Heterogeneous Wireless Access Networks
Architectures and Protocols

T0180622

Ekram Hossain
Editor

Heterogeneous Wireless Access Networks
Architectures and Protocols

 Springer

Editor
Ekram Hossain
Department of Electrical &
Computer Engineering
University of Manitoba
75A Chancellor's Circle
Winnipeg MB R3T 5V6
Canada

ISBN: 978-1-4419-3534-2   e-ISBN: 978-0-387-09777-0
DOI: 10.1007/978-0-387-09777-0

Printed on acid-free paper

springer.com

To my parents and my family

Preface

A Brief Journey through "Heterogeneous Wireless Access Networks"

Ekram Hossain, University of Manitoba, Winnipeg, Canada

Introduction

With the rapid growth in the number of wireless applications, services and devices, using a single wireless technology such as a second generation (2G) and third generation (3G) wireless system would not be efficient to deliver high speed data rate and quality-of-service (QoS) support to mobile users in a seamless way. The next generation wireless systems (also sometimes referred to as Fourth generation (4G) systems) are being devised with the vision of heterogeneity in which a mobile user/device will be able to connect to multiple wireless networks (e.g., WLAN, cellular, WMAN) simultaneously. For example, IP-based wireless broadband technology such as IEEE 802.16/WiMAX (i.e., 802.16a, 802.16d, 802.16e, 802.16g) and 802.20/MobileFi will be integrated with 3G mobile networks, 802.11-based WLANs, 802.15-based WPANs, and wireline networks to provide seamless broadband connectivity to mobile users in a transparent fashion. Heterogeneous wireless systems will achieve efficient wireless resource utilization, seamless handoff, global mobility with QoS support through load balancing and tight integration with services and applications in the higher layers. After all, in such a heterogeneous wireless access network, a mobile user should be able to connect to the Internet in a seamless manner. The wireless resources need to be managed efficiently from the service providers point of view for maximum capacity and improved return on investment.

Protocol engineering and architecture design for broadband heterogeneous wireless access systems is an emerging research area. Load balancing and network selection, resource allocation and admission control, fast and efficient vertical handoff mechanisms, and provisioning of QoS on an end-to-end basis are some of the major research issues related to the development of heterogeneous wireless access networks. The contributed articles in this book from the leading experts in this field cover different aspects of analysis, design, deployment, and optimization of protocols and architectures for heterogeneous wireless access networks. In particular, the topics include challenges and issues in architecture deign and provisioning of QoS for heterogeneous wireless access networks, convergence of heterogeneous wireless and wireline access networks, architectures and protocols for spectrum sharing in heterogeneous wireless networks, cognitive radio techniques for heterogeneous wireless access, radio resource management, admission control, vertical handoff and mobility management, network selection in heterogeneous wireless access networks, modeling and performance analysis of heterogeneous mobile networks, energy saving through heterogeneous wireless access, congestion control in wired-cum-wireless Internet, quality-oriented multimedia streaming in heterogeneous networks, pricing policies in heterogeneous wireless networks, content discovery in heterogeneous mobile networks, and heterogeneous wireless network test-beds. A summary of all of the chapters is provided in the following sections.

Integration of Heterogeneous Wireless Access Networks: Issues and Approaches

As the first chapter in the book, *Chapter 1*, authored by *R. Beaubrun*, introduces the concepts and definitions related to integration of different wireless access networks, different architectural alternatives, and provides examples of signaling exchange protocols for mobility management in a heterogeneous network architecture.

The wireless access networks in a heterogeneous system are likely to be operated by different service providers who will require to operate based on different service agreements to provide seamless services to the mobile users while maximizing their own utilities. The mobile users would like to seamlessly and dynamically roam among the different access networks to maintain the most optimal network connectivity. In this context, intra-technology handoff (or horizontal handoff) and inter-technology handoff (or vertical handoff) will be common phenomena. The inter-technology handoff situations may result in the mobile users moving out of a preferred network or moving in to a preferred network. Mobility over a larger scale is likely to be handled through the Mobile IP (MIP) protocol used in the Internet. Security and authentication operations need to be performed as well during inter-technology handoff. The handoff management protocols should guarantee the network application performances at the expense of tolerable signaling complexity. Different integration architectures should be evaluated based on the above considerations.

Two generic architectures for integrating different access networks are: loosely coupled architecture and tightly coupled architecture. In a loosely coupled architecture, the networks are interconnected independently using a common interworking point. Different mechanisms can be used to handle authentication, billing and mobility management in each network (each of which is presumably operated by different service providers) and MIP can be used for mobility management across the networks. In such a architecture mobile users can use their Subscriber Identity Module (SIM) or User Service Identity Module (USIM) card to access services over different networks. In this case minimal modifications may be required for the existing network components. In a tightly coupled architecture, for example, in a scenario where WLANs and WiMAX networks are integrated with 3G/UMTS networks, the WLANs and the WiMAX networks can be integrated through a 3G logical node and these networks behave as alternate RANs connected with the 3G/UMTS network. The integration can be performed either at the core network level (i.e., Gateway GPRS Support Node (GGSN) level or Serving GPRS Support Node (SGSN) level) or at the access network level (i.e., Radio Network Controller (RNC) level). The WLAN gateways and the WiMAX gateways in this case are required to implement the 3G protocols for mobility management, authentication etc. which are required in a 3G radio access network. Although data traffic can bypass the 3G/UMTS infrastructure, all the signaling traffic would pass through the 3G/UMTS core network. In general, integration through tight-coupling is more complex than the loose coupling approach and it requires modifications in the existing network components (e.g., in SGSN and GGSN).

To this end, the chapter provides examples of signaling protocols required for connection establishment and vertical handoff in an integrated WLAN/3G network where a tightly coupled integration is used at the SGSN level.

Chapter 2, authored by *P. TalebiFard, L. Wong, and V. C. Chang*, first describes the different application and service requirements for communications over a converged heterogeneous network. Then it introduces the architectural elements of IP Multimedia Subsystem (IMS) as a core network technology to integrate heterogeneous wireless networks into a single converged IP-based multi-service network infrastructure. The quality of service (QoS) provisioning and mobility support issues of integrating heterogeneous networks with IMS at the core network are discussed.

An all-IP-based converged network will need to support different conversational services (e.g., voice over IP, video calls), location-based services, multimedia messaging services (MMS), instant messaging services (IMS), content streaming services as well as mobile broadcast and IPTV services. IMS, which was initially defined by the 3rd Generation Partnership Project (3GPP) and 3GPP2, provides an architecture that enables the development of all of the above different types of services in converged fixed, wireless, and mobile networks in a terminal and access technology-agnostic manner. The requirements of the IMS architecture to deliver services to the end users include enabling establishment of IP multimedia sessions, support of QoS negotiation, interworking with Internet and circuit-switched networks, support of subscriber roaming between service domains, policy-based service discovery, rapid service creation, and multiple access technology support. The core network subsystems of IMS include Home Subscriber Server (HSS) and Subscriber Location Functions (SLF) databases, Call Session Control Functions (CSCFs), Application Servers (ASs), Media Resource Functions (MRFs), Breakout Gateway Control Functions (BGCFs), and PSTN gateways.

The layered IMS architecture consists of *interworking and media layer* (to enable the core IP network to interwork with the heterogeneous access networks) and *session control layer* (to initiate and terminate sessions). These two layers together with a *service layer* form a service oriented architecture (SOA) to enable flexible creation and provisioning of application services. IMS is primarily based on MIPv6 and it has chosen SIP (Session Initiation Protocol) as the protocol to perform session control signaling for real-time multimedia applications.

To provide QoS support in an integrated heterogeneous network based on IMS, QoS management and traffic control functions need to be implemented in network nodes. These include flow identification, resource reservation and call admission control (CAC), traffic shaping and traffic scheduling, queue management and QoS routing functions. In this context, provisioning of QoS over an integrated heterogeneous network will involve both horizontal QoS mapping (among different networks) and vertical QoS mapping (among the different layers in a network). To support mobility across heterogeneous wireless access networks for continuous service delivery and access and global roaming, location management, packet delivery, handoff management, roaming and admission control functions will be required. Different aspects of mobility, namely, terminal mobility, user mobility, and service mobility need to be realized in future wireless IP networks. At the end, the chapter presents some of the major design issues and technical challenges towards the integration of heterogeneous networks with IMS and possible directions for future researches in these areas.

Dynamic Spectrum Sharing and Cognitive Radio-Based Architectures and Protocols for Heterogeneous Wireless Access Networks

Dynamic spectrum sharing will be a key technique to enhance the radio spectrum usage efficiency in next generation heterogeneous wireless access networks. *Chapter 3*, authored by *O. Holland et al.*, investigates the usage scenarios, technical requirements, and architectures and protocols for spectrum sharing in the context of dynamic spectrum allocation (DSA) and dynamic spectrum selection (DSS) in heterogeneous wireless access networks. By DSA, the authors refer to the spectrum sharing paradigm where radio spectrum from a spectrum pool is dynamically coordinated and allocated to the radio access networks (RANs) in a centralized manner. In the DSS paradigm, devices/RANs access the radio spectrum as a secondary entity (either in a fully-decentralized or in a network-assisted way) through the use of cognitive radio (CR) technique.

For DSA, the participating networks should exchange information on spectrum usages, predicted traffic load, and network performance requirements (or resource request) with the spectrum controller. The spectrum controller can then perform spatial and temporal spectrum allocation among the participating networks. The major phases of DSA operation are: traffic prediction, resource request, and resource allocation, which are executed in a cyclic fashion. The authors provide a survey on the related works on centralized dynamic spectrum allocation in heterogeneous networks and outlines the major research issues.

For DSS, the major phases of operation are spectrum opportunity (resource) identification, resource access, and resource release, which are executed in a cyclic fashion. DSS in the licensed spectrum domain can be achieved through either *negotiated* or *opportunistic* access depending on whether direct interaction between the primary and secondary system is possible or not. Ideally, spectrum identification should be performed such that there is no tangible interference to primary systems. Compared to DSA, DSS could be faster and more flexible with a potential of better spectrum usage efficiency; however, DSS approaches would be more difficult to implement. A qualitative comparison between the DSA and the DSS approaches is provided in terms of implementation and processing complexity, signaling overhead, scalability, and reactivity to context changes. The IEEE P1900 series of standards under the sponsorship of IEEE Standards Committee (SCC) 41 are developing various mechanisms to support DSA and DSS. The authors also summarize the ongoing regulatory and standardization activities in these domains. To this end, the authors discuss how game theoretical approaches can be used for radio resource optimization in DSA and DSS contexts.

In a spirit similar to that of Chapter 3, *Chapter 4*, authored by *A. Attar et al.*, provides a broader perspective on the design of dynamic spectrum access networks by looking at the protocol design issues at all the layers in the transmission protocol stack. The major functions of a cognitive radio physical layer (PHY) are spectrum sensing, adaptive transmission, spectrum aggregation, and interference mitigation. Spectrum aggregation techniques enable a cognitive radio transmitter to use sporadically idle bands, for example, using multi-carrier modulation techniques such as Orthogonal Frequency Division Multiplexing (OFDM) and Multi-Carrier CDMA (MC-CDMA). Different multi-carrier modulation schemes face different challenges for dynamic spectrum access. For example, interference management is more challenging for OFDM techniques compared to that for MC-CDMA techniques. On the other hand, implementation of MC-CDMA techniques will be more challenging as the number of users and sub-bands increases.

The major issues at the MAC layer are development of mechanisms for resource allocation for coexistence of primary and secondary networks, and resource allocation among secondary users. The resource allocation mechanisms should be designed to minimize interference (if they do not completely avoid interference) caused to primary users and at the same time to maximize the sum of the rate of all of the users. Estimation of interference at the primary receivers, dynamic adaptation of transmission parameters, and synchronization between a secondary transmitter and a secondary receiver are challenging problems. Mobility management, packet routing, and QoS provisioning are among the major network layer tasks in a cognitive radio network. Since the available bands in each node vary from location to location, routing becomes particularly challenging in a multi-hop transmission scenario. The application layer protocols need to be designed to exploit the frequency-agility of cognitive radio. The authors have discussed some of the solutions to address the higher layer protocol design issues in dynamic spectrum access-based heterogeneous wireless access networks. Due to the discontinuous nature of communications, the authors argue that the basic assumptions of a layered protocol stack might not hold true. A cross-layer optimization approach might therefore be more appropriate. Different spectrum access models (e.g., dynamic exclusive use, open sharing, hierarchical access model) would lead to different solutions based on cross-layer optimization. Again, two levels of optimization are possible – local optimization to satisfy application demands, and network-wide optimization to optimize global resource allocations of a dynamic spectrum access-based cognitive radio system. Several example approaches for cross-layer optimization are discussed.

Radio Resource Management and Admission Control in Heterogeneous Wireless Access Networks

Chapter 5, authored by *J. P.-Romero, X. Gelabert, and O. Sallent*, covers the radio resource management (RRM) issues in a heterogeneous wireless access network where different Radio Access technologies (RATs) coexist. Each of RATs could be based on specific multiple access mechanism and local RRM mechanisms would be required for each RAT. The main RRM functions are admission and congestion control, horizontal (or intra-system) handoff, packet scheduling, and power control.

Besides the above local RRM mechanisms, new RRM strategies, which take into account the overall amount of resources in the available RATs, would need to be introduced to achieve a more efficient use of the available radio resources. These RRM algorithms, referred to as Common Radio Resource Management (CRRM) algorithms intend to manage the pool of radio resources belonging to the different RATs in a coordinated way. For example, common admission and congestion control algorithms can be designed which take into account information about several RANs. A specific functionality for RAT selection will be required to decide to which RAT a given service request should be allocated. Also, CRRM algorithms could be used to manage vertical (or inter-system) handoff between two RATs. The current 3GPP standards support some of the envisaged CRRM functionalities.

To support the CCRM functionalities, two different approaches, namely, the CRRM server approach and integrated CRRM approach are proposed. In the former approach, RRM and CRRM entities are located in different physical nodes and are interconnected through an open interface. In the latter approach, advocates a distributed architecture where the CRRM entity may be included in all of the RRM entities or only in a subset of them. Again, several alternatives exist to split the functionalities between CRRM and local RRM, mainly depending on whether it is the RRM or the CRRM entity which acts as the master in taking decisions and also the degree of interactions between them. The highest degree of interaction between CRRM and local RRM is achieved (which in turn gives the highest degrees of freedom to solve the radio resource management problems) by incorporating both long-term functionalities (e.g., RAT selection, vertical handoff) and short-term functionalities (e.g., common admission control, common congestion control, common fast packet assignment) into the CRRM entity.

To this end, the chapter describes specific implementations of CRRM algorithms in the context of RAT selection. Three categories of RAT selection schemes, namely, service-based RAT selection, load balancing-based RAT selection, and interference-based RAT selection schemes are considered. A service-based RAT selection scheme uses a direct mapping between services and RATs. A load balancing strategy distributes the load among different RATs as evenly as possible. An interference-based strategy takes transmitted power level, propagation conditions, interference, and specific network characteristics into consideration. Performance of RAT selection schemes is impacted by the multi-mode terminal availability. Single-mode terminals may degrade the overall network performance in a heterogeneous wireless access environment.

Chapter 6, authored by *F. R. Yu, H. Tang, and H. Ji*, presents a mathematical model for radio resource management for an integrated heterogeneous wireless access network. The objective is to maximize the overall network revenue subject to the QoS constraints in individual access networks. First, the authors describe the *admissible sets* for several individual wireless networks each of which can support several classes of users. In particular, TDMA/FDMA cellular wireless networks, CDMA cellular networks with matched filter receivers, CDMA cellular networks with linear minimum mean square error (LMMSE) multiuser receivers, and IEEE 802.11e-based WLANs are considered. The admissible set gives the number of users in each service class that can be supported in the network simultaneously. Based on the admissible sets of individual wireless networks, the joint radio resource management problem in an integrated heterogeneous network is modeled as a Markov chain. However, due to the problem of large dimensionality, adopting a brute force method to solve the global balance equations for the Markov chain will lead to significant computational complexity. Therefore, the authors consider a set of *coordinate convex schemes* such as the *complete sharing, complete partitioning*, and *threshold* schemes for which a product form of the equilibrium probabilities can be obtained. Using the equilibrium probabilities, different QoS measures (e.g., blocking probability for a particular class of users) can be obtained.

To obtain the optimal admission control scheme that maximizes the radio resource utilization while guaranteeing QoS, a Semi-Markov Decision Process (SMDP) model is formulated. In this formulation, upon arrival of a new call or a handoff call (or session), a decision is made as to whether or not to admit and to which network to admit the call. The performance of the proposed joint radio resource management and optimal admission control scheme is illustrated for an integrated WLAN and CDMA network.

Handoff and Mobility Management in Heterogeneous Wireless Access Networks

Chapter 7, authored by *N. Shenoy and S. Mishra*, provides a survey on the mobility management considerations and related works in the heterogeneous wireless networks. Mobility management is required to provide seamless mobility, which is required for ubiquitous communication and computing. Mobility management embraces three major functionalities - handoff management, location and roaming management. Handoff is a process by which a mobile user (or device) gets handed over from one transceiver station to another. Handoff within one wireless access technology is termed horizontal handoff and handoff across different wireless access technologies is termed vertical handoff. Location management includes paging and location update functions. Roaming includes micro and macro mobility functionalities from an administrative domain perspective. The different categories of roaming and handoff sceanrios are therefore as follows: vetical macro mobility, horizontal macro mobility, vertical micro mobility, and horizontal micro mobility.

Horizontal handoff management is carried out differently in cellular wireless networks and WLANs. When a user roams across several base stations or access points in the same subnet, the handoff can be executed at Layer 2 of the OSI (Open System Interconnection) model. However, if a mobile user crosses several subnets or domains, handoff (horizontal and vertical) management requires involvement of protocols at Layer 3. The handoff decision criteria at Layer 2 include received signal strength (RSS), signal to noise ratio (SNR), and channel availability. In case of vertical handoff, Layer 2 handoff triggers a Layer 3 handoff and the timings of handoff inititaion, execution, and completion at the two layers should be synchronized efficiently to reduce handoff delay. For vertical handoff, additional considerations such as the network traffic conditions, available bandwidth, velocity of the mobile and mobility pattern, and user preference should be taken into account. An optimization can be performed across several handoff criteria to select the best target network for vertical handoff. To simplify the decision process, techniques such as pattern recognition, fuzzy logic and neural networks, and Markov Decision Process (MDP) modeling can be used. The authors have provided a survey on the different handoff decision algorithms proposed in the literature. Handoff implementation aspects, which address mechanisms for message exchanges, have been also addressed. The Media Independent handoff Functions (MIHFs), defined by the IEEE 802.21 Media Independent Handover Working Group, provides a framework for exchange of handoff messages across different wireless access networks.

Macro mobility, which refers to mobility across different administrative domains, is generally handled at Layer 3 by using the Mobile IP (MIP) protocol. MIP is fundamental to most of the vertical handoff solutions in the all-IP paradigm. Micromobility is handled by retaining the handoff decision control in Layer 2 as long as the mobile user does not cross the domain. In the literature, a number of projects involved designing macromobility and micromobility architectures for integrating heterogeneous wireless access networks. The authors have provided a brief survey on these architectures.

Network Selection in Heterogeneous Wireless Access Networks and Performance Analysis

Chapter 8, authored by *W. Song and W. Zhuang*, investigates the problem of network selection in a heterogeneous cellular/WLAN integrated overlay network. Three different network selection strategies, namely, service-differentiated network selection, randomized network selection, and size-based network selection, are proposed and analyzed. These schemes exploit user mobility and multi-service traffic characteristic with an objective to efficiently sharing the overall network resources for QoS provisioning. For location-dependent user mobility, a heavy-tailed user residence time is assumed within a WLAN while an exponential residence time is assumed within the area of a cell. In the multi-service traffic model, conversational voice service (with Poisson distributed call arrival rate and exponentially distributed service duration) and interactive data service (with data file size modeled by a two-stage hyper-exponential distribution) are assumed.

In the service-differentiated network selection strategy, a new voice call first selects the cellular network and it overflows to the WLAN only if its admission control request is rejected by the cellular network. On the other hand, a new data call tries a WLAN first for admission. While a data call can be handed over from a cellular network to WLAN through vertical handoff, a voice call may not be handed over from a cellular network to a WLAN to avoid QoS degradation induced by vertical handoff. With a randomized network selection strategy, an incoming voice (data) call selects the cellular network with probability $\theta_v^c(\theta_d^c)$ and the WLAN with probability $1 - \theta_v^c(1 - \theta_d^c)$. These probabilities are determined as a function of traffic load. In a size-based network selection strategy, a data call selects a cellular network if the call size is not greater than a threshold Φ_d and the cell bandwidth available to data traffic is at least R_d^c; otherwise the data call selects the WLAN. On the other hand, a voice call selects a cellular network as its target network and overflows to the WLAN only if there is not sufficient capacity available in the cell. Also, network re-selection via vertical handoff from WLAN to cellular network is considered for voice calls.

For the above different network selection schemes, analytical models are developed to evaluate the QoS performance and determine the selection parameters. A comparative performance evaluation among the different schemes is carried out in terms of voice call blocking probability and mean data response time. Also, the impacts of mobility and traffic characteristics on selection parameters and resource utilization are studied. It is observed that the size-based network selection strategy outperforms the two other schemes by exploiting the heavy-tailedness of the data call size in initial network selection.

Chapter 9, authored by *R. B. Ali and S. Pierre*, presents performance analysis models and algorithms to evaluate call-level QoS performance measures for voice calls in an integrated heterogeneous wireless access network comprised of WLANs, 3G, and 2.5 G (e.g., GPRS) networks. Assuming a hyper-exponential distribution for cell residence time (CRT) in a WLAN, an Erlang distribution for CRT in a 2G/3G cell, an exponential distribution for CRT in a GPRS cell, and exponential call holding time and a Poisson process for new call arrival, a bi-directional Markov chain model is formulated for voice occupancy in a cell. To generalize the model, N heterogeneous networks on N layers are considered (i.e., the network consists of N heterogeneous cellular layers hierarchically overlaid) and a guard-channel-based admission control scheme is assumed in each layer. The network selection strategy assumes that a new call blocked at a lower layer or a handoff call overflows to cells at higher layers using obligatory vertical handoffs. The Gauss-Seidal iterative method is used to obtain the steady-state probabilities iteratively from the balance equations corresponding to the Markov chain. Then, based on the steady-state probabilities, different QoS performance measures such as the new call blocking probability (at a certain layer k), horizontal/vertical handoff failure probability, call dropping probability, probability of normal call termination, are calculated. Numerical performance results on the analytical models and algorithms are presented for a heterogeneous wireless access network composed of 7 UMTS micro-cells overlaid on 7 WLAN pico-cells and one GPRS macro-cell overlaid on all the cells.

Energy Saving Mechanisms in Heterogeneous Wireless Access Networks

Chapter 10, authored by *G. P. Perrucci, F. H. P. Fitzek, and M. V. Petersen*, focuses on the energy saving potential for mobile devices in a heterogeneous wireless access environment through the exploitation of cooperative communications (i.e., cooperation among mobile devices). Three different architectures using cooperation among mobile devices are considered in each of which a cluster of mobile devices, which are within the transmission range of an access point (AP)/base station (BS), cooperate among each other to share information. Each of the devices in the cluster receives partial information from the AP/BS and then combine the information over short-range communication links. The architectures use three different technology combinations, namely, cellular 3G and IEEE 802.11 WLAN, cellular 3G and Bluetooth v2.0 without broadcast, and cellular 3G and Bluetooth v2.0 with broadcast. Two different application scenarios, namely, *streaming* and *file download*, are considered. Energy consumption performance is analytically investigated for the above different architectures which is complemented by extensive measurement campaigns using commercially available cell phones. It is observed that, for the *file download* scenario, the energy saving performance for the mobile devices can be significantly improved through cooperative wireless networking based on the existing heterogeneous wireless technologies. However, the gain is not significant in the *streaming* scenario.

Routing in Heterogeneous Wireless Access Networks

Chapter 11, authored by *Y. Wu, K. Yang, and H.-H. Chen*, addresses the issues related to designing QoS-aware routing in a heterogeneous wireless access network (HWAN). Based on the network structure of a HWAN, QoS-aware routing can be classified into two categories: inter-working HWAN QoS-aware routing and intra-working HWAN QoS-aware routing. In the former case, routes are constructed through the different heterogeneous network domains while the routes in each separate domain (e.g., UMTS, WiFi, WiMAX) are actually homogeneous. In the latter case, the air interface for each hop is selected from a list of available air interfaces to construct the whole route and the available wireless access networks are tightly coupled with each other. Again, based on the design objectives, QoS-aware routing strategies can be divided into the following four categories: schemes which balance traffic between cellular cells, schemes which extend network transmission range, schemes which guarantee high-bit-rate service, and schemes which offer inter-domain QoS support.

The general structure and procedure of QoS-aware routing in a heterogeneous system can be divided into the following three main parts: source selection, route selection, and destination selection. Source selection involves making decision on the starting point of route discovery - a mobile node which initiates a call request may not have sufficient bandwidth in its home cell to accommodate the request. This selection of a certain source node may require reallocation of bandwidth in the system. Route selection involves consideration of hop number limitation, bandwidth requirement as well as cooperation among heterogeneous nodes and air-interfaces. Destination selection involves selection of one final destination from a list of available destinations. The authors illustrate the QoS-aware routing procedure for a Converged Ad hoc and Cellular Network (CACN).

The general procedure of bandwidth allocation in the source selection procedure (SSP) in CACN is described and three SSPs are presented which aim at reducing the average call request rejection rate in integrated cellular networks such as CACN. Two types of destination selection methods, namely, reactive destination selection and proactive destination selection, are discussed. In an inter-working HWAN system, route selection is performed by collecting the QoS metric information across multiple heterogeneous network domains and considering the QoS requirements of the source. In an intra-working HWAN system, the working air-interface should be decided at each hop. The authors present example algorithms to describe the operational procedure of destination selection and route selection. To this end, a number of research challenges in QoS-aware routing in a HWAN system are mentioned.

Congestion Control Protocols for Heterogeneous Wireless Access Networks

Chapter 12, authored by *S. Chen, X. Hei, J. Zhu, and B. Bensaou*, presents a survey on the issues and challenges related to congestion control for data services (i.e., TCP-based congestion control) in the heterogeneous wired-cum-wireless Internet. The state-of-the-art approaches to address these issues are summarized. Due to high bit error rate and dynamic channel conditions, conventional TCP-based congestion control mechanisms perform poorly and causes underutilization of channel bandwidth. To cope with the channel errors and user mobility, it is desirable that TCP is able to distinguish among congestion losses, wireless random losses, and handoff-related losses. TCP is inefficient in networks with high bandwidth-delay product (e.g., satellite networks) and also in networks where the available bandwidth fluctuates (e.g., due to rate adaptation). Wireless mobile ad-hoc networks pose new challenges to the TCP congestion control due to frequent route changes and failures, hidden and exposed terminal problems, topology-related unequal channel access opportunity, and unpredictability of end-to-end transmission delay. The authors have provided a survey of related work in the literature which addressed the above-mentioned challenges.

To improve TCP performance in presence of wireless random losses, proxy-based, end-to-end, and network cooperation approaches were proposed in the literature. The proxy-based approach relies on a proxy agent which makes the wireless random losses transparent to TCP. In an end-to-end approach, the end systems differentiate wireless random packet losses from congestion losses and take different congestion control actions according to the loss type. The network cooperation approach involves routers to aid the end systems with extra information to improve congestion control efficiency.

Several congestion control protocols were proposed to mitigate the negative effects of handoffs in mobile networks - notable among which are M-TCP (Mobile TCP), Freeze TCP, RCP (Reception Control Protocol), and pTCP (Parallel TCP). pTCP was particularly designed for networks involving multiple wireless interfaces. TCP-Peach, XCP (eXplicit Control Protocol), and REFWA (Recursive, Explicit, and Fair Window Adjustment) protocols were designed for satellite networks. The RA-snoop (Rate Adaptive Snoop) protocol was designed to help TCP to effectively adapt to variable bandwidth. The proposed solutions for congestion control in mobile ad hoc networks include schemes to tackle disconnections due to frequent route changes or failures (e.g., TCP-Feedback, Ad-hoc TCP), schemes to handle persistent packet reordering due to route changes or multi-path routing (e.g., dynamic SACK extension to TCP SACK option, TCP-DOOR), mechanisms which improve fairness by a better sharing of the wireless medium (e.g., Neighborhood RED), and schemes which require network cooperation (e.g., Ad-hoc Transport Protocol).

In designing congestion control schemes for heterogeneous wired-cum-wireless networks, besides performance gain, issues such as deployment cost and TCP friendliness should be also taken into account. Also, since the communication environment is volatile, improved lower layer design, which takes TCP performance into consideration, will be very beneficial to improving the performance of TCP in wired-cum-wireless Internet. Again, the bandwidth mismatch between the wired network and the wireless network requires careful network planning and deployment of new techniques (e.g., content proxies) for improved congestion control performance in heterogeneous wireless Internet access environment.

Multimedia Streaming in Heterogeneous Wireless Networks

Chapter 13, authored by *G.-M. Muntean*, presents a number of solutions for adaptive quality-oriented multimedia streamimg in heterogeneous wireless access networks internetworking with wired networks. Multimedia streaming applications (e.g., digital and interactive TV, video on demand, videoconferencing) over wired and wireless networks are becoming increasingly popular. However, high traffic load in wired networks, low bandwidth and increased number of users in wireless networks and limited mobile device power negatively affect the quality of the transmissions, and hence viewers' perceived quality. This chapter describes four adaptive solutions proposed to improve users' perceived quality when streaming multimedia in various network conditions: Quality Oriented Adaptive Scheme (QOAS), Region-Of-Interest Adaptive Scheme (ROIAS), Priority-based Differentiated Quality of Service in wireless local area networks (pDiffQoS), and Battery power Adaptive Mechanism for wireless multimedia streaming to mobile devices (BAM). Each of these solutions has been described in detail with experimental performance evaluation results.

The QOAS schemes, adaptive decisions on content adjustment during multimedia streaming are based on network performance related QoS parameters such as packet loss rate, delay, and delay jitter rather than the end user perceived quality or Quality of Experience (QoE). QOAS affects equally the entire viewing area of video frames when adjusting the stream quality. On the other hand, ROIAS, when performing adaptive streaming-related adjustments, selectively affects the quality of the regions the viewers are interested in. pDiffQoS is an application-level adaptive scheme for in-home wireless delivery of multimedia-based services. It adapts multimedia content bitrate to existing network conditions by taking the priorities of the multimedia contents as well as device requirements into account. In case of BAM for wireless multimedia streaming to mobile devices, different individual power-saving schemes are employed during data transmission and reception, decoding and multimedia playback. The authors have provided a survey of the existing literature on adaptive multimedia streaming solutions, region of interest video coding, and power saving solutions.

Pricing in Heterogeneous Wireless Access Networks

Chapter 14, authored by *S. Sengupta and M. Chatterjee*, deals with the pricing aspects of mobile data services in heterogeneous networks which are deployed by competing wireless service providers (WSPs). The end users and the WSPs can be considered as buyers and sellers, respectively, in a market environment where the WSPs charge the users so that they are able to maximize their profit and at the same time satisfy the users. Several popular pricing mechanisms for packet data services, namely, the *Paris Metro Pricing*, *Smart Market Pricing*, and *Threat Strategy Pricing* schemes are discussed. These schemes, however, do not consider the interactions among multiple competitive service providers. In a dynamic competitive market based pricing model, a WSP can dynamically adapt the service price for a user based on its revenue and network load. The pricing mechanism should be such that it can successfully compete with other service providers. On the buyer side, an end user would like to select the best (according to some criteria) service provider for a particular service request. The interactions between the service providers and the end users can be modelled formally using non-cooperative games. The duration of such a game is the time the user is connected to a service provider and the concept of *sub-games* in a game need to be introduced. During the game between an end user and a WSP, sub-games are defined over small discretized time intervals within which there is no change of pricing from any one of the end-users or WSPs. Since it may not be possible for the service providers to set the price for the entire service duration beforehand, the service pricing needs to be divided into sub-games. A user and a service provider negotiate price at the beginning of each sub-game.

The authors analyze the solution of the game for both service providers and end users considering voice/video and data services. Assuming that the players are rational, the solution of the non-cooperative game gives the strategies of the players which are mutual best responses to each other. These strategies, which no player would have a reason to deviate from, are referred to as the Nash equilibrium. For non-elastic service user's satisfaction modeled by Sigmoid-type utility function, the authors prove the existence of a unique Nash (price) equilibrium. For elastic services (e.g., data services), the authors show that discretized pricing is undesirable to prevent the service provider to be malicious. From a user's perspective, the equilibrium strategy would be to negotiate price at the beginning for the entire data session such that the net utility of the user is positive and maximized. After the user's equilibrium strategy is established, the service provider's dominant strategy can be defined based on the maximization of its expected net utility. The existence of Nash equilibrium can then be easily proved. Performance evaluation is carried out for a system model with two access networks, namely, 3G network based on CDMA/HDR and 802.11-based Wi-Fi network. The performance evaluation results show that if service providers and users comply with the proposed dynamic pricing mechanism, Nash equilibrium can be achieved which maximizes the net utilities of both users and service providers.

Content Discovery and Dissemination in Heterogeneous Mobile Networks

Chapter 15, authored by *D. Borsetti, C. Casetti, C.-F. Chiasserini, and L. Liquori*, addresses the problem of discovery and sharing of information among users in a heterogeneous wireless mobile network. In a mobile environment, information required by the users can be acquired either from servers in the backbone infrastructure or from other users through cooperative communication mechanisms. The publish/subscribe messaging paradigm, which is based on an asynchronous, many-to-many communication model, is a promising approach to share information among large number of mobile users. In this publish/subscribe system, mobile users can be both information providers (or publishers) and information consumers (or subscribers). While producers publish data to the system and consumers subscribe and specify the type of information they are interested in, the system disseminates the information to the consumers according to their declared interests. There are two types of routing associated with content delivery - topic-based routing and content-based routing. For topic-based routing, all published messages on some topic are broadcast to all users subscribing to that topic. In case of content-based routing, only messages that match the subscriber-preferred attributes are delivered.

To achieve information discovery and retrieval through the publish/subscribe approach implemented via a seamless, geographically distributed open-ended network, an overlay network can be defined, which is built on top of a physical network. One such overlay network, named *Arigatoni*, which was designed in the context of wired computer networks, would be suitable for information delivery and sharing in a mobile environment. The core of the *Arigatoni* overlay network are two main logical entities (the *Agent* and the *Broker*) and two basic protocols (a *Registration* and a *Service Discovery* protocol). The authors provide a brief overview of these main entities and protocols involved in *Arigatoni*.

The authors consider a WiFi and GSM/GPRS (or UMTS)-based heterogeneous mobile network scenario and defines an overlay network based on the publish/subscribe messaging paradigm and content-based routing for this scenario. For this overlay network, referred to as *Arimove*, the main components and their interactions are described. The benefits of using the publish/subscribe system in the *Arimove* system for content retrieval through using multiple wireless technologies are demonstrated by simulation. The simulation study is based on a case study involving WiFi hot spots and UMTS base stations as network infrastructure entities. The differnet performance measures considered for performance evaluation include service discovery time, success probability of a service request, average time to satisfy a service request, and probability of successful retrieval.

Heterogeneous Wireless Network Test-bed

Chapter 16, authored by *A. Botta, A. Pescapé, and R. Karrer*, focuses on planning, deploying, and experimentally evaluating heterogeneous wireless network test-beds. Heterogeneity in the wireless Internet environment arises due to the diversity in data link and physical layer technologies, different protocols implementing the network functionalities, diversity in end-host devices, and wide variety of applications. The major challenges in tackling heterogeneity include efficient control and management of network resources, distribution of control, and implementation of cooperation among the layers in the internet Protocol stack (i.e., cross-layer design and implementation). The authors argue that test-bed implementations, measurements, and evaluations provide fundamental insights into systems and networks design in presence of heterogeneity. In particular, test-beds allow to assess the tradeoff between performance gain and implementation overhead under realistic network conditions.

The authors describe two heterogeneous network test-beds and experimental results obtained from these test-beds. The first test-bed, named as *Magnets*, is for an outdoor multi-hop wireless wide area network which can internetwork with other wireless access technologies such as GPRS, UMTS, and WiMAX. The second test-bed, which has a smaller scale with respect to *Magnets*, consists of a large mix of different devices, operating systems, and access networks. The *Magnets* architecture consists of a high-speed 802.11 backbone (interconncting two facilities in Berlin through a high-speed line-of-sight wireless link), 802.11 mesh networks (which includes an indoor test-bed of 20 nodes and an outdoor test-bed of 100 nodes), and integration points to other wireless access networks. The architecture is characterized by multiple wireless interfaces with diverse link characteristics, heterogeneity in the nodes in terms of processing and storage capabilities, and internetworking among mesh networks using different routing protocols. For experimental performance evaluation of *Magnets*, per-link throughputs of the backbone links are measured. The authors conclude that, even though they are in the same test-bed, the measured link characteristics vary significantly. Significant performance gain was observed through the use of *Turbo* and *Burst* mode of the 802.11 access points.

For the small-scale heterogeneous test-bed several system configuration parameters including operating system, end-user device, access network, transport protocol, and traffic condition are varied. Performance measures such as the average throughput, jitter, round-trip time are obtained with different network configurations.

Conclusion

A summary of the contributed chapters has been provided which will be helpful to follow the rest of the book easily. These chapters essentially feature some of the major advances in the research on heterogeneous wireless networking technology for the next generation wireless communications systems. Therefore, the book will be useful to both researchers and practitioners in this area. The readers will find the rich set of references in each of the chapters particularly valuable.

About the Editor

Ekram Hossain is currently an Associate Professor in the Department of Electrical and Computer Engineering at University of Manitoba, Winnipeg, Canada. He received his Ph.D. in Electrical Engineering from University of Victoria, Canada, in 2000. Dr. Hossain's current research interests include design, analysis, and optimization of wireless communication networks and cognitive radio systems. He is a co-editor for the books *Cognitive Wireless Communication Networks* (Springer, 2007, ISBN: 978-0-387-68830-5), *Wireless Mesh Networks: Architectures and Protocols* (Springer, 2007, ISBN: 978-0-387-68839-8), *Heterogeneous Wireless Access Networks* (Springer, 2008, ISBN: 978-0-387-09776-3), and a co-author of the book *Introduction to Network Simulator NS2* (Springer, 2008, ISBN: 978-0-387-71759-3). Dr. Hossain serves as an Editor for the *IEEE Transactions on Mobile Computing*, the *IEEE Transactions on Wireless Communications*, the *IEEE Transactions on Vehicular Technology*, *IEEE Wireless Communications*, and several other international journals. He served as a guest editor for the special issues of *IEEE Communications Magazine* (Cross-Layer Protocol Engineering for Wireless Mobile Networks) and *IEEE*

Wireless Communications (Radio Resource Management and Protocol Engineering for IEEE 802.16). He served as a technical program co-chair for the IEEE Global Communications Conference (Globecom'07) and IEEE Wireless Communications and Networking Conference (WCNC'08). Dr. Hossain served as the technical program chair for the workshops on "Cognitive Wireless Networks" (CWNets'07) and "Wireless Networking for Intelligent Transportation Systems" (WiN-ITS'07) held in conjunction with QShine'07: International Conference on Heterogeneous Networking for Quality, Reliability, Security and Robustness, during 14-17 August 2007, in Vancouver, Canada. He served as the technical program co-chair for the Symposium on "Next Generation Mobile Networks" (NGMN'06), NGMN'07, and NGMN'08 held in conjunction with ACM International Wireless Communications and Mobile Computing Conference (IWCMC'06), IWCMC'07, and IWCMC'08, and the First IEEE International Workshop on Cognitive Radio and Networks (CRNETS'08) in conjunction with IEEE International Symposium on Personal, Indoor and Mobile Radio Communications (PIMRC'08). He is a Senior Member of the IEEE. Dr. Hossain is a registered Professional Engineer (PEng) in the province of Manitoba, Canada.

Contents

List of Contributors

R. Beaubrun
Université LAVAL, Canada
ronald.beaubrun@ift.ulaval.ca

P. TalebiFard, T. Wong, and
V. C. M. Leung
The University of British Columbia,
Vancouver, Canada
Telus, Toronto, Ontario, Canada

peymant@ece.ubc.ca
vleung@ece.ubc.ca
Terrence.Wong@telus.com

O. Holland, A. Attar, M. Sooriyaban-
dara,T. Farnham,
H. Aghvami, M. Muck, V. Ivanov, and
K. Nolte
King's College London, UK
Toshiba Research Europe Ltd., Bristol,
UK
Infineon Technologies, Munich,
Germany
Intel Corporation, Communications
Technology Lab, St.
Petersburg, Russia
Alcatel-Lucent Deutschland AG, Bell
Labs, Germany

Oliver.Holland@kcl.ac.uk
Ali.Attar@kcl.ac.uk

Hamid.Aghvami@kcl.ac.uk
Mahesh.Sooriyabandara@toshiba-
trel.com
Tim.Farnham@toshiba-trel.com
markus.muck@gmail.com
vladimir.ivanov@intel.com
klaus.nolte@alcatel-lucent.de

A. Attar, O. Holland, T. Farnham,
M. Sooriyabandara,
M. R. Nakhai, and A. H. Aghvami
King's College London, UK
Toshiba Research Europe Ltd., Bristol,
UK

Ali.Attar@kcl.ac.uk
Oliver.Holland@kcl.ac.uk
Reza.Nakhai@kcl.ac.uk
Hamid.Aghvami@kcl.ac.uk
Tim.Farnham@toshiba-trel.com
Mahesh.Sooriyabandara@toshiba-
trel.com

J. P.-Romero, X. Gelabert, and O.
Sallent
Universitat Politècnica de Catalunya
(UPC), Barcelona, Spain

jorperez@tsc.upc.edu
xavier.gelabert@tsc.upc.edu
sallent@tsc.upc.edu

F. R. Yu, H. Tang, and H. Ji
Carleton University, Ottawa, ON,
Canada
Defense R&D Canada, Ottawa, ON,
Canada
Beijing University of Posts and
Telecommunications, Beijing, P.R.
China

richard_yu@carleton.ca
helen.tang@drdc-rddc.gc.ca
jihong@bupt.edu.cn

N. Shenoy and S. Mishra
Rochester Institute of Technology,
Rochester, USA

nxsvks@rit.edu
sxm1145@rit.edu

W. Song and W. Zhuang
University of Waterloo, Canada

wsong@bbcr.uwaterloo.ca
wzhuang@bbcr.uwaterloo.ca

R. B. Ali and S. Pierre
Ecole Polytechnique Montreal, Canada

racha.benali@polymtl.ca
samuel.pierre@polymtl.ca

**G. P. Perrucci, F. H. P. Fitzek, and
M. V. Petersen**
Aalborg University, Denmark

gpp@es.aau.dk
ff@es.aau.dk
mvpe@es.aau.dk

Y. Wu, K. Yang, and H.-H. Chen
University of Essex, UK
National Cheng Kun University, Taiwan

ywud@essex.ac.uk
kunyang@essex.ac.uk
hshwchen@ieee.org

**S. Chen, X. Hei, J. Zhu, and B.
Bensaou**
The Hong Kong University of Science
and Technology,
Clear Water Bay, Kowloon, Hong Kong

chenshan@cse.ust.hk
heixj@ece.ust.hk
csjhzhu@cse.ust.hk
brahim@cse.ust.hk

G.-M. Muntean
Dublin City University, Ireland

munteang@eeng.dcu.ie

S. Sengupta and M. Chatterjee
Stevens Institute of Technology, USA
University of Central Florida, USA
Shamik.Sengupta@stevens.edu
mainak@eecs.ucf.edu

**D. Borsetti, C. Casetti, C.-F. Chi-
asserini, and L. Liquori**
Politecnico di Torino, Italy
INRIA Sophia Antipolis Méditerranée,
France

borsetti@tlc.polito.it
casetti@tlc.polito.it
chiasserini@tlc.polito.it
Luigi.Liquori@inria.fr

A. Botta, A. Pescapé, and R. Karrer
Universitá di Napoli Federico II, Napoli,
Italy
Deutsche Telekom Laboratories, Berlin,
Germany

a.botta@unina.it
pescape@unina.it
roger.karrer@telekom.de

1

Integration of Heterogeneous Wireless Access Networks

Ronald Beaubrun

Department of Computer Science and Software Engineering
Université LAVAL, Canada
ronald.beaubrun@ift.ulaval.ca

1.1 Introduction

The development and proliferation of wireless and mobile technologies have rev-olutionized the world of communications [1]. Such technologies are evolving to-wards broadband information access across multiple networking platforms in order to provide ubiquitous availability of multimedia services and applications [2]. Recent broadband wireless access systems include wireless local area networks (WLAN), broadband fixed wireless access (metropolitan networks) and wireless personal area networks (WPAN), as well as the widely used mobile access technologies, such as General Packet Radio Service (GPRS), Wide Code Division Multiple Access (WCDMA), Enhanced Data Rate for Global Evolution (EDGE), 3G and Beyond 3G (B3G) communications systems, Worldwide Interoperability for Microwave Access (WiMAX), and Bluetooth [2–4].

These wireless access technologies have characteristics that perfectly comple-ment each other [2]. Cellular systems and 3G provide wide coverage areas, full mo-bility and roaming, but traditionally offer low bandwidth connectivity and limited support for data traffic [1]. On the other hand, WLANs provide high data rate at low cost, but only within a limited area, whereas WiMAX can supply mobile broad-band for anyone, anywhere, whatever the technology and access mode [4]. More specifically, WLANs are expected to provide access to IP-based services (including telephony and multimedia conferencing) at high data rates and reduced coverage in public and private areas [5]. In particular, current WLANs offer a bit rate of 54 Mbps with IEEE 802.11g in the 2.4 GHz frequency band [6].

In this context, several types of WLANs are emerging and become profusely used, allowing users to roam inside their home, enterprise or campus without inter-rupting their communication sessions. They are organized in form of hotspots, i.e. relatively small networks covering a particular location providing broadband and easy-to-use Internet access to their customers while supporting high traffic load. Classical hotspot examples are airports, hotels, dense urban areas, campuses, and private offices. Using hot-spots, providers can offer subscribers not only wide-area

E. Hossain (ed.), *Heterogeneous Wireless Access Networks*,
DOI: 10.1007/978-0-387-09777-0_1, © Springer Science+Business Media, LLC 2008

connectivity through the cellular infrastructure, but also increased bandwidth via Wi-Fi access points deployed in high concentration areas [1]. In this context, WLANs are often seen as a suitable complementary technology to the existing cellular radio access networks (2.5G/3G).

In order to fill the gap between WLANs and 3G wireless systems, IEEE 802.16 defines the Mobile WiMAX standard that provides the air interface specification for supporting broadband wireless access (BWA) in the context of wireless metropolitan area networks (WMAN) [7, 8]. The Mobile WiMAX provides the capability to offer new wireless services, such as Voice over Internet Protocol (VoIP), videoconferencing, multimedia streaming, multiplayer interactive gaming, Web browsing, instant messaging, and media content downloading [8, 9]. Such services consume significant bandwidth, require short end-to-end delay, and may be offered with relatively low costs. In this context, the mobile terminal may be a wireless videophone, a laptop, or a PDA, and can be connected with a private or public network, from home to office. Combined with 3G, WiMAX offers high data rate services in addition to original voice services in hotspot areas [4]. As a result, WiMAX and WLAN can be utilized as a powerful complement to 3G/B3G networks.

In order to provide the mobile users with the requested multimedia services and corresponding quality of service (QoS) requirements, these radio access technologies will be integrated to form a heterogeneous wireless access network. Such a network will consist of a number of wireless networks, as illustrated in Fig. 1.1 [10], and will form the 4th generation (4G) or next-generation of wireless networks. Heterogeneous wireless access, extensive support of IP-based traffic and excellent mobility support are among the main drivers for the architecture of such generation [11].

The 4G wireless networks will offer several advantages for both users and network operators [2, 3]. On one hand, users will benefit from the different coverage and capacity characteristics of each network throughout the integrated networks. In this way, a large set of available resources will allow them to seamlessly connect, at any time and any place, to the access technology that offers the best possible quality. For the network operators, the integration of all these technologies provides more efficient usage of the network resources, and may be the most economic and technologically diversified means of implementing the future anywhere, anytime, always-on visions, providing both universal coverage and broadband access.

However, each access network provides different levels of QoS, in terms of bandwidth, mobility, coverage area and cost to the mobile users. As a result, when roaming across heterogeneous wireless access networks and experiencing vertical handoffs, high variability in the required QoS may be introduced, as the most optimal access network can dynamically change [12]. In this case, choosing the correct time to initiate a vertical handoff request and select the best network to connect becomes important. Furthermore, vertical handoffs, i.e. handoffs between radio access networks using different technologies, require additional delay for reconnecting the mobile terminal to the new wireless access network, which may cause packet losses and degrade the QoS for real-time traffic [13]. In this context, roaming and interworking between heterogeneous wireless access networks constitute important issues to the networking community.

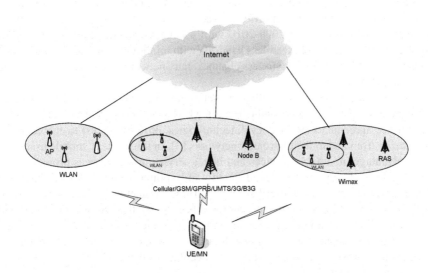

Fig. 1.1. Integration of heterogeneous wireless access networks.

This chapter outlines the most recent works, and describes architectures, as well as signaling exchange, in the area of heterogeneous wireless access networks. More specifically, it presents related definitions and concepts in the next section. After that, architectural proposals for integrating WLAN, WiMAX and 3G/UMTS networks are described. Such proposals are extensions of generic integration methods described in [4, 14, 15]. Finally, we present signaling exchange during vertical hand-off procedures for a specific scenario which considers a WLAN interworking with UMTS.

1.2 Related Definitions and Concepts

In order to take full advantage of future wireless networks (high-speed and low cost of WLAN networks, and large coverage areas of 3G/B3G networks), mobile terminal capabilities need to provide connectivity to a variety of access technologies. In this context, mobile terminals, such as Personal Digital Assistants (PDA) and smartphones, will be equipped with multiple network interfaces connecting to one or more access networks [13]. Such devices are called multi-mode terminals, as they will enable mobile users to seamlessly and dynamically roam between different access technologies in order to maintain the most optimal network connectivity [12].

The Third Generation Partnership Project (3GPP) has identified several issues concerning multi-mode terminals [16]. More precisely, it has specified multi-mode

User Equipments (UE) categories, and described the general principles and procedures for the multi-mode operation. In [17], the parameters for the UE radio access capabilities are addressed, and some reference configurations are provided for utilization in test specifications. It is expected that 2G/2.5G/3G multi-mode terminals, which are able to operate with different Radio Access Technologies (RAT), will be available for most users in 2009-2010 with a penetration rate reaching 90% [18]. Such terminals will enable seamless handoffs with minimum user intervention, while dynamically adapting to the wireless channel state, network layer characteristics and application requirements.

Moreover, the networks building the heterogeneous 4G system will be managed by various wireless providers and operators who have to cooperate and make different service agreements (service level, technologies, profiles, prices) to enable roaming access from and to their partners' networks [3]. In this context, mobile users will take advantage of suitable services and better service price, whereas providers will offer additional services with good quality and optimal network usage. Currently, wireless operators allow subscribers to roam into competing networks in order to offer ubiquitous connectivity. In the future, it is expected that wireless operators will continue to form roaming agreements with their competitors in order to maintain high customer satisfaction [1]. This will enable customers to request advanced services and improved bandwidth when they become available, even if it is through a competing provider.

In general, mobility management addresses two main problems: location management and handoff management [2, 19]. Location management is a process that enables the network to notify the current attachment point of each mobile user for successful information delivery. For this purpose, Mobile IP (MIP) is expected to be the main engine in the next-generation wireless networks. Handoff management is a process that enables mobile terminals to maintain the active connections, as they change their point of attachment to the network.

In the context of heterogeneous wireless networks, both intra-technology handoff and inter-technology handoff can take place [2]. Intra-technology handoff is the traditional Horizontal Handoff (HHO) process where the mobile terminal roams between two Access Points (AP) or Base Stations (BS) using the same radio access technology. On the other hand, inter-technology handoff, or Vertical Handoff (VHO), occurs when the UE changes its point of attachment to a radio access network using different technology. The main difference between VHO and HHO is symmetry [2]. While HHO is a symmetric process, VHO is an asymmetric process in which the UE moves between two networks with different characteristics.

Because of the different capabilities of the networks, Petander et al. [20] consider two kinds of vertical handoffs: upward VHO from the high-speed to the low-speed network, and downward VHO from the low-speed to the high-speed network. Downward VHO are often performed opportunistically, i.e., a UE performs a handoff to a new network even if the current network is still being available. In this case, the handoff timing and duration are often not critical to the network application performance. However, in upward VHO, the UE is typically using the best available network in terms of throughput and cost (e.g., the WLAN), and needs to move outside the cov-

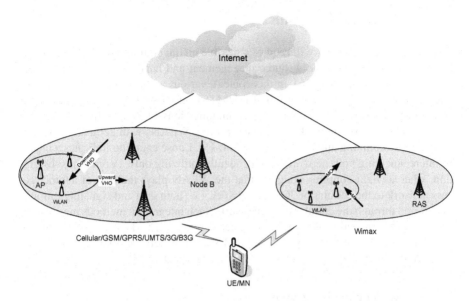

Fig. 1.2. Concepts related to VHO.

erage area of such a network. In this case, the handoff delay and timing become crucial. An early handoff would result in unnecessary costs and low performance from the use of the low-speed network, whereas a late handoff would eventually result in packet loss affecting the application performance. Upward and downward handoffs are illustrated in Fig. 1.2.

These concepts enable to introduce in [2] the concept of preferred network which designates the network that provides better throughput performance at lower cost, even if both networks are available and in good condition for the user. In this context, two main scenarios are defined: moving out of the preferred network (MO) and moving into the preferred network (MI). As long as the preferred network satisfies the user application, it is desirable to associate the UE with the preferred network [2]. This enables to improve both resource utilization of the access networks and the user perceived QoS. MO and MI scenarios are illustrated in Fig. 13.2.

If the MI decision is only based on the preferred network availability, the UE will start the MI process as soon as it detects the high-speed wireless access network [2]. In addition, if more than one high-speed networks are available, the UE will associate itself with the one with the strongest received signal strength (RSS), as it does in traditional HHO. Once it is associated with the preferred network, the UE enjoys all the preferred network advantages (high data rate at low cost) before moving out [2]. In the ideal MO scenario, it only performs one handoff at the high-speed wireless access network edge, i.e. when the preferred network becomes unavailable. However, this ideal MO decision is usually difficult to execute.

1.3 Architectural Proposals

The 4G wireless networks will require a flexible architecture in which the QoS will be closely coupled with the resource management and handoff mechanisms of the wireless access technology [6]. In this context, various architectures have been proposed for integrating WLAN and 3G (or UMTS) networks. Depending on the level of interdependence, the European Telecommunications Standards Institute (ETSI) specifies two generic methods for interconnecting WLANs with cellular or 3G networks: loose coupling and tight coupling [14, 15]. Loose coupling indicates a means of interconnecting both networks independently, utilizing only a common subscription, whereas tight coupling suggests that the WLAN plays the role of another access network to the cellular or 3G core network, i.e., both data and signaling traffic is transferred through the cellular or 3G network. Such integration methods may also be used for interconnecting WiMAX with other wireless networks [4]. As a result, the integration of Mobile WiMAX and 3G/UMTS wireless networks may be considered equivalent to that of WLAN and 3G/UMTS wireless networks.

1.3.1 Loose Coupling Integration

In the loose coupling architecture, the integrated networks exist independently, as they provide independent services [21]. In this configuration, the interworking point is after the interface of the Gateway GPRS Support Node (GGSN) with the Internet Protocol (IP) network, as the networks interconnection uses Mobile IP mechanisms [22, 23]. This approach calls for the introduction of interconnection equipments, such as WLAN gateway and WiMAX gateway. Such gateways support authentication and billing for roaming services, while using the mobile IP to provide mobility between WLAN, WiMAX, and 3G/UMTS networks [21]. Fig. 1.3 illustrates an example of loose coupling integration architecture.

Both gateways are connected with the Internet through the UMTS Authentication Authorization and Accounting (AAA) server, and do not have any direct link to the 3G/UMTS network equipments. As a result, the WLAN and WiMAX data traffic does not pass throughout the 3G/UMTS core network, but goes directly to the Internet. In other words, the WLAN and the WiMAX network are deployed as other access networks complementary to the 3G/UMTS network. In this case, they use the subscriber databases in the 3G/UMTS network to provide the mobile users with direct data access to the Internet. Furthermore, the WLAN and WiMAX network may be owned by a third party, with roaming and mobility enabled via dedicated connections between the operator and the WLAN (or WiMAX), or over an existing public network, such as Internet [15].

In this approach, different mechanisms can be used to handle authentication, billing, and mobility management in each network [24]. Such mechanisms require that the WLAN and WiMAX gateways support Mobile IP functionalities to handle mobility across networks, as well as the services offered by the UMTS AAA server. This enables the 3G/UMTS provider to collect the WLAN or WiMAX accounting records and generate a unified billing statement that indicates usage and

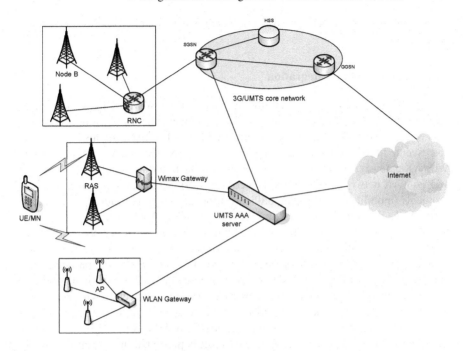

Fig. 1.3. Example of loose coupling integration.

price schemes for both networks. At the same time, the use of compatible AAA services on both networks allows the WLAN and WiMAX gateways to dynamically obtain per-user service policies from their home AAA servers, and to enforce and adapt such policies to both WLAN and WiMAX networks.

There are several advantages in using the loose coupling integration approach. First, it allows independent deployment and traffic engineering of WLAN, WiMAX and 3G networks [4, 24]. This enables 3G operators to take advantage of other providers' WLAN or WiMAX deployments without major investments, as they may continue to deploy 3G networks independently. In this case, simple implementations with minimal enhancements are needed to the existing components. Also, loose coupling integration allows operators to provide their own public WLAN hotspots interoperate through roaming agreements with public WLAN and 3G service providers, or manage a privately installed enterprise WLAN. Based on roaming agreements, subscribers may take advantage of having one service provider for all network access. They can reuse their Subscriber Identity Module (SIM) or User Service Identity Module (USIM) card to access multimedia services over the WLAN or WiMAX network [15].

Currently, the short-term trend is to use the loose coupling approach, specifically in scenarios where the UMTS and the WLAN (or WiMAX) infrastructures belong to different providers. However, some limitations and problems have to be solved to provide continuous services with this approach [25]. In particular, the average handoff duration can reach 400 ms [26], which is not suitable for real-time services and applications.

1.3.2 Tight Coupling Integration

With the tight coupling approach, the WLAN and the WiMAX network are connected with the 3G/UMTS core network in the same manner as any other Radio Access Network (RAN), such as UMTS terrestrial RAN (UTRAN). In this context, they can execute functions that are available in the 3G RAN. The WLAN gateway and the WiMAX gateway, introduced to achieve integration, hides the details of both WLAN and WiMAX network to the 3G/UMTS core network, while implementing all the 3G protocols (mobility management, authentication, etc.) required in a 3G radio access network. As a result, the data traffic from the WLAN or WiMAX users goes through the 3G/UMTS core network before reaching the Internet or other packet data networks (PDN). In this context, each network has to modify its protocols, interfaces and services in order to support the interworking requirements. In particular, this enables to support integrated authentication, accounting and network management.

Moreover, with the tight coupling approach, the interconnection with the WLAN and the WiMAX network can be made at the core network level (i.e., GGSN or SGSN) or at the access network level (e.g., RNC) of UMTS [21, 23]. An example of tight coupling integration at the GGSN level is presented in [27] and illustrated in Fig. 1.4. In this architecture, a logical node called the virtual GPRS support node (VGSN) is introduced to interconnect the WLAN, WiMAX and UMTS networks. Its main functionality is to exchange subscriber and mobility information, and to route packets between the integrated networks. More particularly, the VGSN enables the UMTS, WLAN and WiMAX to handle their own subscribers independently, without the need of Mobile IP functionalities. It can be implemented as an independent node, or may be integrated in the WLAN/WiMAX gateway, or in the SGSN, or in the GGSN. With the VGSN, data traffic can bypass a large part of the 3G infrastructure, which results in less congestion in the 3G/UMTS core network. However, signaling traffic goes through the 3G/UMTS core network, and requires close cooperation between the 3G/UMTS, WLAN, and WiMAX network operators.

The efficiency of coupling at the GGSN level approach, in terms of average bandwidth per user and handoff duration, is similar to that in a loose coupling case where Mobile IP is used [22]. Such approach requires less complicated modifications in the UMTS architecture, since changes only affect high-level protocols. However, two main drawbacks may be mentioned. First, additional routing delays may be caused if the VGSN location is not properly selected. Second, this architecture asks for cooperation stronger than a simple roaming agreement between the operators, since confidential 3G/UMTS network information could be revealed to the service providers, without the 3G/UMTS provider's permission.

Another tight coupling solution, proposed in [15], describes an architecture where the coupling is done at the SGSN. An example of this architecture is illustrated in Fig. 1.5. In this architecture, the WLAN and WiMAX networks are deployed as alternate RAN connected with the 3G/UMTS wireless network. In other words, the

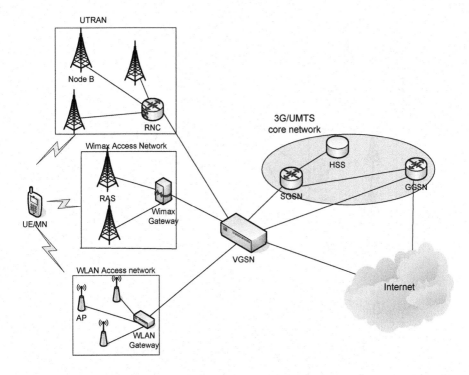

Fig. 1.4. Tight coupling integration at the GGSN level.

3G/UMTS core network does not identify the differences between WLAN/WiMAX
radio technology and 3G radio technology. The key functional elements in the system
are the WiMAX gateway which is connected with the Radio Access Stations (RAS),
and the GPRS Interworking Function (GIF), which is connected with the WLAN
APs. Both gateways are connected with a Serving GPRS Support Node (SGSN) in
order to provide a standardized interface to the 3G/UMTS core network and to virtually hide the WLAN and WiMAX particularities. In this context, performance during
handoff is expected to be better because fewer components located at the lower levels
of the 3G architecture are involved.

In [23], the WLAN and UMTS networks are interconnected in a tight coupling
way at the SGSN level, where the GIF is replaced with an Emulated Radio Network Controller (ERNC) that manages the WLAN network resources similarly to a
typical RNC. More specifically, the ERNC collects information concerning the resources of the attached Access Points (APs), such as traffic load, serving terminals,
received signal strength, and is responsible for the establishment of radio paths with
the User Equipment (UE). In order to provide efficient usage of the radio resources
and advanced handoff decision capabilities, three functional entities are introduced:
the Mobile Terminal Controller (MTC), the Advanced Radio Resource Controller
(ARRC), and the Light Radio Resource Controller (LRRC). The ARRC is an ex-

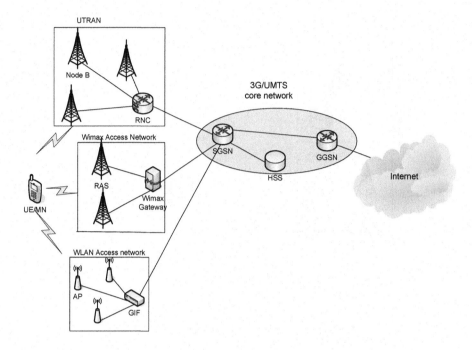

Fig. 1.5. Tight coupling integration at the SGSN level.

tended version of Radio Resource Control (RRC) protocol which handles the radio resource management of the supervising network elements. It manages the resources of the underlying Ues and Node-Bs, while communicating with the LRRC to acquire traffic load information. The LRRC, which is referred to as "light" RRC, is a functional entity that is limited to gather and report this information to the ARRC. Finally, the MTC is integrated into the RRC of the UE, and is responsible for radio link monitoring and processing of terminal parameters.

Finally, tight coupling may be done at the RNC level. This approach focuses on interworking at the UTRAN level and, more precisely, on incorporating RNC or lower UMTS entities' functionality into WLAN/WiMAX components. An example of this architecture is illustrated in Fig. 1.6 [28]. This integration is accomplished with the Interworking Unit (IWU) which is responsible for protocol translation and signaling exchange between the RNC and the other access points (hotspots and WiMAX BS). The critical point in this architecture is the RNC which transmits and receives traffic from both UTRAN and other wireless access technologies. Such architecture is mainly tailored to operators deploying their own WLANs and WiMAX networks, as the 3G/UMTS infrastructure is mostly reused [22].

The main advantage of this approach is the efficient mobility management, based on existing UMTS functionality that ensures at least service continuity, including authentication, authorization, accounting, and billing. This results in reduced handoff latency and seamless service continuation. In particular, the mobile users are able

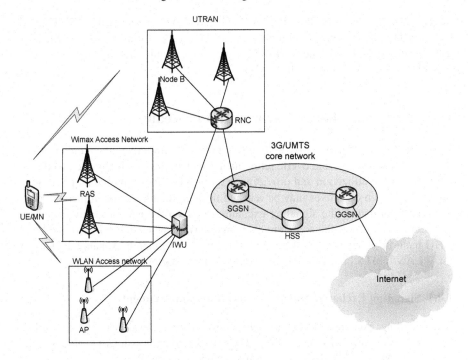

Fig. 1.6. Tight coupling integration at the RNC level.

to maintain their sessions, as they move from a network to another, whereas service continuation is subject to WLAN or WiMAX QoS capabilities. Furthermore, a large part of the 3G/UMTS infrastructure (e.g., core network resources, subscriber databases, billing systems) is reused, minimizing the cost of deployment [15, 22].

However, this approach presents several disadvantages [24]. Since the 3G/UMTS core network is directly connected with the WLAN and WiMAX network, the same operator will typically be required to own both WLAN, WiMAX network and 3G/UMTS networks. More specifically, independently operated WLANs cannot be integrated with 3G/UMTS networks without explicit physical connectivity to the 3G/UMTS core network. In this case, if users have a particular subscription to each network (e.g., SIM-based authentication for UMTS and login/password in WLAN), they will not be able to experience service continuity while roaming. On the other hand, a single subscription to one network with roaming privileges to another access network enables to avoid service interruption, as long as the different authentication entities are closely cooperating. Such cooperation has already been standardized by 3GPP [29].

In general, tight coupling integration is considered more complex than loose coupling approach. By injecting the WLAN and WiMAX network traffic into the 3G/UMTS core network directly, the setup of the entire network, as well as the configuration and design of network elements, have to be modified to support the increased traffic load. Furthermore, several extensions are needed in SGSN and GGSN nodes to support the large amount of data traffic from WLAN or WiMAX users through the UMTS network, while terminals should also include integrated 3G/WLAN/WiMAX functionalities [22].

1.4 Signaling Exchange During VHO Procedures

The main design objectives of a VHO mechanism are minimizing the number of unnecessary handoffs to avoid overloading the network with signaling traffic, maximizing the network utilization and providing active applications with the required level of QoS [2]. Such objectives may be reached through definition and implementation of procedures that define signaling exchange between the network equipments. In this section, we report the signaling exchange during connection establishment and VHO handoff procedures, as described in [23]. Such signaling exchange is defined for WLAN/UMTS integration tightly coupled at the SGSN level.

1.4.1 Signaling Exchange During Connection Establishment

The signaling exchange during an outgoing connection establishment through a WLAN is illustrated in Fig. 1.7 [23]. The UE periodically listens to radio signals from other APs or Node Bs, using passive scanning mode. When a new connection has to be established, it sends an *Activate Packet Data Protocol (PDP) Context Request* message to the SGSN. This message contains several parameters, such as the requested Network Service Access Point Identifier (NSAPI) that identifies the connection, the requested QoS, a PDP address field (e.g., an IP address), and an optional Access Point Name (APN) field that indicates the server for assigning the IP address. The PDP address may already be assigned statically during subscription. Otherwise, the field can be left empty in order to be filled dynamically during this procedure, and the APN server does the allocation.

Upon receipt, the SGSN verifies if the specified APN is valid according to the user's subscription, creates a PDP Context and uses the APN value to find the IP address of the GGSN from a Domain Name Server (DNS). After indicating the appropriate GGSN, the SGSN sends a Create PDP Context Request message to the GGSN, asking it to establish a tunnel between the SGSN and the GGSN. This message contains the Tunnel Endpoint Identifier (TEID) that indicates the signaling tunnel (TEID-C field) for the user plane tunnels in the downlink and the uplink directions (DL-TEID-D and UL-TEID-D, respectively), the DL-TEID-D field, the NSAPI identifying the specific PDP Context and the APN included in the previous message.

After successful tunnel negotiation and IP address allocation, the SGSN receives a Create PDP Context Response message including the IP address requested for the UE. The SGSN, in turn, orders the serving RNC to allocate resources (RAB Assignment Request), by specifying the Radio Access Bearer Identifier (RAB ID) value, the RAB QoS characteristics and the GPRS Tunneling Protocol (GTP) tunnel at the SGSN side. Then, the RNC asks the UE to report its measurements (Measurement

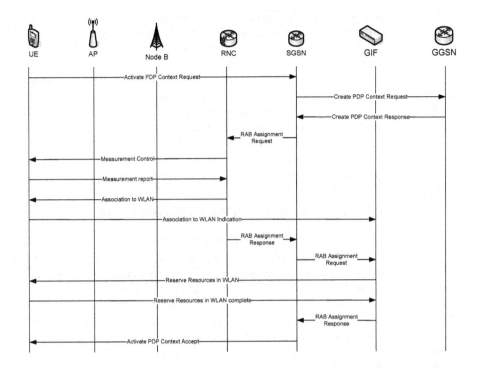

UE   AP   Node B   RNC   SGSN   GIF   GGSN

Fig. 1.7. Signaling exchange during connection establishment.

Control message). This triggers a process that selects the accessible APs or Node Bs according to several parameters, such as signal strength, user, terminal and service profiles. Depending on these parameter values, an ordered list is created and transmitted to the RNC (Measurement Report message). This list is checked to comply with the available network resources and operator's policy, and may be re-ordered by the RNC. If the WLAN network is on top of the list, a message indicates to the UE that it should associate with the WLAN (Association to WLAN message).

After WLAN association finishes, the UE acquires an IP address by a Dynamic Host Configuration Protocol (DHCP) server located in the AP or the GIF. This address is used locally in order to build an IP tunnel between the UE and the GIF, and sent to the GIF with an Association to WLAN Indication message. With this information, the GIF is able to map UMTS messages to WLAN specific ones and reserve resources for the UEs. In order to accelerate the whole process, the WLAN authentication, the WLAN association, and local IP address allocation phases along with the indication to the GIF can be executed in parallel with some UMTS signals, or right after finding WLAN coverage and before the connection establishment procedure.

In the mean time, the RNC answers the SGSN that it cannot serve the current connection (RAB Assignment Response). Then, the SGSN sends a new RAB Assignment Request toward the GIF as a directed retry to the WLAN. This, in turn, triggers a reservation phase in the WLAN (through the *Reserve Resources in WLAN* message). After the reservation procedure, a *Reserve Resources in WLAN Complete* message from the UE informs the GIF that resources have been reserved over the WLAN. The GIF sends a *RAB Assignment Response* to the SGSN in order to report the successful establishment of WLAN resources and the establishment of a tunnel between the GIF and the SGSN. Upon receipt of the previous message, the SGSN sends an *Activate PDP Context Accept* message to the UE, reporting the negotiated QoS and the global routable IP address of the UE assigned by the APN server. After this message, the UE configures the Network Access Session Layer (NASL) to route outgoing packets for this connection over the WLAN interface, which completes the connection establishment procedures [23].

1.4.2 Signaling Exchange During UMTS-WLAN Handoff Procedures

The signaling exchange during handoff from UMTS to WLAN is illustrated in Fig. 1.8 [23]. The UE listens periodically to radio signals from other APs or Node-Bs using a passive scanning mode. When a condition that may initiate a handoff (e.g., signal deterioration, user preferences, network availability) is met for a specific connection, a *Measurement Report* is sent to the RNC. This message contains the RAB ID and its QoS parameters that specify an ordered list of candidate access networks. This list is built according to radio signal measurements, user, terminal, and service profiles.

When the RNC receives this report, it may reorder the reported network list, based on the available network resources and the operator's policy. If the target network on top of the list is the same as the one that serves the specific connection, then the request is simply rejected. Here, the WLAN is assumed to be on top of the list and the serving network for the specific connection is UMTS. In this context, a relocation preparation phase begins, and the RNC sends a *Relocation Required* message to the SGSN. When the SGSN receives this message, it orders the GIF to allocate resources (*Relocation Request*) for the specific RAB, which enables to transfer the necessary information for the establishment of Radio Access Network Application Part (RANAP) signaling bearer and Iu bearer between the SGSN and the GIF [23].

Upon receipt of Relocation Request, the GIF builds the *Inter RAT Handoff from UTRAN* message. This may indicate radio frequency channels or a specific Basic Service Set ID (BSSID) as a recommendation to the UE, and encloses it into the *Relocation Request Ack* sent to the SGSN. Then, the SGSN makes sure that resources have been allocated for the specific RAB ID by sending a *Relocation Command* message to the RNC, including the original *Inter RAT Handoff from UTRAN* message. The RNC informs the UE that it should associate with the WLAN (if not already done) and that resources should be reserved (*Inter RAT Handoff from UTRAN*).

After that, the UE passes through WLAN authentication, WLAN association and local IP address allocation phases, and informs the GIF about its presence. When these procedures are completed, the UE informs the GIF that the resources have been successfully reserved (*Inter RAT Handoff from UTRAN Indication*), and configures the NASL to send all outgoing packets for this connection over WLAN. The GIF,

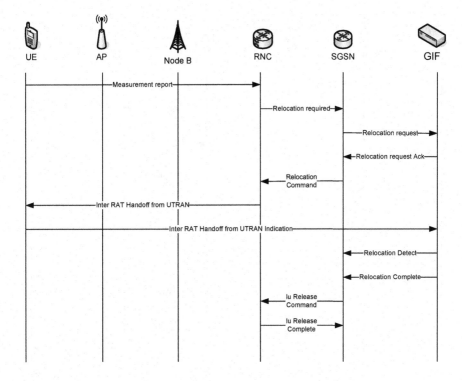

Fig. 1.8. Signaling exchange during UMTS-WLAN handoff.

in turn, reports the detection of the relocated connection to the SGSN (*Relocation Detect*), which switches the data flow in the downlink from UMTS to WLAN. Finally, the completion of the handoff procedure is reported to the SGSN (*Relocation Complete*) and a pair of *Iu Release* messages is exchanged between the RNC and the SGSN in order to release the data bearer for this connection in the old path [23].

1.4.3 Signaling Exchange During WLAN-UMTS Handoff Procedures

The signaling exchange for handoff procedure from WLAN to UMTS is illustrated in Fig. 1.9 [23]. In this case, the UE can use the signaling connection to transmit a *Measurement Report* message to the RNC. Once the UE makes the decision to roam from the WLAN to UMTS, the RNC sends a *Relocation Required* message to the SGSN. This process triggers a Relocation Request message in order to reserve resources in UMTS and to establish a signaling bearer between the RNC and the SGSN. The RNC reports successful allocation of resources to the SGSN and piggybacks the respective information for reservation of resources in UTRAN as an *Inter RAT Handoff to UTRAN* message (*Relocation Request Ack*) [23].

The SGSN orders the GIF to begin relocation (Relocation Command), and the GIF sends an *Inter RAT Handoff to UTRAN* message to the UE, using the WLAN

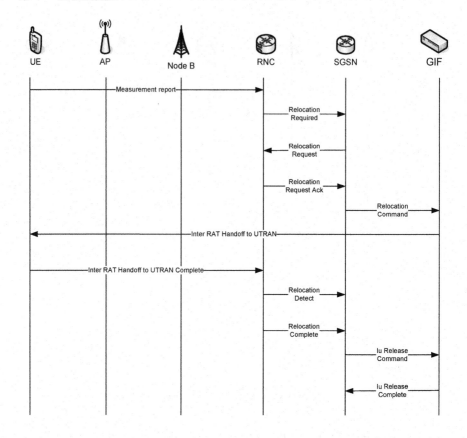

Fig. 1.9. Signaling exchange during WLAN-UMTS handoff.

Management (WLM) protocol. Upon receipt of this message, the UE establishes a RAB with UMTS, and reports successful completion of the handoff to RNC (*Handoff to UTRAN Complete*). After this message, traffic flow can begin over UMTS in the uplink. When the detection of the relocation is reported to the SGSN (*Relocation Detect*), the data flow in the downlink is switched from the WLAN to UMTS. After Relocation Detect, the completion of the handoff procedure is reported to the SGSN (*Relocation Complete*). Finally, the SGSN sends an *Iu Release* Command to the GIF in order to release any resources between WLAN and the core network. Then, an Iu Release Complete message confirms the successful release of the resources in the WLAN.

Conclusion

In order to provide mobile users with real-time multimedia services seamlessly, WLAN, WiMAX and 3G/UMTS networks will be integrated to form a heterogeneous wireless access network. Such integration will enable to provide not only efficient usage of the network resources, but also universal coverage and broadband access. This chapter has presented architectural options for such integration, as well as signalling exchange between the network equipments during VHO procedures. Further work should be oriented towards performance evaluation of the presented architectures, in terms of generated signalling traffic and packet loss.

References

1. R. C. Chalmers, G. Krishnamurthi, and K. C. Almeroth, "Enabling Intelligent Handovers in Heterogeneous Wireless Networks," *Mobile Networks and Applications*, vol. 11, pp. 215-227, 2006.
2. A. H. Zahran, B. Liang, and A. Saleh, "Signal Threshold Adaptation for Vertical Handoff in Heterogeneous Wireless Networks," *Mobile Networks and Applications*, vol. 11, pp. 625-640, 2006.
3. A. Hecker, H. Labiod, G. Pujolle, H. Afifi, A. Serrhouchni, and P. Urien, "A New Access Control Solution for a Multi-Provider Wireless Environment," *Telecommunication Systems*, vol. 29, no. 2, pp. 131-152, 2005.
4. F. Xu, L. Zhang, and Z. Zhou, "Interworking of WiMAX and 3GPP Networks Based on IMS," *IEEE Communications Magazine*, vol. 45, pp. 144-150.
5. A. Bria, F. Geesler, O. Queseth, R. Stridh, M. Unbehaun, J. Wu, and J. Zander, "4th-Generation Wireless Infrastructures: Scenarios and Research Challenges," *IEEE Personal Communications* vol. 8, pp. 25-31, 2001.
6. A. Grilo, M. Nunes, G. Sergio, and N. Ciulli, "Integration of IP Mobility and QoS for Heterogeneous Wireless Access in MOCAINE," in Proc. of *IEEE 14th Conference on Personal, Indoor and Mobile Radio Communications*, 7-10 Sept. 2003, vol. 1, pp. 470-475.
7. W. Jiao, P. Jiang, and Y. Ma, "Fast Handover Scheme for Real-Time Applications in Mobile WiMAX," in Proc. of *IEEE International Conference on Communications*, 24-28 June 2007, pp. 6038-6042.
8. K. H. Teo, Z. Tao, and J. Zhang, "The Mobile Broadband WiMAX Standard," *IEEE Signal Processing Magazine*, vol. 24, pp. 144-148, 2007.
9. B. Li, Y. Qin, and C. L. Gwee, "A Survey on Mobile WiMAX," *IEEE Communications Magazine*, vol. 45, pp. 70-75, 2007.
10. E. Vanem, S. Svaet, and F. Paint, "Effects of Multiple Access Alternatives in Heterogeneous Wireless Networks," in Proc. of *IEEE Wireless Communications and Networking*, 16-20 March 2003, vol. 3, pp. 1696-700.
11. D. O'Mahony and L. Doyle, "Beyond 3G: 4th Generation IP-based mobile networks," in *Wireless IP and Building the Mobile Internet*, Artech House, Norwood, pp. 71-86, 2002.
12. K. Murray and D. Pesh, "Intelligent Network Access and Inter-System Handover Control in Heterogeneous Wireless Networks for Smart Space Environments," in Proc. of *1st International Symposium on Wireless Communication Systems*, 20-22 Sept. 2004, pp. 66-70, 2004.
13. L. D. Chou, J. M. Chen, H. S. Kao, S. F. Wu, and W. Lai, "Seamless Streaming Media for Heterogeneous Mobile Networks," *Mobile Networks and Applications*, vol. 11, pp. 873-887, 2006.

14. *ETSI TR 101 957*, Requirements and Architectures for Interworking between HIPER-LAN/2 and 3rd Generation Cellular Systems, Tech. Rep., 2001.
15. A. K. Salkintzis, C. Fors, and R. Pazhyannur, "WLAN-GPRS Integration for NG Mobile Data Netwoks," *IEEE Wireless Communications*, vol. 9, pp. 112-124, 2002.
16. *3GPP TR 21.910 3.0.0*, Multi-mode UE issues; categories, principles and procedures.
17. *3GPP TR 25.306 5.9.0*, UE Radio Access capabilities definition.
18. X. Gelabert, J. Perez-Romero, O. Sallent, and R. Agusti, "On the Impact of Multi-mode Terminals in Heterogeneous Wireless Access Networks," in Proc. of *2nd International Symposium on Wireless Communication Systems*, 5-7 Sept. 2005, pp. 39-43.
19. B. Liang and Z. J. Haas, "Predictive Distance-Based Mobility Management for Multi-dimensional PCS Networks," *IEEE/ACM Transactions on Networking*, vol. 11, no. 5, pp. 718-732, 2003.
20. H. Petander, E. Perera, and A. Seneviratne, "Multicasting with Selective Delivery: A SafetyNet for Vertical Handoffs," *Wireless Personal Communications*, vol. 43, pp. 945-958, 2007.
21. S. Park, J. Yu, and J. T. Ihm, "A Performance Evaluation of Vertical Handoff Scheme between Mobile WiMax and Cellular Networks," in Proc. of *16th International Conference on Computer Communications and Networks*, 13-16 Aug. 2007, pp. 894-899.
22. L. Lampropoulos, N. Passas, A. Kaloxylos, and L. Merakos, "Handover Management Architectures in Integrated WLAN/Cellular Networks," *IEEE Communications Surveys and Tutorials*, vol. 7, pp. 30-47, 2005.
23. L. Lampropoulos, N. Passas, A. Kaloxylos, and L. Merakos, "A Flexible UMTS/WLAN Architecture for Improved Network Performance," *Wireless Personal Communications*, vol. 43, pp. 889-906, 2007.
24. M. Buddhikot, G. Chandranmenon, S. Han, Y. W. Lee, S. Miller, and L. Salgarelli, "Design and Implementation of a WLAN/CDMA2000 Interworking Architecture," *IEEE Communications Magazine*, vol. 41, pp. 90-100, 2003.
25. M. Inoue, K. Mahmud, H. Murakami, and M. Hasegawa, "MIRAI: A Solution to Seamless Access in Heterogeneous Wireless Networks," in Proc. of emphIEEE International Conference on Communications, 11–15 May 2003, vol. 2, pp. 1033-1037.
26. J. Y. Song, H. J. Lee, S. H. Lee, S. W. Lee, and D. H. Cho, "Hybrid Coupling Schemes for UMTS and Wireless LAN Interworking," *AEU International Journal on Electronics and Communications*, vol. 61, pp. 329-336, 2007.
27. S. L. Tsao and C. C. Lin, "VGSN: A Gateway Approach to Interconnect UMTS/WLAN Networks," in Proc. of *13th IEEE International Symposium on Personal, Indoor, Mobile Radio Communications*, 15-18 Sept. 2002, pp. 275-279.
28. N. Vulic, I. Niemegeers, and S. H. de Groot, "Architectural Options for the WLAN Integration at the UMTS Radio Access Level," in Proc. of *IEEE 59th Vehicular Technology Conference*, 17-19 May 2004, Milan, Italy, vol. 5, pp. 3009-3013.
29. *3GPP TS 23.234 V6.2.0*, 3GPP System to Wireless Local Area Network (WLAN) Interworking; System Description (Release 6), 2004.

2

Integration of Heterogeneous Wireless Access Networks with IP-based Core Networks: The Path to Telco 2.0

Peyman TalebiFard[1]*, Terrence Wong[2], and Victor C. M. Leung[1]

[1] Dept. of Electrical and Computer Engineering, The University of British Columbia
Vancouver, B.C., Canada, V6T 1Z4
{peymant,vleung}@ece.ubc.ca
[2] Telus, 200 Consilium Place Toronto, Ontario, Canada M1H 3J3
Terrence.Wong@telus.com

2.1 Introduction

The Internet is composed of different domains operated by Internet service providers (ISPs) with different capabilities, policies and access networks. Next generation networks (NGNs) will employ standards and architectures that are based on the Internet Protocol (IP) suite. The IP platform can integrate diverse access networks in a common scalable framework, and through IP Multimedia Subsystem (IMS), extend a wide range of multimedia services to subscribers over heterogeneous wireless and wireline access networks. Transitioning to all-IP NGNs builds the foundation of convergence at three levels, namely, application, network, and service.

Application convergence can provide numerous opportunities for carriers because mobile devices are increasingly capable of supporting multiple functions such as voice, video, email, and web browsing.

Network convergence means to integrate several application-specific networks into a single IP-based multi-service infrastructure. This leads to significant reduction of infrastructure costs, capital and operating expenses. It is therefore important that providers develop new services to substitute for legacy applications over the new infrastructure because running old services on new IP networks causes more problems and higher operating costs.

Technical capabilities of IP and IMS are also enabling service convergence. Service convergence delivers more intelligent application level and subscriber level service control. In other words, service providers will be able to bill, operate, and manage their services over a wide range of access networks.

Multi-modal mobile devices that provide several alternate means to access the Internet are becoming widely available. However, mobile users are more interested

---

*This work represents the personal opinion of the authors and not the respective employers.

E. Hossain (ed.), *Heterogeneous Wireless Access Networks*,
DOI: 10.1007/978-0-387-09777-0_2, © Springer Science+Business Media, LLC 2008

in being able to receive services in an access agnostic manner rather than dealing with specific access technologies. They want the technology specifications of the access network to be transparent to them. IP technology is capable of offering service to users in a technology transparent manner. IP-based access networks such as wireless local area networks (WLANs) are becoming important components of public access networks. Furthermore, IP-based wireless access networks can integrate with the Internet more easily and with lower cost than traditional circuit-switched networks.

On the other hand, market and business drivers are also playing a major role in convergence. The telecommunications industry is now in a consolidation phase where the biggest challenge is the integration of heterogeneous networks with IP core networks and IMS. Using an IP backbone to integrate various networks is advantageous because many of the integration issues can be solved with standard network elements and next-generation applications and services can be efficiently and effectively deployed over the resulting NGN.

From a non-technical point of view, convergence is a matter of cost, competition, and regulatory legacy. One of the driving forces of convergence is cost, which has several influencing factors. The first one is accomplished by a shared infrastructure employing a common technology, which leads to reductions in network management costs, support staff costs, and costs associated with service modifications and intra-office calling charges. The second cost driver is capturing of new revenue streams through new business models. The third cost driver is increased productivity and efficiency through introduction of new features. The fourth cost driver is better control of capital that leads to reduction of operational and financial risks through deployment of a single management interface. The fifth driver is client demands for IP-based products.

Another driving force of convergence is competition. An example feature that can create competition is number portability, a feature that allows users to switch between service providers without changing their telephone numbers. This mechanism enhances fair competition among telecommunication service providers. There are three types of number portability:

1. Location portability, which allows a subscriber to move to another location without changing the number;
2. Service portability, which enables a subscriber to keep the same telephone number when changing services;
3. Operator portability, which allows a subscriber to change service provider without changing the telephone number [3].

Bundling is a necessary strategy for service providers to acquire new customers, keep their existing customers, and increase revenue streams. In this manner, customers who are more comprehensively engaged with a service provider are less likely to switch to another provider. Regulatory driving forces are different from the aforementioned, which are driven by economics [1].

During different periods in the evolution of contemporary telecom networks, different layers in the network protocol architecture become converged. Fig. 2.1 shows development of networks from the past to the future. The past approach based on the so-called *stove pipe* model is evolved to the future *service-based* approach. In this approach, resources are being integrated to support a variety of services.

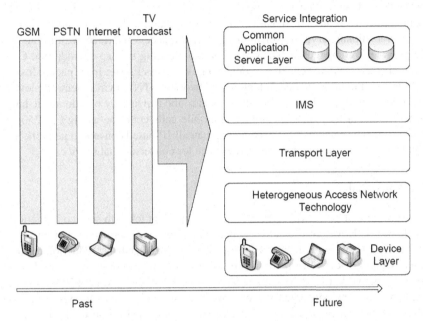

Fig. 2.1. Development of telecom networks from past to future.

Integration of heterogeneous networks may be considered under two different aspects. One aspect is from the perspective of service providers managing the networks according to different transmission media such as cable, satellite, radio that may also implement different protocols such as Asynchronous Transfer Mode (ATM), IP or Multi-Protocol Label Switching (MPLS). Another aspect is from the point of view of users who can access different services with different features, availability and prices [5].

In this chapter, Section 2.2 illustrates several use cases for conversational communications in NGNs, in order to highlight some of the requirements of a converged heterogeneous network environment. In Section 2.3 we introduce IMS, its architectural elements and business aspects as a core network capability to support network and service convergence. Section 2.4 addresses the implementation issues at the service and control layer where the integration of heterogeneous access networks using IMS and functions of the Service Plane to provide users with a converged service and network access experience are described. Furthermore, in Section 2.5 we focus on the development of standards and technical issues of integrating heterogeneous networks with IMS at the core network and access network layer by addressing quality of service (QoS) provisioning and mobility support issues. Section 2.6 outlines some future research directions.

2.2 Use Cases for Conversational Communications over Converged Heterogeneous Networks

Conversational communications consist of interactive voice, video, and other multi-media applications that are sensitive to round trip latency. Users should be able to port conversational services over multiple network access interfaces within the service provider's converged network without experiencing service degradation. Conversational services are sensitive to delays but have less impact from packet losses. In the legacy Public Switched Telephone Network (PSTN) environment employing time division multiplexing (TDM), the typical round trip latency of a domestic land-line call is about 60 ms while mobile-to-mobile calls could be as high as 250 ms due to air interface delays. It is expected that an all-IP-based converged network will provide a similar, if not better, user experience for conversational services.

2.2.1 Voice Call

It is expected that voice will remain the most important telecom service in the fore-seeable future. However, the delivery mechanism for voice service is evolving from TDM to Voice over IP (VoIP) due to value-added services and cost reduction requirements. Although new voice features are easier and more cost effective to implement over an IP network, the TDM-based PSTN network will co-exist with IP for another decade or two. The converged voice network must be able to serve existing PSTN end-points. Service provider should also consider using the IMS network to simulate (e.g., new IP phone equipped with existing PSTN phone features) and emulate (i.e., using IMS and the Session Initiation Protocol (SIP) to replicate PSTN signaling for existing PSTN phones) PSTN functions to support voice service migration. These functions will be described in more detail in Section 2.3.

Use Case – Consumer

Tom, married with 2 kids, would like to have a converged voice service that works for his family. Tom has access to broadband digital subscriber line (DSL) with QoS enabled and Wi-Fi (IEEE 802.11 b/g WLAN) coverage at home. He gives out his mobile number to most people and home number to family members only. He wants all incoming calls (to his mobile and landline) to be delivered to his mobile when he is present on the mobile network. If he is not on the mobile network, mobile calls will be routed to his office phone during office hours or to his home with a special ringing tone after working hours. When he arrives home, his 3G/Wi-Fi dual mode phone will detect the home WLAN Access Point (AP) signal and register his presence in the home network. All incoming mobile calls after this point should be delivered to his landline phone with a special ringing tone or to his mobile using Voice over WLAN (VoWLAN) if the landline phone is busy. If the landline phone rings for 6 times (defined by the user on the service portal) with no answer, the call will be routed to the voice mail system that is common to his mobile and landline services. Tom also wants to route all incoming calls to voice mail from 12:00am-7:00am except those callers/numbers that are on a special list defined by him on the portal. The voice mail system should send notification for message deposit or withdrawal to both landline and mobile devices in a near real-time manner. Mary, Toms wife, stays home most of the time and would like to have both desk top and cordless Wi-Fi phones at home. She has many friends but is not good at memorizing phone numbers. She uses her service providers network-based phone book services to store all contacts. The advantage of a network-based phone book is that Mary can search the contact list with any phone at home. She can also share common contact names such as family members, piazza place, etc., with Toms and her kids network-based phone book. Jeff and Stephanie, Tom and Marys children, are teenagers. They have an active social life and many phone calls. Tom bought Jeff and Stephanie dual mode mobile phones under the family plan. Jeff and Stephanie want their calls to be delivered to the dual mode phone or to a soft phone client on their laptop or personal digital assistant (PDA) if theyre actively on-line. Both of their dual mode phones and soft phone clients should be able to access the network-based phone book.

Use Case – Enterprise

Jennifer is an information technology (IT) manager of a small/medium enterprise (SME) in British Columbia. Most of her companys employees are in a sales role and must meet with clients at the customers premises. Clients complain that they have to make multiple calls to reach their sales person. Most clients end up calling the sales persons cell phone even when they are in the office. Jennifer wants a solution that provides them with a single number and a unified voice mail between landline and mobile services so that most incoming calls will reach the called party instead of voice mail. Jennifer has just upgraded the in-house private branch exchange (PBX) system to support IP calls and SIP signaling and implemented a Wi-Fi network through the office. Her voice service requirements are the following:

- Calls will be delivered to sales desk phone if the sales person is in the office
  - based on location of the mobile device as provided by the mobile network, or
  - based on IP registration through mobile Wi-Fi interface, or
  - a touch icon on mobile for call forwarding.
- If the desk phone rings with no answer, Wi-Fi pages the sales persons mobile device (if the user is registered through the Wi-Fi interface of the mobile device) or calls the sales persons mobile phone;
- If the sales person did not answer the phone, the call is sent to voice mail;
- Voice mail then sends out notifications to the sales persons desk phone, mobile phone and email account. The sales person can use any of these devices to retrieve the voice message. When the message is retrieved, the voice mail system sends out a message retrieved notification to the devices above.

2.2.2 Video Call

With increasing bandwidth in broadband networks and increasing capability of end-user devices, video calls are expected to become a value-added service to pure voice communication. It is expected that many of the devices connected to smart homes and mobile networks will be equipped with a camera and be able to send and receive video calls. This video call service will be built on the all-IP architecture. The calling party can use an E.164 number (i.e., regular phone number) or a SIP Universal Resource Identifiers to reach the called party. In addition to a network-based address book, the user can also find a friends presence information on their devices. Due to multiple end-user devices and access networks requirements, the convergence team should specify the bandwidth and codec requirements for video calls to minimize format conversions. If format conversion is needed, it should be performed in an application server to reduce handset processing load. A typical video call may require up to 128kbps for a 4x4 inch mobile or desktop screen.

Use Case – Video Call with Presence

John subscribes to his service providers converged video call service which allows him to make video calls from his mobile phone, home phone and soft phone in his laptop/PDA. The video call client in these devices can access the buddy list and user profile that contains the users presence and capability information. John, with his video mobile phone, decides to call Kati who is on-line with her laptop. Kati, Johns best friend, has allowed a number of users including John to see her presence and user capability information. John knows that Katis laptop can support video phone function because there is a video call check mark besides Katis name on Johns mobile buddy list. In this case, both users are registered for IMS service in the IP network. When John clicks Katis name for a video call, the bearer traffic carried through the video call server Application Layer then reaches Katis video phone on her laptop. Since the video encoding from Johns mobile phone is different from Katis video client, the video call server is actually performing format conversion to bridge the two devices.

2.2.3 Push-to-X Call

Push-to-X refers to a half-duplex communications service between two or more parties and is usually used for short and quick conversations. Push-to-X described in this chapter is an IP-based system that allows a user to send voice and multimedia contents to other parties. With the recent developments in the Open Mobility Alliance (OMA), it is expected that Push-to-X can be deployed in mobile, desktop (soft client) and gaming devices. Similar to video calls, Push-to-X call utilizes SIP signaling over the IP infrastructure for user registration, and to access presence, user capability and address book information.

Use Case – Push-to-X Call over Mobile and Desktop

John is surfing on-line at home to determine available movies. He finds a movie he likes and would like to invite Kati to see it. John brings up the Push-to-X client on his laptop and sees that Kati is available on her mobile. John clicks on Katis name and chats with her. She is not sure if the movie is good. John wants to convince her by sending her a 3 minute movie trailer file on the same Push-to-X session. The file is about 5 MB and is sent from Johns Push-to-X client to the application server. Because Kati has set up a rule in her user profile that does not allow receipt of files larger than 3 MB on her mobile, the Push-to-X server sends a text message to Katis mobile asking if she wants to switch the Push-to-X session to her laptop. Kati happens to be at home with the laptop connected to DSL and Wi-Fi. She answers Yes on the mobile. The SIP network brings up the Push-to-X client in Katis laptop and switches the session from Katis mobile to her laptop. The movie trailer file arrives almost immediately and is shown on Katis laptop.

2.2.4 Location-Based Service

Location-based technology enables the linkage between physical objects and associated data, effectively turning the world into a geospatial information bulletin board. Location-based data accessed as part of a use case might include:

- Environmental details,
- Cultural information,
- Historical information,
- Mythology,
- Social information about people nearby,
- Geo-demographic information about the local community,
- Micro-local commercial information,
- Specialized enterprise and industry data,
- Safety information based on actuarial data about health, accidents, and crime,
- Political data,
- Facilities information such as telecommunications availability,
- Local public services from government agencies,
- Tying to-do lists to location information to facilitate running errands such as grocery shopping or reminding the user to pick up some spare light-bulbs when passing by a hardware store.

Use Case – Emergency Broadcast

XYZ Chemical is a community minded company and they are concerned about public safety in the event an accident occurs at one of their plants. Should an event occur they want a way to directly notify the public that are in the impacted area. Since the public in these areas may subscribe to fixed or mobile phone services, they have contracted a converged operator as the 911 service provider to provide an emergency broadcast service. This service will notify all customers in the affected area using the location based information associated with the customer devices. If a customer is on a mobile device they will receive a message via the Short Messaging Service (SMS) notifying them of the potential hazards. If the customer is utilizing IP-based television (IPTV) service then the broadcast information is presented on the television set. Notification also occurs over VoIP devices. In addition, the service attempts to notify over legacy voice technologies.

2.2.5 Message-Based Communications

Messaging is becoming an important part of overall communication needs for consumer and enterprise users. Although North American adoption of messaging is not as high as in Europe (due to a flat rate voice plan), it is expected that advanced messaging services proven on other continents will benefit local users by simplifying the services and allowing them to better manage their time. A messaging service is delay insensitive but highly sensitive to packet loss. The bandwidth requirement of each service varies depending on the media and applications. Messaging services can be sent over multiple access interfaces to different end-user devices in the converged network because of its capabilities. Service quality of various access interfaces plays an important role in determining how end-users perceive these services; e.g., 2.5G wireless access supports a lower throughput than DSL, 3G or Wi-Fi. Future home and office desktop phones will be equipped with a 4x4 inch color screen that can display multimedia content, provide touch screen functions and select contact lists. These phones will be IP-based and be connected to home or office gateways.

2.2.6 Multimedia Messaging Service

Multimedia Messaging Service (MMS) technology originated in the mobile environment to deliver multimedia content such as pictures, and small files. The existing MMS delivery mechanism is person-to-person with a store-and-forward mechanism. Future MMS will retain this service characteristic with access to user/device capabilities, presence and location information.

Use Case

Mary had asked Tom to pick up some cereal for the kids before he came home after work. Tom went to the supermarket but was not sure which brand to buy. He took pictures of 2 different cereals with his camera mobile phone and sent them to his wife. The MMS system checked with Marys user profile and presence and saw that Mary was on the mobile network and also registered an active SIP session on her laptop. Based on Marys MM preference, the laptop was her first preference before mobile for MMS. She received the pictures on her laptop screen instantly and replied to Tom with a check mark against one picture. Tom got the reply message on his mobile phone a second later.

2.2.7 Instant Messaging Service

Instant Messaging (IM) is a growing communication tool in the Internet community and is expected to replace part of the voice service. There are a number of IM services, such as MSN, Yahoo and AOL, which dominate the market place. A service providers IM service should be able to inter-work with these solutions to gain market traction. Presence information is the key enabler of IM service and will be shared with other multimedia enablers to create compelling applications. There are two open standard bodies (i.e., OMA and Third Generation Partnership Project 2, 3GPP2) developing presence capability in IMS and mobile environment and they should be considered as the base for presence architecture for this document.

Use Case – IM Whiteboard

Paul is driving to his friend Alans housewarming party. Alan has bought a new house in one of the developing areas north of Toronto. Because the area is under development, street signs are not available and hence, Paul gets lost. Paul decides to call Alan for help. Alan tries to describe how to get to his house but Paul does not feel comfortable because he is not familiar with the area. Alan picks up his laptop in the living room and brings up the IM client. Alan continues to talk with Paul over the phone; Paul, I can see youre on-net; just wait, I will sketch a map for you. Alan quickly draws the map on the IM client and sends over to Paul. Paul receives the map over the IM session on his PDA phone and tells Alan, "I've got it."

2.2.8 Unified Messaging Service

Messaging functions defined in this chapter include SMS, voice mail, email, Multimedia Messaging Service (MMS), and Instant Messaging (IM). We explore the possible integration of these functions into a Unified Messaging (UM) service to enable compelling services and cost reduction in the convergence environment. The recent development of voice Extended Markup Language (XML) [19] technology is a good example of how messaging services can be converged in text, voice, and multimedia.

Use Case – Consumer

Tom had created a user profile on the user portal and indicated that all incoming calls to his mobile and home phones after 12:00am should be routed to the UM system. On Saturday morning at 1:00am, one of Tom's family members from overseas called his mobile phone and left him a voice message. The UM system records the caller's name in text by voice recognition technology and sends a notification message to Tom, voice message from Uncle Sam - marked urgent to Tom's home and mobile phone. Tom gets the notification the next morning and uses his home phone touch screen to retrieve the voice message. After the voice message is retrieved, the UM system sends a notification to Toms mobile phone to cancel the voice messaging waiting indicator.

Use Case – Enterprise

Robert is attending an important meeting and is about to conduct a presentation. He does not want any incoming call to his mobile phone to interrupt him. He sets the mobile to Do Not Interrupt mode. Nicole, Roberts wife, wants to know if Robert wants to go out for dinner. When she picks up the phone, she sees there is a Do Not Interrupt icon besides Roberts name on her phones buddy list. She decides to leave him a voice message anyway, knowing that he will not be returning this call for a while. The system realizes Robert is in Do Not Interrupt mode and searches in Roberts profile for the preferences to handle this message. Robert had instructed the system to send voice messaging to his email account if his mobile phone is in Do Not Interrupt mode. When Robert finishes the presentation and goes back to his office, he sees that there is a message in his email account from Nicole, time deposited, duration, etc. He double-clicks the message and listens to the voice message via the email client.

2.2.9 Content-on-Demand Communications

As increased bandwidth and advanced device capability become available in consumer and enterprise segments, content-on-demand services become feasible. These services allow consumers to access content in real-time over multiple access networks and allow enterprises to push commercial advertisements to user devices based on their locations in order to target specific customers. Telecom operators around the world are implementing an integrated entertainment strategy to defend their positioning with the cable companies. Entertainment is believed to be the 3rd highest area of spending in household budgets, behind education and healthcare, and is one of the major drivers for an all-IP network architecture in the telecom industry.

2.2.10 Browsing Service

Browsing service refers to Internet surfing with various terminals and devices. Internet surfing clients on desktops and laptops are standardized with several available web browsers, while wireless devices use Wireless Application Protocol (WAP) clients to customize the content to be presented on the limited screen. In order to converge the Internet surfing experience, not only does the system have to know all end-user device capabilities but also be able to port services over multiple access interfaces.

Use Case

Paul subscribes to a service providers Internet access service that allows him to surf the Internet over DSL, mobile broadband and hotspot networks. The system stores his user profile in a centralized database so that he can use a common user name and password to authenticate the service over multiple access interfaces. The system also knows the device capability and configuration of his laptop and PDA with Wi-Fi interfaces. Paul arrives at Toronto airport traveling from Vancouver. He wonders if there is a special deal on the web site of the rental car companies in the terminal. He pulls out his PDA and finds that there is Wi-Fi hotspot access. He starts to surf the web and walks toward the rental car pick-up station. Because the hotspot coverage ends at the baggage claim area, the system switches Pauls PDA Internet session from hotspot to EV-DO without service interruption. User and device authentication and authorization are done seamlessly in the backend system.

2.2.11 Content Download

Content download is popular especially in the music and gaming segments. To protect intellectual property rights, the industry is leading toward Digital Rights Management (DRM) mechanisms to restrict users in how they can port or share content between various devices.

Use Case

Calvin uses his mobile device to download a heavy metal rock song from a content providers web site. He likes the song and would like to share it with his buddy Ralph, his grade 10 classmate. After trying the song, Calvin chooses the share icon on the mobile screen and forwards the song to Ralph. Ralph, playing an on-line game at home, receives an IM session from the system that says Calvin wants to share MP3 music with you. Ralph clicks OK and his laptop begins playing the rock song.

2.2.12 Content Streaming

Content streaming refers to a media file that is being played on the server side and streaming through the telecom network. Different from download-and-play, streaming requires QoS support on the network to ensure user experience. Depending on the content type, streaming content can be real-time like news or static like movie trailers.

Use Case

Paul is waiting for a flight back to Vancouver in the Toronto airport. His flight is delayed for one hour because of a maintenance issue. He remembers that he had received a movie ad from his content providers portal for a newly released movie. He brings his laptop with Wi-Fi access to the hotspot and logs in to the content portal to start streaming the movie. When he is half-way done with the movie, the ground crew starts to call for boarding. He is glad that his providers streaming service has a feature to remember where he left the streaming content so that he can continue with the same content on a later day. That evening when he arrives home, he brings up the home Wi-Fi network to finish the movie. The system knows where he left and continues to play the movie from that point.

2.2.13 Content Push

Content push refers to content delivery over various access interfaces based on user demand or network trigger. Typical applications can be location-based advertisements or map-quest services.

Use Case

Jimmy goes to Scarborough Town Center (STC) to shop. He does not have any shopping items in mind but wonders what is on special. He pulls out his mobile phone which is connected to the 3G network. From the shopping menu on his cell phone, he selects shopping discount. The system queries his current location and knows that he is in STC. The content server searches all discounts available in STC and pushes them to his mobile phone.

2.2.14 Mobile Broadcast and IPTV

Broadcast refers to uni-directional content delivery over a wireless or fixed broadband to end-user device or terminal. Different content broadcast technologies exist in landline and mobile environments.

Use Case

Kevin, a subscriber of mobile broadcast and IPTV services, closely follows a TV show. Due to his travel schedule, he cannot be at home to watch the final episode. However, he has received a program guide in his mobile that this episode will be broadcast over the wireless network. He brings up the mobile broadcast client and watches half of the show. The system records the duration Kevin has watched and makes an indication on his IPTV user profile. When Kevin returns home, he brings up the IPTV program guide, and the system continues the TV show where he left off.

2.2.15 Peer-To-Peer Application

Peer-to-peer service refers to direct communications between two end-user devices. Peer-to-peer communications are different from client-server communications in that the service provider likely does not have control of the user traffic in a peer-to-peer communication session. With the Internet experience and young generation adoption of this service, peer-to-peer service will inevitably be requested by end-users.

Use Case

Calvin has just downloaded an online game to his laptop and mobile because his buddy, Ralph, highly recommended the game. After the download is completed, he sees Ralph is online with his mobile. Calvin sends an IM to Ralph and asks if he wants to play the game. Ralph, sitting in the back seat of his parents car traveling on a freeway, is bored and cannot think of a better thing to do.

Use Case

Susan has enabled the information sharing capabilities of her handheld device. This allows the device to communicate with other devices to exchange information. The amount of information shared between devices is under Susans control. The amount of information shared depends on the trust levels between the devices. For example, Susan walks into a store and her handheld device detects and communicates with the store's network access device. Since the store is a trusted entity, her handheld communicates personal information to the server. As a preferred customer, the server responds (via the store's network access device) and provides Susan with personalized promotion information on her handheld device. She decides to purchase one of the promotional items. When she approaches the till, the server notifies the clerk of the promotional items as well as information about Susan. Given this information the clerk is able to provide Susan with personalized service.

2.3 IP Multimedia Subsystem (IMS)

The requirements for enabling ubiquitous and unlimited access for users to networks and to enable and enhance the competition of service providers, have led to the definition of NGNs in which service related functions are made independent from underlying transport related technologies. Furthermore, NGN will enable service providers of any type to deliver their services to end users in a network and terminal agnostic manner. IMS [21] is the enabler of converged communication technologies over NGNs [16].

IMS was initially defined by the 3rd Generation Partnership Project (3GPP) and 3GPP2. 3GPP is in charge of developing specifications for the Global System for Mobile communications (GSM). GSM has two modes of operation: circuit-switched and packet-switched. GSM circuit-switched (CS) mode uses technologies that are commonly used in the PSTN. CS networks have two planes: signaling plane and media plane. The signaling plane is in charge of service invocation that includes protocols to establish a connection between calling and called terminals. The media plane is in charge of user data transmissions between end terminals. General Packet Radio Service (GPRS) is a standard for GSM packet-switched network that forms the base for 3GPP release 4 packet-switched domain. This domain allows users to connect to the Internet via native IP-based packet switching.

IMS provides an architecture that enables ubiquitous cellular access, convergence of fixed and mobile networks, user mobility, access-agnostic application development and a service centric framework that makes the development of new revenue generating services as described in the previous section possible [2].

The aims and motivations of IMS are:

1. To combine the latest trends in technology (i.e., fixed, wireless, and mobile networks convergence);
2. To enable user handover, roaming and mobile Internet;

3. To build a common platform for creating drivers of multimedia applications and possibility of using any terminal type such as personal computer, PDA, mobile telephone, etc.;
4. Convergence of various data communication types.

In order for IMS to deliver services to end users, the following requirements should be met:

1. Enabling establishment of IP multimedia sessions – Audio and video communication services over packet-switched networks are of the most important services that benefit from this.
2. Support of QoS negotiation – IMS shall allow operators to differentiate between groups of customers by controlling the QoS provisioned to each session.
3. Interworking with the Internet and CS networks – Interworking with the Internet expands the potential sources and destinations of multimedia sessions. Interworking with CS networks allows co-existence of IMS-based NGN with the legacy PSTN and enables connections between IMS terminals and ordinary telephones in PSTN domains.
4. Support of subscriber roaming between service domains - Roaming capability enables subscribers to access subscribed service while visiting a foreign service domain. Roaming has always been supported by cellular networks in the second generation and beyond, but limited to service domains employing a common access technology. IMS shall retain this requirement and further extend this capability across service domains employing heterogeneous access technologies.
5. Policy-based service delivery - This allows operators to control services delivered to end users by imposing general policies that apply to all the users in a network and individual policies that apply to a particular user.
6. Rapid service creation – IMS shall reduce the time it takes to introduce new services by standardizing service capabilities so that new services can be readily composed from these capabilities. This obviates the need to pre-establish a large selection of standardized services except for those that are widely subscribed to.
7. Multiple access technology support – IMS is IP-based and independent of the network access technology so long as it supports IP [2]. IMS shall support access to networks other than GPRS, such as WLAN and broadband wireless access.

2.3.1 Architectural Elements

3GPP is also responsible for standardizing a collection of IMS functions linked by standardized interfaces. Fig. 2.2 gives an overview of the IMS architecture as specified by 3GPP.

One of the most important components in the signaling plane is the signaling protocol for session control. SIP [24] is the protocol chosen to perform session control signaling in IMS. The main purpose of SIP is to deliver a session description to the user equipment (UE) at the current location of the end user. SIP is completely independent of the format of the object it transports. It means that SIP can deliver a description of a session written in Session Description Protocol (SDP) or any other

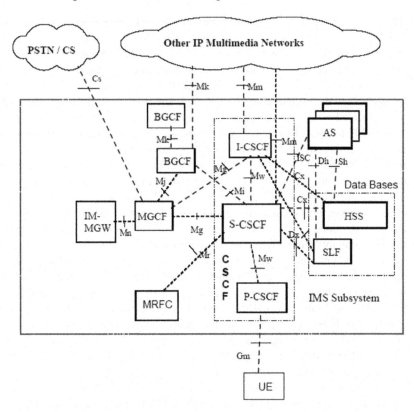

Fig. 2.2. Overview of core elements of the IMS architecture specified by 3GPP.

format. For instance, it can be used to deliver IM. SIP can track the location of a user by means of identifiers such as SIP Universal Resource Identifier (URI) in the Internet.

Core network subsystems of IMS are briefly described as follows [2]:

- *Home Subscriber Server (HSS) and Subscriber Location Functions (SLF) databases* – HSS is the central repository for network provisioning and user-related information that includes user profile data (e.g. identification, contact information, preferences, security information, subscription information, etc.). HSS is technically an evolution of the Home Location Register (HLR) in GSM and GPRS. It incorporates all of the interfaces and functions associated with a mobile networks' HLR. The HSS can serve as the master database and distribute provisioning and user profile data to individual network elements, applications, service enablers, etc. as needed. This not only will enable service provider to converge its wireless and landline provisioning systems but also establish a single interface between an operator's converged network and its IT organization over which provisioning and billing information is communicated. SLF is a simple database that maps users' addresses to entries in the HSS.

- *SIP servers that are collectively known as Call Session Control Functions (CSCFs)*– There are three types of CSCFs: Proxy-CSCF (P-CSCF), Interrogating-CSCF (I-CSCF) and Serving-CSCF (S-CSCF). P-CSCF is the first point of contact between the IMS-capable UE and the IMS network through which all the requests initiated by the IMS UE or destined to the IMS UE traverse. P-CSCF can be located either in the visited network or in the home network. I-CSCF acts as a SIP proxy located at the edge of an administrative domain. It also has an interface to the SLF and HSS to enable it to retrieve user location information from the databases and route the SIP request to the appropriate S-CSCF. A network may include a number of I-CSCFs for scalability and redundancy. It is often located in the home network and in some cases in the visited network. S-CSCF is the central node of the signaling plane, which is located in the home network and acts as a SIP server. It also performs session control, provides SIP routing services, enforces the policy of the network operator and acts as a registrar. The entire SIP signaling to/from the UE traverses the allocated S-CSCF. A network usually includes a number of S-CSCFs for the purpose of scalability and redundancy.
- *Application Servers (ASs)* are SIP entities located either in the home network or in an external third party network. ASs are in charge of executing services.
- *Media Resource Functions (MRFs)* are located in the home network and perform media related functions such as media generation and processing in the home network. MRFs are further divided into a signaling plane mode called Media Resource Function Controllers (MRFC) and a media plane mode called Media Resource Function Processors (MRFP).
- *Breakout Gateway Control Functions (BGCFs)* are in charge of routing functionalities based on telephone numbers. It is used for cases where an IMS UE initiates a call to a CS domain like PSTN.
- PSTN gateways act as interfaces to/from PSTN/CS domains. PSTN gateways are decomposed into a Signaling Gateway (SGW), Media Gateway Control Function (MGCF), and Media Gateway (MGW). SGW interfaces the signaling plane of a CS network. MGCF is the central node of PSTN/CS domain that is in charge of call control protocol conversion and management of resources in a MGW, which provides interfaces to the media plane of the PSTN/CS networks.

IMS UE should meet some prerequisites in order to be able to operate in an IMS environment. Two access levels should be granted to the IMS UE.

1. Access to an IP Connectivity Access Network (IP-CAN).
2. Access to IMS via IMS level registration.

Registrations in the above mentioned layers (IMS and IP-CAN) are independent of each other because in the IMS architecture IP-CAN and IMS application (SIP) are handled by separate layers. The location of the user can be determined at the IMS registration level through the IP address of the UE that is assigned by IP-CAN. In the following an abstract view of the required steps to operate in an IMS environment is summarized:

1. Establishment of IMS service contract through which the IMS service provider will authorize the UE to use the IMS service.
2. Obtaining an IP address and gaining access to IP-CAN.
3. P-CSCF address discovery that can take place as a separate procedure or as a part of IP-CAN connectivity.
4. IMS level registration where the IMS can locate the user through the IP address of the UE, authenticate, authorize session establishments and create security associations.

2.3.2 Business Aspects

Enterprises are moving from centralized data architectures towards content-aware networking, which is a more flexible approach based on Service Oriented Architectures (SOA). In traditional IP networking, packets are routed based on destination IP addresses. Drawbacks of traditional IP networking are complexity in the application layer and increase in the cost of information technology capital and operating expenses. In content-aware networking, messages are routed based on content and context. It embeds business rules in high speed and low latency networks while decreasing the IT capital cost and operational expenses [9]. Deployment of the existing IP infrastructures for content-aware networks brings the following advantages to the enterprises:

• It allows distributed or SOA applications to use the shared infrastructures on a demand basis to filter route and transform information faster than traditional software middleware. Enterprises can invest on this shared infrastructure and reduce their capital and operational costs significantly.
• As executions are done in hardware, it virtually eliminates performance degradations and ensures that data latency is low and predictable in cases where content routing and filtering is required.
• Application servers no longer need to focus on complex business rule executions, event processing and analysis.

Content-aware networks allow service providers to filter, route and transform information to enterprise customers on their behalf as well as to their partners. It therefore raises the position of service providers from providers of bandwidth commodity to strategic suppliers of services [9]. IMS provides an IP-based service centric creation and control framework that leverages common resources for services. It therefore creates the ability for rapid creation of multimedia and enhanced services. Furthermore, IMS application servers are capable of interacting with Web 2.0 applications. Web 2.0 applications are social networking applications that emphasize on services and are based on user participation. One of the significant requirements of enhanced services is application billing and charging functions. IMS features can also be leveraged to offer Fixed Mobile Convergence (FMC) solutions in conjunction with wireless partners and cable operators. Cable operators would then be able to offer enhanced services that provide ease of use and convenience. For instance, users can receive message waiting indications for messaging applications on their TV. Enhanced services offered by IMS can further be used as a distribution channel [8]. For example, in messaging applications and services, mailboxes can be utilized so that users could receive targeted offers of new services and the option to purchase the services. One of the attractive drivers for next generation IMS enhanced services is that operators are able to offer services to small and medium sized businesses via the software-as-a-service (SaaS) business model.

2.4 Service Layer

Integration of heterogeneous access networks via a core IP-networks supporting IMS enables flexible creation and provisioning of application services via the SOA, which can be broadly divided into the following layers from the bottom up:

- Interworking and media layer,
- Session control layer, and
- Service layer.

The interworking and media layer consists of the core-IP network interworking with the heterogeneous access networks to provide various media transport capabilities. SIP signaling is used for initiating and terminating sessions via Call Session Control Functions, media servers and other interfaces such as PSTN interface, under the management of the session control layer. These two bottom layers of the SOA can be implemented using IMS. In this section, the functionalities of the Service Layer are described in details.

As shown in Fig. 2.3, the Service Layer consists of:

1. The Service Delivery Platform (SDP),
2. A centralized policy management function, and
3. Service enablers.

The SDP establishes a set of Application Programming Interfaces (APIs) 'northbound' to the application layer above by abstracting the complexity of the underlying network's infrastructure from the application. The centralized policy management function authorizes or denies requests for network resources made by any application or user based on business rules, user profiles, and/or network resource availability. Service enablers perform tasks common to multiple applications necessary to deliver a service. For example, a centralized charging agent is needed to facilitate consolidated billing and establish a single interface; therefore this charging agent exists as a service enabler in the Service Layer.

2.4.1 Service Delivery Platform

Currently applications are tightly integrated with a network's infrastructure in the form of 'silos'. The converged architecture should utilize common functions in each

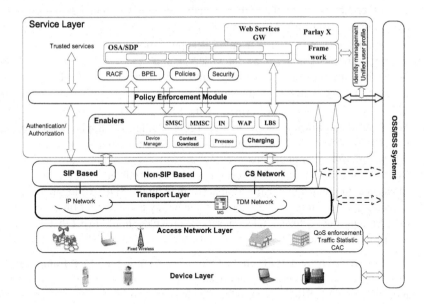

Fig. 2.3. Service layer architecture.

layer to reduce the number of network elements and hence, simplify processes, re-duce service development time and operation cost. The SDP facilitates this strategy by:

1. Abstracting applications from network infrastructure using northbound common APIs, and
2. Interfacing the policy management module and service enablers southbound.

The SDP performs four major functions:

1. First, the Parlay X [20] module within the SDP interfaces with external (non-trusted domain) applications. It enables external applications to securely ac-cess network resources, service enablers (e.g. presence, location-based service (LBS), etc.), and/or the user's end-device. For example, the SDP via the Parlay X module performs bill settlement with external application providers. Simi-larly, Parlay X enables service provider to securely share its network capabili-ties/resources in a controlled manner with enterprise customers, service integra-tors, and/or mobile virtual network operators (MVNOs).
2. Interconnecting internal applications and service enablers with various APIs is the second major function performed by the SDP. The Parlay group has defined a set of APIs for most mobile and landline service enablers. (Note: 3GPP2 has outlined in Multimedia Domain (MMD) Rev A how IMS should interface the SDP through IMS Service Control (ISC) APIs [22]). These APIs will establish an environment in which applications can discover and share information with

each other. For example, the SDP will enable an application that delivers driving directions to a user on her or his mobile to:

- Obtain the user's location from a LBS application,
- Access relevant maps on a content server, and
- Deliver driving directions to the subscriber using a map displayed on the user's device with his/her location identified.

3. The third major function performed by the SDP consists of providing charging information and billing records to a charging enabler. All applications traverse the SDP; therefore, the SDP is optimally located to monitor and collect charging information for all applications.

4. Controlling access to network resources via a centralized policy enforcement function is the fourth major function performed by the SDP. Although a number of leading SDP vendors have integrated the policy enforcement function within their SDP product offering, telecom service provider should insist that this function be separated into a stand alone module for functional development purposes.

Telecom service providers around the world are adopting the SDP framework to reduce the time it takes to bring an application/service to market and reduce operational costs with NTT, SKT, and Vodafone leading the way.

SDP Middleware on Handset

The SDP enables applications to discover and inter-work with each other. This capability must also include discovering clients on user devices so that the SDP can invoke applications. For example, consider an application that enables a user to use his/her mobile device to find a gas station. This application may require downloading, or 'pushing', a client designed to display a map onto the subscriber's mobile device. Determining whether or not a user's device contains the client necessary to complete a service is a function of the SDP middleware. Unfortunately, development of such SDP middleware has not been standardized and is vendor specific. Nonetheless, service provider can implement a common SDP middleware stack on all user devices to facilitate the convergence of its wireless and landline services.

2.4.2 Policy Enforcement

The Policy Enforcement (PE) module enables the service provider to manage and control access to its applications and network resources from a centralized location. The PE module contains service, user, and network policy rules executed by a policy engine (service broker) when users request access to a service or application. These policies/rules can be communicated with other entities using the Generic User Profile (GUP) [18] framework. As shown in Fig. 2.4, when users request access to a service or application, the request is relayed to the policy enforcement function for verification and authorization. If the request is approved, the PE module checks and allocates the underlying network resources.

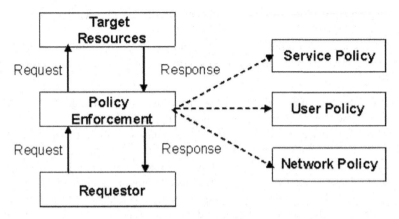

Fig. 2.4. Policy enforcement.

The PE module must also support and track the addition, modification, or removal of a resource (e.g., application, service enabler, and/or network element) as well as the policy associated with that resource. In addition, the PE module must support interrupting a service and/or accepting/rejecting a service request as a result of enforcing a policy. The PE module requires information from other modules within the Service Layer to perform these tasks. This information includes, but is not limited to, the following:

- End user subscriptions,
- End user class (e.g., gold, silver or bronze),
- Service Level Agreements (SLAs) with service provider and 3rd parties,
- End user account status (e.g., online, offline),
- End user personal data and preferences,
- Network resource availability and conditions, and
- Regulatory and legislative (e.g., privacy, security, emergency) variables and conditions.

The PE module may approve, reject, terminate or negotiate changes to any service/resource request from a user or application. If the PE module rejects or terminates a service, it must inform the requestor/user of the 'cause' in real-time. The PE module should retain a log of its policy decisions and record all of the events it observes (e.g., errors, resource request and responses, etc.). This data can then be relayed to the Support System Layer to facilitate trouble shooting and enhance customer service. The 'rules' enforced by the PE module can be sub-divided as shown in Fig. 2.4 into service, user and network policies. Each of these policy categories, along with an explanation of how they relate to the HSS, is explained next.

Service Policy Module

The service policy entity contains a set of business rules that authorize/deny:

- Third party application service providers (ASPs) access to service provider's network resources and
- End users access to other users (within or outside an operator's domain) and services (provided internally by service provider or externally by a third party ASP).

This module also enables third party service provider to exchange service policy information. The service policy module should support a number of security protocols that enable a service provider to securely exchange information with third party ASPs, users, and trusted external service providers. If this exchange is between internal network users or within a trusted domain, a dedicated link or private network may be used. If the exchange is between un-trusted domains or through the Internet, encryption should be used. (The appropriate security mechanisms/protocols/models supported by the SDP should be determined when the service policy function is implemented.) The service policy module should only control the authorization process. It should not perform bill settlement or collect traffic information to facilitate billing. Instead these functions should be performed by a 'charging' enabler and the service policy module should use this charging enabler to determine if:

- A user has good credit before authorizing access to a service/resource and
- A service is billed using an online or offline charging mechanism.

User Policy Module

User policies are determined by the data associated with a user's profile as stored in the HSS. The user profile contains user preferences, service priorities, device configuration information, service subscription data, and network registration information. The user policy module does not authenticate a user; this task is performed by the Control Layer using the HSS. Instead, the user policy module authorizes access to network services/resources after a user is authenticated. A user's profile in existing mobile circuit-switched network is centralized in the HLR. The HSS described above should include HLR functionality. Users will subscribe to both CS and IP-based services for the foreseeable future. A single database that applies a user's profile across both domains is therefore required to deliver services seamlessly across both domains. An HSS that integrates HLR functionality not only reduces the number of profiles associated with a user (thereby eliminating data discrepancies and duplication) but improves the end user's experience by making it consistent.

Network Policy Module

The network policy module enforces the policies associated with the underlying network resources. It should support and interoperate with the charging and service discovery capabilities available within the SDP. Southbound, the network policy function must support a variety of network domains including a heterogeneous mix of access network infrastructures and access network service providers. Northbound, the network policy function must support a variety of internal and external applications. All the while, the network policy module must be transparent to users and applications and not preclude the deployment of any service enablers or limit the scalability of the converged network. The network policy function should have direct access to the Resource and Admission Control Function (RACF) in the transport layer. When a user or application requests a network resource, the network policy module should ask the RACF to determine in real-time if the requested network resource is available. And if the network policy function authorizes a requested service or application (based in part on the response of the RACF), it should secure the resources that satisfy the service's QoS and SLA requirements through the RACF.

Service Enablers

'Service enablers' as shown in Fig. 2.3 link directly with clients on user devices to enable applications. For example, a 'presence-server' is a 'service-enabler' that facilitates the delivery of presence-based call routing services orchestrated by applications hosted on servers within the Application Layer. Some service enablers will be common to both landline and wireless applications (e.g., presence, Intelligent Network (IN), UM, MMS, connection manager, content download); other enablers will be relevant to only landline or wireless applications, but not both (e.g., WAP, SMS, LBS, etc). A centralized charging agent should be deployed as a service enabler as emphasized by the 'Billing' icon shown in Fig. 2.3. This charging enabler/agent must inter-operate with:

1. Any charging 'client' deployed in any application,
2. Other service enablers, and
3. Any charging 'client' deployed in any end-device.

This increases the complexity of the charging agent as explained below.

Charging Agent

The charging agent collects information in near real-time to support online (pre-paid) and offline (post-paid) billing systems. The charging agent also generates billing records that capture volume, duration, application, transaction, etc., information and communicates this information to the billing system via a single interface. Consolidating all communications with the billing system onto a single standardized link that interfaces the charging agent is necessary to abstract the service providers billing system from any changes/upgrades to an application, service enabler, and/or network element. Otherwise such changes/upgrades would require making time-consuming changes/upgrades to the billing system [23].

Fig. 2.5 illustrates the relationship between the charging agent and an application, service enabler, network element, and service providers billing system. The charging client collects user billing information specific to an application, enabler and/or network element. The charging client forwards this information to the charging agent. The charging agent consolidates this information and communicates it to the billing system.

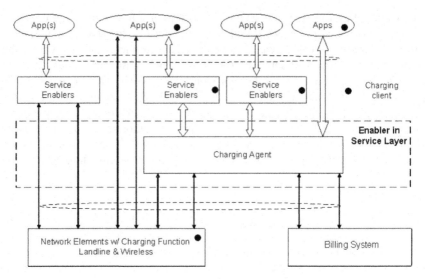

Fig. 2.5. Charging architecture.

2.5 Integrating Heterogeneous Networks Using IMS

2.5.1 Architecture

Packet-switched networks are compatible with IMS and can therefore be integrated with IMS through logical interfaces that implement different instances of SIP. These interfaces have been standardized by 3GPP and shown in Fig. 2.6. In this figure some of the most important logical interfaces are shown with dotted lines [16]. The Gm interface implements signaling communications between the IMS UE and the core network and offers procedures such as user registration, roaming, handover and service charging.

2.5.2 QoS Provisioning

Any application that requires a specific level of quality assurance from the network needs QoS support. QoS depends on type of service and can be categorized based on various characteristics. Examples of service types are best-effort, streaming, inter-active, and conversational [4]. Native IP (IPv4) is connectionless and offers a best-effort service. To identify traffic flows IPv4 offers two ways of traffic flow marking: (1) using the source and destination IP addresses and port numbers, and (2) using the Type of Service (TOS) field in the IPv4 header. Further, methods of Integrated Services (IntServ) and Differentiated Services (DiffServ) were proposed to meet the market demand for QoS. Two service classes are defined in IntServ: Guaranteed Service (GS) and Controlled Load Service (CLS). DiffServ has been proposed to deal with the scalability problem of IntServ. On the other hand, the next generation IP,

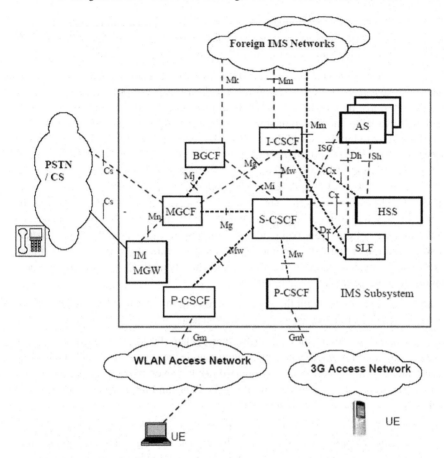

Fig. 2.6. Layout of logical interfaces for integration with IP core networks.

IPv6, uses two fields in the IP header to mark the traffic: flow label field and traffic class field. If used with its full features, IPv6 is very powerful in flow identification. IPv6 can also mark an aggregate of flows and can theoretically identify a single customer. However, packet or flow identification is but the first step in solving the problem of QoS provisioning. QoS management functions and traffic control mechanisms need to be implemented in network nodes for QoS support. In the following a few solutions are briefly described.

Over provisioning of QoS is the act of throwing bandwidth at the problem. This becomes the case when purchasing an oversupply of bandwidth is a simpler and easier solution to the problem of QoS provisioning. Although it can not be classified as a solution as it simply ignores bandwidth optimization and possible future trends of new services, it becomes useful in hazardous situations and military environments.

Flow identification as mentioned earlier is relevant in guaranteeing QoS. Different methods to classify and identify packets are flow label and traffic class fields in IPv6, as well as TOS field, source/destination IP addresses, and port numbers in IPv4.

Resource reservation and call admission control (CAC) are needed to reserve the bandwidth and to allocate the resources based on CAC and policies. CAC decides whether or not a new or handoff call should be admitted or rejected based on the resources available and the applicable Service Level Agreements (SLAs). Each call may be associated with a set of Service Level Specifications (SLS) given in terms of packet loss rate, delay and jitter for each flow or aggregate of flows. However, there is no clear method of identifying the amount of bandwidth and buffer available for a specific flow, except for peak bandwidth assignment. This method has led to many bandwidth allocation problems. The concept of equivalent bandwidth, defined as the minimum rate requirement to guarantee QoS for a single flow, has been introduced in the literature. Equivalent bandwidth techniques are often described by statistical characterization of traffic by parameters such as peak rate, mean rate, and maximum burst size. Equivalent bandwidth techniques are hardly applicable to situations where aggregate flows of traffic are heterogeneous.

Traffic shaping policies limit flow rates to their committed rates. Two basic methods of traffic shaping are leaky bucket and token bucket.

Traffic scheduling determines the order that packets buffered in the transmission queue at an outgoing link of a packet switching node are sent over the link. One of the very basic algorithms is first-in-first-out (FIFO); however, more sophisticated traffic scheduling algorithms are needed to satisfy the delay and jitter QoS requirements of real-time flows such as voice and video.

Queue management is often linked to traffic scheduling where packet dropping strategies can be applied when the transmit buffer is getting full. Other queue management strategies might involve partitioning the available buffer space so that some buffers may be dedicated to an individual traffic flow.

QoS routing is concerned with management of an end-to-end connection by routing it through appropriate intermediate nodes to ensure that the QoS requirements of connection can be met. QoS routing is part of the CAC process, which also takes into account of traffic shaping and scheduling/buffer management in intermediate nodes.

Provisioning QoS over interworking heterogeneous networks can be divided into the following steps:

1. CAC requests with specific QoS requirements should traverse end-to-end the interconnected networks that implement different technologies (e.g., cable, cellular, WLAN) and different protocols (i.e., ATM, TCP/IP).
2. QoS requirements should be interpreted according to the protocol and QoS mechanisms used in each individual network involved in the end-to-end connection.
3. The protocol stack of each of the interconnected networks consists of several layers each having a different scope for QoS provisioning.

4. Mapping should be done between the corresponding layers of the respective protocol stacks employed by the interconnected networks, such that the role played by each layer in provisioning QoS over the corresponding network can be harmonized across adjoining networks and end-to-end QoS can be guaranteed.

In the first two steps the QoS requirements are transferred to different network portions that implement their own technologies and protocols. It is referred to as *horizontal QoS mapping*. Steps 3 and 4 show the concept of *vertical QoS mapping* that is based on the idea that a network is composed of functional layers and the overall QoS depends on the QoS achieved at each layer of the network.

NGNs introduce new requirements for mobility management and provisioning of QoS in heterogeneous networks. QoS requirements include QoS guarantee among different domains/networks and end user perception of QoS. Some examples of mobility requirements are ability to change access point, ability to get access from any network and awareness of user availability. Mobility management with QoS support is of particular importance in NGN that is one of the driving forces of FMC, as it is desirable for users to be able to roam between fixed and mobile networks employing different technologies. There are three requirements to achieve end-to-end QoS in FMC [11]:

1. Capability of network controllers in individual networks to process resource reservations according to aggregate traffic demand, while maintaining fairness across individual traffic flows.
2. Mapping functions to convert the end-to-end QoS requirements to specific levels of QoS support in different types of networks.
3. Fast handover schemes to reduce the handover latency within a single domain and across heterogeneous access domains as a result of user roaming [27].

In access networks employing wireless technologies, QoS management needs to be supported via appropriate radio resource management (RRM) mechanisms. RRM schemes are designed to allocate scarce wireless resources to all users efficiently, fairly and to provide QoS support. The random nature of wireless channels due to path-loss, fading and shadowing brings many challenges in RRM. Due to the statistical nature of the wireless channel, soft or probabilistic QoS schemes may be more applicable in wireless access networks. For instance, RRM algorithms can provide soft QoS guarantees in an opportunistic way [11] so that in the short term users with good channel conditions are favored but in the long term all users receive the required throughput. In [11] QoS management for integrated WiFi and WiMAX services is discussed and a heterogeneous network architecture of FMC is demonstrated. Provisioning of end-to-end QoS in IMS over a WiMAX architecture is described in [13]. In this work a hybrid cross layer QoS scheme based on IMS over WiMAX is discussed. In this scheme, UE is responsible of initiating the QoS negotiation, resource reservation and cancellation. It is based on the following procedure:

1. Application layer QoS authentication at Policy Decision Function (PDF),
2. Vertical mapping of QoS for application layer and MAC layer at UE,

3. MAC layer and IP layer QoS mapping at the Access Service Network (ASN), and
4. Intserv and DiffServ mapping at the Packet Data Gateway (PDG).

NGN's conceptual model for QoS and mobility management consists of two main elements: Domain Policy Manager (DPM) that acts as a policy decision point and PE module which ensures that only authorized IP flows are allowed to use the network resources that are reserved and allocated to them. The framework for end-to-end QoS support is presented in [15] released by 3GPP. Fig. 2.7 shows a conceptual model for policy-based QoS provisioning in IMS.

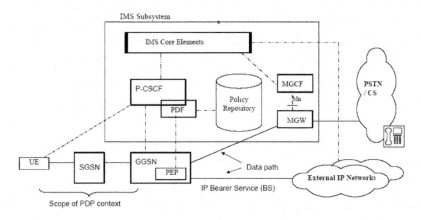

Fig. 2.7. Network architecture for QoS conceptual models.

In this figure policies can be stored and retrieved to/from the policy repository database. PDF is in charge of retrieving policies from policy repository and controlling the PEP. The role of PEP is to take actions based on the decisions made by PDF. PEP is located in the Gateway GPRS Supporting Node (GGSN) because GGSN is in the data path. PEP and PDF communicate in two possible modes of communication, namely, *push mode* and *pull mode*. In the push mode, communication is initiated by PDF and the decision is sent (pushed) to PEP for further actions. In the pull mode, communication is initiated by PEP to request a decision (pulling from PDF).

As shown in the figure, several external IP networks are connected via data paths and may consist of several different domains. In order to provide end-to-end QoS support, it is required to provision QoS within each domain. An IP Bearer Service (BS) manager is used to manage the external IP BSs. As different domains may employ different technologies, translation functions may be needed to communicate QoS requirements to the other external BS managers. When a Packet Data Protocol (PDP) context is set up to enable a UE to access a packet-switched network service, the UE shall have access to the following alternatives:

- Basic GPRS IP connectivity service for which the communication is established based on the users subscription and local operators admission control, policies and roaming agreements.
- Enhanced GPRS based services for which the bearer is used to support application layer services such as IM and would also be in charge of policy control decisions.

In a situation where resources that are not owned by a certain network are required to provide QoS, interworking with external network that controls those resources is required. This interworking can be done in a number of ways [15]:

1. Signaling along the flow path (e.g., using the Resource Reservation Protocol, RSVP),
2. Packet marking or labeling along the flow path (e.g. DiffServ, MPLS),
3. Interaction between policy control and resource management elements, and
4. Border routers to enforce SLAs.

3GPP has also adopted another policy-based QoS solution in which sufficient resources to support a specified QoS are provided to authorized users. In this framework, policy rules are stored in a policy repository and PDF is able to retrieve appropriate policy rules. This action is triggered at the PE modules by contracted QoS-enabled IP services [14].

2.5.3 Mobility Support

Mobility management in NGN should be distinguished from mobility management in previous generations of networks in the sense that vertical handover, heterogeneity and roaming would also be involved. Support of global roaming in NGNs requires interworking of heterogeneous network accesses with different mobility management techniques [27] [25] [28]. Different aspects of mobility can therefore be realized:

Terminal mobility that is the capability of a mobile terminal to continuously remain connected to access the network in range.

User mobility refers to the ability of a mobile user to continuously access services from a network.

Service mobility is the portability of a service which enables a user to access the same service independent of location and terminal.

Future wireless IP networks are expected to support basic mobility requirements such as:

- Support of all forms of mobility and applications,
- Support of mobility across heterogeneous wireless access networks,
- Support of continuous service delivery and access, and
- Support of global roaming.

The most important components of mobility management are location management, packet delivery, handoff management, roaming and admission control. Location management schemes in current cellular networks were capable of the following procedures: location update and location discovery by paging. Location update strategies decide when a mobile terminal should perform location update and what information should be communicated to the network. The decision can be based on time, movement, distance or probability distribution function of time, movement and/or distance. Location discovery by paging is required when the network does not know the exact location of a user. The main goal in design of location discovery strategies is to minimize the paging cost. In the current cellular network the HLR and VLR databases are used to enable mobility management. For the case of handoff management in heterogeneous networks, in addition to triggering handoff and decision criteria, choosing the appropriate access mode based on application requirements and QoS mapping issues are of great importance [26].

Use of IPv6 can enhance the support of terminal mobility. At the time of IMS standardization, IPv6 was also being standardized and 3GPP considered the applicability of IPv6 to be deployed. Deployment of IPv4 in a large scale would lead to the need of Network Address Translations (NAT) as allocations of private IP addresses would be required. As the first IMS product reached the market, IPv6 had not taken off and this caused some efforts towards making IMS compatible with IPv4 via deployment of (IPv4 and IPv6) dual stack implementations of IMS UEs [2]. IMS is mainly based on MIPv6 and uses application layer SIP for signaling and control of real-time multimedia applications. SIP may also be used to support terminal mobility. Support of terminal mobility using SIP attracts many researchers to work in this area. Similar to email users that are identified by their email address, SIP users can be identified by their SIP URIs that works as a unique global Network Access Identifier (NAI). In terms of mobility support, the difference between MIP and SIP-based mobility is that unlike MIP, SIP servers are only in charge of setting up the sessions between users and once session are set up, data traffic will be communicated directly between end users. This also solves the triangular routing problem of MIP. However, when a terminal moves to a new access network, it may need to register its new IP address with the SIP server and it can cause a high load on the home server. One of the solutions to this problem is to use a hierarchical registration similar to MIP. It is also difficult for a roaming terminal to keep a TCP connection alive while changing its IP address. This is a rather challenging problem and one approach is to consider using the Stream Control Transmission Protocol (SCTP) with its multi-homing features for end-to-end reliable transport. Other solutions were also summarized in [6].

2.5.4 Design Considerations for Mobility Support and QoS

There are several design considerations for mobility and end-to-end QoS support [10]. One of the considerations is the interoperability between the networks of different service providers. In a scenario where various administrative domains with different policies, management architectures and traffic mechanisms coexist, a common protocol would be needed for communication of end-to-end QoS requirements of user traffic. It is also necessary to respect the individualities of independent operation of traversed networks at the same time. Different autonomous networks and many proprietary solutions on the other hand, make this task difficult. One solution to make the system interoperable is to separate the signaling protocol from data. In a design proposed in [10] this concept is followed by decoupling the signaling plane from media plane.

Another design consideration targets the choice of signaling protocol. An inter-domain signaling protocol can be used to enable the interaction of different domains, so that they can be integrated in a manner capable of provisioning QoS and supporting mobility functionalities. Solutions and examples that apply this concept were discussed in [10].

Finally, the integration level is also a factor that should be considered in a design. 3GPP defines two levels of integration between 3G cellular networks and WLANs: tightly coupled and loosely coupled integration. In tightly coupled integration, the WLAN access point is connected to the Serving GPRS Support Node (SGSN) and behaves like a node B (i.e., a 3G base station). In loosely coupled integration, the WLAN and 3G cellular networks are autonomous domains but may share a common Authentication, Authorization and Accounting (AAA) server that enables, e.g., a UE to be authenticated for WLAN access using its 3G cellular credential. In this case the two domains interwork via the respective gateways, namely WLAN Access Gateway (WAG) and Packet Data Gateway(PDG).

2.6 Design Issues, Technical Challenges, and Future Research Directions

2.6.1 IMS Session Setup in a Heterogeneous Mobile Environment

As mentioned earlier, SIP is selected as the signaling protocol for multimedia sessions over IP-based mobile networks due to advantages such as simplicity, extensibility, flexibility, and scalability. SIP session setup delay performance is shown to be acceptable over the land-based Internet. However, in a heterogeneous mobile network environment where wireless networks such as 3G and WLAN coexist, session setup may suffer from unexpected delay and performance degradation due to the limited throughput and/or reliability of the wireless channel causing multiple retransmission of SIP request and response messages, and the need to determine which one of several alternative access networks would best serve the session; i.e., the so-called network selection problem [29] [30] [31] [32]. Furthermore, as MIP is used to support when the mobility of users roaming into foreign domains, SIP signaling may suffer from triangular routing [24].

2.6.2 Customer Profile and Identity Management

The network should provide an infrastructure capable of the management of user profiles in terms of their status and locations as well as their identities, taking into account that a user can potentially access application services via one of several heterogeneous access networks, and using one of several UEs. Admission policies in the IMS can be enforced at different layers, including the service layer. Thus, it is important to define an access policy model that describes the elements that form part of a policy and serves as input to a policy (which is, in turn, implemented using a policy language). Profile management is the process of determining the various aspects to be specified in a user's profile and how this information can be applied in order to tailor-make services according to the user's preferences. Another aspect of work in this area is to understand how to securely exchange private subscriber data between the ASPs and the access network operators in order to better adapt and personalize services offered to the users. These are challenging problems to be addressed by future research in this area.

2.6.3 Bandwidth Allocation and Resource Management

Access agnostic framework of IMS has separated the service layers from network and transport layers. The signaling infrastructure therefore, ties these layers together and shall provide the network operators with maximum influence on admission control, efficient bandwidth allocation among users, and applications as well as appropriate choice of session paths. To efficiently utilize network resources, resource allocation and bandwidth management is required to guarantee QoS for certain applications with stringent bandwidth demand. Resource allocation and management is a challenging task given the heterogeneity of access networks and applications. Session control entities in IMS are in charge of allocation of resources. Resource allocation can also be done in the IP-CAN layer. Developing efficient bandwidth allocation mechanisms is therefore a future research direction.

2.6.4 Seamless Mobility Management and Vertical Handover Support

Global roaming is one of the future needs in mobile and wireless communications that enables users to initiate a call from any access network and seamlessly roam across heterogeneous wireless access networks. Another aspect of global seamless roaming is the continuity of a call or a session when a mobile terminal is moving to different coverage areas with different access network. This leads to initiation and performing a vertical handover. IMS should be able to manage various access related constraints at different layers that are imposed by heterogeneous access technologies. Providing a seamless vertical handover for applications that are sensitive to delay and have strict QoS requirement is a challenge and has attracted extensive research work [25].

2.6.5 Location Management – Tracking of User Locations

Mobility of users brings the challenge of location management. It is important that as users move around the coverage area of a specific wireless access network or roam between different access networks, their current locations can be determined for connection management purposes. With recent U.S. E911 requirements to fairly accurately report the location of VoIP and cellular callers, it opens up the opportunity of location-based services and applications. One example is Location-aware PoC (LaPoC) [17]. Other location-aware applications can be built around the existing applications and services enabled by IMS. Some examples are:

- Combination of presence and location where the location of users can be seen by the other end,
- Initiation of messaging based on a defined distance,
- Location based context aware adaptation where the users communication decisions are based on the location (i.e. at work or at home), and
- Multimedia broadcasting based on the end users location.

Various SIP methods for communicating location information are being considered by the Internet Engineering Task Force (IETF) but no preferred standard has been chosen yet. In provisioning of user locations, [17] proposes the introduction of a service enabler entity called IMS Location Server (ILS) to be located at the Service Layer. ILS does not determine the locations of users, but it passes the location requests to positioning systems to determine the users locations. It furthermore acts as a client that interacts with other application servers via a SIP interface. Support of location and presence services at the Service Layer allows the application servers to support location aware applications. Investigation of new applications encompassing location and presence information is very much of interest to both research community and service providers.

2.6.6 Multicast/Broadcast Services Across Multiple Access Technologies

Currently, IMS only supports services based on unicast mode of transmission. Protocols and standards need to be developed in order for IMS to support multicast/broadcast and advanced group management functionalities. In [7], a system architecture for multicast/broadcast in IMS was proposed that is in line with Telecoms and Internet converged Services and Protocols for Advanced Networks (TISPAN) and NGN reference architectures. This design offers a converged logical framework that brings together a diversity of multicast-transport bearers such as Digital Video Broadcast for Handheld devices (DVB-H), Multimedia Broadcast Multicast Service (MBMS), and Broadband Multimedia Communications System (BMCS). Enabling multicast/broadcast services across multiple access technologies opens up many challenges in the areas of network selection, QoS management, session management and advanced multicast group management functionalities.

Conclusion

In the path to Telco 2.0 and transitioning to the future generation of wireless networks, convergence of IP based core networks with heterogeneous wireless access networks is inevitable. IMS aims at providing a standardized solution for multimedia services in an access agnostic manner within a unified framework. While IMS opens up new business and revenue generating perspectives for service providers and network operators, several business and technical issues are still to be solved. NGNs and IMS will enhance existing wireless services such as voice calls, SMS, MMS, and enable new services such as IM, presence, PoC, UM, and multicast/broadcast. In this chapter, we have illustrated the use cases for advanced conversational communications that would be available to the future telecom users over an NGN composed of converged heterogeneous wireless and wireline access networks. IMS plays a major role in enabling these services. We have introduced IMS and discussed the role of IMS and convergence drivers at three levels of network, application and service. The architectural elements and some business aspects have also been addressed. We have discussed the role of the Service Layer in service convergence and revealed some implementation issues. Network architectures and conceptual models for mobility management and QoS provisioning for the case of interworking heterogeneous access networks have been reviewed. Towards the integration of heterogeneous networks with IMS, various challenges such as billing, QoS provisioning, mobility support, bandwidth management, resource allocation, customer profiling, security and location management issues have been presented as possible directions for future research in these areas.

Acknowledgment

This work is partially supported by a grant from TELUS and the Natural Sciences and Engineering Research Council of Canada under grant CRDPJ 341254-06.

References

1. T. Macaulay, *Securing Converged IP Networks*, Auerbach Publications, 2006.
2. G. Camarillo and M. Garcia-Martin, *The 3G IP Multimedia Subsystem (IMS): Merging the Internet and the Cellular Worlds*, Second Edition, John Wiley & Sons, Inc, 2006.
3. Y. Lin, *Wireless and Mobile All-IP Networks*, John Wiley & Sons, Inc, 2005.
4. G. Gomez and R. Sanchez, "End-to-end quality of service over cellular networks: Data services performance optimization in 2G/3G," John Wiley & Sons, Inc, April 2005.
5. M. Marchese, *QoS Over Heterogeneous Networks*, John Wiley & Sons, Inc, 2007.
6. J. C. Chen and T. Zhang, *IP-based next generation wireless networks: Systems, architectures, and protocols*, John Wiley & Sons, Inc, 2004.
7. A. Ikram, M. Zafar, N. Baker, and R. Chiang, "IMS-MBMS convergence for next generation mobile networks," in *Proc. of International Conference on Next Generation Mobile Applications, Services and Technologies, 2007*, pp. 49-56, 12-14 Sept. 2007.
8. I. Moraes, "Exploring drivers for IMS enhanced services platform architectures and applications," *IEC Magazine*, 2007.
9. T. Kourlas, "The evolution of networks beyond IP," *IEC Magazine*, 2007.

10. F. C. de Gouveia and T. Magedanz, "A framework to improve QoS and mobility management for multimedia applications in the IMS," in *Proc. of Seventh IEEE International Symposium on Multimedia (ISM'05)*, pp. 216-222, 2005.
11. W. Peng, C. Takeo, H. Jeyhsin, and Y. J. Hung-Yu, "QoS management and peer-to-peer mobility in fixed-mobile convergence," Fujitsu Sci Tech, Japan (Oct. 2006).
12. V. Y. H. Kueh, R. Tafazolli, and B. G. Evans, "Performance analysis of session initiation protocol based call set-up over satellite-UMTS network," Computer Communications, vol. 28, no. 12, July 2005, pp. 1416-1427.
13. W. Jiao, J. Chen, and F. Liu, "Provisioning end-to-end QoS under IMS over a WiMAX architecture: Research articles," Bell Lab. Tech. pp.115-121, May 2007.
14. A. Elmangosh, M. Ashibani, and F. Shatwan, "Quality of service provisioning issue of accessing IP multimedia subsystem via wireless LANS," New Technologies, Mobility and Security, pp.133-143, Springer Netherlands 2007.
15. "End-to-End QoS Concept and Architecture," 3GPP TS 23.207, v7.0.0 Release 7, June 2007.
16. Spyros L. Tompros, "NGN networks: A new enabling technology or just a network integration solution?," White paper appointed by VITAL consortium, Aug. 2007.
17. M. Mosmondor, L. Skorin-Kapov, and R. Filjar, "Location conveyance in the IP multimedia subsystem," vol. 4003, Springer Berlin/Heidelberg, 2006.
18. "Generic User Profile (GUP)," TS 129.240, 3GPP, Rel. 7, June 2007.
19. "Extensible Markup Language (XML)" ; Jan 2008, http://www.w3.org/XML
20. "Open Service Access (OSA); Parlay X Web Services," TS 29.199-1, 3GPP, Rel. 6, Dec 2005.
21. "IP multimedia subsystem - Stage 2," TS 23.228, 3GPP, Rel. 7, June 2007.
22. "All-IP core network multimedia domain Rev. A," X.S0013, 3GPP2, Sept. 2005.
23. "Charging Architecture," OMA-AD-Charging Draft v1.1, OMA, Oct. 2007.
24. M. Handley, H. Schulzrinne, E. Schooler, and J. Rosenberg, "SIP: Session Initiation Protocol," RFC 2543, March 1999.
25. J. Zhang, H. C. B. Chan, and V. C. M. Leung, "A SIP-based soft-handoff (S-SIP) scheme for heterogeneous mobile networks," in *Proc. of IEEE WCNC'07*, Hong Kong, China, Mar. 2007.
26. J. Zhang, H. C. B. Chan, and V. C. M. Leung, "A location-based vertical handoff decision algorithm for heterogeneous mobile networks," in *Proc. of IEEE Globecom'06*, San Francisco, CA, Nov. 2006.
27. V. C. M. Leung, F. Yu, J. Zhang, and H. C. B. Chan, "SIP signaling for vertical handovers in heterogeneous wireless networks," book chapter in *SIP Handbook: Services, Technologies, and Security*, S. Ahson and M. Ilyas, ed., CRC Press 2008.
28. J. Zhang, E. Stevens-Navarro, V. W. S. Wong, H. C. B. Chan, and V. C. M. Leung, "Protocols and decision processes for vertical handovers," book chapter in *Unlicensed Mobile Access Technology: Protocols, Architectures, Security, Standards and Applications*, Y. Zhang, L. Yang, J. Ma, ed., Auerbach Publications, CRC Press 2008.
29. F. Bari and V. C. M. Leung, "Network selection with imprecise information in heterogeneous all IP wireless systems," in *Proc. of Wireless Internet Conference (WiCon)*, Austin, TX, Oct. 2007.
30. F. Bari and V. C. M. Leung, "Architectural aspects of automated network selection in heterogeneous wireless systems," *International Journal of Ad Hoc and Ubiquitous Computing*, Feb. 2008.
31. F. Bari and V. C. M. Leung, "Automated network selection in a heterogeneous wireless network environment," *IEEE Network*, vol. 21, no. 1, pp. 34-40, Jan.-Feb. 2007.

32. F. Bari and V. C. M. Leung, "Service delivery over heterogeneous wireless systems: Networks selection aspects," in *Proc. of ACM IWCMC'06*, Vancouver, BC, July 2006.

3

Architectures and Protocols for Dynamic Spectrum Sharing in Heterogeneous Wireless Access Networks

Oliver Holland[1], Alireza Attar[1], Mahesh Sooriyabandara[2], Tim Farnham[2], Hamid Aghvami[1], Markus Muck[3], and Vladimir Ivanov[4], and Klaus Nolte[5]

[1] Centre for Telecommunications Research, King's College London, UK
{Oliver.Holland, Ali.Attar, Hamid.Aghvami}@kcl.ac.uk
[2] Toshiba Research Europe Ltd., Bristol, UK
{Mahesh.Sooriyabandara, Tim.Farnham}@toshiba-trel.com
[3] Motorola Labs, Gif-sur-Yvette, France (Markus Muck has since moved to Infineon Technologies, Munich, Germany)
markus.muck@gmail.com
[4] Intel Corporation, Communications Technology Lab, St. Petersburg, Russia
vladimir.ivanov@intel.com
[5] Alcatel-Lucent Deutschland AG, Bell Labs, Germany
klaus.nolte@alcatel-lucent.de

3.1 Introduction

The explosive growth of capacity-hungry applications and the increasing proliferation of mobile and wireless devices are proving challenging for the wireless industry. Worldwide, almost all frequency bands of favorable propagation characteristics for wireless communications are currently allocated to at least one Radio Access Technology (RAT) (see, e.g., [1]– [4]); moreover, only a tiny proportion of that spectrum is allocated to mobile communications and wireless networking means. This leads to an increasing requirement for bit-rate per unit spectrum among the systems and devices accessing it. This chapter focuses on means for dynamic spectrum sharing to enhance spectrum usage efficiency, investigating various architectures and protocols from a number of different perspectives.

Spectrum sharing among wireless systems can occur either horizontally or vertically [5]. If all radios have the same right to access a particular band, as is the case in unlicensed bands, horizontal spectrum sharing is implied. On the other hand, if one system (typically, the owner of the license to the band) has a higher priority to access the band than other systems, the implication is vertical spectrum sharing. Generally, vertical spectrum sharing is envisaged through secondary spectrum access, whereby the licensed band is shared but only with the license owner's consent, between the license owner's network, i.e., the primary system, and lower priority networks, i.e., the secondary networks. Means for vertical spectrum sharing are the main emphasis of this chapter.

E. Hossain (ed.), *Heterogeneous Wireless Access Networks*,
DOI: 10.1007/978-0-387-09777-0_3, © Springer Science+Business Media, LLC 2008

Technically speaking, spectrum can be shared in the realms of time, space, frequency, or a combination of the above. Spectrum sharing in time involves systems using a band at different time intervals, for example, by exploiting a listen-before-talk etiquette. Another possibility here, which is complicated by the fact that it requires synchronization among systems, is to use idle time-slots in a Time Division Multiple Access (TDMA) context. Spectrum sharing in the frequency domain involves transmitting such that no frequency overlap occurs among the coexisting systems. In many such cases, it might be necessary for a system to consolidate several smaller idle frequency bands to create a transmission opportunity, through using multi-carrier modulation approaches such as Orthogonal Frequency Division Multiplexing (OFDM). Spectrum sharing in the spatial domain might be used to benefit from the realization that spectrum occupancy varies from location to location. Hence a system might use a band in a specific location if other systems are not present at that location.

Transmission opportunities arising from spectrum sharing in the domains of time, space, frequency, or a combination of these, are generally called spectrum holes, or sometimes "white spaces" [6]. Spectrum might also be shared in the power domain however. For example, a system might be permitted to transmit with a higher power in any of the above domains (time, space, frequency, or any combination), providing that these transmissions do not breach the interference tolerance threshold at other systems' receivers. Transmission opportunities in this context are known as "grey spaces" [6]. The imposed interference to each receiver here might be minimized by using spread-spectrum techniques, for example through Ultra Wide Band (UWB) or through very wideband Code Division Multiple Access (CDMA). Such methods, comprising open access to other systems' bands with extremely low transmission power spectral density through spread spectrum techniques, are known as underlay transmission [7].

Another way to classify spectrum sharing techniques is based on the consideration of whether the utilized bands are expressly allocated or are (semi-) independently chosen by devices. In the former case, spectrum among operators, or possibly from a spectrum pool, might be dynamically coordinated and allocated to Radio Access Networks (RANs) as triggered by traffic demands. This method is known as Dynamic Spectrum Allocation (DSA). The other approach is that of devices/RANs selecting resources (semi-) independently, usually as a secondary entity in the spectrum. This can be either fully decentralized or network-assisted, and is often envisaged as being achieved through Cognitive Radio (CR) [6], [8]. Several terminologies and approaches to this sharing paradigm might be identified in the literature, such as Dynamic Spectrum Access (DSA) [9] and Opportunistic Spectrum Access (OSA). The umbrella term Dynamic Spectrum Selection (DSS) is employed throughout this chapter.

3.1.1 Demands of Applications and Services

A major cause of the spiraling spectrum demand in recent years is the emergence of novel wirelessly networked applications and services such as video streaming, video telephony, conferencing, high capacity downloads and peer-to-peer transfers, demanding data streaming applications (e.g., Google Earth), and evermore complicated websites with increased active content requiring greater capacity. Another cause of spectrum demand is the increased proliferation and take-up of various wireless technologies among the general public. Unfortunately, as stated by Shannon's Theorem, the laws of physics specify a limit on the amount of data that can be transmitted for a given transmission power and noise (or, under specific assumptions, noise plus interference) in a fixed spectrum bandwidth. Hence to satisfy the demands of pioneering applications and services, as well as the increasing density of transceivers, the need for greater spectrum usage efficiency is ultimately unavoidable.

Whereas in unlicensed spectrum systems/devices might previously have been reasonably happy contending for the available spectrum in a haphazard manner, newer services such as Voice-over-IP and video conferencing, and other services such as high priority data, require a better guarantee of Quality of Service (QoS). To improve service error rate and reduce scheduling delay, this often implies the assurance, with a very high probability, that spectrum will be available when asked for by the system in question. The need for such an assurance further exacerbates spectrum bandwidth needs.

3.1.2 Inefficiencies of Spectrum Usage in the Modern-Day World

The spectrum allocation process as determined by regulators worldwide has historically been ultra-conservative. Each frequency band has been allocated to only one or a small number of RATs, and only those RATs are allowed to access the spectrum. This has commonly led to a situation where the spectrum in question is used, on average, at only a tiny proportion of its theoretical capacity in any one geographical location.

Even within the context of systems serving relatively similar purposes, be they using the same standard or alternative standards, there are often significant differences in the loadings of these systems and their associated spectrum at any one time. Through realizing spectrum sharing, spectrally localized increases in traffic loads can be dissipated across much larger spectrum ranges. The end result of such a paradigm would be a far greater average spectrum utilization across multiple bands, hence vastly improved QoS for a range of wireless users. This is a reasoning behind the globally increased interested in spectrum sharing solutions.

3.1.3 Increased Freedoms and Technological Advances Facilitating Spectrum Sharing

In the light of this interest in spectrum sharing and its potential, industry has been looking at various solutions for it. One significant facilitator for spectrum sharing from a technical point of view is Software-Defined Radio (SDR), which allows characteristics of radio interfaces to be defined in software thereby providing easier adaptation to RATs such as might apply in a different spectrum band or might be more appropriate for the radio characteristics or capabilities of systems in a prospective band. Another facilitator is "Reconfigurability", the objective being to simplify adaptations to a range of layers of the protocol stack, including the associated radio network configuration changes and RAT adaptations. Given such technological advances in tandem with increased interest in spectrum sharing, regulators have begun to look more seriously at making spectrum allocations and usages less viscous. These and similar such advances essentially beg the question of "when" not "if" spectrum sharing paradigms will become more of a reality.

The purpose of this chapter is to investigate the technical requirements, usage scenarios, architectures and protocols for spectrum sharing in the DSS and DSA contexts. This supplements several contributions that present general surveys in the field of spectrum sharing (see, e.g., [6], [7] and [9]). This chapter is organized as follows. In Section 3.2, methods for spectrum sharing in the context of centralized control, through Dynamic Spectrum Allocation (DSA), are discussed. Section 3.3 investigates decentralization of spectrum sharing decisions through Dynamic Spectrum Selection (DSS). Section 3.4 discusses Game Theory as a means for resource optimization in spectrum sharing scenarios, before the chapter concludes.

3.2 Centralized Control: Dynamic Spectrum Allocation

The first classification of spectrum sharing solutions investigated is Dynamic Spectrum Allocation (DSA) which generally implies centralized control in the allocation of resources by an entity or collaborating group of entities.

3.2.1 Basic Concept

DSA aims to manage spectrum in a multi-radio environment by dynamically allocating it among participating networks over time, space, or both. This can serve to exploit both temporal and spatial variations in spectrum requirements among networks in order to achieve higher overall spectrum usage efficiency. Consider the example scenario in Fig. 3.1, depicting normalized traffic loads of two networks at the same location in a 24 hour period. Note that each network operates in its own licensed band. Three regions can be identified in this figure. In region 1, both networks encounter a relatively high traffic load, hence spectrum sharing between them would not be advantageous. In region 2, network 2 is facing a high traffic load while network 1 is relatively idle. Therefore the load on network 2 could be alleviated by dynamically allocating some of the idle bands of network 1 to network 2. Finally, in region 3, both networks are operating with a low traffic demand, hence although idle spectrum exists no gain would be achieved through spectrum sharing. Note that the same scenario is valid if the horizontal axis is space instead of time.

The value of a utilized DSA process can be measured, at least partially, by the DSA gain. The DSA gain is given by the average load increase supported over participating networks when DSA is employed, as compared with the average load supported in fixed spectrum allocation. As implied by the example discussed above, the

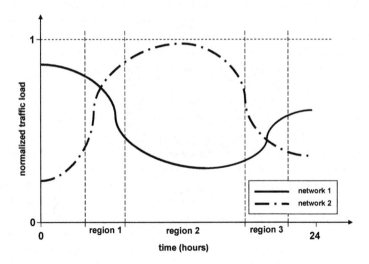

Fig. 3.1. Example of temporal variation in traffic loads for two co-located networks.

DSA gain depends, among other factors, on the correlation in traffic demands among participating networks. A lower correlation in the traffic demands generally indicates a higher DSA gain.

3.2.2 Architectures and Protocols for DSA

The main phases of DSA operation, which are executed in a cyclic fashion, are illustrated in Fig. 3.2. The associated approaches to DSA, with reference to these phases, are discussed in this section. The emphasis is particularly on DSA to satisfy fluctuations in traffic requirements; however, it is noted that in future manifestations there may be other conceivable reasons for spectrum allocations to be reassessed.

The evolutionary steps of DSA liquidity are captured in Fig. 3.3, as inspired by [10]. Fig. 3.3(a) represents a static spectrum allocation case, where clearly no DSA is taking place. This is comparable to the present reality in terms of spectrum usage among systems. Fig. 3.3(b) and (c) represent the contiguous and fragmented DSA schemes, respectively, in both cases of which spectrum sharing is taking place at the RAN level. Finally, Fig. 3.3(d) represents cell-by-cell DSA, where spectrum is shared both in the temporal domain and with a high resolution in the spatial domain. The complexity in achieving DSA increases from Fig. 3.3(a) through to Fig. 3.3(d), whereby DSA architectures are anticipated to progressively aspire towards Fig. 3.3(d). This is the ultimate objective, presenting most DSA gain.

Strategies to address DSA involve joint network coordination and management of resources among multiple operators/systems and/or within meta-operators. In the case of DSA being independently applied through arrangements between distinct entities (e.g., networks), the exact architectures might often be independently developed by those participating entities. The procedures in this chapter (e.g., Fig. 3.2) are

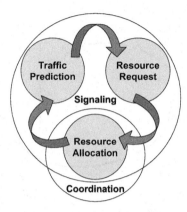

Fig. 3.2. Phases of DSA operation.

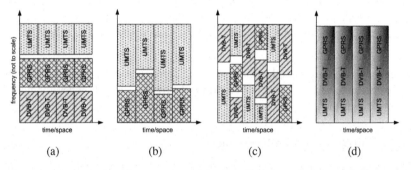

Fig. 3.3. A comparison of spectrum allocation approaches: (a) fixed allocation, (b) contiguous DSA, (c) (highly!) fragmented DSA, and (d) cell-by-cell DSA. Note that this figure might alternatively be represented in terms of RANs (either of the same RAT or different RATs–see [10]).

still necessary, and of course the sharing process must be consistent with local regulations for the use of the associated spectrum. Some specifications applicable to such cases do nevertheless exist: one example of this is the architecture and functional requirements for shared networks in a 3G context [11], in which sharing is specified between operators at the RAN or Core-Network level. Through this approach, multiple operators can maintain independence among their network infrastructures.

Things are more complicated in the case of spectrum being pooled by network operators for DSA purposes, where of course a widely understood architecture/specification for the sharing process is needed. An illustrative architecture for DSA in this scenario is shown in Fig. 3.4, whereby shared spectrum is available to all networks/operators via a Centralized Spectrum Controller (CSC), which dynamically grants Resource Allocations to the networks/operators based on predicted traffic loads and submitted Resource Requests. Associated with this process are the computational and signaling overheads for the coordination and control of DSA,

including the issues such as traffic prediction to appropriately identify required resources (see later sections). Note also that traffic prediction might be a multi-stage
process; e.g., devices might inform the network about their (perhaps anticipated)
traffic requirements (e.g., an anticipated download by an application for an upgrade,
or an anticipated streaming event that the user has signed up for in advance), and
the network might twin this with its own knowledge of traffic requirements based on
experienced loads (using mechanisms such as mentioned in the next section). Given
these requirements, in addition to others such as performing network and device reconfigurations to the new spectrum configuration, DSA is commonly envisaged as
being performed with a relatively low time-resolution, at periodic intervals. Real-
time DSA, where there is the freedom to update allocations at any instance, presents
a number of significant challenges hence is left out of scope of this chapter.

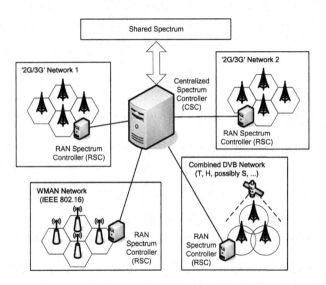

Fig. 3.4. Example DSA architecture in a pooled spectrum environment.

Traffic Prediction for DSA

A first important operation for DSA, as illustrated in Fig. 3.2, is the prediction of
future traffic requirements by networks' RSCs. This prediction is necessary for the
CSC to determine which dynamic spectrum reallocations would be appropriate to
satisfy load variations among networks, and whether the processes, workloads, and
associated temporary detrimental effects on networks and their devices (e.g., due
to signaling and reconfiguration requirements) involved in undergoing a spectrum
reallocation would be justified by longer term gains. Two choices determine the performance of a traffic prediction method: the choice of estimation parameters, and

the choice of estimation algorithm. The most common estimation parameters are the average and peak traffic rates [12]; associated estimation algorithms are discussed as follows.

Future traffic rates are generally predicted based on a set of past observed rates. Two approaches to estimate future traffic are the Autoregressive (AR) and Moving Average (MA) models [13]. Let $Y = \{Y_t | t \in T\}$ denote the predicted traffic series and $X = \{X_1, X_2, \cdots\}$ denote the set of past observed values, where t is the time index. The AR estimate of the series Y, based on p past observed values, is defined by,

$$Y_t = \sum_{i=1}^{p} \phi_i X_{t-i} + W_t$$

where the ϕ_i's are the parameters of the model, and W_t is the error term, commonly taken from a Normal distribution with zero mean and variance σ^2. The AR parameters ϕ can be calculated using Yule-Walker equations [13],

$$\gamma_m = \sum_{k=1}^{p} \phi_k \gamma_{m-k} + \sigma^2 \delta_m$$

where $\gamma_m = E[X_t X_{t-m}]$ is the auto-correlation function of X, and δ_m is an impulse function.

The MA estimate of order q is defined by

$$Y_t = W_t + \sum_{i=1}^{q} \theta_i W_{t-i}$$

where θ are the model parameters (the "weights" of the moving average) and W are i.i.d. error samples, taken from a Normal distribution with zero mean and variance σ^2 as before.

A more general approach is the Autoregressive Moving Average (ARMA) model [13], which combines the AR and MA models to produce

$$Y_t = W_t + \sum_{i=1}^{p} \phi_i X_{t-i} + \sum_{i=1}^{q} \theta_i W_{t-i}.$$

All of the above-mentioned models assume ergodicity and stationarity for the estimated values. More advanced generalizations such as the Autoregressive Integrated Moving Average (ARIMA) and Fractional-Autoregressive Integrated Moving Average (F-ARIMA) models can be used to estimate non-stationary processes [14]. F-ARIMA, and sometimes Fractional Gaussian Noise (FGN), are good for estimating traffic if it exhibits self-similarity [15].

The use of linear or non-linear regression techniques in the form of extrapolation and/or interpolation is also common practice for traffic prediction (see, e.g., [12]). The general equation for linear regression can be written as

$$Y_t = \alpha + \beta X_t + W_t$$

where α is the intercept, β is the slope, and W_t is the error term. To estimate the intercept and slope, different approaches such as Least Square, Maximum Likelihood, or Bayesian methods might be used [13]. For instance, using p samples for estimation, under the LS approach the aforementioned parameters are calculated as

$$\beta = \frac{(\sum_{i=0}^{p-1} X_{t-i} - \bar{X})(\sum_{i=0}^{p-1} Y_{t-i} - \bar{Y})}{\sum_{i=0}^{p-1} (X_{t-i} - \bar{X})^2}$$

and

$$\alpha = \bar{Y} - \beta \bar{X}$$

where $\bar{X}$ and $\bar{Y}$ are the average values of X_i and Y_i ($i = t, t-1, \cdots, t-p+1$). Under non-linear regression techniques, as the name implies, the values of Y_t are estimated from a subset of observed variables using a non-linear function. One example of this is exponential regression [13].

A further traffic estimation approach is to take advantage of historical patterns in the traffic through using past experienced traffic loads at certain times of the day or at certain times of the week/year to predict expected load at that time of day, week or year [12]. In addition to being used as a stand-alone process, this might be combined with the above statistical methods, for example through a weighting mechanism among the results produced by them. Another approach here might be to combine the historical series of statistics from certain periods (e.g., time series from intervals in past busy hours for the same day of each week, perhaps through an averaging method) and use these combined data sets, fed into the above statistical methods, to obtain more precise estimates of future requirements for coming DSA intervals. Finally, it must also be noted that information about future events assisting traffic prediction might come from terminals themselves, or from other entities, e.g., through signaling about subscribed services in future time intervals. Examples of such services might be a streamed sporting event, or a planned operating system upgrade download.

Clearly, there is a tradeoff between the accuracy and complexity of traffic prediction techniques; moreover, since different DSA scenarios need to meet different requirements, it is not practical to select one approach as being optimal for all scenarios. Some comparative studies of performance and complexity covering a limited number of selected traffic estimation approaches are available in the literature (see, e.g., [12], [16], and [17]). There is nevertheless the need for a far more exhaustive study of traffic prediction in a DSA context, encompassing implied requirements from a networking point of view as well as performance and complexity.

Signaling Exchange for DSA

Given a DSA process, resource exchanges should be negotiated between RANs, probably via a centralized server. This corresponds to the Resource Request and

Allocation phases of Fig. 3.2, whereby traffic predictions from networks would be present in Resource Requests. Networks should then inform devices about which bands (and perhaps which RATs) are to be used, before networks' Base Stations (BSs) and devices tune their transceivers, in a coordinated way, to the newly allocated resources (in addition to, perhaps, adapting the RATs applied). Mirroring the architecture for DSA in Fig. 3.4, an associated signaling example is depicted in Fig. 3.5.

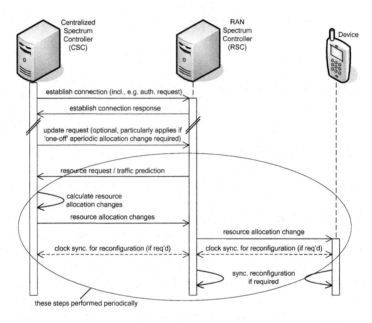

Fig. 3.5. Example signaling exchange for DSA.

In a DSA context, signaling for the purpose of resource coordination between RSCs and the CSC can easily be done over existing wired networks. Such signaling should not be an issue, providing that the participating networks are properly authenticated and the information is appropriately encrypted. With regards to signaling between networks and devices, this might be achieved through one of two options. Either an "in-band" method might be used, whereby information is sent using an existing channel within the network that the device is connected to at that point in time, or alternatively, perhaps in conjunction with an in-band channel (e.g., to provide information to devices when not connected or at start-up), an "out-band" method might be used, whereby a new widely understood (and likely extremely simple) physical channel, outside of operators' spectrum, is defined [18]. For the former, it is a case of simply mapping the information requirement to appropriate channels for the RATs in question, and providing the specified means for that information to be structured. For the latter, a radio interface and prospective spectrum band must be defined. Various ideas have been floated for this, ranging from the use of an entirely new purposefully defined channel in widely available dedicated spectrum (see, e.g., [19]), to the use of a pre-existing Digital Video Broadcasting-Handheld (DVB-H) channel along with the DVB-H RAT [18].

Spatial, Temporal, and Spectral Coordination for DSA

After the estimation of future traffic loads hence required resources, and in conjunction with the necessary signaling operations, an allocation of spectrum will occur as/if assessed appropriate by the CSC. A first critical operation here for the CSC is determining the resolution and appropriate alignment of allocated spectrum slices, and also the appropriateness of performing DSA between certain types of systems with different channel bandwidths. With regards to resolution and alignment, of course this must be compatible with the tunability of the participating systems and devices, as well as the channel bandwidths for each RAT-this is further information which might be conveyed in Resource Requests. Moreover, the channel bandwidths of the participating systems should ideally be divisible; i.e., if three systems are participating in the DSA, and the largest channel bandwidth is A followed by B then C, then ideally, A/B and B/C should both produce integers. In any case, performing DSA among systems with vastly different channel bandwidths might not be worthwhile. To highlight this, reallocation of a channel of spectrum from a system that uses large channel bandwidths to a system that uses small channel bandwidths would render the rest of the larger channel unusable to the larger system, hence would waste a lot of spectrum (as it would be unlikely that the smaller-channel system would be able to make use of the whole reallocated band); conversely, reallocation of a channel of spectrum from a system using smaller channels to a system that uses larger channels would often not be possible as it would require too many of the smaller channels to be reallocated in order to provide one such larger channel. If the channel bandwidths are the same however, spectrum sharing is much simpler. In [10], a DSA method is successfully outlined between a modified Digital Video Broadcasting-Terrestrial (DVB-T) system and a Universal Mobile Telecommunications System (UMTS) network, both of which use 5 MHz channels.

Spatial coordination is also a critical issue for DSA, an important aspect of which is the placement of BSs. In [20], through Monte-Carlo simulations, the possible spectrum sharing gain in a multi-operator network for downlink UMTS-FDD, using speech traffic, was assessed. It was shown that the relative displacement of multi-operator BSs has a significant effect on capacity gain for DSA. Co-located BSs result in better performance compared with displaced BSs, since the near-far effect and interference are minimized.

Further to this context, the coverage overlap among systems clearly affects DSA performance. For example, the average cell radius for DVB-T might be in the tens of km, compared with a cellular system such as the Global System for Mobile communications (GSM) or UMTS which may have an average cell radius of only a km or less. Hence an allocation from GSM to DVB-T requires that allocation to be made for a large number of GSM cells in order to cover one DVB-T cell, which might not be feasible as some of these cells might still require the resource that is being real-located. Conversely, an allocation of one DVB-T cell to GSM might be (partially) wasted as it would likely remain unused in many of the covered GSM cells.

Regarding spatial coordination, [21] proposed a method to group sets of cells/APs among systems using the largest coverage system as a reference, which is useful for the purpose of making spectrum sharing decisions on a cell-by-cell basis; this is de-picted for an exemplary case in Fig. 3.6(a). An alternative depiction in Fig. 3.6(b) highlights that the issue is much more complicated in situations where some of the networks are not spatially aligned. For the case of spectrum sharing in a homogenous environment comprised of only UMTS networks, [22] showed, through simulation, the destructive effect of such coverage misplacements on overall DSA performance. However, no study has yet provided an analytical basis for spatial coordination in a DSA context.

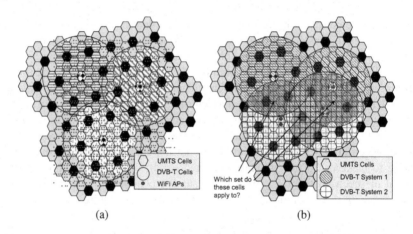

(a)   (b)

Fig. 3.6. Method for spatial coordination (grouping) of network cells/APs for spectrum sharing using the largest coverage system as a reference: (a) case where networks are aligned, and (b) case where networks are not aligned.

Finally, temporal coordination among networks in DSA also has a profound im-pact on performance. Clearly, a synchronization mechanism is required such that all the networks are coordinated, in the time domain, regarding their resource re-quests/traffic predictions, and particularly their spectrum transactions and any re-quired retuning/reconfiguration. This is particularly relevant where the participating spectrum in DSA is being used with a high average load. Such a requirement makes a centralized entity for DSA, as discussed previously, all the more important.

Regarding temporal coordination, there exists the DSA interval τ, acting as the synchronized period of dynamic allocations among networks through the CSC for the time requirements of DSA coordination, traffic prediction, reconfigurations, and so on. If τ is too large, however, the predicted traffic in networks might be outdated at the point in time the reallocation of spectrum is updated, or in a worst case scenario, the spectrum that an operator has been kind enough to allow to be temporarily allocated to another operator could be blocked from the original operator should he require the spectrum back due to an increased traffic load. If τ is too small, issues will usually become apparent, such as computational load, signaling, etc., related to DSA processes, as well as perhaps an increased frequency of unnecessary reconfigurations. In deciding τ and the associated traffic prediction approach, a tradeoff analysis is therefore necessary to address the conflicting requirements of accurate traffic prediction, and computational/signaling and other (e.g., reconfiguration) loads.

In [12] and [23], simulation studies were performed to address the effect of the DSA interval on gain for various approaches to DSA between a DVB-T and a UMTS network. It was observed that increasing τ to above a certain threshold interval of 4-5 hours causes a negative DSA gain (i.e., loss) due to outdated traffic prediction; on the other hand, reducing τ to below 1-2 hours does not provide a significant improvement in DSA gain hence is not justified. Note that such results can only be interpreted in the context of [12] and [23], given the specific assumptions such as the traffic pattern, traffic prediction method, DSA method, etc.

Other Notable Architectures for DSA

Various other architectures and protocols for DSA have been proposed in the literature. One important realization here is the hierarchical arrangement that applies in the structure of wireless/mobile networks, and of radio resources in general. In view of this, [21] and [24] proposed hierarchical spectrum management concepts that can facilitate both DSA and DSS architectures. Another concept, proposed in [25], is the use of a Spectrum Allocation Server (SAS) operating in the Medium Access Control (MAC) layer, where hybrid Frequency Division Multiple Access (FDMA)/Multi-Carrier Code Division Multiple Access (MC-CDMA) systems in a multi-vendor environment were studied. Buddhikot and Ryan [22] developed a coordinated spectrum management concept based on a spectrum broker, with a particular focus on homogeneous CDMA networks. They considered several infrastructure scenarios along the lines of sharing of BSs and the use of collocated antennas. Spectrum sharing can also be directly linked to economic concepts such as auctions. In [26], the authors proposed a token-based sharing protocol which is appropriate in Orthogonal Frequency Division Multiple Access (OFDMA) environments. This method can be applied in both licensed and unlicensed spectrum to create a distributed and real-time channel rental protocol. Various auction mechanisms can apply, whereby this and similar such approaches are in line with activities in the Institute of Electrical and Electronics Engineers (IEEE) 802.22 [27], [28] and IEEE 802.16h [29] standards. Some related works looking at similar approaches include [30] and [31].

In a multi-radio environment with DSA, perhaps the most important issue to consider is the interference effect of systems on each other. An interesting and rather challenging characteristic in this context is that of dynamically changing spectral neighbors, resulting in unforeseen interference effects of systems on each other. One example as to why this might happen is that of power control increasing transmission power thereby affecting the range at which interference on a particular frequency might be relevant, as well as affecting the level of out-of-band emission. As an example, [32] performed a simulation study characterizing the different types of interference present in a two-RAN spectrum sharing scenario involving UMTS and DVB-T type systems.

Reconfigurability [33], [34] and SDR [35] are important practical facilitators for DSA. Through reconfigurability, perhaps as facilitated by SDR, the characteristics of RATs in devices and BSs can be more easily changed to those that are employed in spectrum that is to be switched to: this is another paradigm to spectrum sharing whereby spectrum might be shared without requiring changes to its bounds, responsible party, or deployed technology. Moreover, Reconfigurability and SDR are useful for DSA per se, as through adaptability they allow greater freedom in the definition of spectrum ranges and RATs that might apply to systems/devices as a result of DSA.

Many national and international research initiatives partly or fully addressed various aspects of spectrum sharing, including SDR and Reconfigurability. Studies and consortia in Europe that have focused on flexible spectrum usage and related concepts include IST-DRiVE [36] and OverDRiVE [37], IST-SCOUT, IST-WINNER Phases I and II [38], IST-E2R Phases I and II [39], and Mobile VCE [40], no doubt among others. Various other studies in the US and the Far East have complemented the European interest in DSA. In the US, the SPEAKEASY [41], JTRS [42], and XG projects [43], among others, have assisted progress in related areas to spectrum sharing, and in Japan a number of Projects, many of which are driven by the National Institute of Information and Communications Technology (NICT) [44], have made a significant impact in such fields. In the US, much progress on SDR and related technologies is facilitated by the SDR Forum [45]. Some further interesting reading on DSA and related architectures includes [10], [12], [23], [46], [47] and [48] among others references in this chapter.

Summary of Architectural Requirements for DSA

Finally, the requirements for successful DSA operation are summarized as follows:

- DSA must be performed only in the context of agreement between participating networks, and particularly with the agreement of the networks that own spectrum that is to be dynamically reallocated.
- From a network architecture point of view, it is anticipated that there should be some form of centralized spectrum controller, which must know about the spectrum usages and requirements of each network across the shared spectrum at each location (e.g., in an ultimate realization of DSA, in each cell). Such information should be exchanged between the spectrum controller and the networks in question using predefined signaling means.
- The centralized controller should coordinate, spatially and temporally, the spectrum allocations among participating networks. In most conceptualizations, these allocations must be updated at understood periodic intervals. The centralized

controller must allocate spectrum using intelligent algorithms, in a way that minimizes interference (of course within the bounds of regulations) between participating networks, and satisfies traffic requirements conveyed in resource requests as best as possible. Allocations must be made in a fair and efficient manner and made in accordance with prearranged QoS policies, and of course must give preference to the owner of the spectrum in question through policies agreed by that owner.

- Each network, through some form of traffic/load prediction, must estimate its spectrum requirement for at least the coming DSA interval, and report this to the centralized controller. This estimation must be done in a fair way among participating networks, likely using a method that it agreed among networks.
- There is a short time interval during which the actual spectrum reallocation takes place across multi-operator networks/systems. The DSA process should guarantee no service interruption to active users during such intervals, e.g., through appropriate synchronization/coordination mechanisms.
- There should be sufficient provision/backup in case of any failure in a participating network or the spectrum manager. Mirroring spectrum controllers might be provided, or automated mechanisms might be employed in networks to revert to their default (licensed) spectrum allocations should the DSA process crash. The algorithms employed in the spectrum manager should be robust against failure, stable, and glitch-free.

3.2.3 Regulatory and Standardization Viewpoints on DSA

In realization of the importance of novel spectrum management techniques such as DSA and their potential to improve system performances, regulators around the world are introducing new consultations, measures and rules to the context. The Office of Communications (Ofcom) in the UK has proposed the concept of "Spectrum Usage Rights (SURs)" [49]. Under this idiom, spectrum usages by license holders are no longer necessarily tied to specific RATs. SURs also intend to specify acceptable interference limits for geographical, in-band and out-of-band interference, thereby taking an approach more in line with allowing novel spectrum usages so long as no unacceptable effects result. Moreover, it was planned through SURs for the involved parties in spectrum decisions to be able to interact directly, with minimal involvement of the regulator. These are all promising developments leading closer to DSA being employed among collaborating networks. Other Ofcom reports have proven to be heading in a similar direction regarding spectrum liberalization (e.g., [50] and [51]), as has been recommended by influential studies such as the "Review of Radio Spectrum Management" ("The Cave Report") [52] and the "Independent Audit of Spectrum Holdings" [53] in the UK.

Through means such as SURs, the Ofcom vision is to replace command and control methods with market-based mechanisms in 71.5% of bands by 2010 [50]. Moreover, Europe in general appears to be heading in a similar direction, with the passing of the non-legislative resolution "Towards a European Policy on Radio Spectrum" [54], based on an initial submitted report "A Market-Based Approach to Spectrum Management in the European Union".

Other international DSA-related regulatory developments include proposals such as a "Real-Time Spectrum Auctioning Model" presented to the Federal Communications Commission (FCC) in the US by Google Inc., and the release by the FCC of key documents such as "Promoting Efficient use of Spectrum Through Elimination of Barriers to the Development of Secondary Markets" [55]. In Ireland, the Commission for Communications Regulation (ComReg) has recently published a consultation on Dynamic Spectrum Access [56], much of the content of which also relates to wider visions for enhanced spectrum sharing including DSA. Many other regulators worldwide are following similar paths to greater dynamicity and freedom in spectrum attributions.

Standardization is an area that is currently waking up to the possibilities concerning DSA. One particular effort in this direction is the IEEE P1900 series of standards [57]. Established in 2005 and now under the sponsorship of IEEE Standards Coordinating Committee (SCC) 41, these standards are developing various mechanisms in support of SDR, CR, and improved spectrum operation paradigms. The IEEE P1900.4 standard has been concerned with developing a generic architecture to facilitate mechanisms such as DSA and DSS [56].

3.2.4 Future Outlook for DSA

Technically, some forms of DSA, such as sharing among operators with the same RAT, are already achievable. Specification of processes/architectures for DSA operation in systems often still need to be addressed, for example, providing means to coordinate DSA within a composite network or among networks, and providing means to inform devices of changes in spectrum allocations. Nevertheless, given an agreement to allow DSA between distinct networks/operators, those operators might use independently agreed mechanisms (of course, in alignment with regulatory requirements—which might have to be validated by the regulator in some cases), and/or might use specifications to assist where they exist (e.g., [11]).

There are technical issues to be addressed in the realization of some advanced forms of DSA, such as ensuring appropriate reconfigurability and tunability of networks/devices. For the most part however, challenges in the DSA context have for some time been largely political rather than technical. First, these political challenges exist between operators: why should an operator who has paid a significant price to acquire prime spectrum allow competitors to use it? Operators do nevertheless foresee mechanisms to achieve approaches to DSA through charging the "secondary" operator, akin to spectrum leasing. Second and perhaps most significantly, political challenges to DSA exist for regulators. This is because regulators are responsible for assuring that spectrum allocations are reliable, and want to ensure that any new technical approaches will not lead to a situation where those who have paid for spectrum are disadvantaged by interference from other systems. Nevertheless, as indicated in the prior section, there are significant, perhaps irreversible signs that the present conservative situation within regulation is rapidly changing.

In the light of the observations that have been presented, it seems that DSA processes are certain to become widely employed-the question is simply how long it will take. To this end, DSA mechanisms will be realized long before the event of full "Mitola" CR, which can be considered the ultimate and most challenging achievement in spectrum sharing. Given recent advances, various forms of DSA can be expected to be widely used long before the end of the 2010s decade.

3.3 Decentralization of Spectrum Sharing

Next investigated are architectures and protocols for the decentralization of spectrum sharing decisions, whereby the term "decentralization" is used in the sense of secondary devices or systems independently (or perhaps semi-independently) choosing spectrum to use based on usage opportunities they have identified. Such decentralization is assumed to be achieved through mechanisms such as Opportunistic Spectrum Access / Dynamic Spectrum Access [6], [9], whereby for simplicity these and all similar such terms are collected under one generic term "Dynamic Spectrum Selection (DSS)". A degree of centralization in decisions will often nevertheless remain; for example, DSS might be performed only within constraints/policies as conveyed by the network/operator of the primary system or an associated entity. Such constraints/policies might specify, for example, the maximum allowable inter-sensing period, the maximum allowable secondary access transmission power, the bands (perhaps dynamically) that are allowed to be opportunistically used, and an acceptable opportunistic usage mechanism (e.g., a MAC).

3.3.1 Basic Concept of DSS

DSS, similarly to DSA, is a basis for secondary access to spectrum of a licensed system. DSS might, however, also be performed in the context of unlicensed spectrum, akin to what is already achieved through IEEE 802.11 networks. There are fewer challenges to the realization of DSS in unlicensed spectrum, particularly because it is not controversial to the spectrum license owner (as there is none!), transmission power is usually limited (to minimize interference), and there is no imposed RAT which must be adhered to. Hence only DSS in the licensed spectrum domain is considered here.

DSS, unlike DSA, is based on a device-centric (or sometimes secondary-system centric) approach, whereby secondary devices/systems choose which bands they will use rather than being allocated them. In making this choice, a transmitting DSS-enabled secondary radio must operate such that the imposed interference to all primary radios is below a required threshold. If such a condition is met, for instance through the distance and/or physical obstacles between the secondary transmitter and primary receiver(s) being significant enough to cause sufficient path loss/shadowing, or though the DSS radio intelligently using opportunities within the RAT of the primary system (e.g., idle frequency divisions, timeslots, codes, etc.), the DSS device is allowed to transmit in the primary band.

Two categories of DSS are *negotiated* and *opportunistic* access. In the former case, a signaling channel, likely to be wireless in nature, provides an interaction medium between the primary and secondary system, whereby several signaling solutions for DSS have been proposed in the literature, e.g., [19], [58] and [59], and are under investigation as part of the IEEE P1900.4 standardization effort [57]. Note that such a signaling channel might also play a part in a controlled form of opportunistic access, where identification of access opportunities is directed/constrained by signaling from the primary system or an associated entity. In pure opportunistic access however, direct interaction (i.e., signaling) is not possible between the primary and the secondary systems. One example of such is TV bands, which are considered for secondary access within the IEEE 802.22 standard [27], [28]. In such cases, it is the responsibility of the secondary system to reliably identify secondary usage opportunities, and to vacate the band as soon as the primary system reappears or interference to it is anticipated to be an issue [48]. The alternative of using "underlay transmission" is extremely controversial.

3.3.2 Architectures and Protocols for DSS

The main phases involved in DSS operation, which are employed cyclically, are illustrated in Fig. 3.7. The associated approaches to DSS, with reference to these phases, are discussed in this section. Note also that such a representation might be extended to the context of the Cognition Cycle, introduced by Mitola [60].

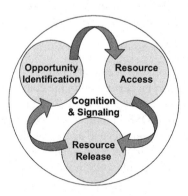

Fig. 3.7. Phases of DSS operation.

Opportunity Identification

A first major requirement in DSS is the identification of transmission opportunities, i.e., white or grey spaces in the spectrum. This is particularly important given the need to avoid interference to primary systems.

The goal of finding secondary spectrum access opportunities can be achieved through "spectrum sensing", a classification of techniques for which is provided in [61]. There are two general approaches to spectrum sensing: *energy detection, and feature detection.* Energy detection is easier to perform, but is hampered because sources of energy other than the primary system, i.e. noise, are present in the channel. Feature detection is able to more accurately detect the presence of signals, although has a notable processing requirement and latency. In any case, it has been suggested that the link budget analysis for accurate spectrum sensing, given the noise, interference and channel (i.e., shadowing and fading) uncertainties makes it almost impossible for individual CRs to identify opportunities reliably [61]. Simply considering a fade margin in the link budget analysis to solve this is not sufficient, due to uncertainty of fading distribution models among other reasons.

Two architectural approaches to tackle this problem are cooperative sensing [63], and to decouple the sensing task from devices through using an infrastructure of sensors [64]. The cooperative sensing approach reduces the error margin in the sensing task by collating measurements of CRs which are at somewhat different locations and times (hence experiencing different fades and, dependent on purpose, different shadowing characteristics). The degree of sensing complexity (sensitivity) at each node, and the channel characteristics (e.g., coherence time and bandwidth) determine the degree of cooperation required [65].

Cooperative sensing faces several challenges. Specifically, cooperation is not possible without a reliable signaling channel between secondary devices, whereby this channel would most commonly have to be achieved using wireless means. A frequent proposal is for dedicated spectrum to exist for such signaling (see, e.g., [18] and [19]), spectrum which has not been allocated by any regulatory body so far (indeed, a related proposal has had to be somewhat watered down in the World Radiocommunications Conference (WRC) 2007 [66]). Some dubious alternatives include using the opportunistic access itself for signaling, thereby reducing the inherent reliability, and to use of underlay transmission. Further challenges to cooperative sensing are the sheer amount of signaling overhead likely to be required, and the question of reliability of sensed information. As noted in [63], cooperative sensing gain is affected by the presence of failing/malicious nodes.

Despite such issues, a form of cooperative sensing architecture is already close to commercial realization within the IEEE 802.22 standard. One reason that this is possible is because of the static nature of primary and secondary transmitters in the context of IEEE 802.22, thus reducing required sensing overhead. Sensing in this case is performed in two stages. First, a fast sensing algorithm is used by all Customer Premise Equipments (CPEs) (i.e., home APs), where the results of this are aggregated at the centralized 802.22 BS. This fast sensing algorithm typically takes less than 1ms per channel. Next, the BS decides if more sensitive, $\sim$ 25ms per channel, fine sensing is necessary on secondary access channels to look for signatures of primary systems. This ensures that the primary system experiences no interference [28].

Considering the approach depicted in Fig. 3.8 of using a dedicated infrastructure of sensors as the spectrum sensing means, a number of technical challenges exist.

First, there is the question of who would implement and maintain such a sensing network. Next, the question of reliability of information also applies here. Furthermore, there is the question of how the gathered information is signaled among sensors and to/from devices, that of sensor energy, and also the issue of network roll-out time and cost, given the necessity to cover large areas and to optimize sensors for geographic locale.

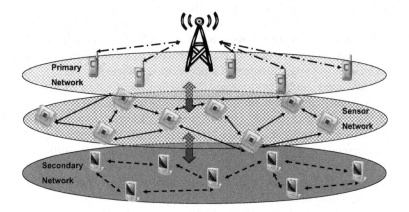

Fig. 3.8. Multi-layered network architecture decoupling spectrum sensing from the secondary spectrum access devices.

Various alternative methods to prevent interference to primary receivers, other than spectrum sensing, exist. Consider the case where the secondary transmitter is much closer to the primary transmitter than the primary receiver. In such circumstances, it is possible for the CR to know the primary system's interference characteristics and benefit from using a dirty paper coding technique to transmit without causing detectable interference to the primary receiver. It has been shown that, through this "cognitive radio channel" strategy, rate regions of somewhere between the interference channel and Multiple-Input and Multiple-Output (MIMO) channel can be achieved [67]. Of course, issues of delay (reactivity) and processing load are a challenge here.

As another approach, secondary transmitters might try to determine the locations of primary receivers. Through this approach, [62] proposed to introduce "No Talk Zones" for secondary devices within a certain range of each primary receiver. To account for channel uncertainty (e.g. shadowing and fading), this range should be significantly larger than the range of the primary system. An analysis of trade-off was performed in [62] to address different options in this scenario.

Another means for identifying the presence of the primary system is the use of dedicated wireless signaling channels per se. Different approaches here include the advertising of available resources by the primary network [8], [18], or broadcasting of beacons associated with occupied bands (e.g., [58]). Along similar lines, a dedicated signaling channel termed the Resource Awareness Channel (RAC) has been proposed in [19], utilized by primary receivers to broadcast information about the occupancy of the bands they are receiving data on. In accordance with reciprocity, the secondary transmitters might also glean some limited information about the channel conditions to primary receivers to enhance opportunity characterization.

Finally, given any opportunity identification mechanism, its performance must be continually monitored, likely with the involvement of a regulatory body. Technical bounds for the operation of such a mechanism must also be specified in its selection or development. To serve such purposes, a quantitative metric for induced interference is of importance. The Interference Temperature, proposed by the FCC in the US [68], might be such a metric; however, the FCC has recently dropped (with strong reservations) this proposal due to a lukewarm reception of the concept from some commenting parties [69]. Another recent interference-based suggestion for spectrum usage, albeit not directly aimed at secondary spectrum usage, has been that of SURs proposed by Ofcom in the UK [49].

Resource Access and Release

Having identified an appropriate spectrum transmission opportunity, the next task for DSS devices/systems is to efficiently use it. This phase comprises definition of an appropriate networking and RAT solution for the secondary system, both of which may feed on solutions that have already been developed in other contexts. Because of cross-layer effects, the definition of suitable higher layers might also apply. One consideration in the choice of an appropriate RAT is that, given use of a sensing mechanism for opportunity identification, the spectrum must frequently be rescanned to ensure that the primary system has not reappeared. Hence time periods must be set aside for the scanning procedure to be re-performed, during which the secondary system must remain silent in the band. IEEE 802.22 [27], [28] gives a first realization of an appropriate RAT within this context.

Regarding networking for the secondary system, different approaches such as CORVUS [70] and CogMesh [71] were proposed, and IEEE 802.22 might again be referred to as an example of a centralized networking approach. [70] proposed a random access procedure similar to that used in unlicensed spectrum access, whereby the secondary users can create "Secondary User Groups (SUG)" to coordinate communication. Members of a SUG might communicate with each other in an ad-hoc fashion, or alternatively might access a fixed infrastructure via a dedicated access point. Through using the adaptive clustering technique, [71] proposed to create a secondary ad hoc network by joining such clusters. Each cluster is formed based on the availability of a local coordination channel in a shared spectrum scenario. The problems of neighbor discovery, cluster formation, network formation and network topology management in this scenario are addressed.

With regards to Resource Release, secondary access to spectrum must curtail as soon as it becomes apparent that the primary system is present, or interference to it is likely to be above a threshold. Moreover, to minimize the probability of interference to the primary system, through coordination mechanisms, maximum usage of primary bands that are already being opportunistically used must be made by the secondary system(s), before any new bands are accessed.

Cognition

A final important aspect of DSS, particularly in the case of CR, is that of the cognition level of devices. A quick mention of this applies here: issues such as CR and cognition are expanded upon as the main purpose of the following chapter.

The cognitive capabilities of DSS enabled devices might differ considerably, where the complexity of a CR is in direct relation to the level of cognition in each phase of the Cognition Cycle: Observing, Orienting, Planning, Deciding, Learning, and finally Acting [60]. Policies, which assist in identifying spectrum opportunities and how they must be used, are necessary to bound/guide the cognition of CRs. Definitions of policies in the form of machine-understandable policy description language are provided in [72], [73]; moreover, the creation of policies is further simplified by an accurate representation of context such as the Radio Knowledge Representation Language (RKRL) [60]. Generic methods that can be used to form policies include the Common Information Model (CIM) [74], Ponder [75], and DEN-ng [76].

In general, for purposes such as intelligent spectrum opportunity identification and intelligent RAT/MAC definition, the cognition level of a CR goes in direct relation to the performance improvement it commands. Clearly however, a higher level of cognition is associated with more complex software/hardware requirements, hence increases the cost of the device. As inspired by [77], the level of cognition for different purposes, versus complexity, is depicted in Fig. 3.9.

Other Notable Architectures for DSS

A number of further notable architectures for DSS have been published in the literature, some examples of which are briefly discussed as follows. OFDMA is concentrated upon initially, as this represents the future of multi-user access in a vast array of wireless systems and is capable of shared spectrum usage through different sub-carriers being allocated to different systems. A DSS method in this context was developed in [78], where through using binary allocation vectors to mark the utilized sub-carriers of the OFDMA-based system, as well as an efficient physical layer detection mechanism termed the "boosting protocol", spectral coexistence of primary and secondary systems is possible. Other topics discussed in [78] include MAC issues, synchronization, and scheduling algorithms. QoS in such OFDMA-based DSS scenarios was investigated in [79], which has an emphasis on the tradeoff between fairness and QoS of participating nodes in a shared spectrum pool.

The DARPA NeXt Generation (XG) program has aimed to develop and expand technologies and system concepts for implementing dynamic spectrum sharing through opportunistic access [43]. Recent field demonstrations have shown communicating radios through XG protocols to be effective at avoiding occupied frequency bands [80]. The XG project also aims to develop a framework for machine-understandable policies managing radio behavior, facilitating devices to make intel-

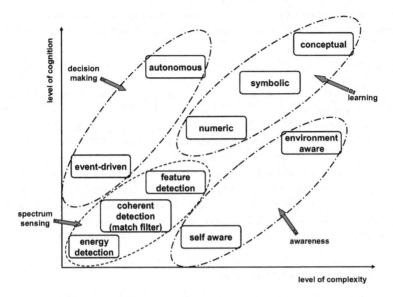

Fig. 3.9. Levels of cognition and complexity.

ligent opportunistic decisions. More about the XG framework and protocols can be found from [73], [81] and [82].

Further interesting literature on architectures for DSS is obtained by referring to the aforementioned references in this chapter.

Summary of Architectural Requirements for DSS

Finally, the requirements for successful DSS operation are summarized as follows:

- In licensed spectrum, the reliable identification (through, e.g., spectrum sensing) of transmission opportunities must apply, such that there is no tangible interference to primary systems. This is a challenge because of uncertainties such as noise, fading and shadowing. Cooperative sensing might be a solution in opportunistic DSS scenarios, the various overheads (signaling, computational etc.) of which should be carefully addressed. Another option might be to decouple the sensing task from secondary systems through the use of a sensor network. Among further options, secondary devices may be warded off of utilized bands by transmissions from primary systems.
- Secondary transmitters in licensed spectrum must keep (periodically) checking that the primary system has not appeared. Under commonly assumed sensing means, this has implications for the MAC/RAT as employed by the secondary system, as the secondary system must be silenced during each sensing period.
- In licensed spectrum, in the vast majority of foreseeable scenarios, DSS must operate within constraints as specified by the owner of that spectrum. Various

mechanisms for conveying these constraints must be specified, such as are currently being conceived through network-device distributed policies within the IEEE P1900.4 working group [57].

- Both structured and unstructured/ad-hoc network architectures for secondary systems can be envisaged. A structured architecture might be more suitable if the primary system's spectrum usage characteristics in the channel are not frequently changing (in space and time), for example in the IEEE 802.22 architecture for secondary access of TV bands. Unstructured secondary system architectures are more suitable to benefit from localized (in time/space) changes to the spectrum usage of the primary system.
- A reliable signaling channel to coordinate the usage of resources and for cooperative spectrum sensing is a challenge in DSS due to the unavailability of dedicated bands for such signaling. Such signaling means would likely need to be created.
- Finally, the networking problem in secondary systems is another issue to be considered. Different means for this have been outlined in the literature.

3.3.3 Arguments for and Against DSS

Numerous researchers have identified potential uses for DSS, some of the most wide-ranging of which are within the CR idiom as introduced by Mitola [8]. Other less ambitious uses simply revolve around devices being able to communicate directly, bypassing networks and thus making the communication far more efficient, user-friendly and cheaper. For instance, if you are calling or sending a text message to someone across the street, why is the communication not sent directly with one transmission over a local link through a DSS mechanism, instead of being transmitted to a BS, through one or more networks, and then transmitted again by a BS to the recipient? Such observations represent very strong arguments for DSS.

There are also various other arguments as to why DSS might be used. DSS is faster and more flexible, potentially achieving far better spectrum usage efficiency in the sharing mechanism. For instance, through DSS mechanisms, the loads (i.e., signaling and computational) and other issues in performing network-wide decisions can be largely avoided, as decisions are made at the "grass-roots" of local radio transmission links. This leads to decisions being made in a far more timely manner, thus improving reactivity to dynamic changes in spectral loads. Moreover, through the spectrum detection or other awareness mechanisms in DSS, far more detailed local information can be leveraged, achieving a much enhanced spatial accuracy in the optimality of spectrum sharing decisions.

On the other side of the coin, centralized approaches are usually easier to develop and maintain. DSA takes complexity away from devices thus making them simpler and cheaper to produce, whereas in the DSS case, complexity is brought to the edge of the network, i.e., to the access network and to devices. Dependent on the range of tasks and nature of the DSS scenario, such complexity might vary significantly on a case-by-case basis. For example, if a sensor network is used for spectrum sensing as illustrated in Fig. 3.8, the complexity will mostly be related to adaptive transception capabilities. On the other hand, if spectrum sensing is left to devices and secondary systems, the complexity and cost of devices and systems will increase dramatically.

As another argument against DSS, of course it is controversial because of the increased potential for interference to primary systems. This is discussed in other parts of this chapter.

3.3.4 Regulatory and Standardization Viewpoints on DSS

Regulatory and standardization viewpoints are both progressing towards DSS becoming a reality. In unlicensed spectrum, through IEEE 802.11 and more recently through emerging IEEE 802.16 networks, means for what might be considered a form of DSS have already been existence for some time. DSS in licensed spectrum is both technically and politically more complicated, because of the need to avoid interference to the owner of that spectrum. Standards in this context are nevertheless developing: some of these are the IEEE 802.22 standard which allows secondary access to locally unused TV bands [27], [28], IEEE 802.11y which is intending to extend Wi-Fi to the scope of secondary access to licensed spectrum [83], and the IEEE P1900.4 standard which aims to provide a generic architecture to assist decision making and information provision in both DSS and DSA [57]. Moreover, efforts such as DARPA XG are looking to standardize their developing CR technology [43], [73], [81], [82].

Because of the need to avoid interference, DSS in the regulatory domain is far more challenging than DSA, hence regulators are far more cautions about accepting it. Nevertheless, the FCC in the US has recently announced a rule change as regards CR, allowing secondary access to TV bands [84]. An imposed requirement is that all devices/systems certified to use locally vacant TV bands must employ CR technology to avoid interference to primary systems. This is among other developments such the "Cognitive Radio Technologies Proceeding" [85], and a document considering the promotion of secondary spectrum markets [55]. Moreover, in mid-2006 the Senate Commerce Committee in the US adopted the "Advanced Telecommunications and Opportunity Reform Act of 2006". This required the FCC to continue rulemaking procedures regarding the opening of Very High Frequency (VHF) and Ultra High Frequency (UHF) TV bands (54 MHz-698 MHz) to use by wireless broadband services and other DSS enabled devices. Furthermore, interesting proposals have been submitted to the FCC, including those related to using the FM band for mass-communication in trains [86], and one submitted to Google Inc. related to a secondary spectrum auctioning model [87].

The outlook in other counties is similar as regards progress towards DSS. This is evident from various studies in the UK and Ireland [51], [56], [88], as well as the support in Japan and Europe of a number of DSS related projects [39], [44]. Moreover, the implication of recent consultations and decisions is that regulators worldwide are heading in a broadly similar direction towards the acceptance of DSS, although the rate of progress, aside from secondary access to TV bands, is likely to be slow [88].

3.3.5 Future Outlook for DSS

DSS by secondary systems in radio and TV bands is much easier to achieve, as primary radio/TV transmissions are generally with a high power and fixed location so are much easier to detect and avoid. Moreover, though the situation may change, the kinds of secondary systems that have thus far been envisaged to use such bands in a secondary access fashion are relatively stationary. Hence it is perhaps no surprise that these bands are expected to provide some of the earliest commercial realizations of DSS through CR, possibly even as early as in 2009-10 in the US.

In situations where either or both the primary and secondary systems employ moving transmitters, and particularly if the primary systems are of relatively low transmission power, progress towards the realization of DSS is expected to be slow. This is because it must ensured that the primary systems experience no interference: there would be massive opposition from owners of primary spectrum, especially in cellular bands, should secondary access degrade their services. Given the need for greater checks in such contexts, complexities result on the regulatory side: e.g., the question of what duration of interference being marginally above a threshold for a spectrum band is considered unacceptable (this is important as regards the frequency of spectrum sensing by the secondary system), and what constitutes an acceptable threshold, for each specific kind of allocated band. Furthermore, how is it *ensured* by the regulator that mechanisms employed are able to avoid causing interference (i.e., which mechanisms should the regulator enforce as being acceptable for this), and how it is known that they will remain reliable and won't be 'hacked' to misbehave? It is likely that all these answers must be provided by the regulator before DSS can become a reality, and the types of answers will likely be very different for different types of primary bands and potentially also for different types of secondary systems using those bands. Hence there are many unknowns and challenges in the DSS context, some of which technical solutions are yet to be provided for, and some of which the eventual solutions for will almost certainly be controversial to some stakeholders.

To summarize, the development of DSS is expected to be significantly delayed in many contexts. Ofcom in the UK has anticipated the realization of a full "Mitola" CR, the "ultimate" in advanced DSS-related capability, as being as late as 2030 [88]. Finally, significant aspects of DSA and DSS are compared in Table 3.1.

3.4 Resource Optimization in Spectrum Sharing Using Game Theoretical Approaches

In this section, analytical approaches to radio resource optimization in both DSA and DSS domains are discussed, based on Game Theory. Game theory is chosen because it provides a set of powerful mathematical tools for analysis of "conflict and cooperation between intelligent rational decision makers" [89]. Radio resource management in DSA and DSS contexts can be modeled by game theory, whereby the networks are either cooperating (as in DSA or negotiated DSS), or are competing for resources

Table 3.1. Characteristics of DSA and DSS compared.

	Dynamic Spectrum Allocation (DSA)	Dynamic Spectrum Selection (DSS)
Management	Centralized	Centralized/Decentralized
Intelligence	Centrally located within the network	Relies on locally distributed information gathered/processed by one or several devices/BSs and/or a sensor network? Decision making might be centralized within the DSS network, dependent on the architecture
Implementation Complexity	Changes required mainly at the network side; an operator upgrade may be necessary. Greater reconfigurability/tunability for both networks and devices might be required for advanced solutions	Newer devices required, often based on SDR, and often with a capability to sense and access spectrum in different bands according to the defined specification
Processing Complexity	Traffic prediction algorithms necessary	Spectrum activity detection and adaptive transmission required (in devices and BSs if employed). In many cases will comprise SDR or a related solution (implying excessive digital signal processing)
Signaling Overhead/Information Exchange	Some signaling required between the centralized spectrum coordinator and networks, and to convey policies to devices	Likely cooperation between multiple intelligent devices; must be done with a high probability of reliable detection of spectrum holes
Scalability	As system size grows, centralized decision maker becomes heavily loaded. A hierarchical architecture might improve this	Amount of signaling overhead might become prohibitive as network dimension (particularly number of nodes per area) increases
Reactivity to Context Changes	Non real-time and to some extent real-time	Usually must be real-time

hence creating conflict (as in opportunistic DSS). The players, which by game theory definition are the decision makers, are the centralized spectrum management entities in the DSA case. In the DSS case, devices, or sometimes the transmission-reception pairs, are the players. The DSA domain is discussed first.

3.4.1 The DSA Case

The multi-operator architecture in Fig. 3.4 can be used as the baseline for analyzing the DSA case, whereby different RAN Spectrum Controllers (RSCs) interact via the Centralized Spectrum Controller (CSC) to satisfy their traffic demands for a specific resource allocation period. The CSC resembles a marketplace, where different players (i.e., the RSCs) bargain to obtain goods (i.e., spectrum) in order to maximize their payoffs, which might be throughput or percentage reduction in blocking/dropping probability. The assumption of rationality of the players implies that if using strategy A (e.g., using a specific band or transmission power level) results in payoff X, and using strategy B results in payoff Y, and $Y > X$, the rational player chooses strategy B.

Since in a DSA context different players are cooperating to achieve their goals, cooperative game theory is appropriate. Assume there are M different networks co-operating in a contiguous DSA case, as depicted in Fig. 3.3(b). Suppose the goal of each player (i.e., network) is to maximize its revenue from the available spectrum, where each network has N_m ($m = 1, 2, ..., M$) units of spectrum. Each participating network predicts the need for D_m units of spectrum (as calculated from anticipated traffic load) in the next DSA interval, where, in this case, the GSM channel size of 200 kHz is chosen as the spectrum unit. If $D_m > N_m$ for network m, that network needs to borrow $D_m - N_m$ units of spectrum from another network. We let the borrowing of each unit of spectrum have a cost C_b, the revenue from providing service per unit of spectrum be R_s, and the initial wealth of network m be $W_m(0)$. Depending on the traffic pattern of each network, the player should decide how many spectrum units to use, rent or borrow such that its wealth at the DSA interval i

$$W_m(i) = W_m(i-1) + D_m R_s - \frac{\frac{D_m}{N_m} - 1}{\left|\frac{D_m}{N_m} - 1\right|}(D_m - N_m)C_b$$

is maximized. Consider a simple scenario where an operator has 10 units of spectrum (i.e., $N_m = 10$), where borrowing one spectrum unit costs \$0.5, and where the initial wealth of the operator, $W_m(0) = \$1$. Fig. 3.10 plots the total operator revenues in the next period based on the different amounts of required spectrum (D_m) and the revenue per spectrum unit (R_s). Also represented is the plane corresponding to zero revenue in order to identify the domains (strategies) that improve revenue.

The choice of strategy would likely be based on long-term goals, such as to maximize revenue after a specific (high) number of DSA periods. Other utility functions might also be defined, depending on the optimization goal. Moreover, since each player is rational and he/she knows that other users are also rational, the players will only select a strategy that maximizes their utility among the set of possible strategies.

If a dominant strategy exists, the game will finally reach an equilibrium point, known as a Nash Equilibrium (NE) [89]. A strategy is said to be Pareto-optimal if the payoff of no user can be improved without decreasing the payoff of some or all of the other users. Depending on the characteristics of the game, the NE might be Pareto-optimal or not. The game might also have multiple NEs or Pareto-optimal solutions.

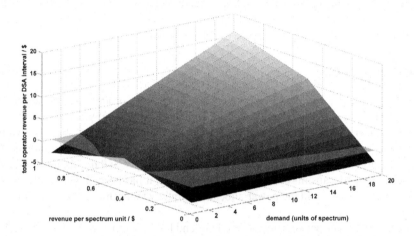

Fig. 3.10. Total operator revenue given the revenue per spectrum unit and spectrum demand per DSA interval.

There are different cooperative game solutions that can be used to analyze this scenario; for example, Nash Bargaining Solutions (NBS) provide a fair, unique, and Pareto-optimal result [89]. If it is assumed that without any cooperation between the players they can achieve the payoff denoted by the vector $\mathbf{v} \in \Re^M$, then by coordination of resource usages through the NBS the payoff of the players increases proportionally to their minimum payoff requirement vector $\mathbf{v}$. Therefore, as argued in [90], the NBS provides a proportional fairness solution paradigm.

Denote the bounded set of all feasible payoff allocations by $\mathbf{F}$, and the allocation vector (in $\Re^M$) for the bargaining problem $(\mathbf{F}, \mathbf{v})$ by $\phi(\mathbf{F}, \mathbf{v})$. It was shown that there exists a unique NBS for a bargaining problem, which can be calculated by [89]

$$\phi(\mathbf{F}, \mathbf{v}) \in \arg\max_{x \in F, x > v} \prod_{i=1}^{K} (x_i - v_i)$$

or equivalently by [90]

$$\phi(\mathbf{F}, \mathbf{v}) \in \arg\max_{x \in F, x > v} \sum_{i=1}^{K} \ln(x_i - v_i).$$

Since the solution obtained is unique and Pareto-optimal, any set of rational players will choose this same result.

The players in the above DSA game will coexist for a long period of time (of the order of months to many years), hence they can learn from history (i.e., from the outcomes of the DSA operations in previous periods) as to optimal coexistence strategies. For example, if in a certain DSA period, one RSC tries to increase its payoff by misguiding the other networks regarding the resources available or required, other players can punish such behavior by stopping the cooperative approach and by using a similar greedy strategy. Such greedy behavior will, clearly, result in the reduction of payoff for all players. Hence, rational players will avoid misleading one another in a cooperative game, and will behave honestly.

3.4.2 The DSS Case

For the DSS case, it is possible to extend the same lines of argument presented for DSA. In [91], a scenario is outlined for negotiated DSS, whereby an ad hoc network of CRs is trying to access the resources of a primary cellular OFDM-based network. The primary system guarantees a level of QoS for its users, defined as a minimum bit rate and target BER. Using the achievable rates as payoffs, and the choice of OFDM sub-channel and transmission power as the strategy, it has been shown in [91] that a NBS solution exists and the power allocation strategy that achieves this solution was derived.

The differences between negotiated DSS and DSA in terms of resource allocation strategy are as follows. First, to avoid interference, the resource allocation period in the DSS case is generally much shorter than that in DSA, of the order of milliseconds in some extreme cases. This DSS interval is generally based on the primary system's resource allocation period. A second difference is the choice of utility function (payoff). In the DSA case, the utility could be a long term goal such as reducing dropping or blocking probability; however, in DSS the utility is usually a short term goal such as maximizing the achievable data rate at any given time.

Further reading on the use of Game Theory for spectrum sharing can be obtained from, e.g., [92], in addition to previously mentioned references.

Conclusion

Spectrum is in significant demand, but evidently it is used with extremely low efficiency. To fulfill the requirements of pioneering wireless services and applications, novel technical solutions, as well as more flexible spectrum regulations, are needed. The challenge of improving spectral efficiency through secondary spectrum access has been studied in this chapter, the focus being on Dynamic Spectrum Allocation (DSA) (centralized decision making) and Dynamic Spectrum Selection (DSS) (largely decentralized decision making). Architectures and protocols for spectrum sharing has been investigated in the paradigms of DSA and DSS (particularly opportunistic DSS), and major challenges and open issues have been identified in each case. As a final contribution, radio resource optimization based on Game Theory has been discussed, of relevance to both the DSA and DSS contexts.

It is clear that by improving spectrum utilization through spectrum sharing, more resources can be made available to wireless technologies; moreover, spectrum sharing might be a source of cost reduction for operators and service providers, and therefore a way to increase profit. Sound economic and business models are nevertheless required for spectrum sharing, in addition to solutions to various technical challenges, in order to maximize its benefits. Moreover, the question of what types of services can be sufficiently provided in various shared spectrum environments must be addressed, and business cases must be investigated.

Clearly, far more work is needed to overcome various issues, technical and political, and particularly in the DSS domain, before the full potential of spectrum sharing can be realized. Nevertheless, significant strides are currently being made towards spectrum sharing in the technical, regulatory, and standardization domains. Given such progress, it is entirely reasonable to expect simplified solutions for both DSA and DSS to be widely employed, in a number of countries, within a decade.

Acknowledgment

Some of the work reported in this chapter formed part of the Delivery Efficiency Core Research Programme of the Virtual Centre of Excellence in Mobile & Personal Communications, Mobile VCE (http://www.mobilevce.com). This research has been funded by EPSRC and by the Industrial Companies who are Members of Mobile VCE. Fully detailed technical reports on this research are available to Industrial Members of Mobile VCE. Some of this work has been performed within the project E2RII, which has received funding from the European Community's Sixth Framework programme. This work reflects only the authors' views; the Community is not liable for any use that may be made of the information contained herein. The contributions of colleagues from the E2RII consortium are hereby acknowledged.

References

1. US Frequency Allocation Table, accessible from `http://www.fcc.gov/oet/spectrum/table/`, accessed Jan. 2008.
2. UK Frequency Allocation Table, accessible from `http://www.ofcom.org.uk/radiocomms/isu/ukfat/`, accessed Jan. 2008.
3. Germany Frequency Allocation Table, accessible from `http://www.bundesnetzagentur.de/media/archive/1820.pdf`, accessed Jan. 2008.
4. Japan Frequency Allocation Table, accessible from `http://www.tele.soumu.go.jp/e/search/share/plan.htm`, accessed Jan. 2008.
5. N. Devroye, P. Mitran, and V. Tarokh, "Cognitive Decomposition of Wireless Networks," in *Proc. CROWNCOM 2006*, Mykonos, Greece, Jun. 2006.
6. S. Haykin, "Cognitive Radio: Brain-Empowered Wireless Communications," *IEEE J. Sel. Areas Commun.*, vol. 23, no. 2, pp. 201-220, Feb. 2005.
7. A. Tonmukayakul and M. Weiss, "Secondary Use of Radio Spectrum: A Feasibility Analysis," in *Proc. Research Conference on Communication, Information, and Internet Policy*, Arlington, VA, USA, Oct. 2004.

8. J. Mitola, "Cognitive Radio for Flexible Mobile Multimedia Communications," in *Proc. IEEE Int'l. Workshop on Mobile Multimedia Commun.*, San Diego, CA, USA, Nov. 1999, pp. 3-10.

9. I. Akyildiz, W. Lee, M. Vuran, and S. Mohanty, "NeXt Generation/Dynamic Spectrum Access/Cognitive Radio Wireless Networks: A Survey," *Computer Networks*, vol. 50, no. 13, pp. 2127-2159, Sep. 2006.

10. P. Leaves *et al.*, "Dynamic Spectrum Allocation in Composite Reconfigurable Wireless Networks," *IEEE Commun. Mag.*, vol. 42, no. 5, pp. 72-81, May 2004.

11. 3GPP, "Network Sharing; Architecture and functional description (Release 6)," 3GPP TS 23.251 V6.6.0, Mar. 2006.

12. IST-DRiVE Deliverable D09, "Dynamic Spectrum Allocation Algorithm Including Results of DSA Performance Simulations," Jan. 2002.

13. A. Papoulis and S. U. Pillai, *Probability, Random Variables and Stochastic Processes (4th Edition).* New York: McGraw-Hill, Jan. 2002.

14. T. C. Mills, *Time Series Techniques for Economists.* New York: Cambridge University Press, 1990.

15. W. Leland, M. Taqqu, W. Willinger, and D. Wilson, "On the Self-Similar Nature of Ethernet Traffic (Extended Version)," *IEEE/ACM Trans. Netw.*, vol. 2, no. 1, pp. 1-15, Feb. 1994.

16. H. Feng and Y. Shu, "Study on Network Traffic Prediction Techniques," in *Proc. Int'l. Conference on Wireless Communications, Networking and Mobile Computing*, Wuhan, China, Sep. 2005, pp. 995-998.

17. S. Fernandes, V. Teichrieb, D. Sadok, and J. Kelner, "Time Series Applied to Network Traffic Prediction: A Revisited Approach," in *Proc. Applied Modeling and Simulation (AMS 2002)*, Cambridge, MA, USA, Nov. 2002.

18. O. Holland *et al.*, "Development of a Radio Enabler for Reconfiguration Management within the IEEE P1900.4 Working Group," in *Proc. IEEE DySPAN 2007*, Dublin, Ireland, Apr. 2007, pp.232-239.

19. O. Holland *et al.*, "A Universal Resource Awareness Channel for Cognitive Radio," in *Proc. IEEE PIMRC 2006*, Helsinki, Finland, Sep. 2006, pp. 1-5.

20. M. Pereirasamy *et al.*, "Dynamic Inter-Operator Spectrum Sharing for UMTS FDD with Displaced Cellular Networks," in *Proc. IEEE WCNC 2005*, New Orleans, LA, USA, Mar. 2005, pp. 1720-1725.

21. A. Attar and A. H. Aghvami, "A Framework for Unified Spectrum Management (USM) in Heterogeneous Wireless Networks," *IEEE Commun. Mag.*, vol. 45, no. 9, pp. 44-51, Sep. 2007.

22. M. Buddhikot and K. Ryan, "Spectrum Management in Coordinated Dynamic Spectrum Access Based Cellular Networks," in *Proc. IEEE DySPAN 2005*, Baltimore, MD, USA, Nov. 2005, pp. 299-307.

23. P. Leaves, J. Huschke J, and R. Tafazolli, "A Summary of Dynamic Spectrum Allocation Results from DRiVE," in *Proc. IST 2002*, Thessaloniki, Greece, Jun. 2002, pp. 245-250.

24. O. Holland, Q. Fan, and A. H. Aghvami, "A Dynamic Hierarchical Radio Resource Allocation Scheme for Mobile Ad-Hoc Networks," in *Proc. IEEE PIMRC 2004*, Barcelona, Spain, Sep. 2004, pp. 1005-1010.

25. R. Kulkarni and S. Zekavat, "Traffic Aware Inter-Vendor Dynamic Spectrum Allocation: Performance in Multi-Vendor Environment," in *Proc. Int'l. Conf. on Commun. and Mobile Computing*, Vancouver, Canada, 2006, pp. 85-89.

26. D. Grandblaise *et al.*, "Credit Token based Rental Protocol for Dynamic Channel Allocation," in *Proc. CROWNCOM 2006*, Mykonos, Greece, Jun. 2006.

27. IEEE 802.22 Homepage, `http://grouper.ieee.org/groups/802/22/`, accessed Jan. 2008.
28. C. Cordeiro *et al.*, "IEEE 802.22: An Introduction to the First Wireless Standard based on Cognitive Radios," *Journal of Communications*, vol. 1, no. 1, pp. 38-47, Apr. 2006.
29. IEEE 802.16's License-Exempt (LE) Task Group Homepage, `http://grouper.ieee.org/groups/802/16/le/index.html`, accessed Jan 2008.
30. J. Huang, R. Berry, and M. Honig, "Auction-Based Spectrum Sharing," *Springer Mobile Networks and Applications*, vol. 11, no. 3, pp. 405-408, Jun. 2006.
31. V. Rodriguez, K. Moessner, and R. Tafazolli, "Market Driven Dynamic Spectrum Allocation over Space and Time among Radio-Access Networks: DVB-T and B3G CDMA with Heterogeneous Terminals," *Springer Mobile Networks and Applications*, vol. 11, no. 6, pp. 847-860, Dec. 2006.
32. C. Hamacher, "Spectral Coexistence of DVB-T and UMTS in a Hybrid Radio System," in *Proc. IST 2001*, Barcelona, Spain, Sep. 2001, pp. 203-208.
33. M. Mehta *et al.*, "Reconfigurable Terminals: An Overview of Architectural Solutions," *IEEE Commun. Mag.*, vol. 39, no. 8, pp. 82-89, Aug. 2001.
34. D. Grandblaise, K. Moessner, P. Leaves, and D. Bourse, "Reconfigurability Support for Dynamic Spectrum Allocation: From the DSA Concept to Implementation," in *Proc. Workshop on Mobile Future and Symposium on Trends in Communications*, Bratislava, Slovakia, Oct. 2003.
35. W. Tuttlebee (ed.), *Software Defined Radio: Enabling Technologies.*   London: Wiley, May 2002.
36. IST-DRiVE Project Website, `http://www.ist-drive.org`, accessed Jan. 2008.
37. IST-OverDRiVE Project Website, `http://www.comnets.rwth-aachen.de/$\sim$o\_drive`, accessed Jan. 2008.
38. IST-WINNER Project Website, `http://www.ist-winner.org`, accessed Jan. 2008.
39. IST-E2R Project Website, `http://e2r.motlabs.com`, accessed Jan. 2008.
40. Mobile VCE Consortium Website, `http://www.mobilevce.com`, accessed Jan. 2008.
41. R. Lackey and D. Upmal, "Speakeasy: The Military Software Radio," *IEEE Commun. Mag.*, vol. 33, no. 5, pp. 56-61, May 1995.
42. Joint Tactical Radio System (JTRS) Homepage, `http://enterprise.spawar.navy.mil/body.cfm?type=c\&category=27\&subcat=60`, accessed Jan. 2008.
43. DARPA XG Program Homepage, `http://www.darpa.mil/ATO/programs/XG/`, accessed Jan. 2008.
44. National Institute of Information and Communications Technology (NICT) Website, `http://www.nict.go.jp`, accessed Jan. 2008.
45. SDR Forum Website, `http://www.sdrforum.org`, accessed Jan. 2008.
46. I. Akyildiz, S. Mohanty, and J. Xie, "A Ubiquitous Mobile Communication Architecture for Next-Generation Heterogeneous Wireless Systems," *IEEE Commun. Mag.*, vol. 43, no. 6, pp. S29-S36, Jun. 2005.
47. S. Souissi and E. Callaway, "Method and Apparatus for Dynamic Spectrum Allocation," US Patent 6553060, Apr. 2003.
48. J. Peha, "Approaches to Spectrum Sharing," *IEEE Commun. Mag.*, vol. 43, no. 2, pp. 10-12, Feb. 2005.
49. Ofcom, "Spectrum Usage Rights: Technology and Usage Neutral Access to the Spectrum," Consultation, `http://www.ofcom.org.uk/consult/condocs/sur/`, 2006.

50. Ofcom, "Spectrum Framework Review," Consultation, `http://www.ofcom.org.uk/consult/condocs/sfr/`, 2004-2005.
51. Ofcom, "A Study into Dynamic Spectrum Access–Final Report," `http://www.ofcom.org.uk/research/technology/overview/emer\_tech/dsa/dsafinal.pdf`, Mar. 2007.
52. M. Cave *et al.*, "Review of Radio Spectrum Management," for UK Dept. Trade and Industry and H. M. Treasury, Mar. 2002, accessible from `http://www.ofcom.org.uk/static/archive/ra/spectrum-review/index.htm`.
53. M. Cave *et al.*, "Independent Audit of Spectrum Holdings," for UK H. M. Treasury, Dec. 2005, accessible from `http://www.spectrumaudit.org.uk`.
54. European Parliament Resolution, "Towards a European policy on radio spectrum," ITRE/6/37236, Feb. 2007.
55. FCC ET Docket No. 04-167, "Promoting Efficient Use of Spectrum Through Elimination of Barriers to the Development of Secondary Markets," 2004.
56. ComReg, "Dynamic Spectrum Access," Document No. 07/22, Apr. 2007.
57. IEEE SCC41/P1900 Website, `http://www.scc41.org`, accessed Jan. 2008.
58. A. P. Hulbert, "Spectrum Sharing Through Beacons," in *Proc. IEEE PIMRC 2005*, Berlin, Germany, Sep. 2005, pp. 989-993.
59. D. Raychadhuri and X. Jing, "A Spectrum Etiquette Protocol for Efficient Coordination of Radio Devices in Unlicensed Bands," in *Proc. IEEE PIMRC 2003*, Beijing, China, Sep. 2003, pp. 172-176.
60. J. Mitola III, PhD dissertation, "Cognitive Radio: An Integrated Agent Architecture for Software Defined Radio," Royal Institute of Technology, Sweden, 2000.
61. S. Shankar, C. Cordeiro, and K. Challapali, "Spectrum Agile Radios: Utilization and Sensing Architectures," in *Proc. IEEE DySPAN 2005*, Baltimore, MD, USA, Nov. 2005, pp. 160-169.
62. A. Sahai, N. Hoven, and R. Tandra, "Some Fundamental Limits on Cognitive Radio," in *Proc. Allerton Conf. on Commun., Control, and Computing*, Champaign, IL, USA, Oct. 2004.
63. S. Mishra, A. Sahai, and R. Brodersen, "Cooperative Sensing Among Cognitive Radios," in *Proc. IEEE ICC 2006*, Istanbul, Turkey, Jun. 2006, pp. 1658-1663.
64. A. Sahai, S. Mishra, R. Tandra, and N. Hoven, "Sensing for Communication: The Case of Cognitive Radio," in *Proc. Allerton Conf. on Commun., Control, and Computing*, Monticello, IL, USA, Sep. 2006.
65. A. Sahai, R. Tandra, S. Mishra, and N. Hoven, "Fundamental Design Tradeoffs in Cognitive Radio Systems," in *Proc. Int'l. Workshop on Technology and Policy for Accessing Spectrum (TAPAS)*, Boston, MA, USA, Aug. 2006.
66. ITU World Radiocommunications Conference 2007 (WRC 2007), Geneva, Switzerland, Oct.-Nov. 2007, accessible from `http://www.itu.int/ITU-R/index.asp?category=conferences\&link=wrc-07`, accessed Jan. 2008.
67. N. Devroye, P. Mitran, and V. Tarokh, "Achievable Rates in Cognitive Radio Channels," *IEEE Trans. Inf. Theory*, vol. 52, no. 5, pp. 1813-1827, May 2006.
68. FCC ET Docket No. 03-237, Interference Temperature, Notice of Inquiry/Notice of Proposed Rulemaking, 2003.
69. FCC Order FCC-07-78A1, "In the Matter of Establishment of an Interference Temperature Metric to Quantify and Manage Interference and to Expand Available Unlicensed Operation in Certain Fixed, Mobile and Satellite Frequency Bands," May 2007.
70. D. Cabric *et al.*, "A Cognitive Radio Approach for Usage of Virtual Unlicensed Spectrum," in *Proc. IST 2005*, Dresden, Germany, Jun. 2005.

71. T. Chen *et al.*, "CogMesh: A Cluster-Based Cognitive Radio Network," in *Proc. IEEE DySPAN 2007*, Dublin, Ireland, Apr. 2007, pp. 168-178.
72. L. Berlemann, S. Mangold, and B. Walke, "Policy-Based Reasoning for Spectrum Sharing in Cognitive Radio Networks," in *Proc. IEEE DySPAN 2005*, Baltimore, MD, USA, Nov. 2005, pp. 1-10.
73. BBN Technologies, "XG Policy Language Framework," Version 1.0, 2004.
74. Distributed Management Task Force (DMTF) Website, http://www.dmtf.org, accessed Jan. 2008.
75. N. Damianou, N. Dulay, E. Lupu, and M. Sloman, Book Chapter: "The Ponder Policy Specification Language," in *Lecture Notes in Computer Science*.   Springer, 2001.
76. J. Strassner, "DEN-ng: Achieving Business-Driven Network Management," in *Proc. IEEE/IFIP NOMS 2002*, Jeju, Korea, Apr. 2002, pp. 753-766.
77. A. Attar, A. Sheikhi, H. Abiri, and A. Mallahzadeh, "A New Method for Communication System Recognition," *Iranian Journal of Science & Technology, Transaction B, Engineering*, vol. 30, no. B6, pp. 775-788, 2006.
78. T. Weiss and F. Jondral, "Spectrum Pooling: An Innovative Strategy for the Enhancement of Spectrum Efficiency," *IEEE Commun. Mag.*, vol. 42, no. 3, pp. 8-14, Mar. 2004.
79. T. Weiss, M. Spiering, and F. Jondral, "Quality of Service in Spectrum Pooling Systems," in *Proc. IEEE PIMRC 2004*, Barcelona, Spain, Sep. 2004, pp. 345-349.
80. F. Seeling, "A Description of the August 2006 XG Demonstrations at Fort A.P. Hill," in *Proc. IEEE DySPAN 2007*, Dublin, Ireland, Apr. 2007, pp. 1-12.
81. BBN Technologies, "The XG Vision," Version 2.0, 2004.
82. BBN Technologies, "The XG Architectural Framework," Version 1.0, 2003.
83. IEEE 802.11y, http://grouper.ieee.org/groups/802/11/, accessed Jan. 2008.
84. FCC ET Docket No. 04-186, "Unlicensed Operation in the TV Broadcast Bands," 2004.
85. FCC ET Docket No. 03-108, "Cognitive Radio Technologies Proceeding," http://www.fcc.gov/oet/cognitiveradio/, accessed Jan. 2008.
86. FCC ET Docket No. 06-161, Request for Comments on a Request for a Waiver of Part 15 of the Commissions Rules, 2006.
87. FCC DA 07-2197, Comments Sought on Google Proposals Regarding Service Rules for 700 MHz Band Spectrum, May 2007.
88. Ofcom Cognitive Radio Study, accessible from http://www.ofcom.org.uk/research/technology/research/emer_tech/cograd, accessed Jan. 2008.
89. R. Myerson, *Game Theory: Analysis of Conflict (3rd Edition)*.   Cambridge: Harvard University Press, 1999.
90. H. Yaiche, R. Mazumdar, and C. Rosenberg, "A Game Theoretic Framework for Bandwidth Allocation and Pricing in Broadband Networks," *IEEE/ACM Trans. Netw.*, vol. 8, no. 5, pp. 667-678, Oct. 2000.
91. A. Attar, M. R. Nakhai, and A. H. Aghvami, "Cognitive Radio Game: A Framework for Efficiency, Fairness and QoS Guarantee," in *Proc. IEEE ICC 2008*, Beijing, China, May 2008.
92. M. Halldórsson, J. Halpern, L. Li, and V. Mirrokni, "On Spectrum Sharing Games," in *Proc. ACM Symposium on Principles of Distributed Computing*, St. Johns, Canada, Jul. 2004.

4

Cognitive Radio Techniques in Heterogeneous Wireless Access Networks

Alireza Attar[1], Oliver Holland[1], Tim Farnham[2], Mahesh Sooriyabandara[2], and M. Reza Nakhai[1], and A. Hamid Aghvami[1]

[1] Centre for Telecommunications Research, King's College London, UK
{Ali.Attar, Oliver.Holland, Reza.Nakhai, Hamid.Aghvami}@kcl.ac.uk
[2] Toshiba Research Europe Ltd., Bristol, UK
{Tim.Farnham, Mahesh.Sooriyabandara}@toshiba-trel.com

4.1 Introduction

Since the late 1990's, when Mitola coined the term Cognitive Radio (CR) [1], [2], this relatively new branch of wireless communications has enjoyed tremendous research activity. At the last count, there were two dedicated conferences on CR, namely, the IEEE International Symposium on Dynamic Spectrum Access Networks (DySPAN) and the International Conference on Cognitive Radio Oriented Wireless Networks and Communications (CrownCom), along with numerous workshops, seminars, and sessions on CR within almost all major communications conferences. There are also several dedicated or related standardization efforts on CR technologies.

This degree of excitement within the wireless communications research community might be related to the freedom that the prospect of CR will bring about. After all, a CR is an intelligent radio that can realize the futuristic dreams of users, based on sensing the Radio Frequency (RF) environment, gathering information, and performing autonomous operations to fulfill the goals and meet the preferences of end-users through sophisticated learning capabilities. In other words, your CR will tailor itself to meet your requirements/preferences, which might be different from everyone else's. This injects a sense of individuality (referred to as brain-empowered radio [3]) into formerly soulless devices which were controlled (at least concerning their main functionalities) by operators and treated all users in the same way.

This chapter presents an in-depth, detailed look at the concept of CR. Whereas the majority of CR research effort in the literature has thus far been focused on the PHY and MAC layers, this chapter strives to provide the reader with a broader perspective by balancing its content to cover all relevant layers of the OSI reference model. To set the scene, the vision for CR is introduced in Section 4.1.1, and current activities within the field of CR, including standardization efforts, are briefly covered in Section 4.1.2. Section 4.2 is devoted to PHY and MAC layer issues in CR, whereby two exemplary scenarios are elaborated on to highlight the open problems in the PHY and MAC layers. Section 4.3 proceeds to networking scenarios, whereby

E. Hossain (ed.), *Heterogeneous Wireless Access Networks*,
DOI: 10.1007/978-0-387-09777-0_4, © Springer Science+Business Media, LLC 2008

the sub-sections of Section 4.3 deal with signaling and routing considerations for a network of CRs. A list of open issues and challenges in cognitive networking is also provided at the end of Section 4.3. In Section 4.4, the higher layers of CR are discussed, presenting several application scenarios along with the introduction of open problems. Finally, in Section 4.5, cross-layer optimization techniques for CR in a heterogeneous wireless context are investigated, before the chapter concludes.

4.1.1 Vision for Cognitive Radio

In his dissertation, Mitola envisions a new intelligent radio by "integration of model-based reasoning with software radio", creating the CR, which would be "trainable in a broad sense, instead of just programmable" [1]. He introduces the cognition cycle, which can loosely be interpreted as the map of a CR's brain, showing how its interactions with the outside world are processed, and outlining cognitive functions such as learning, orienting, deciding, etc. Mitola does not use a State Machine description for the cognition cycle, thereby broadening the vision of a cognitive system to a higher level of intelligence than pre-programmed, event-driven devices. In fact, he introduces different levels of cognition, ranging from pre-programmed radios at level zero, to protocol adapting radios at level eight (the highest level), which are radios that can autonomously propose and negotiate protocols. To facilitate such radios interacting among themselves as well as with stimuli from the outside world, he introduces the Radio Knowledge Representation Language (RKRL) [1].

As it is clear from the above, the major capabilities envisioned for CR are "reasoning" and "autonomous" operation. The decisions of CRs involve the aggregation of vectors representing radio-self (resources, capabilities, etc.), the RF scene (channel, propagation, etc.), and the user (preferences, requirements, etc.). The grand vision for CR has been to create a "society" of "individual" radios that can cooperate and coexist with each other, given the rules of that society. Interestingly enough, such rules can be proposed by the intelligent radios themselves, as discussed earlier. To this end, the initial step is for the radio to be aware of "itself". Although cognitive capability can reside within any radio entity within a network, e.g., the handsets or the Base Station (BS), let's consider the case of cognitive handsets as follows.

Today's handsets are an amalgamation of numerous hardware chipsets with software applications. Most applications (whether available in a specific handset and whether hardware or software based) are operating in isolation within the handset, e.g., different radio front ends, audio/video file players, digital cameras, internal and external memories, the Global Positioning System (GPS), games, and so on. The handset is usually not intelligently aware of these capabilities, hence cannot operate with optimum efficiency.

An intelligent handset might employ the user's location information, obtained from its built-in GPS, to assist in transmission by selecting the radio front end that provides the best coverage (or any other user preference such as cost, Quality of Service (QoS), etc.) in that region. Moreover, a higher degree of efficiency is achieved if handsets in the environment are also intelligent and capable of interacting. For example, a cognitive handset that can recognize that its loss of communication is because the user's car has entered a tunnel (through using information from its own camera and/or light sensor, perhaps partially assisted by its road mapping database in conjunction with location awareness) might be able to switch to multi-hop communication using the handsets in neighboring cars to reach an access point outside the tunnel.

Two major shifts in the focal point of CR research and development are occurring. The first concerns a narrower focus of CR, particularly concentrating on intelligent spectrum utilization through opportunistic/secondary spectrum access. For instance, the FCC in the US has recently defined the CR as "a radio that can change its transmitter parameters based on interaction with the environment in which it operates" [4], and many recent studies have used the term "cognitive" in this same sense [3], [5]. The other major shift is from autonomous CR operation, which is more suitable for ad hoc networking scenarios, to cooperative approaches and structured networking architectures, e.g., as manifested in the IEEE 802.22 standard [6].

4.1.2 Current Trends and Activities

Regulators around the world are introducing new consultations, measures and rules to the context of novel spectrum management techniques, including CR, in realization of their importance and potential to improve system performances. The FCC in the US have released several key documents, for instance the "Cognitive Radio Technologies Proceeding (CRTP)" [4], "Unlicensed Operation in the TV Broadcast Bands" [7], and another on the development of secondary spectrum markets [8]. Moreover, several proposals in this regard have been submitted to the FCC, which in turn have been published for consultation. One example is the US Rail Network's proposal to use the 88-108 MHz FM broadcast spectrum for informing passengers in mass transit rail cars [9]. Another example is the Google Inc. proposal of a "Real-Time Airwaves Auction Model" for 700 MHz TV bands, geared at the deployment of interactive broadband services in those bands [10].

In Europe, the European Parliament adopted a resolution in February 2007 which encourages the European Commission (EC) to consider allocating license-free bands, given the freeing of spectrum resulting from the migration of terrestrial television from analog to digital [11]. In the UK, the Office of Communications (Ofcom) has also shown genuine interest in liberalizing spectrum usage [12]. The Ofcom vision is to replace command and control methods with market-based mechanisms in 71.5% of bands by 2010 [13]. In Ireland, the Commission for Communications Regulation (ComReg) has recently published a consultation on Dynamic Spectrum Access [14]. Many other regulators worldwide are pursuing similar paths.

Within standardization, significant activities have recently been devoted to CR and associated technologies. The IEEE, through the IEEE P1900 standardization effort (which has now been given new powers as Standardization Coordinating Committee 41-SCC41), is a leading initiative in this direction [15]. Also within the IEEE, the first drafts of the IEEE 802.22 standard have recently been introduced, the objective being to improve spectrum utilization and wireless coverage by realizing Wireless Regional Area Networks (WRANs) through CR [16]. In the US, these WRANs will opportunistically use the VHF and UHF TV bands between 52 and 862 MHz, the way for which has been paved by [7]. Other groups, for example IEEE 802.16h [17] and IEEE 802.11y [18], are working on different proposals to improve resource management and reduce interference. The Software Defined Radio (SDR) Forum [19] has been leading active discussions aimed at promoting SDR and redirecting regulation to be in favor of SDR, whereby SDR is an important facilitating technique for advanced forms of CR. Through such methods, the SDR Forum is accelerating the proliferation of SDR in wireless networks.

Several proof-of-concept demonstrations of CR device prototypes have recently been reported. In the US, the DARPA XG program [20] is investigating novel ways to achieve flexible spectrum management. Recently, a CR demonstration has been conducted using a six node network of XG radios, whereby these secondary XG radios sensed 225-600 MHz bands, detected existing military and civilian systems within these bands, and avoided interfering with them while configuring and communicating in several wireless network formations using the remaining available bands [21]. Another prototypical example is in the recent filing of a device with the FCC, designed for secondary access to TV bands, by a coalition including Dell, Google, Hewlett-Packard, Intel, Microsoft, and Philips, among others [22]. A recent reported proof-of-concept demonstration is the Motorola Labs experimental CR implementation for DySPAN 2007 [23], in which Motorola exhibited a dynamic OFDM PHY layer for CR, along with spectrum sensing and neighbor discovery, achieving a live video communication link.

4.2 Lower Layers in Cognitive Radio

It is not an exaggeration to state that the majority of CR research currently carried out is focused on the PHY and MAC layers. This section therefore discusses the techniques that have been developed at these two layers; it then elaborates on an example from each of the layers in order to provide the reader with an in-depth view of the typical problems that should be addressed in the lower layers of a CR system.

4.2.1 Physical Layer Techniques

The major responsibilities of the CR's PHY layer in a heterogeneous wireless environment consist of spectrum sensing, adaptive transmission, spectrum aggregation, and interference mitigation. Two categories of secondary access cognitive communication are *negotiated* and *opportunistic* spectrum access. In the former case, a signaling channel provides an interaction medium between the primary and secondary systems; signaling channels for CR operation are discussed later in Section 4.3.2. However, it is sometimes assumed that direct interaction (i.e., signaling) is not possible between the primary and the secondary systems. This assumption is valid, for instance, if the primary system is based on a legacy radio technology; one example of such is TV transmission, which is considered within the IEEE 802.22 standard [6].

Under such circumstances, it is the responsibility of the secondary system to identify secondary usage opportunities and vacate the band as soon as the primary system starts using the band. The goal of finding opportunities for secondary spectrum usage can be achieved through spectrum sensing, as described in the previous chapter.

Some primary systems might use a pilot signal or training sequences to aid detection. This will increase the chance of detection for secondary users, for example, by using coherent detection via a matched filter [24]. Furthermore, pilots allow users to measure the local SNR of the primary signal, which can then be employed as a proxy for distance from the primary transmitter. Armed with this information, CRs can adjust their transmit power accordingly.

Note that for real-time secondary spectrum access, the optimum choice of sensing duration is calculated based on a trade-off. The allocated time for spectrum sensing should be sufficient to allow scanning of the band, identifying the existence of primary users, possible cooperation with other secondary devices to exchange channel information, and so on. On the other hand, the sensing time should be sufficiently short such that the channel condition in terms of the primary system's activity is detected very quickly should there be a change in the situation, particularly if the utilized channel has become occupied by the primary system. A more detailed discussion on spectrum sensing techniques is provided in the previous chapter.

After reliable information of the wireless environment is gathered by CRs, the next step is to use this information to gain maximum benefit from available resources. In the case of white spectrum holes, the transmission of CRs should be confined to the boundaries of the detected spectrum hole. If the CR uses a single carrier modulation, the only way it can use such spectrum holes without causing interference to the primary system is to identify an idle chunk of spectrum equal to or bigger than its bandwidth (the possible alternative of using gray spectrum holes will be discussed shortly). It is, however, possible that multiple smaller bands of spectrum might be idle at the same time/location when the CR is looking for spectrum opportunities. Hence, the cognitive system should be able to use sporadically idle bands through a "spectrum aggregation" technique, e.g., using multi-carrier modulation. In [25], Orthogonal Frequency Division Multiple Access (OFDMA) has been exploited for spectrum sharing. Using an allocation vector, i.e., a vector of zeros and ones for the bins of the IFFT block at the OFDM transmitter, desired portions of the band can remain void to mitigate interference. Moreover, [26] proposed an OFDM-based design for cognitive operations, namely, discontiguous OFDM. Both [25] and [26] avoid the fixed-bandwidth sub-bands overlapping with the primary system in order to not cause interference. However, even after sub-band deactivation, there is the possibility of leaked interference from the active sub-bands into the nulled sub-bands. One solution to this is to use a filter after the IFFT process in the transmitter; however, this will cause performance degradation due to the loss of orthogonality of OFDM sub-bands [26].

Consider a heterogeneous wireless networking scenario where the primary system and the secondary system, i.e., the cognitive network, both use OFDM modulation. In [27], the problem of mutual interference in OFDM-based spectrum pools[3] was analyzed. One of the draw-backs of OFDM spectrum pools is the interaction of the licensed system with secondary users, due to the non-orthogonality of their respective transmit signals. Two methods to overcome the side effects of this interaction are windowing the OFDM signal in the time domain or by adaptive deactivation of adjacent sub-carriers. Unfortunately, both methods are based on sacrificing valuable bandwidth which might otherwise be used by the secondary system. A quantitative comparison of both approaches was presented in [27].

Another multi-carrier modulation approach is to use DS CDMA to spread data over a series of disjoint carrier frequencies, to create MC-CDMA [28]. Interference in MC-CDMA systems can be avoided using the sub-band deactivation technique, similarly to the OFDMA case mentioned above. A comparison of Non-Contiguous OFDM (NC-OFDM) with Non-Contiguous MC-CDMA (NC-MC-CDMA) was investigated in [29]. It was shown that as the number of deactivated sub-bands increases, the Bit Error Rate (BER) performance degradation of NC-MC-CDMA becomes higher than the NC-OFDM approach. Another example is [30], which investigated MC-CDMA and OFDM modulation in conjunction with DS-CDMA and TDMA. That work suggests the use of multi-carrier technologies in general, and Carrier Interferometry (CI) signaling in particular. Note that the authors of this chapter have also proposed a novel cognitive DS MC-CDMA system, which will be discussed in Section 4.2.3.

For the gray spectrum hole scenario, the complexity of CR opportunity detection increases greatly. Fewer works have focused on such scenarios. The studies in [31] and [32], were based on the concept of Interference Temperature (IT) to quantify the interference effect of CRs on primary receivers. However, since the FCC has abandoned the concept of IT [33], this chapter will not elaborate on such results.

4.2.2 Medium Access Control Techniques

Two major MAC topics of study in the CR context are the development of mechanisms for coexisting primary and secondary networks (mostly from the radio resource allocation point of view), and MAC considerations in a network of CRs.

The former topic is first discussed here. Obviously, if secondary users of spectrum use the same multiple access scheme as primary users, interference avoidance is relatively straight forward. In TDMA (e.g., see [34]) and FDMA, or a combination of them, spectrum opportunities for secondary users (i.e., idle time slots or carrier frequencies, or a combination) have to be equal or larger in size than the secondary user's needs.

Given the capabilities of OFDMA, which has made it a strong candidate for many next generation wireless communications systems, several proposals in the literature focus on moulding OFDMA as a MAC in CR scenarios. As mentioned before, the

---

[3]Note that in [25], the term "spectrum pooling" was employed with a slightly different meaning than its common usage in the context of Dynamic Spectrum Access (DSA), as discussed in the previous chapter.

authors in [25] proposed the concept of spectrum pooling, which can potentially be used in a heterogeneous wireless network, where the CRs might also be part of the spectrum pool. Primary and secondary systems can coexist in this scheme by intelligent usage of the flexibility of OFDM modulation, where the secondary system can deactivate any number of its sub-channels matching the occupied bands of primary users. From MAC layer point of view, an OFDMA MAC can be utilized based on the available resources for the secondary system. Therefore, the proposed method in [25] can essentially be interpreted as a spectrum-variable OFDMA MAC for CRs.

Among various other mechanisms making coexistence of primary and secondary systems possible by adaptive resource allocation, [35] formulated a game between a primary and secondary network using a cellular OFDM architecture, which optimally allocates power such that a level of QoS can be guaranteed for the primary users of the band. The proposed resource allocation scheme in [35] is a network-centric approach where both primary and secondary users inform the primary BS of their channel conditions and QoS requirements. The primary BS determines U_n, the set of primary and secondary users which can coexist in sub-channel n, according to an algorithm in each of N OFDM sub-channels. Using the fairness axioms of Game Theory, i.e., Nash Bargaining Solutions (NBS), a set of weights is assigned to all the users in U_n. Finally, the power allocation is determined and broadcast to all users by solving the relevant optimization problem, i.e., the weighted sum power minimization problem. Since such optimization problems are non-linear and NP-hard to solve, a method using Sequential Quadratic Programming (SQP) was proposed [35].

If the primary system uses CDMA, the secondary CRs have two choices. They can either employ orthogonal codes with respect to the primary radios' transmissions, which in turn means that they should be able to recognize unused codes in the primary system's code book. Another possibility is underlay transmission, so as not to cause a harmful interference level to the primary users. For example, coexistence of Ultra Wide Band (UWB) systems with a wideband system based on CDMA is possible through underlay transmission [36]. Such an underlay system might use CDMA (direct sequence or another method), OFDM with a sufficient number of carriers, or pulse shaping.

A challenging consideration in CR contexts is what happens if the primary and secondary systems are using different MAC mechanisms. For example, consider the case where the primary system uses a TDMA MAC. This means that the frequency band used by the primary system is fully utilized but only for discrete intervals of time, so the secondary system can only access spectrum in a time sharing mode regardless of its own MAC scheme. That is, the secondary system might still use, e.g., FDMA or CDMA, but only in conjunction with TDMA. For other combinations of MAC schemes of the primary and secondary systems, similar arguments can be provided [37].

Now let's discuss proposed solutions in the second thread of MAC layer studies for CR, i.e., designing a MAC for a network of CRs. Currently, there are few exhaustive studies in this field. Nevertheless, there is a MAC protocol proposed for a mesh network of CRs in [38], which is a combination of TDMA (guaranteeing access to the medium) and a Random Access Period (RAP), mainly used for exchanging control messages intra- and inter-cluster. Given that several different frequency bands might be idle at the location of a specific cluster, simultaneous transmissions are allowed, where in each channel TDMA is used. The authors also suggested the use of spread spectrum communications by using a global spreading code for general parts of the MAC superframe, such as a beacon transmission or inter-cluster control message exchange, and a private spreading code for data transmissions within each cluster, which should be unique to that cluster.

The concept of CORVUS, introduced in [39], [40], is another example of a network of CRs, where the need for an efficient MAC is of utmost importance. The authors, however, did not propose any specific MAC mechanism, and for simple networking scenarios rely on conventional random access methods. It is also worth mentioning that [41] proposed a multiple access scheme for secondary access to a specific band, based on the IT concept, namely Interference Temperature Multiple Access (ITMA). As a further example, authors in chapter 10 of [42] elaborated on the design principles for a cognitive MAC in opportunistic spectrum access scenarios. The focus of their study is reliable and real-time spectrum sensing, with the access policy being based on sensing results. The effects of inaccuracy of sensing in the PHY layer, and its result on the access policies in the MAC layer are investigated. As a final example from the literature, [43] developed a decentralized MAC scheme which can be used in ad hoc network of CRs.

4.2.3 Exemplary Scenarios for the PHY and MAC Layers

An Example PHY Layer: Cognitive MC-CDMA [44]

Many approaches to increasing spectrum efficiency in heterogeneous wireless environments have been proposed in the literature, such as overlay mechanisms [45], and Dynamic Spectrum Allocation (DSA) [46]. Alternatively, using the CR paradigm might increase resource usage and system efficiency dramatically. Consider a heterogeneous wireless environment where a number of legacy systems are operating in different bands of a chunk of spectrum, and a cognitive system (using MC-CDMA) wishes to utilize the idle bands in that part of spectrum. It is assumed that the legacy system's usage of spectrum varies temporally (and spatially), but at much longer intervals compared with the resource allocation period of the cognitive MC-CDMA system. It is also assumed that the bandwidth and the frequency location of the legacy systems are available at the cognitive MC-CDMA transmitter. In this scenario, such information can be obtained in an efficient centralized fashion, i.e., by the Base Station (BS) of the cognitive system, based on the measurements of the cognitive users, similar to IEEE 802.22 approach [6].

Within this context, instead of deactivating the fixed-bandwidth sub-bands as discussed in Section 4.2.1, a cognitive MC-CDMA is proposed that can adaptively change its transmission parameters. The transmitted signal of this cognitive MC-CDMA system in the downlink can be formulated as

$$S_{MC}(t) = \sum_{k=1}^{K} \sum_{\substack{s=1 \\ s \notin G}}^{S} \sum_{n=-\infty}^{+\infty} \sum_{i=1}^{N_s} A_s d_k(n) c_{k,s}(i) h_s(t - iT_{c,s}) \cos(\omega_s t + \theta_s)$$

where $d_k(n)$ represents the nth data symbol of the kth user ($k = 1, 2, \cdots, K$), $c_{k,s}(i)$ is the ith chip of the sth sub-band spreading code (signature sequence) ($i = 1, 2, \cdots, N_s$) and G denotes the set of sub-bands occupied by the legacy signals. The number of chips per symbol is denoted by N_s, assuming BPSK signaling. The data symbols, spread by the signature sequence $c_{k,s}(i)$, modulate an impulse train. Then, chip wave-shaping filters, with impulse response $h_s(t)$ for the sth sub-band, are used to create a Nyquist pulse shape for the modulated impulse train. By adaptively modifying the above parameters, the characteristics of the transmitted signal, including the bandwidth and power of sub-bands, can be tailored to the interference pattern in that band. This approach is shown in Fig. 4.1. Compared with the sub-band deactivation method, this method increases spectrum usage efficiency considerably, because it only excludes the portion of the band affected by interference. Also, the proposed method does not suffer the OFDM interference leakage problem [26].

In [44], an algorithm to compute the values of transmission parameters adaptively for such a cognitive environment has been derived, and resulting system performance has been analyzed in terms of error probability. Comparing system performance under different circumstances, it was shown that cognitive MC-CDMA outperforms conventional MC-CDMA, especially when the power and number of interfering signals is increased. While a conventional MC-CDMA system becomes unreliable in such conditions, the cognitive MC-CDMA system performs robustly in the presence of external interferers [44].

An Example MAC Layer: Interference-Limited (Aware) Scheduling [47]

Consider two OFDM-based cellular systems using co-located BSs covering almost the same area, i.e., an isolated cell, where the focus is only on the downlink of both systems. Sharing sites for BS installations is a common practice among operators/service-providers, where generally each operator has its own licensed spectrum band. Assuming users are not too close to BS antennas, the two channels between any one user in either of the systems and the two co-located BSs can be considered to be reasonably similar. This gives the opportunity for various interesting spectrum sharing solutions.

Vertical spectrum sharing (through secondary access in a CR sense, to OFDM sub-channels) can help increase the effective capacity of each operator. This is desirable in the sense that in acquiring new cell sites, the cost of licenses for new bands and cell planning are challenging issues. Hence in a specific band, the assumption here is that the licensed system operates as a primary and the other one acts as a secondary spectrum user. While the secondary operators benefits from accessing the primary system's spectrum to satisfy application and traffic demands, the primary operator might benefit from using the licensed bands of the secondary system at other times. Another possible option benefiting the primary operator is simply to charge

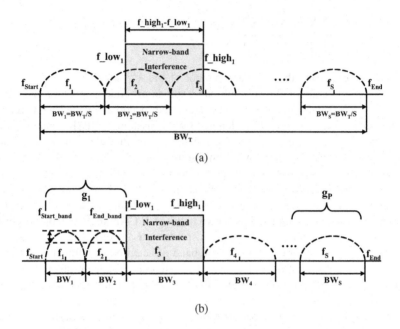

Fig. 4.1. Comparison of spectrum partitioning for (a) conventional systems, and (b) cognitive MC-CDMA, in the presence of an interfering signal from a legacy system.

a flexible fee based on the spectrum usage of the secondary system. Although this possibility is not considered specifically in the analysis, the basis of the analysis (i.e., scheduler design) is not affected by the choice of business model.

Note that this considered scenario, for a two-operator case, can be modeled as if a single BS of an OFDM-based system with fixed spectrum allocation (i.e., the licensed band) operates with maximum transmit power $P_{max}^{1} + P_{max}^{2}$, where P_{max}^{i}, $i = 1, 2$, is the maximum transmit power of each BS at that site. There is also a need for a signaling channel to feedback Channel State Information (CSI) to the BS.

In this context, a resource allocation scheme was developed to maximize system throughput (i.e., the sum of the rates of all users), while limiting received interference level at each user [47]. It must be noted that the throughput here is comprised of primary plus secondary system throughputs. The threshold for acceptable interference level is allowed to be different for each user; this makes it possible to define a higher protection for primary users, as likely is required.

It was shown through extensive simulations that performances in terms of overall throughput of the proposed allocation scheme are reasonable, while the scheme provides a high degree of fairness amongst users [47].

4.2.4 Open Issues and the Way Forward for the PHY and MAC Layers

There remain numerous obstacles and challenges to be addressed before a cognitive network in a heterogeneous wireless environment can be practically implemented. First discussed here are the challenging issues at the PHY layer.

A consideration in implementing a network of CRs is the use of multi-carrier modulation, as discussed in Section 4.2.1. First, it should be mentioned that no comparison of the performances of secondary spectrum aggregation methods with a single-carrier cognitive system is available in the literature. Also, each of these multi-carrier modulation schemes faces different challenges for secondary spectrum access. In the MC-CDMA case, the size of sub-bands is in direct relation with the chip size of the spreading code. Therefore, in many cases, due to the bandwidth and location of available opportunities in the spectrum, different chip rates for each sub-band should be used. To preserve the orthogonality of spreading codes, variable rate spreading means such as Orthogonal Variable Spreading Factor (OVSF) codes are necessary. Given that the codes should also be unique for different users in each sub-band, a large codebook is required, which might impose a challenge as the number of users and sub-bands increases. These modifications are not readily available in conventional MC-CDMA systems.

The OFDM case differs in the sense that spectrum aggregation does not require any modification to the modulation. Different numbers of sub-carriers can be grouped together to use different sized available opportunities. However, the issue of interference management is more challenging than in the MC-CDMA case. The imposed interference from any sub-channel of MC-CDMA on the primary systems, for example, as Co-Channel Interference (CCI) in cellular primary-secondary systems, is not considerable due to the spread spectrum nature of this method. The same argument is true for leakage of power to adjacent bands due to imperfect filters of transmitters, i.e., Adjacent-Channel Interference (ACI). On the other hand, any form of imposed interference in secondary OFDM systems might have a considerable effect on the primary system. For the ACI case, a simple solution is to leave one or more sub-channels idle between the operating band of secondary OFDM system and the primary system. However, this wastes bandwidth in an opportunity, and CCI still remains an issue which must be addressed.

The primary system architecture can notably affect the performance of secondary users. The major issues for consideration are whether the primary system is fixed or mobile, whether the primary system is capable of interaction (in the form of cognitive capabilities or signaling mechanisms) with secondary systems, and whether there is a reuse pattern in the spectrum (e.g., frequency reuse in cellular systems) or the primary system is a Single Frequency Network (SFN). In the former case, CCI should be mitigated, while in the latter case, Inter-Channel Interference (ICI) should be mitigated.

Depending on the type of application that the CR requires spectrum for, in addition to the architecture of the cognitive network, different duplex mechanisms can be exploited. For point-to-point communications with other secondary users, without support of infrastructure, half-duplex or full duplex mechanisms can be used. The ad hoc network structure for secondary users encourages random access schemes where the transmitter tries to establish a link to the receiver only for the duration of its transmission. If the receiver wants to reply it might need to establish a new link. This might be acceptable for some applications (best-effort, non-real-time) but this is certainly not appropriate for real-time voice or multimedia communication. Time Division Duplex (TDD) mechanisms are less appropriate in such conditions due to the need for very accurate synchronization mechanisms. Also, time duplexing strategies for secondary spectrum users might increase holding time of the channel and, hence, increase the chance of contention with primary users. On the other hand, Frequency Division Duplex (FDD) mechanisms need more of the already scarce bandwidth for the reverse communication link. It seems that thus far, no study in the literature has considered the effects of duplex mechanisms and their effects on the performance of secondary systems; this issue is still an open question, which is particularly crucial for secondary communications utilizing infrastructure support. Another important and related issue for selecting a duplex mechanism is its effect on interference. For example, if the secondary system exploits TDD, while the resource it is using is part of spectrum in downlink band of the primary system, does it have the same interference characteristics when using FDD under the same conditions? What are the implications of using half duplex mechanisms?

There is no doubt about the need for a metric that facilitates interference management, especially for secondary spectrum access. To efficiently benefit from such a metric a number of issues need to be addressed. First, the relationship between the interference metric and parameters such as bandwidth, power, and system capacity should be investigated [48], so that suitable interference thresholds can be identified. The open questions in this area include that of what degree of accuracy is necessary for practical use of interference measurements, and how often should interference be measured, or in other words, what correlation in the interference pattern exists between neighboring areas, or in one area at different times. This would have an enormous effect on mobile secondary users, since if a powerful correlation exists, it is possible to reduce the required rate of spectrum measurements for secondary devices, which in turn reduces the power consumption of secondary users' devices. In light of this, [49] proposed a joint coordination and power control (JCPC) algorithm with the objective of maximizing the secondary system capacity. To this end, they assume all users adopt a spread spectrum signaling format, in which the transmitted power is evenly spread across the entire available band, controlled by a manager. They formulate an iterative and distributed joint coordination and power control algorithm, which only requires users to obtain limited local information in order to converge to a Nash Equilibrium (NE), which guarantees that the received power at the measuring point will not exceed the interference temperature constraint.

Another issue here is how to regulate the received power, and more specifically interference, rather than transmission power. Should regulators consider variable thresholds for different times/locations? Should the same interference threshold used in a city center be used in rural areas? Also, how should one define interference thresholds for each band, considering different transmission power and coverage area of different Radio Access Technologies (RATs) and bands? What will happen at the boundaries of different bands, i.e., are there step transitions from one interference threshold level to another or should another mechanism be considered? Furthermore, should there be interference thresholds for license-exempt bands? All such questions will have to be addressed in the eventual realization of lower layers for CR solutions.

4.3 Networking for Cognitive Radios

This section elaborates on networking considerations for CRs. As mentioned in Section 4.2.2, few studies thus far have focused on the interactions among CRs in a network; instead, most studies have tried to address primary-secondary device interactions. It might at one time have been commonly assumed that CR implies an ad hoc networking architecture; however, IEEE 802.22, among other work, has recently paved the way for other cognitive network architectures.

4.3.1 From a Network of Cognitive Radios to Cognitive Networks

CRs were originally envisioned for autonomous operation [1]. However, the complexity of tasks, for instance in spectrum sensing, has shifted the CR domain towards cooperative approaches. Also, to facilitate profitable business models, recent cognitive architecture designs are changing from ad hoc scenarios towards structured methods, for example the cellular architecture as manifested in the IEEE 802.22 standard [6]. That is, standalone CRs are being grouped and evolving towards "cognitive networks". Among the limited number of studies that have thus far proposed cognitive network designs, two examples are [38], [39].

To provide a reliable cognitive communications network, clearly suitable PHY and MAC layers are required, as discussed in the previous sections. In the network layer of a cognitive system, several tasks should be satisfactorily accomplished, including packet routing and QoS provisioning. Normally, a mapping process from network traffic classes to link it to QoS classes using tags (such as provided by the IEEE 802.1p extended MAC header), and also mapping of network address to (next hop) link address to facilitate these tasks are involved. The network layer also handles mobility and security aspects of mobile terminal devices. For instance, mobile IP, hierarchical mobile IP and cellular IP are examples of this; although it is also possible to support mobility at higher layers (such as through the Session Initiation Protocol (SIP)), the network layer is more appropriate for the purpose. Network layer security is required for authentication prior to allocation of network resources, and to enable secure connections through networks. Finally, the network layer host configuration of the end devices is supported by various methods, such as IP auto-configuration or the Dynamic Host Configuration Protocol (DHCP).

The modifications required to the network layer of a cognitive system, compared with legacy networks, are mainly to the QoS provisioning mechanism, mobility management, and routing. More specifically to address QoS in cognitive networks, efficient, flexible and dynamically reconfigurable resource reservation, mobility management and scheduling algorithms must be developed. Such QoS-aware schedulers should allow for the evolution of existing and emerging air interfaces, e.g., IEEE 802.11, UMTS (HSDPA, HSUPA), and mobile WiMAX, towards cognitive solutions which may support several radio technologies simultaneously. Hence link selection (routing), mobility management and scheduling need to be closely coupled. Also in a cognitive network, upper layers will face an additional challenge regarding the possibility of changing the underlying spectrum band during communication based on activity of the primary system.

Generally, QoS is provisioned through a scheduling algorithm based on differentiated service classes [50], which will not be sufficient in these highly dynamic network environments. For these different service classes, different QoS requirements should be met such as throughput, jitter, delay and packet loss, but if the CR configuration changes dynamically simple service classes will likely not correlate well with the QoS obtained. This poses the questions of how to define traffic service classes and corresponding QoS parameters for cognitive network layers, and how to support the QoS measures by efficient resource allocation. A step towards this goal has been made within the Open Mobile Alliance (OMA) with proposals for the definition of protocol managed objects and service QoS profiles to support reconfigurable devices, but this approach is very much a centralized solution which is not appropriate for ad-hoc networks. Existing approaches based on differentiated class-based or per-flow based solutions for QoS provisioning are often suboptimal for ad-hoc networking scenarios, since they do not take the dynamics of environment changes into account. To address this issue, the IETF and IRTF have been looking into policy based management to realize flexible QoS approaches that adapt to changing requirements. Still, these policies are normally associated with network parameters and user needs only.

The issue of routing is particularly challenging in cognitive networks, especially in multi-hop scenarios. The reason for this is that available bands in each node vary from location to location. Therefore, it is necessary to define a multi-dimensional routing scheme, based on available spectrum bands as well as the locations of source, destination, and intermediate hop nodes, as illustrated in Fig. 4.2. Hence, a challenging goal of network layer research for CR is to identify efficient and practical routing solutions for such cognitive network scenarios. This is different from ad hoc QoS routing, e.g. [51], [52], in the following aspects. First, as stated above, the available channels vary from node to node, and also throughout time. Hence, reservation techniques (for instance bandwidth or time slot reservation) are not easily extendible to cognitive networking. Also, a globally available common control channel (used for synchronizing different nodes in traditional ad hoc networks [51]) in the cognitive network is not realistically always available. Neighboring nodes might therefore instead use locally available bands for signaling. The issue of signaling is investigated in the following section.

Note that the extra dimensions of the routing algorithm for cognitive networks might also include the Radio Access Technology (RAT). For example, in [53] the concept of Multi-Radio Multi-Hop (MRMP), based on the Generic Link Layer (GLL), was introduced. This notion can readily be applied to cognitive network-

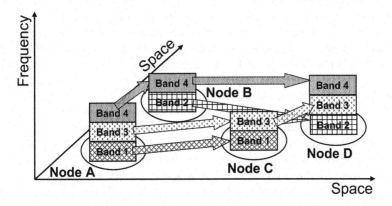

Fig. 4.2. Multi-dimensional routing mechanisms are required for successful cognitive networks.

ing scenarios in a heterogeneous wireless environment, whereby the traffic routing in each hop of the cognitive network might be based on a different RAT.

4.3.2 Signaling Considerations

The availability of a signaling mechanism facilitates several aspects of cognitive networking. First, it helps in identification of available resources for the use of CRs, discussed as follows.

In general, there are two main paradigms for spectrum information acquisition. Some methods put the responsibility in the hands of primary (licensed) users of the band to either broadcast occupancy of a specific band [54] or advertise availability of some bands [2]. Another approach is to make CR responsible for reliably acquiring this information using different mechanisms, for example independent spectrum sensing by each CR, interacting with a network of CRs sensing their local environment (distributed intelligence) [55] or through using some kind of real-time resource database accessible by CRs [56].

The possibility for interference to be caused by a transmission resource selection of a CR must be ascertained using information conveyed to that radio by each other radio that is within interference range. This should be achieved not through radios sending information about which resources (frequency bands, time-slots, etc) they are transmitting upon (such information is useless, as it does not accurately indicate the potential for interference in received signals), but by these radios (i.e. terminals, base stations, and any transceivers using the spectrum range over which CR applies) transmitting information about the resources being used by the traffic that they are receiving.

[57] proposed a signaling mechanism for conveying this information, namely the Resource Awareness Channel (RAC)-a single common channel upon which all terminals must broadcast (with equal power) information about their resource usages,

and to which all cognitive radios must listen to assess the effects of a prospective transmission resource usage change. Through this approach, a cognitive radio can infer a very rough estimate of the power of interference at other radios of a prospective resource usage change, from the powers of these radios' transmissions on the RAC. Hence it is potentially possible for a cognitive radio to intelligently mould its resource choices and transmission powers according to phenomena such as large-scale shadowing, through the use of the RAC scheme.

As further examples, in [58] and [59] an efficient signaling method for spectral resources and synchronization algorithms in the context of a spectrum pool [25] was introduced. Exchanging information on the radio scene can increase the accuracy of spectrum opportunity recognition on the one hand, and can cause spectrum occupation and create overhead on the other. A signaling method termed the "boosting protocol" was therefore devised in [58], in order to help efficiently detect, collect and broadcast available resources. In [59], preamble and synchronization issues of a spectrum pool was considered. It was shown that existing methods are insufficient, and frequency and frame synchronization in the presence of multiple narrow-band interferers (i.e., licensed users of the band) were discussed.

Another important usage of signaling channels is to exchange neighboring node IDs and resource information for routing metrics (such as available channel state information and queue size), to facilitate operation of a routing algorithm in a cognitive network. Both [38] and [40] propose to use locally available channels for such signaling purposes.

4.3.3 Open Issues and the Way Forward for Cognitive Networks

Since networking considerations for CRs have not been investigated in the literature as extensively as PHY and MAC layers have, there are still many open questions and problems to be addressed in this field. For instance, the issues and evaluation of QoS provisioning, routing, and mobility management in cognitive networks have largely not been investigated.

Furthermore, the evaluation of the performance of different networking architectures for CR has rarely been investigated in the literature. As an example of proposed networking solutions for CR, [40] proposed that secondary users create a "Secondary User Group (SUG)" to coordinate their communication. Members of a SUG might communicate with each other in an ad-hoc fashion or, alternatively, might access a fixed infrastructure via a dedicated access point which is part of an existing networking infrastructure.

In [60], Akyldiz et al. discussed the requirements and challenges for secondary networks (under the xG title in that article, referring to x-generation) for both licensed band and unlicensed band operations. The xG network components are the xG users, xG base station, and spectrum broker. The latter is a centralized entity that can be connected to all networks to facilitate spectrum sharing, for example, via spectrum information exchange. An important contribution of this paper is collaboration between routing and spectrum management as a fundamental design feature of a cognitive radio network. Proving the case, in [61], [62] the authors demonstrate that a cross layer solution for joint route and spectrum selection can outperform a solution that selects routes independently of the spectrum. More specifically, the authors in [63] argued that using network topology information, such as the geometric relation of nodes, can have a direct impact on the connectivity and capacity estimates of cognitive networks. Using spatial correlation functions and statistics can facilitate more intelligent network optimization, for example through power control and locality-based protocols.

The issue of routing is particularly challenging in a cognitive network context. Given limited availability of transmission opportunities for secondary users, simple routing schemes, such as flooding, are wasteful of resources. Hence, any proposed routing algorithm should address the trade-off between efficiency, complexity, and agility in such scenarios. Another challenge is that each node might only have limited knowledge of a subset of its immediate neighbor nodes, depending on the capabilities and frequency bands of operation at each point in time. Also, the capabilities and configuration of these devices may change rapidly over time.

Many CR applications are based on ad hoc networking methods. The flexibility and adaptivity of ad hoc networks make them an ideal candidate for the unstable, constantly changing environment of CR. For such ad hoc cognitive networking scenarios, important considerations are the mobility and the different capabilities of CRs within a network. A mobile secondary user will enter/exit the coverage area of different Radio Access Networks (RAN) while trying to find access to spectrum. The main challenge here will be finding a method to perform all the network related routing and resource, QoS and mobility management, without network assistance. For instance, distributed solutions are inevitably often less reliable than centralized solutions, due to the lack of complete knowledge of the network and device configuration and capabilities, but can nevertheless react faster to dynamic environmental or network changes. This will be a very important consideration for cognitive networks in which each CR is more flexible and dynamic than conventional radio devices, and can also operate in different radio bands. Very few studies have considered this issue; however, some have provided solutions as an integral part of their plan. In non ad hoc cognitive networks, such as using a cellular architecture, the issue of Hand Over (HO) should also be addressed. Note that, given the reconfigurability of CRs, HO operations can be horizontal (e.g., from one cell to another cell) or vertical (e.g., from one RAT to another). The issue of HO has not been studied vigorously thus far. In [40], the spectrum pool was divided into sub-channels where it is suggested that "in order to keep a continuous QoS, Secondary Users should always have a redundant amount of sub-channels". Obviously, such a strategy can be easily modified to support horizontal HO.

Finally, the area of security for networks is potentially very challenging, as there is an obvious benefit in facilitating cooperation between CR devices in order to effectively share spectrum resources. However, the very principle of cooperation requires a level of trust, and opens up possibilities for exploitation by misbehaving or malicious CR devices. This is of particular concern to networks that must provide guaranteed QoS to CR users, such as in emergency service scenarios. Although security issues in various networking scenarios have been under investigation for some time now, the specific security requirements brought about by deployment of CRs as secondary devices have not been vigorously addressed.

4.4 Higher Layer Solutions for Cognitive Radio

As discussed in the previous sections, few studies have focused on higher layer aspects of CR, despite the importance of higher layers. This section attempts to redress the balance by consideration of Application and Transport layer characteristics for CR.

4.4.1 Application Scenarios

In the previous sections, two well-known definitions for CR have been presented. One is based on the FCC definition, referring to intelligent radio devices that employ dynamic spectrum access techniques to utilize radio resources more efficiently. Another is based on Mitola's definition which places more emphasis on the intelligence at the higher layers. The cognition associated with both types of CR systems will provide new opportunities especially to application services providing better experience to the end user. Nevertheless, higher layers will be presented with a number of challenges when CR devices are used in the context of heterogeneous wireless access networks, for instance, errors, delays, and delay variation patterns created by dynamic spectrum access and scheduling techniques associated with the CR systems.

CRs are characterized by their ability to understand and reason about the context they are in and act accordingly. Therefore, the benefit of CR can only be realized fully if application requirements and context are determined or inferred in order to know what modes of operation and spectrum bands to use at each point in time. To facilitate this, basic application communication requirements can be characterized by attributes such as:

- Urgency,
- Latency,
- Data rate,
- Security, and
- Reliability.

The above list is not exhaustive, but rather are examples of the sort of information that is needed in order that a CR device can know how to plan, decide and act on behalf of the user. A CR device must also understand its own context and the user's situation, and can use auxiliary context data such as:

- Location,
- Movement,
- Battery charge state,
- Connectivity (modes and networks available),
- Spectrum availability and state, and

• Devices (including neighbouring devices) and their capabilities.

The combined context can now be used to determine the most likely application scenario and consequently how the device should behave to satisfy the user's requirements. Gathering, managing and interpreting context related information is a complex process that has been the subject of research in recent years (see [78], [65] and [66]). One of the main issues is to abstract context information in a generic and meaningful manner so that it is possible for the cognitive process or the user to comprehend. There is also need for universal methods of determining and accessing context information that can be scaled to different devices with diverse capabilities. One such Application Programming Interface (API) that has been developed for this purpose is the Unified Link Layer API (ULLA), see [67]. The purpose of the API is to allow cognitive applications to gather context information in a link technology and device platform independent manner and to subsequently control the configuration of the available links. Thus, it is possible for cognitive applications to interact directly with radio links in order to optimize the use of resources on different terminal devices regardless of the Operating System (OS) or underlying link technologies supported. The API provides smart triggers and handlers for accessing context information to determine when significant context change events occur. This has the benefit of reducing the processing required within the cognitive application to interpret the context, which can be significant within multi-mode devices supporting many options and possible operational frequency channels within each of these modes.

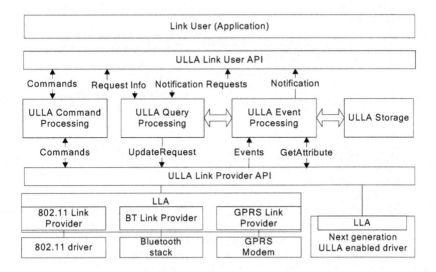

Fig. 4.3. ULLA architecture.

In addition to the retrieval and interpretation of context information it is further necessary for cognitive devices to interpret the policies governing their behavior in order to ensure correct operation and to facilitate resource management at a network level. The definition of policies has been the subject of standardization activities within the IEEE SSC41 (P1900.4) group and can be either of the typical event, condition action form or goal or utility based. The policy is provided by the operator in order that the network resource use can be optimized, for instance by re-allocating resources at busy periods (as illustrated in the example in Table 4.1). There is also a further level of optimization that can take place within the device as depicted in Fig. 4.4 within these constraints. For instance, the P1900.4 framework can provide the means to access extended network and terminal context information (in addition to context information already available within the local device). As is common with all radio communication, the most useful measurements are collected at the receiving end of a link and so often it may be necessary to utilize mechanisms (such as will be developed within P1900.4) to obtain measurement from the remote end of a link even for local resource optimization purposes.

Table 4.1. Example event - condition - action policy.

Event	Conditions	Actions
Terminal state =connected mode on a particular RAN (e.g. GPRS)		
	Location = London AND Terminal class = any AND Time of day = daytime AND Terminal capabilities include, WiFi.	Change permitted RAN (e.g. include WiFi) OR Change frequencies/channels set AND Perform specific measurements report before opening a new radio link
Terminal state =connected mode on a particular RAN (e.g. GPRS)		
	Location = London AND Terminal class = any AND Time of day = evening	Change permitted RAN (e.g. exclude WiFi) AND Perform specific measurements report before opening a new radio link

An example scenario is described to illustrate the need for the application and transport layer cognition and context awareness.

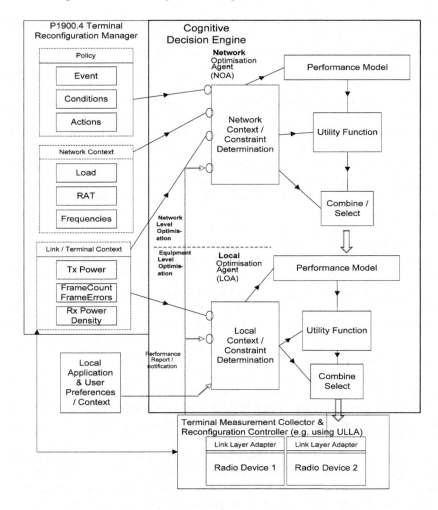

Fig. 4.4. Integration of cognition within P1900.4 policy management framework.

Exemplary Cognitive Radio Application Scenario

A Business Manager (BM) travels to a customer office for the first time with his CR device. He walks into building and the CR device identifies the wireless services available via a local information point (via a cognitive pilot channel or radio enabler) in the unmanned reception area. The CR device interrogates the service and determines that it is able to use the corporate WLAN with visitor credentials (obtained via the information point) and automatically logs onto the network and sets up a VPN to BM's corporate network and opportunistically downloads e-mails and voice mails that it knows are waiting in the mail server. The device also determines the devices approximate position in building, via wireless location sensor network within the office. The CR device shows the route from his current position to the meeting room which he is about to attend. The CR device automatically reports BM's arrival to meeting host's CR via an e-mail or instant message and the host unlocks the door.

Meanwhile, BM's mother-in-law (ML) who is recovering from heart surgery has constant electronic monitoring of her health condition. This is done by several monitoring devices which take periodic measurements and are wirelessly connected to the CR device and sends daily reports to the medical center by opportunistically using a secondary spectrum access channel. ML's heart rate increases beyond the critical threshold level set by the doctor. The monitoring application takes additional health readings from several other monitoring devices and uses the CR to set-up an urgent, high data rate and reliable link over an emergency primary user channel to send reports to the medical center. BM and ML's doctor are informed directly on their CR devices of the situation.

BM initiates a video conference with his ML using his CR terminal (which has already negotiated an emergency priority service on the customer's wireless LAN due to the nature of the situation) to comfort ML until the doctor can join in the conference call. Half-way through the conversation BM's terminal battery power becomes lower than an operational level that the CR determines is suitable for this situation. BM is notified by an alert and alters QoS preferences in the video application which reduces video quality (or switches to audio only).

4.4.2 Transport Layer Scenario

Many Internet applications including email, file transfer, remote access and especially web browsing require a reliable data delivery service. They often rely on Transmission Control Protocol (TCP) at the transport layer to guarantee end-to-end reliability.

TCP provides flow control, congestion control and loss recovery functions and employs an acknowledgement-based scheme to realize a reliable packet delivery service. The flow control uses a sliding window mechanism to allow receiving node to control transmission window size to avoid potential buffer overflows. The congestion control scheme uses two basic algorithms called as Slow Start and Congestion Avoidance. Transmission node probes the available capacity by gradually increasing and decreasing a variable called "congestion window" to reflect the packets outstanding in the network, based on the acknowledgements received. Many variants of congestion control algorithms are described in the literature. In the simplest case, congestion window will be increased, exponentially during slow start phase and, linearly during congestion avoidance phase, for each acknowledgement received. Consequently, throughput of TCP has an inverse relationship with the round trip time or delay of the path. The loss recovery of TCP is generally based on duplicate/selective- acknowledgements or retransmit-timers. In addition, when loss is detected, TCP sender infers congestion. It then multiplicatively decreases the TCP sender rate by halving the congestion window variable. TCP assumes loss as a congestion indication although in practice loss may occur from other sources including wireless errors and outages.

This behavior of TCP has been seen as a major drawback for wireless systems where there packet errors and outages are more common than in wired networks.

TCP specifications provide a number of techniques and a wide-range of implementations considering different types of networks. However, traditionally, TCP has shown a number of vulnerabilities when used over wireless links. A number of research studies have shown that they perform poorly in wireless networks employing dynamic radio resource management schemes since such networks breaks the transport layer assumptions, e.g. see references [68], [69] and [70]. There are also a number of studies on performance related adaptations of TCP for wireless links covering a wide range of network and dynamic system scenarios, such as in [71] that propose state aware handling of TCP packets in order to avoid unnecessarily triggering adaptive behavior and also unfairness between flows. This is a well known problem that occurs when scheduling is based on serving the longest queue first (or shared queuing disciplines), resulting in unfair distribution of throughput between TCP flows. TCP flows in the slow start state do not get enough resource to compete with established flows, which prolongs the slow start period. If the slow start period is never exited it is often referred to as TCP resource starvation, and will be more prevalent in dynamic spectrum access networks which utilize simple etiquettes (such as listen before talk) without consideration of transport protocol type or state.

Real time transport protocols used for multi-media streaming such as the Real Time Protocol (RTP) are sensitive to excessive latency and latency variations. This results in packet dropping due to deadlines being missed and necessitates that delivery latency targets are considered when opportunistically utilizing dynamic spectrum resources. In this case the transport protocol is not state dependent but the RTP stream could be transporting application data that has different reliability requirements, such as compressed video I and P frames, which have different impacts on the perceived video quality. It is also generally the case that RTP transport does not provide any delivery guarantee or congestion indication or avoidance mechanism, and so buffer overflow is a major issue, which will be harder to manage in dynamic spectrum access networks.

Exemplary Cognitive Radio Transport Scenario

A web browsing session is started on a CR device. In order to rapidly initiate the TCP connection, the CR obtains a spectrum resource opportunity that enables the TCP protocol state to rapidly exit the slow start state. The CR device can then be more power efficient or selective in the choice of spectrum opportunities but ensures that the TCP adaptation mechanisms are not triggered unnecessarily.

Now a multi-media session is initiated using RTP and the CR searches for spectrum opportunities that satisfy the latency and resource requirements of the stream to avoid buffer overflow. The video I frames are given higher importance (and delivery reliability precedence) over P frames in order to ensure the best perceived quality to the user. When the opportunistic spectrum resources become limited it is necessary for the CR for switch to a different spectrum access scheme (such as exclusive use) to maintain video quality.

4.4.3 Novel Methods and Solutions

Having briefly introduced the issues relevant to higher layers in communication systems with an eye on cognitive radio in particular, this section proceeds to discuss some novel solutions to address such higher layer considerations.

Application Layer

Currently, applications traffic flows are not normally associated with a dynamic urgency or precedence level in commercial devices and no generic classification mechanism or API to facilitate this exists. Also, it is often difficult to infer this information from the traffic characteristics when the application itself is adaptive to different situations. This hampers the process of determining how a CR device should handle the traffic from different applications. For instance in the exemplary scenario it would be difficult for a CR using current technologies to determine that there is a need to change its behavior. To facilitate this requires an API and associated mechanisms to allow applications to inform the device of its needs (even if they change part way through a session), such as in [72]. Alternatively, inferring application requirements by considering device context and traffic flow information (such as IP port numbers and addresses, video frame types) is possible, but may be very processor intensive and not sufficiently reliable. Even so cognitive resource management is necessary to determine the best possible way to satisfy all application requirements.

Also, currently applications may be adaptive to network performance variations, but are not cognitive in the sense of being able to reason and configure links to best suite the user requirements. This shift in level of reasoning and context awareness (as depicted in the exemplary application scenario in Section 4.4.1) requires that applications can determine the urgency and importance of service they are providing and the options available to support them. When multiple applications and services must coexist or share resources then optimization is possible based on the knowledge of the context and service requirements (as described in more detail in Section 4.5). However, to support the arbitration between applications (or agents) performing optimization requires a hierarchical approach consisting of different privilege levels (as depicted in Fig. 4.4). This is supported by ULLA with the link manager plug-in concept (see [67]), which permits a higher privilege level plug-in to constrain and override the decisions made at lower levels (of privilege) when necessary (for instance when there are conflicting actions being performance by different applications). In the determination of the role and privilege level of each application or agent is therefore a key consideration for any cognitive radio system. Indeed, it is also normal in policy based management to associate each policy (event-condition-action) with a role in order that policies can be interpreted correctly depending on the context and privilege level (or role) of the corresponding application. Therefore, in the exemplary application scenario the BM's CR device is able to escalate the level of service provided by the WLAN because of the change in privilege from that of a visitor application level to an emergency video call. In this case the application triggers the change of role (by initiating the emergency video conference session), but this may not automatically result in a change in privilege level unless it is acceptable to do so by the higher privilege application (or agent).

It is also necessary to have mechanisms which police the use of roles or privilege levels and subsequently the use of the spectrum and network resources by application services. These types of approach are used within network nodes (for instance selective dropping of packets within switches or routers), but there has been very few proposals specifically for CR device based policing mechanisms. It is currently assumed that it will be necessary to have mechanisms both within the CR devices and also in the network nodes so that it is always possible to ensure the best allocation of network resources to different CR applications, and particularly in emergency or specific (such as congestion or failure) situations.

Transport Layer

Research on TCP enhancements for wireless networks has been going on for over a decade and already there are several standard mechanisms to enhance performance when used over wireless links. In addition, there are several variants of TCP (known as TCP flavors) arguably more suitable for dynamic spectrum access networks or any wireless links which exhibit variable bandwidth properties and vulnerable to non-congestion related losses [73]. One example of such protocol enhancement is the TCP Westwood proposal [74], which dynamically estimate the available bandwidth to set the congestion window and slow start threshold after a loss. Apart from that, there are also middleware solutions such as performance enhancing proxy (PEP) which could rectify some of the problems mentioned above [75]. Moreover, there are also cross layer methods, such as Explicit Loss Notification (ELN), to make TCP stack aware of packet losses due to link errors [76].

State aware scheduling of TCP traffic is a necessary solution to avoid unfairness and resource starvation in dynamic spectrum networks. This relies on the ability to be able to have protocol state awareness and prioritize access opportunities of different TCP flows depending on state. Alternatively, per flow resource allocation could be exploited if it is possible to allocate spectrum resources to individual TCP flows.

4.4.4 Open Issues and the Way Forward for Higher Layers

As mentioned previously, so far most of the studies on CR have focused on the PHY and MAC layers of the protocol stack. For any wireless communications system, higher layers assume a reliable service from their corresponding lower layers. However, given the discontinuous nature of cognitive communications (jumping from one band to another, or one access technology to another etc) the basic assumptions of a layered protocol stack might not remain valid. This issue is probably best addressed in a cross-layer approach, as will be discussed in the next section.

As an example consider the case of transport layer using TCP as discussed in Section 4.4.2. It is widely known that TCP, which was originally developed for wired networking, is not efficient in wireless communications. The problems only worsen as wireless communications evolve towards cognitive communications. Since a CR needs to change its operating frequency band from time to time, as to meet the primary and the secondary users' requirements, the end-to-end communication link is disrupted several times in the life-time of a specific session. Given the fact that at each disruption stage, the TCP will start a new end-to-end communication link with the minimum congestion window size, the effective achievable data transfer rate might be much lower than legacy radio technologies. While the discourteous nature of available resources for secondary radios is partly to be blamed, the legacy transport layer protocols are also involved in creating this inefficiency. Hence, there is an urgent need for cross-layer approaches in CR domain. Is it, for instance, possible to create a "virtual end-to-end link" from transport layer point of view, while the PHY layer is operating discontinuously? Then, how would flow control be optimized in such circumstances?

Cognitive application can exploit the flexibility provided by multi-radio and frequency agile CR devices. However, a number of open research challenges exist to ensure that the applications are not only aware of the context of the device (which facilitates application level adaptation), but also that the applications coexist and cooperate effectively to configure CR devices and share the radio resources. This necessitates generic context awareness mechanisms and API's to permit applications to interact more intelligently, together with policies (policy language), roles and privilege levels and the associated policing mechanisms to ensure that the cognitive applications use the resources in an optimal manner. This includes resource sharing mechanisms and detection, prevention or reporting of observed exceptions and policy infringements in different manners, for instance in distributed or centralized ways (depending on the situation). This clearly involves multiple stakeholders (including operators, device manufacturers, application developers, and operating system vendors) and requires that consensus is formed to ensure that CR enabled applications are not only beneficial to the end user, but provide value and added revenue to the other stakeholders. The P1900.4 standardization project is just one piece in this jigsaw, providing the policy language and framework for supporting interaction between CR devices and the network, and in the future this will also develop the required protocol specifications.

4.5 Cross-Layer Optimization for Cognitive Radio

Due to the volatility in the underlying spectrum resource available to CRs affecting higher layers, among other factors, cross-layer implications are also relevant to CR design. Cross-layer optimization for CR is investigated in this section.

4.5.1 Cross Layer Optimization in the Cognitive Radio Domain

There has been much research within the umbrella of Cross Layer Optimization (CLO) for wireless networks. Much of this prior work focused on the exploitation of opportunistic channel scheduling by considering the classical resource (rate) allocation and congestion control optimization problem in conjunction with knowledge of the channel (radio) state and application requirements (e.g., see [77]). This approach yields performance benefits by taking advantage of dynamic channel variations that are present in all radio systems and the utility for allocating different rates to different application flows. The goal of the CLO is, therefore, to maximize the overall system utility based on the service (application) utility functions, which are dependent on knowledge and characterization of the application flows. One outstanding research challenge is to provide efficient CLO algorithms that can be applied to in-elastic application flows which have non-concave utility functions. This type of application flow is typical of voice and video traffic and leads to a combinatorial NP-hard optimization problems. There are heuristic methods (such as in [71]) of tackling this problem, which avoid discontinuities in the utility function and, thus, prevent instabilities or non-convergence of the solution.

The flexibility provided by CR systems adds an additional dimension to the CLO problem, which is in the dynamic spectrum resource assignment dimension. CR also allows different ways of performing dynamic spectrum access, based on dynamic exclusive use, open sharing or hierarchical access models, as described in [78]. These different models lead to different opportunities for CLO solutions, which can be centralized, distributed or decentralized. The open sharing and hierarchical models have a common characteristic, which is that there is uncertainty in spectrum opportunities due to the fact that different secondary and primary systems must coexist in the same spectrum allocation, but may not be able to directly interoperate. This reduces the predictability of channel availability determination, which can result in the need for channel access etiquette, based on listen-before-talk principles or a spectrum underlay approach which can use physical layer interference avoidance techniques. This section focuses on the open sharing and hierarchical access overlay models applied to CR CLO, as the other approaches suit a more traditional CLO problem with variable resource availability. Two levels of optimization are possible in these types of CR systems, one being local to devices in order to satisfy application demands and the second network-wide optimization to optimize a network of cooperating CR devices as a whole in the presence of multiple radio systems.

Local Optimization

This section examines the probabilistic analysis of CLO in more detail, when considering local optimization from the CR device perspective. For this, measurements (such as signal strength) are required to determine channel access opportunities. It is then possible to reformulate a revised CLO utility maximization expression based on the maximum utilization of CR opportunities as an additional dimension to the channel condition (or state). For instance, the observed probability of a frequency resource being available (P_t) can be defined as the probability of a spectrum resource not being occupied (or would not cause interference to any of the transmissions on the channel) for a certain period of time (t which can be a set of discrete interval values in multiples of a base unit, $t = n \times T$, where $n = 1, 2, \cdots, N$) within an observation period of duration $N \times T$. In addition, it is desirable to accumulate observation data over successive periods within a window of M successive observation periods (i.e. a sliding window). The reason for this is to improve reliability of the opportunity estimates given that the accuracy of the channel measurement (such as signal strength) normally has low accuracy and, so, requires compensation (for instance using a calibration offset based on long term observations) and also to account for time varying channel fading. In this manner, P_t can be computed for each channel by observing transmission (passive monitoring) and the most appropriate channel selected for the particular application flow using cross layer information. For this to be achieved, knowledge of application traffic characteristics is also a necessary part of a CLO approach and these can include aspects, such as:

- Protection requirements (for instance, for I frames and P frames in video traffic),
- Periodicity characteristic,
- Latency requirements,
- Protocol state awareness (for instance, in TCP streams —slow start, congestion avoidance),
- Size of application transfer units (objects), and
- Burst/interactivity characteristics in case of data traffic such as web browsing.

The above application characteristics can be abstracted into a single performance requirement metric, which is the reliability target for achieving successful delivery of a data unit s of size S_s within a latency target (d) denoted by $R_{d,s}$. Clearly, this formulation does not place any restriction on the variation in d and S_s or $R_{d,s}$ between successive transmissions and so the values can change depending on the application requirements or flows. It is also possible to adjust the observation (and channel dwell) period so that it equates to the target delay (i.e. $d = N \times T$), by either adjusting N or T.

In addition to application requirements are the device specific optimization requirements, which are typically concerned with minimization of the power consumption (and other terminal resource usage). For power consumption, the energy required to transmit a given individual data unit s of size S_s can be expressed as $E^c{}_s(r_l)$ at a given average link rate r_l. Therefore, the link and channel with the minimum energy $min_{\forall c}[E_s^c(r_1)]$ will be selected, if the probability of meeting the delivery objectives are fulfilled within the acceptable level of confidence (probability). Selection of optimal link rate under time varying channel conditions is a well known problem, as described in [79]. The problem has an information theoretic derivation (provided by Shannon) and can be solved in time varying channels with lazy scheduling using water filling in time, channel inversion or other techniques. The common characteristic is that if larger delay can be tolerated, lower energy is consumed by exploiting channel variability, as given by the theoretical minimum energy expression:

$$min[E_s^c(r_1)] = S_s(aP + P_0)[S_s B_c^{-1} / \log_2(1 + PW) + T_0]$$

where P_0 is the constant (quiescent) power consumption, T_0 is the constant fixed time overhead for transmission/reception of data unit s, S_s is the number of bits in the data unit s, B_c is the bandwidth of channel c, P is transmit power, a is related to the power amplifier efficiency and W is path gain (reciprocal of path loss) to noise ratio.

Therefore, the minimum theoretical per channel power consumption is dependent on the bits per second per Hertz transmitted on the channel and on the transmitter (and quiescent) power consumptions. It is also possible to split individual data units (s) across multiple channels (or subcarriers in multicarrier modulation schemes and MIMO channels in multi antenna systems) to provide optimal channel usage, so we consider this to be catered for by considering energy versus average link rate for logical channels (or composite channels which includes multicarrier and MIMO schemes) and derive a CR local optimization utility function for using a particular logical channel c to send an individual data unit s as:

$$U_{s,1}^c(r_1) \propto 1/E_s^c(r_1), \quad \text{iff} \quad F_c(S_s/r_1) > R_{d,s} \quad \text{and} \quad 0, \quad \text{otherwise} \quad (4.1)$$

or

$$U_{s,1}^c(r_1) \propto 1/E_s^c(r_1) \quad \text{for latency insensitive traffic}$$

where r_l is the average application level data rate achievable on link l (from device i to j) considering the link related overhead, bandwidth, modulation scheme, link scheduling and state (signal to noise ratio), including retransmissions and coding overhead. F_c is the complementary cumulative distribution function of the transmission opportunity probability P_t, with duration t. Therefore, it is possible to calculate F_c from the observed transmission opportunity probability for channel c as follows:

$$F_c(x) = \sum_{t > x} P_t. \quad (4.2)$$

The expressions in (4.1) and (4.2) do not take into account overhead caused by channel switching and access time, which reduces the usefulness of the opportunities by a scalar amount. This can be considered by a substitution of expression (4.2) with (4.3), where o is the overhead time and is considered to be zero if no channel changing is required.

$$F_c(x) = \sum_{t > (x+o)} P_t. \quad (4.3)$$

If it is assumed that o is a constant independent of the link technologies and frequency, then the computation of expression (4.3) is simplified. The final consideration in the CR CLO utility expression is the combination of opportunity considering all the recipient devices, in case of multicast or broadcast communication. This is important from the point of view of selecting the optimal opportunities as seen by all recipient devices. Therefore, transmission opportunity probabilities considered in expression (4.3) should be collective as depicted by the new expression (4.4) below to represent the least occurrences (lowest probability) of transmission opportunity observed by all the recipients. The collection of the transmission opportunity data $P_{t,j}$ could be performed centrally or in a distributed manner. It is also possible that a device estimates this from its own observations rather than observations made by the intended recipient device.

$$F_c(x) = \sum_{t > (x+o)} \min_{over\ j} [P_{t,j}]. \quad (4.4)$$

In (4.4), $P_{t,i}$ is the transmission opportunity probability of duration t, seen by recipient device j. If the opportunity durations $P_{t,j}$ follow exponential distributions (which is often the case for user initiated sessions, such as file transfers see [80]), then the $\min_{over\ j}[P_{t,j}]$ is also exponentially distributed. Therefore, it is feasible to approximate the distribution $P_{t,j}$ and consequently $\min_{over\ j}[P_{t,j}]$, using the maximum likelihood estimate based on the knowledge that for the exponential distribution, $E[_{Pt,j}] = 1/\lambda$. In this way it is possible to only measure the mean channel occupancy during each observation period, which equal to $N \times T - E[P_{t,j}]$ to form an approximation for $F_c(x)$ with reduced signaling overheads (by at least a factor of N).

In order to illustrate the CR local optimization process, a set of five channels with linearly increasing bandwidth it considered. The formulation of $F_c(x)$ assumes an exponential opportunity duration distribution as shown in Fig. 4.5 (which corresponds to a uniform occupancy bandwidth ratio on each channel). The utility for utilizing each channel (reciprocal of energy consumption) as given in expression (4.1) is assumed to be proportional to bandwidth as is the case for radio devices with constant transmit power and quiescent power consumption regardless of bandwidth. Then, the optimal channel bandwidth selection for applications is based on the trajectories in Fig. 4.5 for different target latencies and reliability. For instance, if an application requires 95% confidence that a $5 \times T$ duration opportunity to transmit will be obtained (within the period $N \times T$, in this case $N = 30$), then channel 1 (lowest bandwidth) must be used, whereas if only 80% reliability is necessary channel 5 (highest bandwidth) should be used. It is also necessary to consider the computation of S_s/r_l to dynamically derive the optimal x value, which will be different for each of these channels (given the different bandwidths and variability). Ideally the average link rate (r_l) required should be minimized to minimize overall power consumption, but this requires longer channel opportunity duration, which is less likely to occur in the required time interval. Therefore, a trade-off occurs to balance power consumption with delivery reliability and this leads to the need to dynamically adapt link rates depending on the actual observed channel conditions (if they turn out to be different from that expected). As the channel opportunity trajectories are probabilistic, it is possible to periodically compute values in a look-up table and only update them when significant changes occur.

Network Optimization

In order to optimize global resource allocations of a CR system, each local optimization process must be combined to determine the correct schedule (or link and channel allocation) at each point in time. If it is assumed that only one link and channel can be used at a time, the problem is simplified, but this is not generally the optimal case. For instance, typically a network will consist of several CR devices each capable of switching independently with only the need to be on the same channel as the intended recipient or multiple recipients in the case of multicast and broadcast transmissions. It is further acceptable that broadcast transmissions are duplicated on different channels, if necessary, to ensure that each device has the opportunity to receive them.

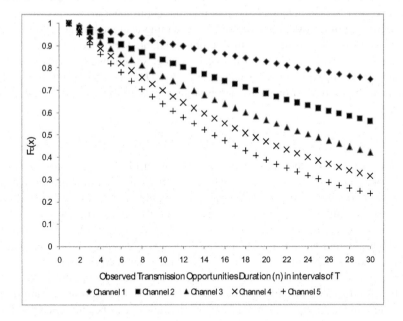

(a)

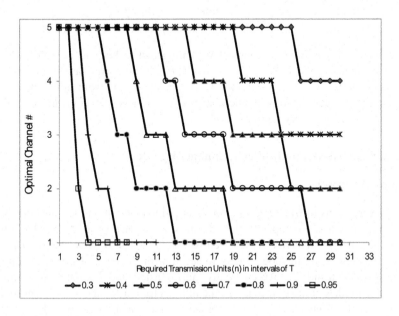

(b)

Fig. 4.5. (a) Example channel opportunity complementary cumulative distribution function for different channels, and (b) optimal channel vs. required transmission time units (n) for different reliability $R_{d,s}$ targets with latency sensitive traffic.

This leads to a much more complex optimization problem in which each local utility expression for each channel and link must be coordinated to achieve an optimal approach from the network perspective. By assuming a single hop network scenario (i.e. no multi-hop forwarding), with a common reliable logical control channel, i.e., the default channel for devices to listen on for requests to change to a particular channel and channel opportunity information, such as $E[P_{t,j}]$ or channel utility $U_{s,l}{}^c(r_l)$ then the utility optimization expression is straightforward to formulate and is shown in (4.5).

$$\max[\sum_{all\ c,s\ \&1} U^c_{s,1} \ ].$$
$$\text{such that all l are non interfering}$$
$$(4.5)$$

However, solving the optimization expression in (4.5) involves comparison of many combinations for different c, s and l alternatives and determination of whether the different links are mutually interfering. It also potentially incurs a high signaling overhead to communicate dynamically changing utility values. This can be simplified by considering mutual exclusion within groups of links for each link l, given as G_l, for which a common transmitter (i) or receiver (j) exists and, so, only one link in group G_l can be scheduled at a time. This assumes that each radio device can only operate on a single channel and link at a time. A maximum group utility expression can be used to determine the optimal link schedule within a group. In addition, a minimum channel frequency separation requirement f can be introduced to avoid adjacent channel interference between links within different groups and so the overall network utility is:

$$\max \quad [\sum_{all\ l,c} W_{l,c} \times \max_{over\ all\ l \in G_l} [U^c_{l,s}(r_1)]]$$
$$\text{such that all the frequency separations between } c \text{ are greater than } \delta f$$
$$(4.6)$$

where $W_{l,c}$ is a per link and channel weighting factor that can be used to favor particular channels and link combinations. This can be particularly important for network scenarios in which different links and channels may incur different costs or revenue, for instance as would be the case in different license exempt and secondary spectrum usage regulatory regimes.

4.5.2 Cross Layer Optimization Techniques for CR

Example Approach

Taking expressions (4.6) and (4.1) as starting points and solving the network optimization problem is a complex task and can be performed using different solution approaches, from exhaustive search, heuristic approaches to game theory. The expression also does not include a factor to account for the backlog congestion problem (as discussed in [77]), which ideally must consider the amount of buffering at each device to determine how to adjust the utilities for elastic latency tolerant traffic. The derivation of these expressions is based on individual application data units s, which implicitly assumes that the latency requirement for each unit copes with congestion control by dropping units which do not meet the delay objectives. However, most application traffic is in fact latency sensitive in nature even if it is hard to determine the exact reliability and latency requirements, but this is eased with cross layer approaches.

Now let us consider an exemplary approach to solve the CR CLO problem, and its performance evalation in a technology specific scenario. The scenario considers a mixture of in-elastic multimedia (video or voice) traffic and elastic data traffic within a dynamic open sharing model and wireless LAN technology. The solution approach uses an observation period ($N \times T$) aligned to the multimedia traffic latency target requirement and the scheduling of transmissions is controlled in two possible ways. Firstly, if the two traffic types are on links corresponding to different link groups G_l then two different channels are selected by network optimization processes (separated by at least δf) and the transmissions continue in parallel. Otherwise, for links within the same group (G_l) with an overlay protocol, as described in [81], that either assigns time slots within logical time channels by way of dynamic resource reservation or by a distributed contention based approach. In either case the assignment opportunity probabilities are weighted by the locally perceived channel state (based on the most recent received signal strength indicator from the last received frame). Therefore, performance depends on the reliability of the channel state estimation and on the type of scheduling (distributed or centralized). An intuitive conclusion, that has been reinforced by simulation results, is that as the channel state becomes more dynamic (shorter coherence time), the distributed resource assignment approach becomes more attractive, due to the shorter time between channel state feedback and exploitation potential. To capitalize on this it should also be possible for different CR devices within the same network to use distributed or centralized resource assignments as determined by the network optimization process.

In either of the above cases the local optimization process selects from the available channels and links (radio access) options provided by the network optimization process and utilizes the channels in a locally optimal manner, for instance by determining at which rate and which data unit (s) to send in each channel opportunity. The local optimization process also adapts the rate of the multi-media application transmissions in response to changes in the observed average link rate, which is also itself adapted to try to meet delivery reliability for the target latency with minimum energy. Experimental testing using wireless LAN technology has shown that this is a viable optimization approach for this type of scenario (see [82]), but it may not be suitable for other types of application traffic or dynamic spectrum access model.

The above example highlights an obvious issue with CLO techniques for CR, which is that network resource assignment techniques are typically designed to provide determinism in the allocation of resources using knowledge of queue size or other measures of resource requirements (such as utility). However, CR overlay opportunities are often non-deterministic, and so we assume a reliance on probabilistic estimation of resource availability. At a certain point (depending on the channel coherence time and transmission opportunity distribution), it becomes inappropriate to attempt to perform CLO in this manner.

Another aspect of CLO for CR that requires consideration is how the performance degrades and stability is affected as the radio resources become more fully utilized. For instance, as more CR devices are deployed the traffic patterns and observed interference by CR devices will become dominated by other CR devices rather than the primary users. As CR devices are expected to behave in an opportunistic (spectrum overlay and open sharing) way, then the transmission opportunities will become shorter and also potentially follow a different distribution to the primary users traffic. Consideration of this situation requires more thorough analysis of the transmissions and some promising techniques were described in the previous Chapter. With more detailed knowledge and characterization of the type of transmissions would result in a better CLO solution.

Reinforcement Learning

Considering the above example a suitable way to use cognition in a CR CLO solution can be to exploit reinforcement learning techniques to adapt to the changes in the channel opportunity distributions and prediction of the utility for the different applications to utilize the available channels. For instance, this could use the back-propagation technique to train a neural network with the channel opportunity distributions presented to the input nodes and corresponding expected utilities for each of the possible opportunity durations at the output nodes (with a number of weighting layers in between, but normally two). Once trained with expected utilities (for instance, derived from analytical or simulation models or even empirical data) the neural network can be continuously updated (retrained) with actual utility values obtained by measurements taken when the channels are selected. For cases in which the channels have significantly different characteristics (coherence time, power consumption, and noise levels) it may further be necessary to categorize the channels and use separately trained networks. This approach can be applied to both the network level optimization and also to the local optimization processes (as depicted in Fig. 4.4 in Section 4.4). The channels in this case can be physical frequency channels as well as logical (composite frequency or time domain based) channels, such as proposed in the above example approach. The advantage of this approach over simple lookup tables is that the reinforcement (or training) can generalize between different (but similar) opportunity distributions and, thus, not rely on interpolation between specific fixed sample data point or statistical approximation of the channel opportunity characteristics (such as with the exponential distribution). Also, as the exact relationship between channel utility value and opportunity distribution can be complex to compute directly, it can lead to a more efficient (lower complexity) implementation. A candidate network and training error pattern are shown in Fig. 4.6, which is fed channel opportunity distribution at the input nodes, in discrete steps of T, and the utility for different links and discrete average link rates at the output, the training data used for the network is based on the analytical data in Fig. 4.5.

The use of neural networks with reinforcement learning does raise a question as to how much processing is required to retrain the network as the situation changes. The results in Fig. 4.6 are indicative of the number of training cycle iterations required to converge from an arbitrary (random) starting point. Therefore, if the dynamic nature of the channel characteristics (as quantified by the coherence time and

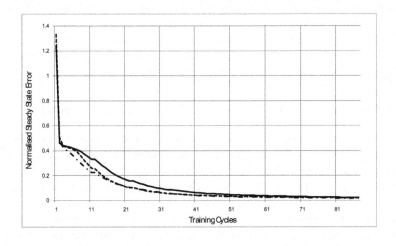

(a)

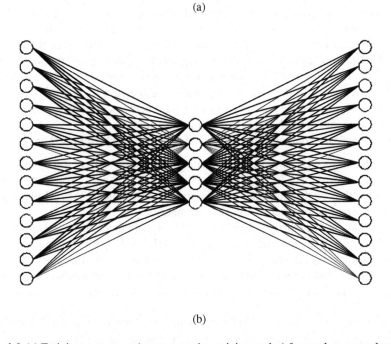

(b)

Fig. 4.6. (a) Training error curve (over successive training cycles) for two layer neural network using back-propagation and random weight initialization, and (b) example two layer network with five hidden nodes.

opportunity distribution) is not itself changing rapidly then retraining is not a significant processing burden. However, when for instance changing from a stationary to a fast moving scenario, the channel characteristics suddenly change and it may be necessary to utilize separately trained networks for these different scenario contexts. It could further be possible to utilize sensor data, such as movement detection, to select the most appropriate networks (or network weights) to use.

4.5.3 Open Issues and the Way Forward for Cross Layer Optimization

This section has addressed the problem of CLO for CR systems. It has highlighted the problems and formulated optimization expressions. However, to exploit CLO requires application knowledge and this necessitates direct interaction (such as using a suitable application programming interface, see [72]) or by inferring application requirements from user or application behavior. With local application and radio device knowledge, it is possible to perform local optimization, but to achieve a network-wide optimization requires a further and potentially much more complex step as the possible combinations of options explodes. A simplification is proposed that is consistent with proposed frameworks to support dynamic optimization [83], [84], which splits local and network optimization in a hierarchical manner. This framework can also utilize fuzzy classification of application requirements, when specific detailed knowledge or inference is not possible. To support the download of optimization algorithms to different CR device (in a device agnostic manner) requires additional execution environment functionality (for instance using Java or Javascript), as described in [81], that ideally have support for neural network or other reinforcement learning and cognitive (such as reasoning) mechanisms.

Some open issues that need to be resolved to enable CLO within CR are:

- Reliable and power efficient determination of channel state that may combine different sensory data (such as signal strength and link performance) to derive an estimation of the transmission opportunity distribution (P_t) and efficiently characterize this for efficient signaling exchanges.
- Determination of realistic generic application requirement characterization that can be easily obtained or inferred, such as reliability $R_{d,s}$ for meeting a latency target (for latency sensitive traffic) or buffer size limit for elastic data.
- Radio access technologies independent abstractions for performance and power consumption to permit calculation of a channel energy (vs rate) relationships $E^c_s(r_l)$.
- Cross layer optimization for multi-hop cognitive radio networks in which the dynamic optimal routing dimension must also be considered, which can be regarded as the optimal allocation of data units (s) to links (l) for minimum total energy consumption.
- Simple near-optimal algorithms to select channels and schedules in distributed scenarios that have realistic implementation complexity and overhead. Use of reinforcement learning to adapt to changing environments and the associated complexity evaluation.

- Radio devices that are able to flexibly trade-off power consumption for link transmit rate, unlike, for example, off-the-shelf wireless LAN devices which tend to have constant transmit power and constant quiescent power consumption regardless of modulation scheme.

There has recently been some progress in addressing many of the open issues (such as [72], [79], [81], [82] and [84], particularly within single technology solutions. However, it remains to be seen whether a generic CR CLO approach can be realistically realized to support heterogeneous technology and device agnostic CLO using a general framework, such as proposed in [83], for different multi-technology scenarios and dynamic spectrum access models.

Conclusion

The emergence of CR technology presents opportunity and challenge in equal measure. Through the decentralized approach to spectrum usage and control in CR, far greater average spectral utilization can be achieved, thereby providing users with levels of QoS which in many respects might previously have been associated only with wire lines. CR, however, has the potential to greatly disrupt conventional users of the spectrum, should it not be carefully planned and regulated. Hence, secondary spectrum access through CR is perhaps one of the most pertinent research areas at present, presenting enormous potential reward but also significant obstacles which must be proven to have been reliably addressed before it emerges into the market place.

CR research should not only concern lower layers. Numerous higher layer implications for CR exist; moreover, CR approaches can have significant cross-layer implications. For example, the necessary cautionary/interrupted radio access of CRs in a secondary usage scenario will likely filter through to transport protocol performance implications, particularly in terms of congestion control and loss recovery. This chapter has therefore attempted to provide a balanced view on CR, considering its implications across the ISO reference model, rather than taking the more restricted (MAC/PHY only) view of CR that has perpetuated in some past works.

This chapter has also highlighted, through example scenarios and approaches, how the flexibility and dynamic opportunities provided by CR can be exploited. There is clearly a very careful trade-off to be addressed in all future CR systems, to balance the complexity and overhead associated with the methods of coordinating the use of different CR device configurations against the resource utilization and efficiency gains that are able to be made. As these gains depend not only on the radio environment, but also on protocols and applications, the overall approach to managing them must also be adaptive and all-encompassing across these layers. The grand challenge for CR systems is, therefore, to combine the decision making intelligence in a sufficiently flexible, adaptive and cooperative way, such that different scenarios can be catered for regardless of device types and application requirements in order to always provide the optimal resource utilization.

As a final comment, a successful CR system must satisfy the various stakeholders' interests (such as operators, manufacturers, regulators, software or operating system vendors, and end users), which may well be conflicting. Resolving conflicts and roles of the stakeholders and associated applications and agents is therefore a major challenge that has implications on various aspects of CR devices. As CR devices of different capabilities and types are likely to emerge, CR will provide a great opportunity for product differentiation and optimization above that possible with conventional technologies.

Acknowledgment

Some of the work reported in this chapter formed part of the Delivery Efficiency Core Research Programme of the Virtual Centre of Excellence in Mobile & Personal Communications, Mobile VCE (http://www.mobilevce.com). This research has been funded by EPSRC and by the Industrial Companies who are Members of Mobile VCE. Fully detailed technical reports on this research are available to Industrial Members of Mobile VCE. A part of this work has been performed within the projects E2RII, GOLLUM and Aragorn, which have received funding from the European Commission's Sixth and Seventh framework programmes. The ideas presented here only reflect the authors' views; the community is not liable for any use that made of the information contained herein. The contributions of colleagues from E2RII, Gollum and Aragorn are hereby acknowledged.

References

1. J. Mitola III, "Cognitive Radio: An Integrated Agent Architecture for Software Defined Radio," Doctorate of Technology Dissertation, KTH, Sweden, May 2000.
2. J. Mitola III, "Cognitive Radio for Flexible Mobile Multimedia Communications," in *Proc. Sixth Int'l. Wksp. on Mobile Multimedia Communications*, San Diego, CA, USA, Nov. 1999, pp. 3-10.
3. S. Haykin, "Cognitive Radio: Brain-Empowered Wireless Communications," *IEEE J. Select. Areas Commun.*, vol. 23, no. 2, pp. 201-220, Feb. 2005.
4. FCC NPRM, *Cognitive Radio*, ET Docket No. 03-108, Sep. 2003.
5. N. Devroye, P. Mitran, and V. Tarokh, "Achievable Rates in Cognitive Radio Channels," *IEEE Trans. Inf. Theory*, vol. 52, no. 5, pp. 1813-1827, May 2006.
6. C. Cordeiro *et al.*, "IEEE 802.22: An Introduction to the First Wireless Standard based on Cognitive Radios," *J. of Commun.*, vol. 1, no. 1, pp. 38-47, Apr. 2006.
7. FCC press release, ET Docket No. 04-186, Dec. 2004.
8. FCC press release, *Promoting Efficient Use of Spectrum through Elimination of Barriers to the Development of Secondary Markets*, ET Docket No. 04-167, 2004.
9. FCC press release, ET Docket No. 06-161, 2006.
10. FCC press release, *Comments Sought on Google Proposals Regarding Service Rules for 700 MHz Band Spectrum*, FCC DA 07-2197, May 2007.
11. The European Parliament, *Towards a European Policy on Radio Spectrum*, ITRE/6/37236, 2007.

12. Ofcom, *Spectrum Usage Rights: Technology and Usage Neutral Access to the Spectrum*, Apr. 2006.
13. Ofcom, *Spectrum Framework Review*, Consultation, Nov. 2004.
14. Ireland's Commission for Communications Regulation, *Dynamic Spectrum Access*, Document No. 07/22, 2007.
15. IEEE P1900/SCC41 website: http://www.scc41.org/
16. IEEE 802.22 website: http://grouper.ieee.org/groups/802/22/
17. IEEE 802.16's License-Exempt (LE) Task Group website: http://grouper.ieee.org/groups/802/16/le/index.html
18. IEEE 802.11y standards activities: http://grouper.ieee.org/groups/802/11/
19. SDR Forum website: http://www.sdrforum.org/
20. DARPA XG program website, http://www.ir.bbn.com/projects/xmac/working-group/
21. F. Seelig, "A Description of the August 2006 XG Demonstrations at Fort A.P. Hill," in *Proc. IEEE DySpan 2007*, Dublin, Ireland, Apr. 2007, pp. 1-12.
22. FCC press release, *Notice of Ex Parte Communication*, ET Docket No. 04-186, 02-380.
23. D. Taubenheim *et al.*, "Implementing an Experimental Cognitive Radio System for DySPAN," in *Proc. IEEE GLOBECOM 2007*, Washington D.C., USA, Nov. 2007.
24. N. Hoven and A. Sahai, "Power Scaling for Cognitive Radio," in *Proc. Int'l Conf. on Wireless Networks, Communications and Mobile Computing*, Maui, HI, USA, Jun. 2005, pp. 250-255.
25. T. A. Weiss and F. K. Jondral, "Spectrum Pooling: An innovative Strategy for the Enhancement of Spectrum Efficiency," *IEEE Commun. Mag.*, vol. 42, no. 3, pp. 8-14, Mar. 2004.
26. J. D. Poston and W. D. Horne, "Discontiguous OFDM Considerations for Dynamic Spectrum Access in Idle TV Channels," in *Proc. IEEE DySPAN 2005*, Baltimore, MD, USA, Nov. 2005, pp. 607-610.
27. T. Weiss *et al.*, "Mutual Interference in OFDM-Based Spectrum Pooling Systems," in *Proc. IEEE VTC-Spring 2004*, Milan, Italy, May 2004, pp. 1873-1877.
28. S. Kondo and L. Milstein, "Performance of Multicarrier DS CDMA Systems," *IEEE Trans. on Commun.*, vol. 44, no. 2, pp. 238-246, Feb. 1996.
29. R. Rajbanshi *et al.*, "Quantitative Comparison of Agile Modulation Technique for Cognitive Radio Transceivers," in *Proc. IEEE CCNC 2007*, Las Vegas, NV, USA, Jan. 2007, pp. 1144-1148.
30. Z. Wu *et al.*, "The Road to 4G: Two Paradigm Shifts, One Enabling Technology," in *Proc. IEEE DySPAN 2005*, Baltimore, MD, USA, Nov. 2005, pp. 688-694.
31. T. Clancy, "Achievable Capacity Under the Interference Temperature Model," in *Proc. IEEE INFOCOM 2007*, Anchorage, AK, USA, May 2007, pp. 794-802.
32. T. Clancy, "Formalizing the Interference Temperature Model," *Wiley Journal on Wireless Communications and Mobile Computing*, vol. 7, no. 9, pp. 1077-1086, Nov. 2007.
33. FCC Order, FCC-07-78A1, May 2007.
34. F. Capar and F. Jondral, "Spectrum Pricing for Excess Bandwidth in Radio Networks," in *Proc. IEEE PIMRC 2004*, Barcelona, Spain, Sep. 2004, pp. 2458-2462.
35. A. Attar, M. R. Nakhai, and A. H. Aghvami, "Cognitive Radio Game: A Framework for Efficiency, Fairness and QoS Guarantee," in *Proc. IEEE ICC 2008*, Beijing, China, May 2008.
36. J. Lansford, "UWB Coexistence and Cognitive Radio: How Multi-band OFDM Lead the Way," in *Proc. Wireless Networking Symposium (WNCG 2004)*, Austin, USA, Oct. 2004.

37. A. Attar, S. A. Ghorashi, M. Sooriyabandara, and A. H. Aghvami, "Challenges of Real-Time Secondary Usage of Spectrum," *Computer Networks*, vol. 52, no. 4, pp. 816-830, Mar. 2008.
38. T. Chen *et al.*, "CogMesh: A Cluster-Based Cognitive Radio Network," in *Proc. IEEE DySPAN 2007*, Dublin, Ireland, Apr. 2007, pp. 168-178.
39. R. W. Brodersen, A. Wolisz, D. Cabric, S. M. Mishra, and D. Willkomm, "CORVUS: A Cognitive Radio Approach for Usage of Virtual Unlicensed Spectrum," *UC Berkeley White Paper*, Jul. 2004.
40. D. Cabric, S. M. Mishra, D. Willkomm, R. Brodersen, and A. Wolisz, "A Cognitive Radio Approach for Usage of Virtual Unlicensed Spectrum," in *Proc. 14th IST Mobile and Wireless Communications Summit*, Dresden, Germany, Jun. 2005.
41. T. Clancy, "Interference Temperature Multiple Access," in *Proc. Int'l Wksp. on Technology and Policy for Accessing Spectrum (TAPAS 2006)*, Boston, MA, USA, Aug. 2006.
42. Q. Zhao, Y. Chen, and A. Swami, "Cognitive MAC Protocols for Dynamic Spectrum Access," in *Cognitive Wireless Communication Networks*, Eds. E. Hossain and V. K. Bhargava, Springer, 2007.
43. Q. Zhan, L. Tong, A. Swami, and Y. Chen, "Decentralized Cognitive MAC for Opportunistic Spectrum Access in Ad Hoc Networks: A POMDP Framework," *IEEE. J. Select. Areas Commun.*, vol. 25, no. 3, pp. 589-600, Mar. 2006.
44. A. Attar, M. R. Nakhai, and A. H. Aghvami, "Cognitive Radio Transmission Based on Direct Sequence MC-CDMA," to appear in *IEEE Trans. Wireless Commun.*, 2008.
45. H. Vincent Poor, "Active Interference Suppression in CDMA Overlay Systems," *IEEE J. Select. Areas Commun.*, vol. 19, no. 1, pp. 4-20, Jan. 2001.
46. P. Leaves *et al.*, "Dynamic Spectrum Allocation in Composite Reconfigurable Wireless Networks," *IEEE Commun. Mag.*, vol. 42, no. 5, pp. 72-81, May 2004.
47. A. Attar, O. Holland, M. R. Nakhai, and A. H. Aghvami, "Interference-Limited Resource Allocation for Cognitive Radio in OFDM Networks," to appear in *IET Communications, Special Issue on Cognitive Spectrum Access*, 2008.
48. T. Clancy and W. Arbaugh, "Iterative Techniques for Interference Temperature Measurement," in *Proc. Virginia Tech MPRG Symposium on Wireless Personal Communications*, Blacksburg, VA, USA, Jun. 2006.
49. Y. Xing and R. Chandramouli, "QoS Constrained Secondary Spectrum Sharing," in *Proc. IEEE DySPAN 2005*, Baltimore, MD, USA, Nov. 2005, pp. 658-661.
50. J. G. Andrews, A. Ghosh, and R. Muhamed, "Fundamentals of WiMAX: Understanding Broadband Wireless Networking," *Prentice Hall Communications Engineering and Emerging Technologies Series*, 1st ed, Feb. 2007.
51. C. R. Lin and J. S. Liu, "QoS Routing in Ad Hoc Wireless Networks," *IEEE J. Select. Areas Commun.*, vol. 17, no. 8, pp. 1426-1438, Aug. 1999.
52. C. Zhu and M. S. Corson, "QoS Routing for Mobile Ad hoc Networks," in *Proc. IEEE INFOCOM 2002*, New York, NY, USA, Jun. 2002, pp. 958-967.
53. K. Dimou *et al.*, "Generic Link Layer: A Solution for Multi-Radio Transmission Diversity in Communication Networks Beyond 3G," in *Proc. IEEE VTC Fall 2005*, Dallas, USA, Sep. 2005, pp. 1672-1676.
54. A. P. Hulbert, "Spectrum Sharing Through Beacons," in *Proc. IEEE PIMRC 2005*, Berlin, Germany, Sep. 2005, pp. 989-993.
55. M. Gandetto *et al.*, "A Distributed Approach to Mode Identification and Spectrum Monitoring for Cognitive Radios," *SDR Tech. Conf. & Product Expansion*, San Diego, CA, USA, Nov. 2005.
56. B. Krenik and C. Panasik, "Cognitive Radios for Unlicensed WANS," in *Proc. Berkeley Wireless Research Center Cognitive Radio Wksp.*, Nov. 2004.

57. O. Holland *et al.*, "A Universal Resource Awareness Channel for Cognitive Radio," in *Proc. IEEE PIMRC 2006*, Helsinki, Finland, Sep. 2006, pp. 1-5.
58. T. Weiss *et al.*, "Efficient Signaling of Spectral Resources in Spectrum Pooling Systems," in *Proc. 10th Symposium on Communications and Vehicular Technology (SCVT 2003)*, Eindhoven, Netherlands, 2003.
59. T. Weiss *et al.*, "Synchronization Algorithms and Preamble Concepts for Spectrum Pooling Systems," in *Proc. IST Mobile and Wireless Communications Summit*, Aveiro, Portugal, Jun. 2003, pp. 788-792.
60. I. F. Akyildiz, W. Lee, M. C. Vuran, and S. Mohanty, "NeXt generation/dynamic spectrum access/cognitive radio wireless net-works: a survey," *Computer Networks*, vol. 50, no. 13, pp. 2127-2159, Sep. 2006.
61. Q. Wang and H. Zheng, "Route and Spectrum Selection in Dynamic Spectrum Networks," in *Proc. IEEE CCNC 2006*, Las Vegas, NV, USA, Jan. 2006.
62. C. Xin, "A Novel Layered Graph Model for Topology Formation and Routing in Dynamic Spectrum Access Networks," in *Proc. IEEE DySPAN 2005*, Baltimore, MD, USA, Nov. 2005.
63. P. Mahonen, M. Petrova, and J. Riihijarvi, "Applications of Topology Information for Cognitive Radios and Networks," in *Proc. IEEE DySpan 2007*, Dublin, Ireland, Apr. 2007, pp. 103-114.
64. J. Zhao, H. Zheng, and G. Yang, "Spectrum Sharing Through Distributed Coordination in Dynamic Spectrum Access Networks," *Wiley Wireless Communication and Mobile Computing Journal*, vol. 7, no. 9, pp.1061-1075, Nov. 2007.
65. J. Salo *et al.*, "Context Provisioning in the WWI System Architecture," in *Proc. WWRF 17*, Nov. 2006, Heidelberg, Germany.
66. S. Hrishikesh and P. Balamurali, "A Context Interpretation Architecture for Cognitive Network Devices," *SDR Forum Technical Conference*, Denver, CO, USA, Dec. 2007.
67. Unified Link Layer API (ULLA) Specification, www.ist-gollum.org/deliverables.html
68. P. Karn *et al.*, "Advice for Internet Subnetwork Designers," IETF RFC 3168, 2004.
69. H. Inamura *et al.*, "TCP over Second (2.5G) and Third (3G) Generation Wireless Networks," IETF RFC 3481, 2003.
70. M. Sooriyabandara and G. Fairhurst, "Dynamics of TCP over BoD Satellite Networks," *Int'l Journal of Satellite Communications and Networking*, vol. 21, no. 5, pp. 427-449, 2003.
71. J. Singh *et al.*, "TCP Dynamics and Link-Layer Adaptation Based Optimization Methods for Wireless Networks," *IEEE Trans. on Wireless Commun.*, vol. 6, no. 5, pp. 1864-1879, May 2007.
72. M. Sooriyabandara *et al.*, "Unified Link Layer API: A Generic and Open API to Manage Wireless Media Access," *Journal on Computer Communications*, Elsevier Science, to appear, 2008.
73. Y. Tian, K. Xu, and N. Ansari, "TCP in Wireless Environments: Problems and Solutions," *IEEE Commun. Mag.*, vol. 43, no. 3, pp. S27-S32, Mar. 2005.
74. C. Casetti *et al.*, "TCP Westwood: Bandwidth Estimation for Enhanced Transport over Wireless Links," in *Proc. ACM Mobicom*, Jul. 2001, pp. 287-297.
75. J. Border, M. Kojo, J. Griner, G. Montenegro, and Z. Shelby, "Performance Enhancing Proxies Intended to Mitigate Link-Related Degradations," RFC 3135, Jun. 2001.
76. H. Balakrishnan and R. H. Katz, "Explicit Loss Notification and Wireless Web Performance," in *Proc. IEEE GLOBECOM 1998*, Sydney, Australia, Nov. 1998.
77. X. Lin, N. Shroff, and R. Srikant, "A Tutorial on Cross-Layer Optimization in Wireless Networks," *IEEE J. Select. Areas Commun.*, vol. 24, no. 8, pp. 1452-1463, August 2006.

78. Q. Zhao and B. M. Sadler, "A Survey of Dynamic Spectrum Access: Signal processing, networking and regulatory policy," *IEEE Signal Process. Mag.*, vol. 24, no. 5, pp. 79-89, May 2007.

79. B. Bougard *et al.*, "Transport Level Performance-energy Trade-off in Wireless Networks and Consequences on the System-level Architecture and Design Paradigm," in *Proc. IEEE Wksp. on Signal Processing Systems (SIPS 2004)*, Austin, Texas, USA, Oct. 2004, pp.77-82.

80. V. Paxson and S. Floyd, "Wide-Area Traffic: The Failure of Poisson Modeling," *ACM/IEEE Trans. Netw.*, vol. 3, no. 3, pp. 226-244, Jun. 1995.

81. T. Farnham, M. Sooriyabandara, and C. Efthymiou, "Terminal Centric Optimisation for the Wireless Local Area," in *Proc. IEEE PIMRC 2007*, Athens, Greece, Sep. 2007, pp. 1-5.

82. T. Farnham, M. Sooriyabandara, and C. Efthymiou, "Enhancing Multimedia Streaming Over Existing Wireless LAN Technology Using the Unified Link Layer API," *Wiley Int'l. Journal of Network Management*, vol. 17, no. 5, pp. 331-346, Aug. 2007.

83. Q. Wei *et al.*, "A Hierarchical General Purpose Optimisation Framework for Wireless Networks," in *Proc. 11th European Wireless Conference*, Nicosia, Cyprus, Apr. 2005.

84. T. Farnham and S. Zhong, "Dynamic Optimisation of Reconfigurable Equipment," in *Proc. IST Mobile Communications Summit*, Mykonos, Greece, May 2006.

5

Radio Resource Management for Heterogeneous Wireless Access Networks

Jordi Pérez-Romero, Xavier Gelabert, and Oriol Sallent

Universitat Politècnica de Catalunya (UPC), Barcelona, Spain
{jorperez, xavier.gelabert, sallent}@tsc.upc.edu

5.1 Introduction

Technological advances and market developments in the wireless communications area have been astonishing during the last decade; they were mainly driven by the successful deployment of GSM (Global System for Mobile communications) networks from the European perspective. The extension of GSM to GPRS (General Packet Radio Service) including packet transmission capabilities in the radio interface was a first milestone (commonly named 2.5G) in the evolution path from Second Generation (2G) cellular systems towards the Third Generation (3G), with UMTS (Universal Mobile Telecommunications System) being one of its most significant representatives. Although GPRS penetration probably was not as successful as could be originally expected, its compatibility with the popular GSM technology together with the fact that initial UMTS releases make use of the same GSM/GPRS core network infrastructure, provide the basis to believe that the co-existence and interaction between UMTS and GSM/GPRS technologies constitutes one of the key points for the success of 3G technologies. In addition to this, wireless technologies are rapidly evolving in order to allow operators to deliver more advanced multimedia services. In this sense, GPRS evolved towards EDGE (Enhanced Data rates for GSM Evolution) providing higher bit rates comparable to those of UMTS. As for UMTS, HSPA (High Speed Packet Access) for uplink and downlink is seen as intermediate evolutionary step since the first wave of UMTS networks rollout, while E-UTRAN (Evolved UMTS Terrestrial Radio Access Network) is the long term perspective for the 3GPP (Third Generation Partnership Project) technology family. Similar paths are drawn from the 3GPP2 around the evolution of CDMA2000. Moreover, the IEEE 802 is in turn producing an evolving family of wireless standards, such as 802.11 local, 802.15 personal, 802.16 and 802.20 metropolitan, and 802.22 regional area networks, some of which have already experienced a widespread use and deployment.

As a result of the above evolutions, leading to different standards coexisting in time, it is generally acknowledged today that beyond 3G encompasses network heterogeneity. A plethora of different network topologies will have to co-exist or be inter-connected. These network topologies ought to be inter-connected in an opti-

E. Hossain (ed.), *Heterogeneous Wireless Access Networks*,
DOI: 10.1007/978-0-387-09777-0_5, © Springer Science+Business Media, LLC 2008

mum manner with the ultimate objective to provide the end-user with the requested services and corresponding QoS (Quality of Service) requirements. The heterogeneous network concept for beyond 3G systems is intended to propose a flexible and open architecture for a large variety of different wireless access technologies, applications and services with different QoS demands, as well as different protocol stacks. Fig. 5.1 shows an example of such heterogeneous networks scenario. It is constituted by several Radio Access Networks (RANs) interfacing a common core network. Radio access networks include cellular networks, e.g. UTRAN R99, UTRAN-HSPA, GERAN (GSM/EDGE Radio Access Network), etc., which may in turn be subdivided into different cellular layers (e.g. macro, micro or picocells) depending on the expected coverage area, and also other wireless local area (e.g. WLAN) and wireless metropolitan area (e.g. WiMAX) networks. The core network infrastructure is typically subdivided into the circuit switched (CS) and packet switched (PS) domains providing access to external networks, e.g. PSTN or Internet. The scenario assumes the existence of multi-mode terminals, possibly with reconfigurability capabilities, which can be connected to multiple access networks either in different time instants or even simultaneously.

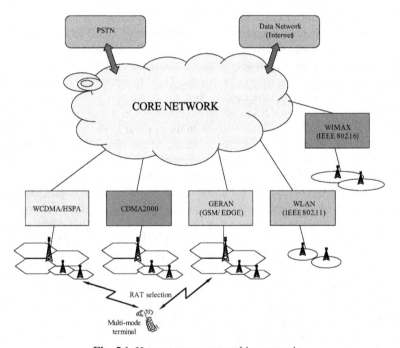

Fig. 5.1. Heterogeneous networking scenario.

Rather than being an inconvenience, these new heterogeneous scenarios must indeed be regarded as a new challenge to offer to the users an efficient and ubiquitous radio access, by means of a coordinate use of the available Radio Access Technologies (RATs). In this way, not only the user can be served through the RAT that fits better to the terminal capabilities and service requirements, leading to the 'connected everywhere, anytime, anyhow' experience, but also a more efficient use of the available radio resources can be achieved from the operator's point of view. In this way, the heterogeneous network becomes transparent to the final user and the so-called ABC (Always Best Connected) paradigm [1], which claims for the connection to the RAT that offers the most efficient radio access at each instant, can be achieved.

The realization of the ABC concept in a heterogeneous scenario where several RATs coexist calls for the introduction of new radio resource management (RRM) strategies operating from a common perspective that takes into account the overall amount of resources in the available RATs, and therefore are referred to as CRRM (Common Radio Resource Management) algorithms . In this scenario the provision of radio resources can be seen as a problem with multiple dimensions, since every RAT is based on specific multiple access mechanisms exploiting in turn different orthogonal dimensions, such as frequency, time and code. Then, local RRM mechanisms are needed for every considered RAT, as shown in Fig. 5.2(a). In addition to that, CRRM is based on the picture of a pool of radio resources, belonging to different RATs but managed in a coordinated way, as shown in Fig. 5.2(b). Then, the additional dimensions introduced by the multiplicity of available RATs provide further flexibility in the way how radio resources can be managed and, consequently, overall improvements may follow thanks to the resulting trunking gain [2]. Notice that the CRRM vision allows also considering different amounts of radio resources spatially available, because in terms of current network deployment, different spatial availabilities are found for the existing RATs. For example, GERAN tends to be the most widespread RAT, while UTRAN is not yet deployed everywhere with full capabilities and in turn WLAN hotspots with reduced coverage areas are also widespread around cities.

The achievement of these goals requires not only the definition of appropriate CRRM algorithms but it also has implications that affect the network architectures in terms of interworking and coupling between RANs depending on how they are interconnected to the core network. These new network architectures must face requirements like mobility management for seamless handover, including vertical handover between RATs, as well as authentication and billing mechanisms. Furthermore, convergence between radio access networks is necessary, which requires a standardization effort and business commitment to support it.

5.2 Interworking and Coupling among RANs

The provision of heterogeneous network topologies is conceptually a very attractive notion; however, it is certainly a challenge to the network designer. For a proper support of CRRM algorithms, the currently existing network architectures must be modified accordingly to ensure the desired interworking capabilities among the different technologies [3, 4]. Here, coupling between the networks of possibly different characteristics can be provided. The stronger the coupling the better resources can

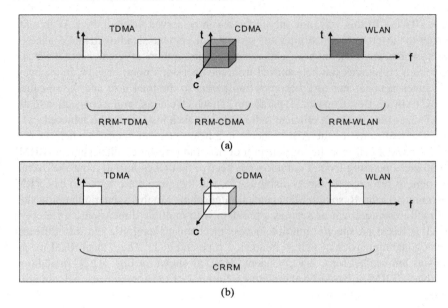

(a)

(b)

Fig. 5.2. (a) RRM at single RAT level managing orthogonal multiple access dimensions and (b) CRRM managing a pool of orthogonal multiple access dimensions.

potentially be utilized leading to an optimum of performance. However, this comes along with an increased effort in the definition and implementation of required interfaces. A suitable trade-off for specific systems thus ought to be determined.

5.2.1 3GPP CRRM Functional Model

The functional model assumed in 3GPP for CRRM operation considers the total amount of resources available for an operator divided into radio resource pools. Each radio resource pool consists in the resources available in a set of cells, typically under the control of a Radio Network Controller (RNC) or a Base Station Controller (BSC) in UTRAN and GERAN, respectively. Two types of entities are considered for the management of these radio resource pools [4], as shown in Fig. 5.3:

- The RRM entity, which carries out the management of the resources in one radio resource pool of a certain radio access network. This functional entity involves different physical entities depending on the specific considered functions, although for representation purposes it is usual to assume the RRM entity residing in the RNC or the BSC.
- The CRRM entity, which executes the coordinated management of the resource pools controlled by different RRM entities, ensuring that the decisions of these RRM entities also take into account the resource availability in other RRM entities. Each CRRM entity controls several RRM entities and may communicate with other CRRM entities as well, thus collecting information about other RRM entities that are not under its direct control.

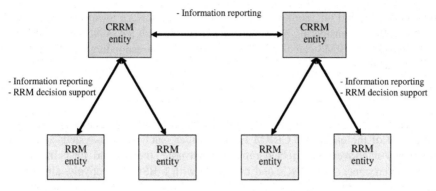

Fig. 5.3. CRRM functional model.

The interactions among RRM and CRRM entities mainly involve two types of functions:

1. *Information reporting function*: The information reporting function allows the RRM entity to indicate to its controlling CRRM entity either static or dynamic information like the cell capacity or different types of measurements. The reporting is controlled by the CRRM entity, which can request it either at given instants or according to periodical or event-triggered reports. The exchange of information is also possible among different CRRM entities in order to know the status of their corresponding RRM entities.

2. *RRM decision support function*: This function describes the way how the CRRM entity affects the decisions taken by the RRM entities under its control. Depending on how the CRRM is implemented in the network, there exist several possibilities for the RRM decision support function. For example, it is possible that the CRRM simply advises the RRM entity, so that the RRM remains as the master of the decisions, and, on the contrary, it is also possible that the CRRM is the master so that its decisions are binding for the RRM entity. Similarly, there exist several degrees of coupling or interaction between the CRRM and the RRM entities, ranging from the case in which the CRRM is involved in any RRM decision (e.g. in every intersystem handover) to the case in which the CRRM simply dictates policies for RRM operation and the RRM entity takes decisions according to these specific policies.

With respect to the network topologies to support the previous CRRM functional model, there exist two different approaches, namely the CRRM server and the integrated CRRM solutions, which impact the way the CRRM functions are realized in practice, as explained in the following.

CRRM server

This approach introduces the CRRM functionality in a stand-alone node, denoted as CRRM Server (CRMS), and common to all RANs, thus constituting a centralized approach, as depicted in Fig. 5.4. RRM and CRRM entities are then located in different physical nodes and interconnected through an open interface towards the RNC and the BSC. When considering heterogeneous networks with RATs other than GERAN or UTRAN, the CRMS can also be connected to the corresponding RRM entities in these networks.

Integrated CRRM

This approach is based on the fact that the current 3GPP standards already support most of the envisaged CRRM functionalities, such as intra-system and inter-frequency handovers or directed retry. Furthermore, the Iur and Iur-g interfaces, which interconnect two RNCs and one RNC with a BSC respectively, already include almost all the necessary functions to support the CRRM procedures. Because of that, a natural approach consists in integrating the CRRM functionality in the existing UTRAN and GERAN nodes, leading to a distributed CRRM architecture as depicted in Fig. 5.5. Notice that the CRRM entity may be included either in all the RNC/BSCs or only in a subset of them. In the first case, only the reporting information function between different CRRM entities must be standardized, because the RRM decision support function between CRRM and RRM is done locally at the same physical entity. On the contrary, if only a sub-set of RNC/BSCs include the CRRM entity, the RRM decision support function must also be standardized [4].

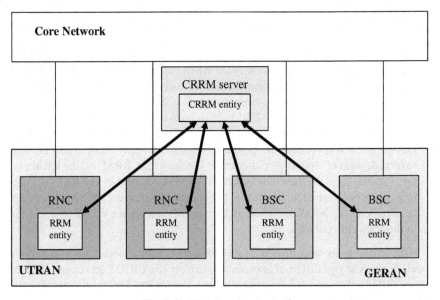

Fig. 5.4. CRRM server approach.

Different procedures can be used in the integrated CRRM approach to exchange the common measurements and static information on cells controlled by the distant entities (RNC or BSC), depending on the considered interface. The options provided

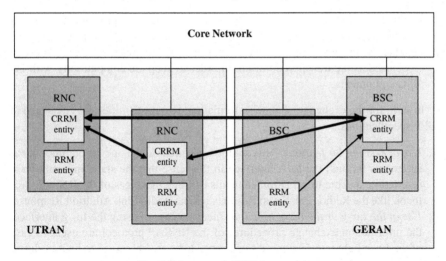

Fig. 5.5. Integrated CRRM approach.

by 3GPP recommendations for the exchange of common measurements are the following ones:

1. *Common measurements over Iur-g*: This is the adopted solution when the BSC supports an Iur-g interface to interwork with the RNC. In this case, the measurement of common resources functionality of the RNSAP (Radio Network Subsystem Application Part) protocol is used (see [5] for details).

2. *Common measurements over Iu/A interfaces*: This solution is used when the BSC does not support the Iur-g interface. In this case, the communication between the BSC and the RNC is done through the MSC of the core network, by using the A interface between BSC and MSC together with the Iu interface between the MSC and the RNC. Measurements are transmitted through the A interface in the messages of the BSSMAP (Base Station Subsystem Management Application Part) protocol handover procedures, inserted in some transparent information elements. Similarly, through the Iu interface, measurements are inserted in messages of the RANAP (Radio Access Network Application Part) relocation procedure. Consequently, the availability of common measurements is limited by the periodicity of the above handover and relocation procedures in the network. For example, during high traffic situations, there are more frequent inter-RAT handovers thus having more frequent measurements and more reliable information. In any case, this solution is not as efficient as the use of the Iur-g interface.

3. *Common measurements over a Iur-g light interface*: This solution is also used when the BSC does not support the Iur-g interface. It consists in the introduction of a new interface between the BSC and the RNC specifically designed to support the exchange of common measurements and information on cells controlled by distant entities. In order to ensure the compatibility with the Iur-g interface, it is proposed that the Iur-g light interface is based on a RNSAP based protocol

denoted as IRNSAP (Inter Radio Network Subsystem Application Part) using the measurement of common resources functionality (see [5]). Furthermore, the IP (Internet Protocol) stack defined in release 5 for the Iur interface is proposed to be used as the transport solution, in order to keep the compatibility with 2G BSC equipments.

In turn, for the exchange of static information, two possibilities are envisaged in 3GPP specifications:

1. *Use of existing Iur features*: This solution is similar to the one for common measurements through the Iu/A interface in the sense that the static information is inserted in specific information elements of some messages of the RNSAP protocol, like the Radio Link Setup Response or the Radio Link Addition Response.
2. *Use of the Iur-g or the Iur-g light interfaces*: In case of using the Iur-g interface, the information exchange procedures of the RNSAP protocol are used. In case of the Iur-g light interface these same procedures are envisaged to be introduced in the new IRNSAP protocol.

5.2.2 Interworking between 3GPP and WLAN

The interworking between 3GPP networks (e.g. UTRAN) and the IEEE family of standards (e.g. WLAN) is devised from a different perspective because WLANs are being mainly deployed by parties not belonging to 3GPP, and consequently they do not follow the same networking architectures. Nevertheless, the successful deployment of WLAN systems worldwide and the high data rates offered by such systems make them an interesting alternative to increase wireless coverage of cellular systems in hotspots locations. Consequently, there exist efforts in the standardization bodies addressing the integration of 3GPP and WLAN technologies. Such interworking should take into account both technical and non-technical aspects, because the environments where both systems co-exist (i.e. public, corporate or residential environments) may involve different administrative domains and different WLAN owners, thus leading to e.g. different security, billing or authentication requirements. In that respect, in [6] some initial scenarios were identified in 3GPP for WLAN interworking with different degrees of requirements. From those initial works, current specifications consider mainly two different approaches, as detailed in the following:

Interworking WLAN (I-WLAN)

This approach is defined in [7], where the 3GPP WLAN subsystem is assumed to provide bearer services for connecting a 3GPP subscriber via WLAN to IP based services compatible with those offered via the PS domain of the core network. This is enabled with the introduction of three new functional elements in the UMTS PS core network: the 3G AAA (Access, Authentication and Authorization) server, the WLAN Access Gateway (WAG), and the Packet Data Gateway (PDG), as illustrated in Fig. 5.6. The WLAN must also support similar interworking functionality to meet the access control and routing enforcement requirements. The 3G AAA server in the

UMTS domain terminates all AAA signaling originated in the WLAN corresponding to UMTS roamers. This signaling is securely transferred across the Wa interface, which is typically based on Radius or Diameter. The 3G AAA server interfaces with other 3G components, such as the WAG, PDG, and Home Subscriber Server (HSS), which stores information defining the subscription profiles of 3G subscribers. In turn, the data is transferred from the WLAN to the WAG through the Wn interface, while the connection from the PDG to the external data network is done through the Wi interface. The interconnection between the UE (User Equipment) and the WLAN Access Network is outside the scope of 3GPP specifications.

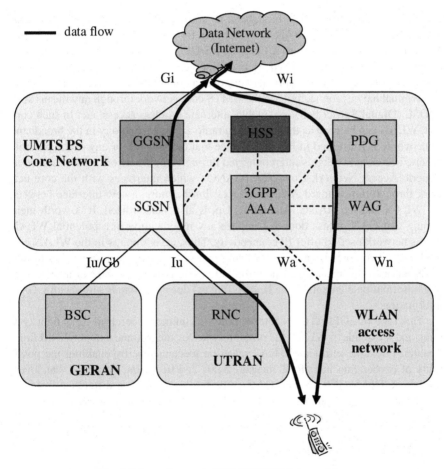

Fig. 5.6. Loosely coupled WLAN/3GPP interworking.

This kind of interworking is commonly referred to as loose coupling [8], in which the 3GPP and WLAN networks are connected at the Gi interface, but the user data flow does not go through the GGSN or SGSN core network elements. As a result, the WLAN UE does not deal with Non Access Stratum (NAS) protocols in the PS domain (e.g. Packet Data Protocol signaling). Instead, plain IP connectivity services are offered either through the WLAN network directly or through the PS Domain. Then, legacy AAA mechanisms deployed in the WLAN are connected to those of the cellular network. It is worth mentioning that I-WLAN solution does not specify any support for RRM within the WLAN Access Network and, in the same way, no support for CRRM is included in terms of inter-RAT system information, inter-RAT UE measurements and inter-RAT mobility management.

Generic Access Network (GAN)

A generic access network is a broadband IP network providing access to A/Gb interfaces of the GERAN/UTRAN Core Network for CS/PS services, respectively [9]. Thus, it is really an extension of GSM/GPRS/UMTS mobile services into the customers premises that is achieved by tunneling certain NAS protocols between the customer's premises and the core network over a broadband IP network. This allows the terminal having connectivity to the UMTS core network through any means such as DSL (Digital Subscriber Line), cable, alternate wireless access, etc. In such context, WLAN can be used as the underlying radio access technology in the broadband IP network, as illustrated in Fig. 5.7. Notice that the WLAN (or any of the possible generic access networks) is interconnected through a new network node, denoted as Generic Access Network Controller (GANC), which interfaces with the core network through the standard A/Gb interfaces. Furthermore, a new interface between the WLAN and the GANC, denoted as Up, is also standardized. It is worth mentioning that GAN access does not impose any modification in a potential WLAN access network used to offer IP connectivity. Thus, access points in the WLAN network can be completely unaware of being part of the infrastructure used to offer a GAN service. This means that the information sent through the 802.11 air interface (e.g. information elements in the beacon frames) does not contain any specific GAN information.

This type of 3GPP/WLAN interworking is commonly referred to as tight coupling and it extends to WLAN the possibility to execute to some extent CRRM functionalities dealing with simple RAT selection mechanisms, by enabling the possibility of exchanging inter-RAT measurements and triggering inter-RAT handovers between UTRAN/GERAN and GAN cells. Furthermore, as a difference from the loose coupling approach, access to both CS and PS domains can be provided.

5.3 RRM/CRRM Functionalities

The main RRM functions arising in the context of a single RAN are [10]: admission and congestion control, horizontal (intra-system) handover, packet scheduling and power control. When these functionalities are coordinated between different RANs in a heterogeneous scenario, they can be denoted as "common" (i.e. thus having common admission control, common congestion control, etc.) as long as algorithms

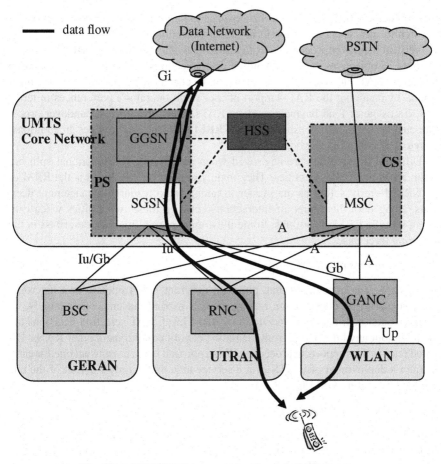

Fig. 5.7. 3GPP/WLAN interworking through GAN function.

take into account information about several RANs to make decisions. In turn, when a heterogeneous scenario is considered, a specific functionality arises, namely the RAT selection (i.e. the functionality devoted to decide to which RAT a given service request should be allocated).

After the initial RAT selection decision, taken at session initiation, vertical (inter-system) handover is the procedure that allows switching from one RAT to another. The successful execution of a seamless and fast vertical handover is essential for hiding to the user the underlying service enabling infrastructure. Issues related to vertical handover comprise scanning procedures for the terminal to discover available RATs, measurement mechanisms to capture the status of the air interface in the different RATs, vertical handover triggers (i.e. the events occurring in the heterogeneous network scenario that require the system to consider whether a vertical handover is actually required or not), vertical handover algorithm (i.e. the criteria used to decide whether a vertical handover is to be performed or not) and protocol and architectural aspects to support handover execution.

Vertical handover procedures from one RAT to another may be useful to support a variety of objectives, such as avoiding disconnections due to lack of coverage in the current RAT, blocking due to overload in the current RAT, possible improvement of QoS by changing the RAT, support of user's and operator's preferences in terms of RATs usage or load balancing among RATs. Thus, the vertical handover procedure enables another dimension into the CRRM problem and provides an additional degree of freedom in rearranging traffic.

Different possibilities are envisaged when considering the functional split between RRM and CRRM entities. They mainly depend on whether it is the RRM or the CRRM entity acting as the master in taking radio resource management decisions, along with the degree of interactions between these two entities when executing specific RRM algorithms. Some illustrative possibilities are described in the following three cases.

No functionalities at CRRM entity

In this case, it is considered that, although different RATs operate in a heterogeneous scenario, no coordination among them is carried out and, consequently, no specific functionalities are associated to CRRM level. RAT selection decisions are taken without any knowledge from the radio network conditions in other RANs. Directed retry is the supporting procedure. This approach is the most usual one for early operator's deployment of UMTS over a service area that is only a subset of the one provided by mature GSM/GPRS network.

Long-term functionalities at CRRM entity

In this situation shown in Fig. 5.8, the RAT selection procedures are associated with the CRRM entity. The local RRM entities provide measurements including the list of candidate cells for the different RATs and cell load measurements, so that the CRRM can take all these inputs into account for the corresponding decision.

This approach is the first step towards the exploitation of improved efficiency in heterogeneous RANs scenarios. We note that, comparatively, RAT selection is the functionality operating at the longest time scale. In what initial RAT selection concerns, it is in the order of call/session set-up interarrival processes. For vertical handover (on-the-fly RAT re-selection), it is related to the transit time elapsed within the same RAN. This approach would, comparatively, offer the fewest constraints in terms of RRM/CRRM interactions regarding measurement reporting, decision support and execution.

A further step in this direction would consist in moving congestion control functions, which also operate on a rather long-term basis, from RRM entities to CRRM entity. Then, the Common Congestion Control algorithm, would take advantage of RAT diversity to cope with congestion problems in a specific RAT, then providing higher degrees of freedom for the congestion resolution actions. Clearly, vertical handover would be the supporting procedure to alleviate congestion in the affected RAT.

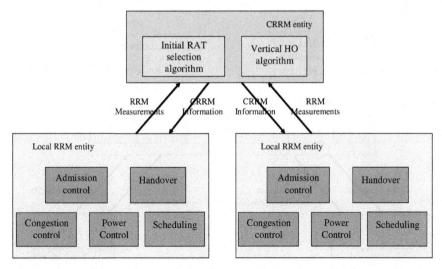

Fig. 5.8. CRRM/RRM functional split possibilities in the case of CRRM having only long-term functionalities.

Long and short-term functionalities at CRRM entity

This approach provides the highest degree of interaction between CRRM and local RRM by executing also common fast packet assignment algorithms in the CRRM entity, as shown in Fig. 5.9. The local RRM functionality would remain at a minimum, limited to the transfer of the adequate messages to CRRM and some specific technology dependent procedures that occur in very short periods of time (e.g. inner loop power control in case of UTRAN, which occurs with periods below 1 ms). This solution would require for CRRM decisions to be taken at a very short time scale (in the order of milliseconds). Clearly, future envisaged multi-homing capabilities supporting more than one simultaneous connection with different RATs would favour shifting these long-term functionalities to CRRM entity.

5.4 RAT Selection

RAT selection becomes a key CRRM issue to exploit the flexibility resulting from the joint consideration of the heterogeneous characteristics offered by the available radio access networks. This RAT selection can be carried out considering different criteria (such as, service type, load conditions, etc.) with the final purpose of enhancing overall capacity, resource utilization and service quality. The scenario of heterogeneity is also present from the customer side, because users may access the requested services with a variety of terminal capabilities (e.g. single or multi-mode) and different market segments can be identified (e.g. business or consumer users) with their corresponding QoS levels. Then, selecting the proper RAT and cell is a complex problem due to the number of variables involved in the decision-making

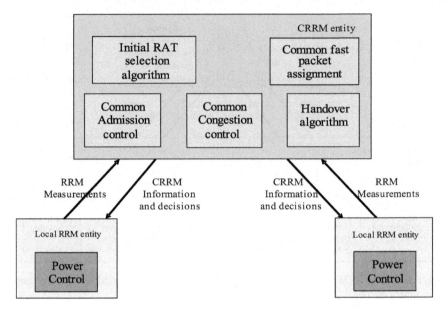

Fig. 5.9. CRRM/RRM functional split possibilities in the case of CRRM having both long and short-term functionalities.

process, as reflected in Fig. 5.10 with some possible inputs. Furthermore, some of these variables may vary dynamically, which makes the process even more difficult to handle.

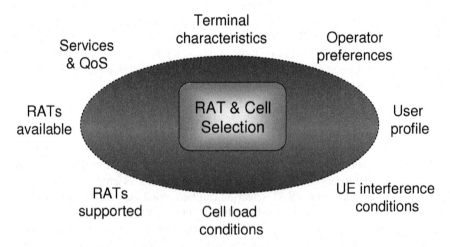

Fig. 5.10. Factors influencing the RAT and cell selection.

In this context, CRRM in general and RAT selection mechanisms in particular have received a lot of attention in recent years, clearly acknowledging the key role that these strategies will have for a full realization of Beyond 3G (B3G) scenarios. Research efforts have been oriented either to propose and assess the performance of heuristic algorithms [11] [12] [13] [14] [15] [16] [17] [18] or to identify architectural and functional aspects for CRRM support [3] [4] [19] [20]. From an algorithmic point of view, in [13] and [14], mechanisms to balance the load in different RATs by means of vertical handover decisions were analyzed. However, the service-dimension is not captured in the problem because only real-time services are considered. In turn, in [15] the authors compared the load balancing principles with respect to service-based CRRM policies, as described in [11]. Similarly, the CRRM problem was discussed from a more general perspective in [16] [17] and references therein, comparing several substitution policies and including the multi-mode terminal dimension with speech and data services. Finally, in [18] the authors proposed a RAT allocation methodology that considers the specific radio network features of a CDMA network to reduce the interference by allocating users to RATs depending on the total measured path loss and capturing also the service dimension but considering that all terminals support the available RATs and that the involved RATs support the same services.

The abovementioned studies usually approached the problem of RAT selection from a system-level simulation point of view. The analytical approach to the RAT selection problem, however, has received less attention in the literature. Fewer analytical proposals have been developed up to date, among those e.g. [21] [22] [23]. In [21], Lincke et al. proposed an analytical approach to the problem of traffic overflowing between several RATs using an M-dimensional Markov model. Nonetheless, in order to derive a closed form solution by means of applying independence between service types, Markov states in this model indicate the number of sessions of each service that are being carried in whole composite network, but not on which RAT each session is being carried out. In [22] a flexible framework for evaluating generic RAT selection policies in CDMA/TDMA scenarios with two different services was built by using a 4-dimensional Markov model. Given a total offered traffic to the network, the fractional traffic arriving to each RAT was dependant on the chosen RAT selection scheme which is fully embedded in the state transitions of the Markov chain. In this way, the model allows the evaluation of different RAT selection schemes accounting for different principles, ranging from the simplest ones like service-based selection or load balancing, up to more sophisticated schemes accounting for the amount of multi-mode terminals in the scenario or trying to minimize the resulting congestion probability in each RAT.

Clearly, an advanced RAT selection algorithm may integrate several of the above principles. Then, it is very important to devise the role of the above principles, which of them has a predominant effect on the overall performance, which are the aspects related to the scenario, services, service mix and RATs characteristics influencing the relative importance, etc. After this, and even though a single optimum algorithm can not be claimed to exist for the wide variety of operation conditions in practice, a robust and advanced algorithm showing very promising performance in many representative situations can be firmly stated. In the following sub-sections, some basic RAT selection solutions are presented aiming at identifying the different role played by each component in the decision. For that purpose, a first classification of RAT selection schemes according to service principles, load balancing and interference aspects is considered.

5.4.1 Service-Based RAT Selection

A service-based RAT selection policy is based on a direct mapping between services and RATs [11]. As an example, in a scenario including voice and interactive service assuming all terminals have multi-mode capabilities (i.e. they can work either with UTRAN or GERAN), two possibilities would be:

- VG (voice GERAN) policy: This policy allocates voice users into GERAN and interactive services into UTRAN.
- VU (voice UTRAN) policy: This policy allocates voice users into UTRAN and interactive services into GERAN.

If no capacity is available in the primary RAT, the other RAT is selected instead. If no capacity is available in the alternative RAT, the service request gets blocked (at service set-up) or dropped (during service life-time).

Table 5.1 compares the performance in terms of aggregated throughput (i.e. including both voice and www users) when basic policies VU and VG are considered in a scenario with seven omnidirectional cells with radius 1 km for UTRAN and GERAN assuming that the cells of both systems are co-sited. Notice that, in all the cases, VG policy outperforms VU, revealing the suitability of allocating voice users in GERAN. The main reasons are two-fold. First, with respect to www users, a higher throughput can be obtained in UTRAN as long as dedicated channels are used while in GERAN www users use shared channels and therefore they are subject to a scheduling algorithm. In turn, from the voice users' point of view, no significant differences between VU and VG are observed provided that the cell radius is small. However, when increasing the radius, a higher degradation is observed in VU because UTRAN users at the cell edge experience some erroneous transmissions due to power limitations and the interference-limited nature of WCDMA.

Table 5.1. Total throughput (Mb/s) for VU and VG policies.

Users		VU		VG	
Voice	www	UL	DL	UL	DL
	200	2.08	2.17	2.14	2.22
400	600	2.88	3.09	2.95	3.15
	1000	3.64	3.96	3.76	4.08

5.4.2 Load Balancing-Based RAT Selection

Load balancing (LB) is another possible guiding principle for resource allocation in which the RAT selection policy will distribute the load among all resources as evenly as possible. Specifically, the selected RAT will be the one having the lowest load. Therefore, an influential run-time parameter in a load balancing decision-making procedure is the load metric. For UTRAN, an average of the cell load factor can be used, while in GERAN a useful way to measure the data load is to measure the average amount of time slots utilized by GSM/EDGE services.

Table 5.2 compares the LB RAT selection algorithm against the service-based VG policy in a scenario with seven omnidirectional cells for UTRAN and GERAN co-sited with cell radius 500m. Particularly, it shows the voice call dropping probabilities (in %) for both algorithms. Up to 600 voice users, dropping values are kept sufficiently low. For 800 voice users however, VG results in higher dropping values than policy LB. The higher dropping rates experienced by VG policy is explained bearing in mind the load distribution in GERAN induced by VG and LB policies (see Fig. 5.11). In particular, for VG the load is at its maximum value most of the time which implies a lack of flexibility in order to accommodate handover users being redirected to GERAN. Therefore, VG may incur in more potential dropping situations than in the case of LB policy appliance which presents more fluctuations in the load values and can provide resources to incoming handover users if necessary.

Table 5.2. Dropping probability (%) for service-based and load balancing policies.

Users		VG	LB
www	Voice		
	200	0.0	0.0
600	400	0.0	0.0
	600	0.051	0.0
	800	3.343	0.0

5.4.3 Interference-Based RAT Selection

This category of RAT selection mechanisms take into account the different resource consumption that a given user may have in one or other RAT depending on the specific network characteristics. One example is the so-called Network Controlled Cell Breathing (NCCB) algorithm [18], developed for CDMA/TDMA heterogeneous scenarios taking into consideration that CDMA technology is more sensitive to interference than TDMA technology. The concept is illustrated in Fig. 5.12 for a situation where CDMA and TDMA cells are co-sited (although the algorithm is also applicable in other situations). For a given service, the TDMA cells with radius R_T ensure coverage in the whole area. In turn, by an appropriate control of the effective cell radius in CDMA cells R_C (e.g. in the figure by changing from R_{C2} to R_{C1}) through

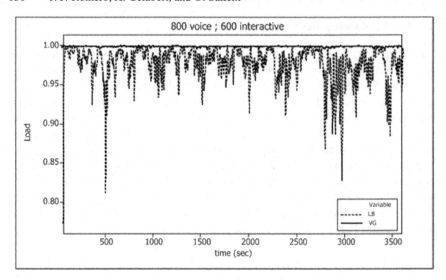

Fig. 5.11. Load evolution in GERAN considering service-based and load balancing policies.

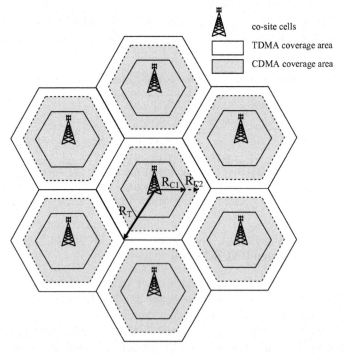

Fig. 5.12. Network controlled cell-breathing when CDMA and TDMA cells are co-sited.

RAT selection strategies, the required transmitted power levels and the inter-cell interference will be reduced, thus improving the capacity for the considered service in CDMA. In practice, due to the shadowing effects, the cell radius is controlled by setting the maximum propagation loss that can be allowed for a given RAT. Taking this into account, this strategy allocates users to RATs according to their propagation losses. Then, users with low propagation loss will be allocated to the CDMA cells and users with high propagation loss will be allocated to the TDMA cells. In this way, by setting a suitable maximum path loss threshold, the intercell interference in CDMA is reduced with the corresponding capacity increase. An illustrative result of the performance in terms of total throughput can be observed in Fig. 5.13, which plots the comparison between NCCB and LB strategies in a scenario with only voice users. Throughput improvements of around 13% in the uplink and 24% in the downlink can be appreciated.

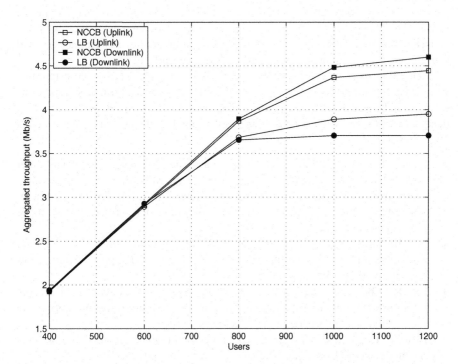

Fig. 5.13. Comparison between NCCB and LB throughput in uplink and downlink.

5.5 Impact of Multi-Mode Terminal Availability on RAT Selection

Within the previous context, and in order to take full advantage of heterogeneous networks, existing mobile terminal capabilities need to be extended. Particularly, to provide connectivity to a variety of underlying access technologies is a must. In this sense, multi-mode terminals, which are able to operate via different RATs, are devised [24]. The assumption that 2G/2.5G/3G multi-mode terminals are available for most users in 2009-2010 with a penetration reaching 90% is still valid [25]. Furthermore, it is expected that the penetration of multi-mode 2G/2.5G/3G/WLAN terminals in the same timeframe will reach 50% of the population [25]. Therefore, in the short term, both single-mode (2G only) and multi-mode terminals will co-exist. On the other hand, the increasing complexity of multi-mode terminal devices may in turn result in a price increase. Consequently, some users may prefer simpler, smaller and cheaper devices for their basic needs such as, e.g., voice and short messaging service (SMS). Then, single-mode 2G or 2.5G terminals may not become extinct. All these facts, together with many other factors will cause differentiation in terminals and will cause segmentation in the terminal market to grow even further.

According to [26], a multi-mode UE in 3GPP is considered to be a terminal with at least one UMTS radio access mode (FDD and/or TDD) that supports one or more additional RATs, e.g., GSM, GPRS/EDGE, WLAN, etc. Moreover, [26] defines several types of UEs, namely Type I through Type IV, which are described in the following:

- Type I: This type of UE, when in a given mode, does not monitor any other mode behaving as a single-mode terminal. Change of mode is based on manual selection with user interaction or automatic selection with timers to avoid ping-pong effects between modes. This type of UE will switch off from the old mode and then switch on in the new mode.
- Type II: This type of UE can, when in a given mode, perform monitoring of other modes and report it through the current mode. It does not support simultaneous reception or transmission through different modes.
- Type III: This type of UE can receive simultaneously in more than one mode, but not transmit simultaneously in more than one mode. It performs monitoring and both manual and automatic mode selection.
- Type IV: This type of UE can receive and transmit simultaneously in more than one mode and it also supports features available in types I to III.

In the following, we study the impact of multi-mode terminal availability on RAT selection procedures. In particular, it is assumed that multi-mode terminals are those with connectivity to GERAN and UTRAN radio interfaces. On the contrary, single-mode terminals are those that support GERAN RAT only. The assumed multi-mode terminal type in our study is a Type II terminal.

For illustrative purposes, a service-based policy will be used (see Section 5.4.1) to evaluate the impact of multi-mode terminal availability. Specifically, the adopted policy first attempts to assign voice users to GERAN and interactive users to UTRAN will be used. If no capacity is available in GERAN, voice users try admission in UTRAN. Similarly, rejected interactive users in UTRAN will attempt admission in GERAN. If no capacity is available in any of the RATs, the user gets blocked. Note that all this will apply provided the terminal has the required capabilities to operate with the suitable RAT, otherwise GERAN is selected as the default RAT.

Fig. 5.14 shows the throughput degradation for different multi-mode terminal availabilities (measured in %) and different number of users requesting service, namely voice users (VU) and interactive users (WU). This throughput degradation is measured as the ratio between the achieved throughput for a particular multi-mode terminal availability and the throughput obtained if all terminals were multi-mode (i.e. 100% multi-mode availability). In general, as the multi-mode terminal availability increases so does the throughput degradation, which is, on the other hand, an expected behaviour. In particular, as long as GERAN is able to manage voice and single-mode terminals, i.e. at low voice and interactive loads, throughput degradation is unnoticed. The increase of users requesting to be served is translated into a bigger degradation in terms of throughput as multi-mode terminal availability decreases given GERAN has to manage both voice and single-mode users.

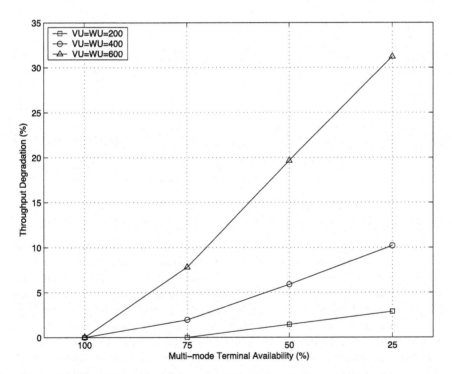

Fig. 5.14. Uplink throughput degradation due to single-mode terminals.

Fig. 5.15 shows the uplink average packet delay for interactive users being served through GERAN for different mixings of multi-mode terminals and service-classes. Note the increasing packet delay when multi-mode terminal availability decreases for VU=WU= 400. As for VU=WU= 200, the average packet delay remains almost constant with multi-mode terminal availability and also at an acceptable level, therefore exhibiting no degradation in this sense.

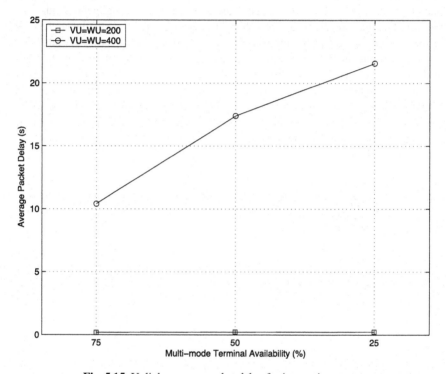

Fig. 5.15. Uplink average packet delay for interactive users.

These illustrative results indicate degradation in terms of throughput and delay introduced by the limited operation of single-mode terminals. In order to compensate the limitations imposed by non-multi-mode terminals, in [27] it is suggested to actuate over GERAN by using a resource reservation scheme for interactive users. In this sense, Fig. 5.16 shows the average packet delay for interactive users when considering 3 dedicated EGPRS slots for interactive users. Clearly, these users benefit from the dedicated slots exhibiting lower packet delays than in the case of not having any reservation scheme (see Fig. 5.15). In addition, for low multi-mode terminal availabilities this reservation scheme improved both aggregated throughput and packet delay figures, particularly for high number of interactive users, as detailed in [27].

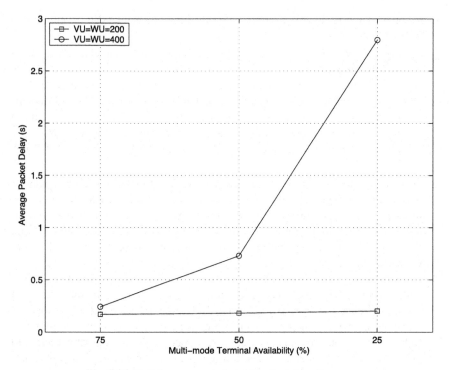

Fig. 5.16. Uplink average packet delay for interactive users.

Other approaches dealing with the lack of flexibility of single-mode terminals include the re-definition of RAT selection policies that account for terminal capabilities which will be addressed in the next section.

5.6 A Generalized Framework for RAT Selection

Trying to combine the different principles influencing on the RAT selection, a generic framework based on the so-called fittingness factor concept was presented in [28] [29]. Specifically, in order to cope with the multi-dimensional heterogeneity reflected in Fig. 5.10, the following main levels are identified in the RAT selection problem:

1. *Capabilities*: A user-to-RAT association may not be possible for limitations in e.g. the user terminal capabilities (single-mode terminal) or the type of services supported by the RAT (e.g. videophone is not supported in 2G).
2. *Technical suitability at the radio part*: A user-to-RAT association may or may not be suitable depending on the matching between the user requirements in terms of QoS and the capabilities offered by the RAT (e.g. a business user may require bit rate capabilities feasible on HSDPA and not on GPRS or these capabilities can be realized in one technology or another depending on the RAT

occupancy, etc.). There is a number of considerations, which can be split at two different levels:

a) *Macroscopic*: Radio considerations at cell level such as load level or, equivalently, amount of radio resources available.

b) *Microscopic*: Radio considerations at local level (i.e. user position) such as path-loss, intercell interference level. This component will be relevant for the user-to-RAT association when the amount of radio resources required for providing the user with the required QoS significantly depends on the local conditions where the user is (e.g. power level required in WCDMA downlink).

3. *Technical suitability at the transport part*: A user-to-RAT association may or may not be suitable depending on the matching between the user requirements in terms of QoS and the capabilities offered by the transport network, that interconnects the different base stations with the access network controllers and the core network, mainly depending on the current load existing in the different links.

4. *Operator/user preferences*: Specific user-to-RAT associations may be preferred without any specific technical criterion but responding to more subjective and economic-related aspects (e.g. due to the investment carried out by an operator to deploy a given technology it can be preferred to serve the traffic through this technology so that investments can be recouped faster, the operator prefers to give some precedence of a service over another one depending on market strategies, etc.).

5.6.1 Fittingness Factor Definition

The above concepts can be captured in the so-called *fittingness factor*, which reflects the degree of adequacy of a given RAT to a given service requested by a given user. Although each RAT has its own particularities, it is possible to make a general definition of the fittingness factor by grouping the different terms under some commonalities general to any RAT. Specifically, the fittingness factor is defined with respect to each cell of a given RAT for each user belonging to a certain profile (e.g. business/consumer) and requesting a specific service, as the product of four different terms ranging from 0 to 1, which are:

Terminal and RAT capabilities

The first term reflects the hard constraints posed by the capabilities of either the terminal or the technology, and therefore it simply takes the value 1 if both the terminal and the RAT support the requested service and 0 otherwise.

Technical suitability in the radio part

This factor reflects the user-centric suitability of the RAT to support the service requested by the user. It basically accounts for the bit rate that can be achieved by the corresponding user/service/profile in the RAT normalized with respect to the maximum bit rate that could potentially be achieved by this user/service/profile in any of the existing RATs. It consists mainly of two multiplicative factors, dealing with the microscopic and the macroscopic dimensions. The macroscopic dimension accounts for the load in the corresponding RAT, as well as for the degree of precedence with respect to other services/profiles, which eventually impacts on the bit rate that can be achieved by the user due to user multiplexing considerations. In turn, the microscopic dimension accounts for the bit rate available per single channel (e.g. one time slot in GERAN, one dedicated channel in UTRAN, etc.) depending on the propagation and interference conditions experienced by the user as well as on the specific characteristics of the RAT (e.g. in case of UTRAN will account for the load factor and/or power availability, in GERAN and WLAN will depend on the link adaptation mechanisms, etc.). Specific definitions of the technical suitability factor in the radio part for UTRAN/GERAN RATs and different services can be found in [28].

Network-centric suitability

This term intends to capture the suitability from an overall RAT perspective, then to provide further flexibility on the fittingness factor definition. For that purpose, this term can include tunable parameters to allow the enforcement of specific operator policies arising from the trade-off between the degree of QoS to be provided to different types of traffic. Just as an example, let us define the non-flexible traffic, which is the traffic that can only be served through one specific RAT and therefore it does not provide flexibility to CRRM (e.g. in UTRAN it could be the videocall users assuming they cannot be served through other RATs such as GERAN, WLAN, etc., while in GERAN, it could be the mono-mode terminals, which cannot be served through other RATs). In such a situation, the operator may decide to give more precedence to the non-flexible traffic depending on the desired policy. Then, the network-centric suitability term is a function that reduces the fittingness factor of flexible traffic depending on the amount of non-flexible load. The idea is that if there is a high amount of non-flexible load in a given RAT, this RAT is made less attractive for flexible load, thus leaving room to non-flexible users. It is worth mentioning that the definition of the network-centric suitability can include tunable parameters to allow the enforcement of specific operator policies arising from the trade-off between the degree of QoS to be provided to the non-flexible traffic with respect to the flexible traffic [29].

Transport network capabilities

This term will account for the bit rate available for this user/service/profile in the transport network of the RAT in accordance with the bottleneck link utilization for a given path between a base station/access point and its corresponding controller. Consequently, this term reflects the amount of load existing in the transport network, in the sense that a value close to 1 means that the transport network has very low load and therefore it does not introduce limitations in the service bit rate, while a value close to 0 means that the transport network is overloaded and as a result it can limit the achievable service bit rate.

5.6.2 Session Set-up and Vertical Handover Algorithm

Based on the above definition of the fittingness factor, two different RAT selection procedures can be identified depending on whether the selection is done at session set-up (i.e. initial RAT selection algorithm) or during an on-going connection (i.e. vertical handover algorithm) for a user requesting a given service s, as detailed in the following:

Session set-up case

Step 1: Measure the fittingness factor for each candidate cell k_j of the j-th detected RAT. The fittingness factor should be computed separately for uplink and downlink of each RAT. Then, a weighting between the two values can be carried out depending on the service characteristics.

Step 2: Select the RAT J having the cell with the highest fittingness factor among all the candidate cells. In the case that two or more RATs have the same value of the fittingness factor, then select the less loaded RAT.

Step 3: Try admission in the RAT J.

Step 4: If admission is not possible, try with the next RAT in decreasing order of fittingness factor, provided that its fittingness factor is higher than 0. If no other RATs with fittingness factor higher than 0 exist, block the call.

On-going connection case

For on-going connections, the proposed criterion to execute a VHO algorithm based on the fittingness factor would be as follows, assuming that the terminal is connected to the RAT denoted as "servingRAT" and cell denoted as "servingCell".

Step 1: For each candidate cell and RAT, monitor the corresponding fittingness factor. Measures should be averaged during a period T.

Step 2: If the fittingness factor of a given RAT in a specific cell is greater than the fittingness factor of the current serving RAT and cell for a period T_{VHO} then a vertical handover to the new RAT and cell should be triggered, provided that there are available resources for the user in this new RAT and cell.

5.6.3 Illustrative Results

The developed framework has been analyzed through exhaustive simulations in a variety of scenarios with different traffic mixes to reveal the ability to adapt to the conditions in each case. In the following some results are given to illustrate the benefits of the proposed framework. They consider a scenario with UTRAN R99 and GERAN cells with EDGE capabilities. Voice at 12.2 Kbps and videocall at 64 Kbps services are considered as representative of the conversational traffic class while a www browsing service with two different profiles, namely consumer (with bit rate up to 128 Kbps in UTRAN and low priority in GERAN) and business (with bit rate up to 384 Kbps and high priority in GERAN), have been selected as representative of the interactive traffic class.

In order to illustrate how the fittingness factor algorithm affects the traffic splitting among the two RATs, Fig. 5.17 plots the fraction of traffic served through GERAN for the voice, interactive consumer and business profiles when increasing the total load coming from videocall users (which are always served through UTRAN given GERAN does not support this type of traffic). For comparison purposes, the distribution according to the load balancing case (LB) is also shown, in which the less loaded RAT is selected at session set-up. It can be observed how LB does not make significant distinctions among the considered services, with the general trend that, by increasing the load of videocall users, more traffic of the other services should be derived to GERAN in order to keep similar load levels in the two RATs. On the contrary, the fittingness factor based algorithm is able to split the traffic according to the peculiarities of each service. In particular, most of the interactive business traffic is served through UTRAN, where this type of traffic can achieve a higher bit rate. Only in the case that the videocall load is very high there is a certain interactive business traffic that should be moved to GERAN. In turn, when looking at the voice and interactive consumer users, as a result of the increase in videocall load, the algorithm tends to move to GERAN mainly the voice traffic, while it keeps a significant fraction of interactive consumer traffic still in UTRAN.

The different traffic split impacts over the QoS observed by each service, as it is reflected in Fig. 5.18, which compares the packet delay of the interactive consumer and business users with the fittingness factor based algorithm and with LB when increasing the load of voice users in the scenario. It is observed that the performance from a user point of view is better with the fittingness factor-based algorithm than with LB for the two user profiles. Although it is not plotted here for the sake of brevity, the total throughput achieved in the scenario in this case is approximately the same for the two approaches, which reflects that the fittingness factor is able to improve the user QoS perception without degrading the overall capacity.

The impact of the videocall users, which are a non-flexible type of traffic, is plotted in Fig. 5.19 and Fig. 5.20 in terms of packet delay and total throughput in the scenario, respectively. To illustrate the ability of the algorithm to reflect different operator criteria, two different settings have been considered in the network-centric suitability component in the fittingness factor definition. Setting 1 reflects the case in which the operator aims at improving the QoS of the interactive business users. In this case, as reflected in Fig. 5.19, the delay for this traffic is the smallest one among the considered approaches, which is achieved by keeping as much as possible this traffic in UTRAN, even if videocall load is high. However, this improvement occurs at the expense of a reduction in the throughput of non-flexible traffic, because the interactive users leave less room in UTRAN. As a result, there is some reduction in the total throughput in the scenario, as shown in Fig. 5.20. In turn, setting 2 reflects the situation in which the operator prefers to keep more capacity for the videocall traffic in UTRAN, which is ensured by allowing that some interactive business users are served through GERAN. Notice in the two figures that, with this setting, the total throughput in the scenario is increased with respect to setting 1 at

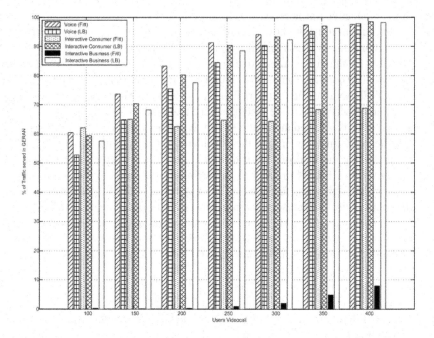

Fig. 5.17. Fraction of traffic served through GERAN as a function of the videocall users for the fittingness factor and the load balancing strategies.

the expense of some delay degradation. Nevertheless, the delay is still much better than that achieved with LB.

Conclusion

This chapter has covered the radio resource management issues that arise in heterogeneous network scenarios, where different RATs coexist and cooperate in a coordinated way, in order to to provide the end-user with the requested services and corresponding QoS requirements. When jointly considering different RATs, a more efficient use of the available radio resources can be achieved with the introduction of CRRM algorithms that take into consideration the overall resources in all the available RATs.

The envisaged architectures for interworking and coupling between UMTS, GERAN and WLAN has also been discussed. For a proper support of CRRM algorithms, the currently existing network architectures must be modified accordingly to ensure the desired interworking capabilities among the different RATs. In this sense, the 3GPP has identified some interactions and topologies among RRM and

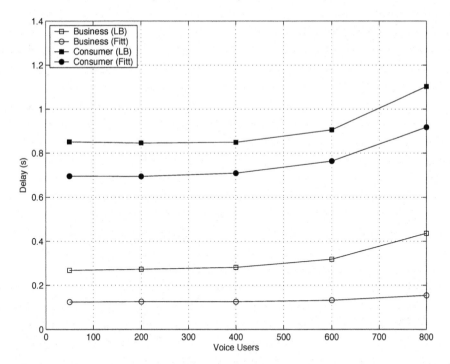

Fig. 5.18. DL packet delay for interactive business and consumer users according to the fit-tingness factor and the load balancing case.

CRRM entities for an efficient management of the total radio resource pools. Of special interest is the integration of 3GPP technologies with WLAN standards which provide high data rates along with increasing wireless coverage in hotspot locations. In this sense, two different approaches were presented, the Interworking WLAN (I-WLAN) and the Generic Access Network (GAN), which provide different levels of coupling between the involved entities. For the GAN interworking scheme, it enables the execution of CRRM functionalities given it supports the exchange of inter-RAT measurement information between UTRAN/GERAN and GAN cells.

The well-known RRM functions that arise in the context of a single RAN (i.e. admission control, congestion control, packet scheduling, etc.) have been redefined in the context of heterogeneous networks where information about several RANs are used to take decisions. We refer to them as "common" RRM functions, thus having common admission control, common congestion control, etc. In addition, the RAT selection functionality appears both at session initiation (initial RAT selection) or during session lifetime (i.e. vertical or inter-system handover). Different possibilities have been envisaged when considering the functional split between RRM and CRRM entities, mainly depending on whether it is the RRM or the CRRM entity acting as the master in taking decisions along with the degree of interactions between them. Some illustrative alternatives have been described in this sense, ranging from no

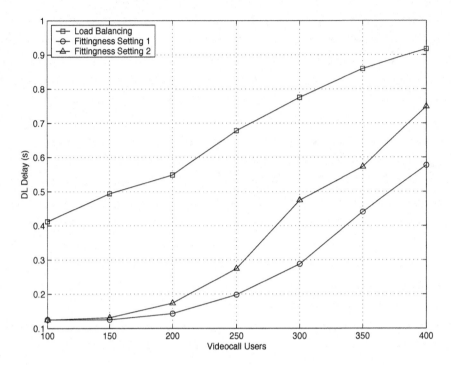

Fig. 5.19. DL packet delay of business interactive users.

functionalities present at the CRRM entity to only long-term functionalities and, finally, long and short-term functionalities to be carried out at the CRRM entity.

Some specific implementations of CRRM algorithms in the context of RAT selection have also been addressed. In particular, a first classification of RAT selection schemes according to service principles, load balancing and interference aspects has been considered. In the case of service-based RAT selection, results indicate the suitability of allocating voice users to GERAN while www users to UTRAN (i.e. policy VG), mainly due to the improved throughput obtained by www users in UTRAN, which use dedicated channels, as opposed to shared channels in GERAN. As for the load balancing (LB) RAT selection, this scheme was compared to the service-based VG policy. It was shown that LB provided a higher flexibility in allocating resources as compared to VG which incurred more potential user droppings. The presented interference-based RAT selection, namely the Network Controlled Cell Breathing (NCCB), revealed throughput improvements as compared to the LB strategy, by effectively reducing the inter-cell interference in CDMA which in turn results in a capacity increase. The impact of multi-mode terminal availability in RAT selection procedures has been also addressed and results showed the degradation introduced by the limited operation of single-mode terminals. In this sense, a resource reservation scheme for www users in GERAN was applied in order to mitigate the high packet delay caused by the excessive load of single-mode UEs in GERAN.

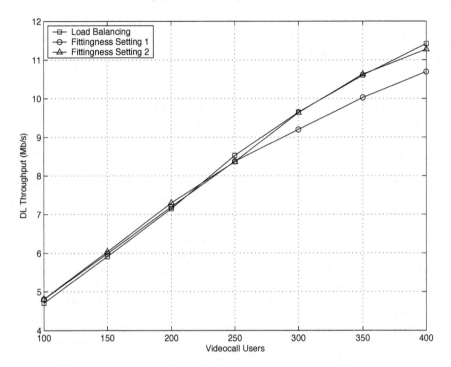

Fig. 5.20. Total scenario DL throughput when increasing the videocall load.

Finally, trying to combine the different parameters influencing the RAT selection, a generic framework based on the so-called *fittingness factor* concept has also been introduced in order to capture the degree of adequacy of a given RAT to a given service requested by a given user. Then, the fittingness factor can be appropriately adopted for the case of RAT selection at session set-up stage (i.e. initial RAT selection) as well as during the session lifetime stage (i.e. vertical handover). So as to illustrate how the fittingness factor affects RAT selection, representative results have been provided in a variety of scenarios with different traffic mixes revealing the ability to adapt to the conditions in each case. Compared to the LB strategy, the fittingness factor approach exhibits a better behavior in terms of guaranteeing several QoS levels without degrading the overall capacity. In addition, a better performance of the fittingness factor in managing non-flexible traffic has been observed and also the ability to respond to different operator-based criteria has been assessed.

References

1. G. Fodor, A. Eriksson, and A. Tuoriniemi, "Providing quality of service in always best connected networks," *IEEE Communications Magazine*, vol. 41, no. 7, pp. 154-163, Jun. 2003.

2. A. Tölli, P. Hakalin, and H. Holma, "Performance evaluation of common radio resource management (CRRM)," in *Proc. of IEEE International Conference on Communications 2002 (ICC'02)*, vol. 5, New York, NY, USA, 2002, pp. 3429–3433.
3. 3GPP TR 25.881 v5.0.0, "Improvement of RRM across RNS and RNS/BSS".
4. 3GPP TR 25.891 v0.3.0, "Improvement of RRM across RNS and RNS/BSS (post rel-5) (release 6)".
5. 3GPP TS 25.423, "UTRAN Iur interface RNSAP signalling".
6. 3GPP TR 22.934 v6.2.0, "Feasibility study on 3GPP system to Wireless Local Area Network (WLAN) interworking".
7. 3GPP TS 23.234, "3GPP system to Wireless Local Area Network (WLAN) interworking".
8. A. Salkintzis, C. Fors, and R. Pazhyannur, "WLAN-GPRS integration for next-generation mobile data networks," *IEEE Wireless Communications*, vol. 9, no. 5, pp. 112-124, Oct. 2002.
9. 3GPP TS 43.318, "Generic Access Network (GAN)".
10. J. Pérez-Romero, O. Sallent, R. Agustí, and M. Díaz-Guerra, *Radio Resource Management Strategies in UMTS*. John Wiley & Sons, 2005.
11. J. Pérez-Romero, O. Sallent, and R. Agustí, "Policy-based initial RAT selection algorithms in heterogeneous networks," in *Proc. of 7th IFIP International Conference on Mobile and Wireless Communication Networks (MWCN'05)*, Marrakech, Morocco, Sept. 2005.
12. G. Fodor, A. Furuskär, and J. Lundsjo, "On access selection techniques in always best connected networks," in *Proc. of 16th ITC Specialist Seminar on Performance Evaluation of Wireless and Mobile Systems*, Antwerp, Belgium, 2004.
13. A. Tölli and P. Hakalin, "Adaptive load balancing between multiple cell layers," in *Proc. of IEEE 56th Vehicular Technology Conference Fall 2002 (VTC Fall'02)*, vol. 3, Vancouver, BC, Canada, Sept. 2002, pp. 1691-1695.
14. A. Pillekeit, F. Derakhshan, E. Jugl, and A. Mitschele-Thiel, "Force-based load balancing in co-located UMTS/GSM networks," in *Proc. of IEEE 60th Vehicular Technology Conference Fall 2004 (VTC Fall '04)*, vol. 6, Los Angeles, CA, USA, 2004, pp. 4402-4406.
15. X. Gelabert, J. Pérez-Romero, O. Sallent, and R. Agustí, "On the suitability of load balancing principles in heterogeneous wireless access networks," in *Proc. of Wireless Personal Multimedia Communications Symposium (WPMC'05)*, Aalborg, Denmark, Sep. 2005.
16. S. Lincke, "Vertical handover policies for common radio resource management," *International Journal of Communication Systems*, vol. 18, no. 6, pp. 527-543, Aug. 2005.
17. ——, "The benefits of load sharing when dimensioning networks," in *Proc. of the IEEE Annual Simulation Symposium*, Arlington, VA, United States, 2004, pp. 115-124.
18. J. Pérez-Romero, O. Sallent, and R. Agustí, "Network controlled cell breathing in multi-service heterogeneous CDMA/TDMA scenarios," in *Proc. of IEEE 64th Vehicular Technology Conference Fall 2006 (VTC Fall'06)*, Montreal, QC, Canada, Sept. 2006, pp. 1305-1309.
19. J. Pérez-Romero, O. Sallent, R. Agustí, P. Karlsson, A. Barbaresi, L. Wang, F. Casadevall, M. Dohler, H. Gonzalez, and F. Cabral-Pinto, "Common radio resource management: Functional models and implementation requirements," in *Proc. of IEEE International Symposium on Personal, Indoor and Mobile Radio Communications, PIMRC*, vol. 3, Berlin, Germany, 2005, pp. 2067-2071.

20. W. Zhuang, Y.-S. Gan, K.-J. Loh, and K.-C. Chua, "Policy-based QoS-management architecture in an integrated UMTS and WLAN environment," *IEEE Communications Magazine*, vol. 41, no. 11, pp. 118-125, Nov. 2003.
21. S. Lincke and C. Hood, "Integrated networks that overflow speech and data between component networks," *International Journal of Network Management*, vol. 12, no. 4, pp. 235-257, Jun. 2002.
22. X. Gelabert, J. Pérez-Romero, O. Sallent, and R. Agustí, "A 4-dimensional Markov model for the evaluation of radio access technology selection strategies in multiservice scenarios," in *Proc. of IEEE 64th Vehicular Technology Conference Fall 2006 (VTC Fall'06)*, Montreal, QC, Canada, Sept. 2006, pp. 1-5.
23. A. Furuskär and J. Zander, "Multiservice allocation for multiaccess wireless systems," *IEEE Transactions on Wireless Communications*, vol. 4, no. 1, pp. 174-184, Jan. 2005.
24. R. Schuh, P. Eneroth, and P. Karlsson, "Multi-standard mobile terminals," in *Proc. of 11th IST Mobile and Wireless Telecommunications Summit 2002 (IST'02)*, Thessaloniki, Greece, June 2002, pp. 174-178.
25. P. Karlsson (editor) et al., "Target scenarios specification: Vision at project stage 2" Deliverable D13 of the EVEREST IST-2002-001858 Project. [Online]. Available: http://www.everest-ist.upc.es, February, 2005.
26. 3GPP TS 21.910 3.0.0, "Multi-mode UE issues; categories, principles and procedures".
27. X. Gelabert, J. Pérez-Romero, O. Sallent, and R. Agustí, "On the impact of multi-mode terminals in heterogeneous wireless access networks," in *Proc. of International Symposium on Wireless Communication Systems (ISWCS'05)*, Siena, Italy, Sep. 2005.
28. J. Pérez-Romero, O. Sallent, and R. Agustí, "A novel metric for context-aware RAT selection in wireless multi-access systems," in *Proc. of IEEE International Conference on Communications 2007 (ICC '07)*, Glasgow, Scotland, Jun. 2007, pp. 5622-5627.
29. O. Sallent, J. Pérez-Romero, R. Ljung, P. Karlsson, and A. Barbaresi, "Operator's RAT selection policies based on the fittingness factor concept," in *Proc. of 16th IST Mobile and Wireless Communications Summit 2007 (IST'07)*, Budapest, Hungary, July 2007, pp. 1-5.

6

Radio Resource Management and Admission Control in Heterogeneous Wireless Access Networks

Fei Richard Yu[1] and Helen Tang[2], and Hong Ji[3]

[1] The Department of Systems and Computer Engineering, Carleton University, Ottawa, ON, Canada
richard_yu@carleton.ca
[2] Defense R&D Canada, Ottawa, ON, Canada
helen.tang@drdc-rddc.gc.ca
[3] Key Lab of Universal Wireless Communications, Beijing University of Posts and Telecommunications, Beijing, P.R. China
jihong@bupt.edu.cn

6.1 Introduction

There has been significant growth in the use of wireless communication services. Currently, there exist disparate wireless systems, such as Bluetooth [1] and ultra-wideband (UWB) radios [2] for personal areas, wireless local area networks (WLANs) for local areas, WiMAX for metropolitan areas, 3G cellular networks for wide areas, and satellite networks for global networking. The complementary characteristics of different wireless networks make it attractive to integrate these heterogeneous wireless access technologies. With the proliferation of wireless networking standards, increasingly portable electronic devices are being equipped with multiple transceivers to enable them to access different wireless networks. For example, 3G cellular smart phones are available with built-in WLAN and Bluetooth interfaces. In next generation wireless networks, wireless devices are equipped with multiple wireless transceivers for different air interface standards to satisfy the different needs for wide area, metropolitan area, local area and personal area wireless networks. These heterogeneous wireless systems will cooperate with each other to provide ubiquitous "always best connection" to mobile users [3]. Users would benefit from the lower overall cost and the enhanced performance of the integrated services.

Moreover, in recent years, there has been an explosion of interest in cognitive radio for next generation wireless networks [4]. As an extension of software defined radio [5], cognitive radio can support multiple air interfaces and protocols for enabling high spectrum utilization, facilitating the convergence of heterogeneous wireless networks, providing more reliable radio services, and reducing harmful interference. Wireless devices with cognitive radio can sense their environment and adapt to many frequency bands with heterogeneous transmission protocols so as to better communicate in the sensed environment. With the capabilities of awareness of and

E. Hossain (ed.), *Heterogeneous Wireless Access Networks*,
DOI: 10.1007/978-0-387-09777-0_6, © Springer Science+Business Media, LLC 2008

adaptation to the environment, cognitive radio holds the promise of a new frontier in wireless communications [4].

Next generation heterogeneous wireless networks present significant technical challenges that require advancements in wireless networks architecture, protocol, and management algorithms. There are several networking protocols at different layers to integrate heterogeneous wireless networks. In addition, managing radio resources among multiple heterogeneous wireless systems is important to preserve system robustness. How to fairly and efficiently share radio resources among multiple networks is one of the major challenges in integrating multiple networks together. The traditional radio resource management schemes optimized for an individual network may result in unsatisfactory performance of the overall integrated systems. Therefore, in order to fully utilize the best of different wireless technologies, it is important to consider the radio resource jointly. Moreover, an efficient and effective admission control scheme is crucial to guarantee the quality of services (QoS) and to maximize the radio resource utilization simultaneously.

In this chapter, we present a framework for radio resource management in heterogeneous wireless access networks considering vertical handovers between these networks. In addition, we propose an optimal admission control scheme to decide whether or not to admit and which network to admit a new or vertical handover session arrival. Some numerical examples illustrate the effectiveness of the schemes.

The rest of the chapter is organized as follows. Section 6.2 describes heterogeneous wireless networks and the networking protocols to integrate them together. Section 6.3 describes the radio resource management schemes. The optimal admission control scheme is presented in Section 6.4. Some numerical examples are given in Section 6.5 before we conclude this chapter.

6.2 Heterogeneous Wireless Networks

In this section, we present the integrated heterogeneous wireless networks considered in this paper. We introduce the networking protocols at different layers that enable the integration of these heterogeneous wireless networks.

6.2.1 Integrated Heterogeneous Wireless Networks

Generally speaking, there are two different ways of integrating heterogeneous wireless networks, defined as tight coupling and loose coupling interworking. Fig. 6.1 shows the architecture for the integration. In this figure, we have two different networks, Network 1 and Network 2. A 3G cellular network is used as an example of Network 1, and a WLAN can be an example of Network 2. In reality, the number of heterogeneous networks is not limited to two.

In a tightly coupled system, a network is connected to another network in the same manner as other radio access networks. Note that many wireless networks, such as Universal Mobile Telecommunication System (UMTS), have the interfaces to other radio access networks. In this approach, these two networks could use the same authentication, mobility, and billing infrastructure, and the gateway of a network needs to implement the protocols required in the other network. The main advantage of this approach is that the existing mechanisms for authentication, mobility and QoS can be reused directly over another network. However, this approach requires modifications of the design to accommodate the increased traffic from the other network.

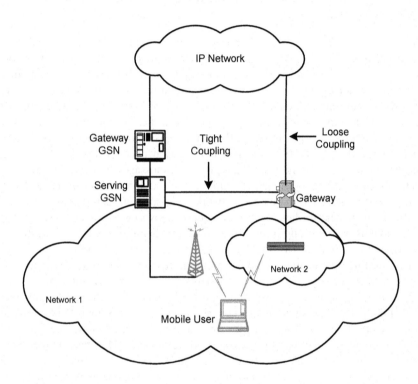

Fig. 6.1. Integrated heterogeneous networks.

In a loosely coupled system, the heterogeneous wireless networks are not connected directly. Instead, they are connected to the Internet. In this approach, different mechanisms and protocols are used to handle authentication, mobility and billing, and the traffic from one network would not go through the other network. Nevertheless, as peer IP domains, they can share the same subscribe database for functions such as security, billing and customer management.

Since wireless mobile users are free to move in integrated heterogeneous wireless networks, the support of handover between these networks, which provides ongoing service continuity and seamlessness, is needed in this integration. Handover in a heterogeneous network environment is different from that in a homogeneous environment, where it occurs only when a mobile user moves from one base station (or access point) to another. While handover within a homogeneous system is called *horizontal handover*, handover between different wireless access technologies is referred to as *vertical handover*. There are several networking protocols at different layers that enable the integration of these heterogeneous wireless networks. We present them in the following.

6.2.2 Networking Protocols Enabling the Integration of Heterogeneous Wireless Networks

Network Layer Solutions: IETF proposes a protocol called Mobile IP (MIP) to keep the IP consistence across heterogeneous networks by adding some transferring nodes [6], home agent (HA) and foreign agent (FA). HA stores the registration information of all terminals originated from the current network. When a mobile node (MN) roams into a foreign network, it will first obtain a Care of Address (CoA) from the FA. The MN will further notify the HA in the home network of the CoA using a registration message. In MIP, MN should register to the HA every time when it changes the IP address. If MN moves frequently between different subnets of a foreign network, it will send back the registration messages frequently to the HA, which will generate heavy signaling traffic as well as long handover delay if HA is far away from the MN. In general, network layer solutions can provide universal mobility management for heterogeneous wireless networks. However, there are two main problems with this approach. It requires changes on networks, such as the set up of HA and FA in all networks. It also requires the modification of CN to implement the routing optimization. In addition, it requires to assign a static IP address to each MN, which is very costly in the IPv4 network that short of IP addresses.

Transport Layer Solutions: Transport layer solutions follow the end-to-end principle [7] in the Internet: anything that can be done in the end system should be done there. Since the transport layer is the lowest end-to-end layer in the Internet protocol stack, it is a natural candidate for vertical handover support. Moreover, in the transport layer solutions, no third party other than the endpoints participates in vertical handover and no modification or addition of network components is required. Authors in [8] proposed a new set of migrate options for TCP to support mobility. In this protocol, the MN and CN will determine a unique token number to identify the TCP connection when the connection is set up. Such token numbers are calculated according to the addresses/ports number of peer nodes. The main drawback of these approaches is that TCP has been globally deployed so that it is practically infeasible to change it. A new IETF-standardized transport layer protocol called Stream Control Transmission Protocol (SCTP) [9], which can be used in place of both the TCP and UDP, has gained significant attention as a candidate transport protocol for the next generation Internet. The multi-homing, multi-stream and partially reliable data transmission features of SCTP are especially attractive for applications that have stringent performance and high reliability requirements. In [10], a MN can be configured with multiple addresses in a connection with one of the addresses as the primary address. Using MSCTP to enable vertical handovers has many advantages, including simpler network architecture, improved throughput and delay performance, and ease of adapting flow/congestion control parameters to the new network during and after vertical handovers.

Application Layer Solutions: As a signaling protocol in application layer, Session Initiation Protocol (SIP) has been widely adopted by IETF and 3GPP to help establish sessions in packet-switched networks [11]. The session is multiple data communications among multiple participants. SIP also affords various mobility supports and allows a connection to be set up in the middle of a session. It is generally envisioned that the SIP should be the main signaling protocol in next generation wireless networks. In SIP, the MN is identified with a unique logical SIP address in the format of email address. When the MN roams into the foreign domain, it obtains a contact address with the domain name in the foreign domain. In comparison with the mobility solutions at other layers, SIP can be easily deployed in the network without requiring the modification of network entities or end-user protocol stacks. Moreover, SIP can be easily extended or escalated due to the operation at the highest layer and the use of text-based signaling messages. Having been accepted by a standard protocol in both Internet and tele-network, SIP is regarded as an attractive candidate to support mobility in NG networks. Apart from all those advantages, SIP also suffers from the long handover delay due to its operation at the highest layer. A "make-before-break" handover procedure was proposed in [12] to reduce the handover delay.

6.3 Radio Resource Management Scheme

Radio spectrum is one of the most important resources in wireless networks. Designing an efficient and effective radio resource management scheme is challenging in the integration of heterogeneous wireless networks. In this section, we first describe the admissible sets for some example individual wireless networks. Then, we present the joint radio resource management scheme.

6.3.1 Admissible Sets for Individual Wireless Networks

TDMA/FDMA Cellular Wireless Networks: Assume that there are J classes of traffic and L wireless networks. Let $n_{l,j}$ denote the number of active class j users in network l. For TDMA/FDMA circuit-switched cellular networks, such as GSM, there is a fixed number of channels, C_l, in network l. The number of channels used by a class j session is $c_{l,j}$. Define a vector

$$x_l = (n_{l,1}, n_{l,2}, \ldots, n_{l,J}). \tag{6.1}$$

The admissible set in network l can be expressed as

$$X_T = \left\{ x_l \in Z_+^J : \sum_{j=1}^{J} n_{l,j} c_{l,j} \leq C \right\}. \tag{6.2}$$

The above admissible set is derived under the assumption that a session will be admitted to the network as long as a free channel is available.

CDMA Cellular Networks with Matched Filter Receivers: For CDMA cellular networks with matched filter receivers and variable spreading gain, an important physical layer QoS requirement for class j users is the signal-to-interference ratio, $\text{SIR}_{l,j}$, which should be kept above the target value $\omega_{l,j}$ [13]. Let W denote the total cell bandwidth. The average bit rate of a class j user is $R_{l,j}$ in network l. The orthogonality factor is ρ. The ratio between intercell interference and total intracell power is γ. The path loss of user i is PL_i. The downlink capacity can be evaluated using cell load factor, which is defined as

$$\eta_l = \sum_{j=1}^{J} \sum_{i=1}^{n_{l,j}} \frac{(\rho + \gamma)}{\frac{W}{\omega_{l,j} R_{l,j}} + \rho}. \tag{6.3}$$

The transmit power needed at the base station to guarantee the SIR requirements is

$$P_T = \frac{P_p + P_N \Lambda}{1 - \eta_l} \tag{6.4}$$

where P_p is the power used by common control channels, P_N is the background noise power, and $\Lambda = \sum_{j=1}^{J} \sum_{i=1}^{n_{l,j}} PL_i / ((W/\omega_{l,j} R_{l,j}) + \rho)$. The admissible set in the CDMA cellular network can be expressed as [13]

$$X_{CM} = \left\{ x_l \in Z_+^J : P_T \leq P_T^{MAX} \right\} \tag{6.5}$$

where P_T is defined in (6.4) and P_T^{MAX} is the maximum available power of the base station.

CDMA Cellular Networks with Linear Minimum Mean Square Error (LMMSE) Multiuser Receivers: Unlike the conventional matched filter receivers, the LMMSE receivers take into account the structure of the interface from other users when demodulating a user, and therefore significantly outperform the conventional matched filter receivers [14]. The admissible set can be expressed as

$$X_{CL} = \left\{ x_l \in Z_+^J : \inf_s \left[s \left(\sum_{j=1}^{J} n_{l,j} \alpha_{l,j} - 1 \right) \right] \leq \log T P^{\text{out}} \right\} \tag{6.6}$$

where $\alpha_{l,j}$ is the scaled logarithmic moment generating function of the instantaneous work load of a class j session, and the SIR outage probability constraint $P^{\text{out}} \leq T P^{\text{out}}$ will be satisfied. The above equation is derived using the concept of cross-layer effective bandwidth in multimedia CDMA networks with LMMSE receivers. Please refer to [15] for details.

Wireless Local Area Networks (WLANs): Throughput and packet delay are important QoS metrics in WLANs. There are several models in the literature that can be used to formulate throughput and packet in WLANs. For example, authors of [16] derived throughput and packet delay in IEEE 802.11e. An optimal operating point for WLANs regulated by applying rate control at the access point was considered in [17] to derive throughput and packet delay. Using these models, we have the WLAN admissible set

$$X_W = \{x_l \in Z_+^J : B_{l,j} \geq TB_{l,j}, D_{l,j} \leq TD_{l,j}\} \tag{6.7}$$

where the throughput constraint $B_{l,j} \geq TP_{l,j}$ and the packet delay constraint $D_{l,j} \leq TD_{l,j}$ will be satisfied.

6.3.2 Joint Radio Resource Management Scheme

In the integrated heterogeneous wireless networks, define the state vector of the system as

$$x = (x_1, x_2, \ldots, x_l, \ldots, x_L). \tag{6.8}$$

The state space X of the system can be derived using the admissible sets of individual wireless networks, such as (6.2), (6.5), (6.6), (6.7). The state space X of the integrated system is given as

$$X = \{x = [x_1, x_2, \ldots, x_l, \ldots, x_L] \in Z_+^{LJ}, \\ : \text{The constraints in individual networks are satisfied}\}. \tag{6.9}$$

For each given state $x \in X$, an action $a(x)$ is performed by a joint radio resource management scheme. The action space is a set of all possible actions. The action is done according to a joint radio resource management scheme $u \in U$, where U is defined as

$$U = u : X \to A. \tag{6.10}$$

$\{x(t), u\}_{t \in R_+}$ is a Markov process under each radio resource allocation scheme. Let $\pi_u(x)$ denote the equilibrium probability that the system is in state x under scheme u. Define $e_{l,j} \in \{0, 1\}^J$ as a row vector containing only zeros except for the jth component, which is 1. $x + (-)e_{l,j}$ corresponding to an increase (decrease) of the number of class j connections in network l by 1. The global balance equations for the Markov Chain under scheme u are [18]

$$\sum_{j=1}^{J} \sum_{l=1}^{L} [\pi_u(x - e_{l,j})\lambda_{l,j}a(x - e_{l,j}) + \pi_u(x + e_{l,j})\mu_{l,j}(n_{l,j} + 1)$$

$$= \sum_{j=1}^{J} \sum_{l=1}^{L} [\lambda_{l,j}a(x) + \mu_{l,j}(n_{l,j} + 1)]\pi_u(x), x \in X \tag{6.11}$$

where $\lambda_{l,j}$ and $1/\mu_{l,j}$ are class j connection arrival and departure rates in network l, respectively. These global balance equations can be solved by using any linear equation procedure, such as Jacobi and Gauss-Seidel methods. Once the equations are solved, network layer blocking probability, can be directly calculated. The blocking probability for a class j connection is

$$P_{l,j}^b = \sum_{i \in X_{l,j}} \pi(i) \tag{6.12}$$

where $X_{l,j} \subseteq X$ is the set of states that system will move out of X with addition of one connection of class j in network l. This approach is general enough to be applicable to a variety of radio resource allocation schemes.

The above approach may have a problem. The computation complexity of solving the global balance equations is extensive when the cardinality of the state space is large. Feasible solutions are difficult to obtain in real networks due to the problem of large dimensionality. We can consider a set of *coordinate convex schemes* that have a product form of the equilibrium probabilities. The coordinate convex schemes form several important resource allocation schemes, such as *complete sharing, complete partitioning*, and *threshold* schemes. The name of coordinate convex scheme comes from the concept of coordinate convex set. A coordinate convex scheme is characterized by a coordinate convex set, which is any nonempty set $\Delta \subseteq X$ with the following property: if $x \in \Delta$ and $n_{l,j} > 0$ then $x - e_{l,j} \in \Delta$. In a coordinate convex scheme associated with coordinate convex Δ, a connection arrival is admitted to the system if and only if the system state remains in Δ after the admission. The equilibrium probabilities of the system can be obtained from the the the theory of multiservice loss networks.

$$\pi(n) = \begin{cases} \pi_0 \prod_{j=1}^{J} \prod_{l=1}^{L} \frac{(\lambda_{l,j}/\mu_{l,j})^{n_{l,j}}}{n_{l,j}!}, & \text{if } n \in \Delta \\ 0, & \text{otherwise} \end{cases} \tag{6.13}$$

where π_0 is a normalization constant given by

$$\pi_0 = \frac{1}{\sum\limits_{n \in \Delta} \prod\limits_{j=1}^{J} \prod\limits_{j=1}^{J} \frac{(\lambda_{l,j}/\mu_{l,j})^{n_{l,j}}}{n_{l,j}!}}. \tag{6.14}$$

6.4 Optimal Admission Control Scheme

Using the above joint radio resource management scheme, we may not be able to derive an optimal admission control scheme that maximizes the radio resource utilization and guarantee the QoS simultaneously. An optimal admission control scheme is presented in this section with this goal in mind. The mathematical tool we will use is semi-Markov decision process (SMDP) [19]. In the SMDP formulation, when a new or a handover session arrives, a decision must be made as to whether or not to admit and to which network to admit the session request based on the current state of the system. The same state information can be used as that in (6.9). The decision time instants are called *decision epochs*. The long-run reward is the optimality criterion for the SMDP.

6.4.1 SMDP Formulations

The decision epochs can be chosen as the set of all session arrival and departure instances. At each decision epoch t_k, $k = 0, 1, 2, \ldots$, a decision will be made in the time interval $(t_k, t_{k+1}]$, which is referred to as an *action*. Action $a(t_k)$ is defined as

$$a(t_k) = [a_1(t_k) \in \{0,1\}^J, \ldots, a_l(t_k) \in \{0,1\}^J, \ldots, a_L(t_k) \in \{0,1\}^J] \quad (6.15)$$

where $a_l(t_k) \in \{0,1\}^J$ is defined as follows. Define row vector

$$a_l(t_k) = [a_{l,1}(t_k), a_{l,2}(t_k), \ldots, a_{l,J}(t_k)]$$

where $a_{l,j}(t_k)$ denotes the action for class j session arrival in network l. If $a_{l,j}(t_k) = 1$, a class j session that arrives in network l is admitted to the network. If $a_{l,j}(t_k) = 0$, a class j session that arrives in network l is rejected. For a given state $x \in X$, a selected action should not result in a transition to a state that is not in X. Moreover, action $(0,0,\ldots,0)$ should not be a possible action in state $(0,0,\ldots,0)$. Otherwise, the system cannot evolve.

The action space of a given state $x \in X$ is defined as

$$A_x = \{a \in A_x : a_{l,j} \neq 1 \text{ and if } [x_1, \ldots, (x_l + e_j), x_L] \notin X,$$
$$\text{and } a \neq (0,0,\ldots,0) \text{ if } x = (0,0,\ldots,0)\} \quad (6.16)$$

where $e_j \in \{0,1\}^J$ denotes a row vector containing only zeros except for the jth component, which is 1. $x_l + e_j$ corresponds to an increase of the number of class j sessions by 1 in network l.

The state transition probabilities of the embedded chain and the expected sojourn time $\tau_x(a)$ for each state-action pair can be used to characterize the dynamics of the system.

$$\tau_x(a) = \left[\sum_{l=1}^{L} \sum_{j=1}^{J} (\lambda_{l,j} a_{l,j} + \mu_{l,j} n_{l,j}) \right]^{-1}. \quad (6.17)$$

The state transition probabilities of the embedded Markov chain are

$$p_{xy}(a) = \begin{cases} \lambda_{l,j} a_{l,j} \tau_x(a), & \text{if } y = x + e_{l,j}, \ l = 1, 2, \ldots, L; j = 1, 2, \ldots, J \\ \mu_{l,j} n_{l,j} \tau_x(a), & \text{if } y = x - e_{l,j}, \ l = 1, 2, \ldots, L; j = 1, 2, \ldots, J \\ 0, & \text{otherwise.} \end{cases}$$
$$(6.18)$$

Based on the action a taken in a state x, a cost $c(x,a)$ occurs to the network. The blocking probability can be expressed as an average cost criterion in our setting. We define the cost as

$$c(x,a) = \sum_{l=1}^{L} \sum_{j=1}^{J} w_{l,j} (1 - a_{l,j}) \quad (6.19)$$

where $w_{l,j} \in R_+$ is the weight associated with class j in network l.

The blocking probability for class j in network l can be defined as the fraction of time the system is in a set of states $X_{l,j}^b \subset X$ and the chosen action is in a set of actions $A_{x_{l,j}^b} \subset A$, where $x_{l,j}^b \in X_{l,j}^b$ and $A_{x_{l,j}^b} = \{a \in A : a_{l,j} = 0\}$,

$$P_{l,j}^b = \frac{\sum_{x \in X_{l,j}^b} \sum_{a \in A_{x_{l,j}^b}} \tau_x(a)}{\sum_{x \in X} \sum_{a \in A} \tau_x(a)}. \quad (6.20)$$

The blocking probability constraints can be expressed as

$$P_{l,j}^b \leq \phi_{l,j}, l = 1, 2, \ldots, L; j = 1, 2, \ldots, J \tag{6.21}$$

which can be addressed in the linear programming formulation by defining a cost function related to these constraints

$$c_{l,j}^b(x,a) = (1 - a_{l,j}), l = 1, 2, \ldots, L; j = 1, 2, \ldots, J. \tag{6.22}$$

6.4.2 Optimal Solution

The following linear program can be used to derive the optimal policy.

$$\min_{z_{xa} \geq 0, x \in X, a \in A_x} \sum_{x \in X} \sum_{a \in A_x} \sum_{j=1}^{J} \sum_{l=1}^{L} w_{l,j}(1 - a_{l,j})\tau_x(a)z_{xa}$$

subject to

$$\sum_{a \in A_y} z_{ya} - \sum_{x \in X} \sum_{a \in A_x} p_{xy}(a)z_{xa} = 0, y \in X$$
$$\sum_{x \in X} \sum_{a \in A_x} z_{xa}\tau_x(a) = 1, \tag{6.23}$$
$$\sum_{x \in X} \sum_{a \in A_x} (1 - a_{l,j})z_{xa}\tau_x(a) \leq \phi_{l,j}, l = 1, 2, \ldots, L; j = 1, 2, \ldots, J.$$

In the above linear program, the term $z_{xa}\tau_x(a)$ can be interpreted as the steady-state probability of the system being in state x and a is chosen. The first constraint is a balance equation and the second constraint can guarantee that the sum of the steady-state probabilities to be one. The blocking probabilities of class j sessions in network l are expressed in the third constraint to guarantee the blocking probabilities. The decision variables are $z_{xa}, x \in X, a \in A_x$.

6.5 Numerical Results and Discussions

Using some numerical examples, we illustrate the performance of the proposed optimal joint radio resource management scheme. We use two wireless networks, a WLAN and a CDMA network in the numerical examples. We compare the performance of the proposed scheme with another scheme, where radio resource management is done independently in individual networks and there is no interaction between these two networks. This scenario represents current systems in practice where WLANs and cellular networks are isolated networks.

One class of video traffic is considered in a system with a single WLAN cell and a single CDMA cell. Each video flow is 1.17 Mbps in the WLAN, which is generated by a constant inter-arrival time 10 ms with a constant payload size of 1464 bytes. It corresponds to a traffic-shaped constant bit rate (CBR) video flow. IEEE 802.11b physical layer is used and the average bit rate is assumed to be 11 Mbps. Because of the scarce bandwidth available in cellular networks and the small screen

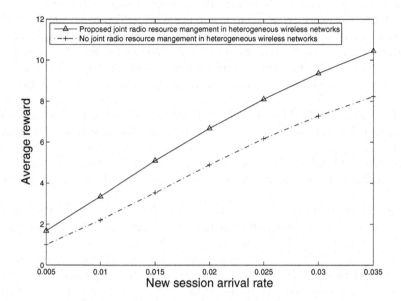

Fig. 6.2. Average reward in different schemes.

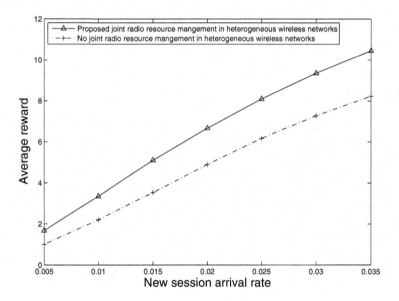

Fig. 6.3. Average reward in different schemes.

of handsets, we assume that the bandwidth consumption of a video session is smaller in the CDMA network compared to that in the WLAN. This bandwidth adaptation can be achieved by adjusting the compression parameters and coding techniques such as layered coding [20]. Specifically, the transmission rate for the video traffic in the CDMA network is 240 Kbps and correspond to an equivalent spreading gain $N = 16$, which can be interpreted as multiple code transmission using a higher spreading gain (say, $N = 256$) to ensure the accuracy of the asymptotic approximation of CDMA system with LMMSE receivers. The new session arrival rates in the CDMA and the WLAN areas are $\lambda_{c,n}$ and $\lambda_{w,n}$, respectively. The total new session arrival rate is $\lambda_n = \lambda_{c,n} + \lambda_{w,n}$. $\mu_{c,t}$ and $\mu_{c,h}$ are the session termination rate and the session vertical handover rate, respectively, in the CDMA network. $\mu_{w,t}$ and $\mu_{w,h}$ are the session termination rate and the session vertical handover rate, respectively, in the WLAN.

Fig. 6.2 shows the average rewards earned in different schemes. In this example, 40% of the total new session arrivals in the system occurs in the WLAN area. $\mu_{c,t} = \mu_{w,t} = 0.005$, $\mu_{c,h} = 0.004$ and $\mu_{w,h} = 0.0005$. $w_{c,n} = w_{c,h} = 2$. $w_{w,n} = w_{w,h} = 1$. From Fig. 6.2, we can see that the reward earned in the proposed scheme is always more than that in the other scheme. The percentage of reward gain is shown in Fig. 6.3. It is observed that the higher the new session arrival rate, the less the percentage of reward gain. This is because the system becomes saturated when the arrival rate is high, and the proposed scheme does not have much room to select sessions to admit based on the reward rate. Nevertheless, the reward earned in the joint radio resource management scheme is about 27% percent higher than that in the scheme without joint radio resource management scheme even when the system is in high load.

Conclusion

In next generation heterogeneous wireless networks, the complementary characteristics of different wireless technologies make it attractive to integrate them. There are several networking protocols that enable the integration of heterogeneous wireless networks. Designing efficient and effective radio resource management schemes is one of the major challenges in heterogeneous wireless networks. In this chapter, we have presented a framework of joint radio resource management scheme. The global balance equations and the blocking probabilities were described. Then, we presented an optimal admission control scheme that can maximize the overall network revenue subject to several QoS constraints in individual networks. Numerical examples have been used to illustrate the performance of the proposed scheme.

References

1. P. Johansson et al., "Bluetooth: An enabler for personal area networking," IEEE Networks. Mag., vol. 15, no. 5, pp. 28-37, Sep. 2001.

2. L. Yang and G. B. Giannakis, "Ultra-wideband communications: An idea whose time has come," *IEEE Signal Processing Mag.*, vol. 21, no. 6, pp. 26-54, Oct. 2003.
3. E. Gustafsson and A. Jonsson, "Always best connected," *IEEE Wireless Commun. Mag.*, vol. 10, no. 1, pp. 49-55, Feb. 2003.
4. S. Haykin, "Cognitive radio: Brain-empowered wireless communications," *IEEE J. Select. Areas Commun.*, vol. 23, no. 2, pp. 201-220, Feb. 2005.
5. M. Milliger *et al.*, Eds, *Software Defined Radio: Architecture, Systems and Functions*, Wiley, New York, 2003.
6. C. E. Perkins, *IP Mobility Support for IPv4*, RFC 3220, Jan. 2002.
7. J. H. Saltzer, D. P. Reed, and D. D. Clark, "End-to-end arguments in system design," *ACM Trans. on Comp. Syst.*, vol. 2, no. 4, pp. 278-288, Nov. 1984.
8. A. C. Snoeren and H. Balakrishnan, "An end-to-end approach to host mobility," in *Proc. of ACM MobiCom'00*, pp. 155-166, Sep. 2000.
9. R. Stewart and Q. Xie, *Stream Control Transport Protocol*, IETF RFC2960, Oct. 2001.
10. L. Ma, F. Yu, V. C. M. Leung, and T. Randhawa, "A new method to support UMTS/WLAN vertical handover using SCTP," *IEEE Wireless Comm. Mag.*, vol. 11, no. 4, pp. 44-51, Aug. 2004.
11. M. Handley *et al.*, *SIP: Session initiation protocol*, IETF RFC 2543, Mar. 1999.
12. J. Zhang, H.C.B. Chan, and V.C.M. Leung, "A SIP-based handoff scheme for heterogeneous mobile networks," in *Proc. of IEEE WCNC'07*, pp. 3946-3950, Mar. 2007.
13. H. Holma and A. Toskala, *WCDMA for UMTS: Radio Access for Third Generation Mobile Communications*, Wiley, NY, 2004.
14. J. Evans and D. N. C. Tse, "Large system performance of linear multiuser receivers in multipath fading channels," *IEEE Trans. Info. Theory*, vol. 46, no. 6, pp. 2059-2078, Sept. 2000.
15. F. Yu and V. Krishnamurthy, "Effective bandwidth of multimedia traffic in packet wireless CDMA networks with LMMSE receivers - a cross-layer perspective," *IEEE Trans. Wireless Comm.*, vol. 5, no. 3, pp. 525-530, Mar. 2006.
16. Y.-L. Kuo, C.-H. Lu, and G.-H. Chen, "An admission control strategy for differentiated services in IEEE 802.11," in *Proc. of IEEE Globecom'03*, pp. 707-712, Dec. 2003.
17. H. Zhai, X. Chen, and Y. Fang, "A call admission and rate control scheme for multimedia support over IEEE 802.11 wireless LANs, *Wireless Networks*, vol. 12, no. 4, pp. 451-463, July 2006.
18. K. W. Ross, *Multiservice Loss Models for Broadband Telecommunication Networks*, Springer-Verlag, 1995.
19. M. Puterman, *Markov Decision Processes*, John Wiley, 1994.
20. D. Wu and Y. T. Hou, and Y. Q. Zhang, Scale video coding and transport over broadband wireless networks, *Proceedings of IEEE*, vol. 89, no. 1, pp. 6-20, Jan. 2001.

7

Vertical Handoff and Mobility Management for Seamless Integration of Heterogeneous Wireless Access Technologies

Nirmala Shenoy and Sumita Mishra

Networking, Security and Systems Administration Department
Rochester Institute of Technology, Rochester, USA
{nxsvks, sxm1145}@rit.edu

7.1 Introduction

Communications networks today are a combination of different wireless and wired networks, that are based on an enriched set of technologies. Wired networks are preferred when the 'communicating' user's mobility or freedom to roam is not a prime concern. However, given the busy schedules of humans today and the tremendous growth rate in travelling professionals, wireless networks are gaining popularity and are becoming an integral part of our lives. Research in innovative wireless technologies continues in the academia and industry as we seek to provide enriched wireless services to cater to the requirements of the variety of roaming users. Restricting the movement of a roaming user to a certain zone or service provider or interrupting a service because the user moved out of the administrative domain of a wireless network will no more be attractive, especially since wireless networks today span the globe and offer a rich variety of services. Hence, providing continued connectivity as users roam across different wireless networks with acceptable quality of service continues to be a topic of intense interest.

One of the major requirements for continued call connectivity is that the user's call or session be handed over seamlessly as he traverses the different wireless networks. It would be beneficial however if the capabilities of the existing wireless networks be leveraged for this purpose. Secondly, it would be highly preferable if the solutions can cater to roaming across several types of wireless networks that exist today, while keeping in mind that new wireless technologies will emerge and the solution should evolve with the new technologies, networks and growing user's demands. The process by which a user gets handed over from one wireless network to another 'near seamlessly' or with the least disruption to the service is called *vertical handoff*. Handover and handoff are synonymous. In this chapter we use the term handoff predominantly except when quoting work from other authors who preferred to use the term handover.

E. Hossain (ed.), *Heterogeneous Wireless Access Networks*,
DOI: 10.1007/978-0-387-09777-0_7, © Springer Science+Business Media, LLC 2008

We have qualified "seamlessly" in the later part of the last paragraph with the term 'near' and described this process as causing least disruption to the ongoing services of the roaming user. It may not be possible to provide a truly seamless handoff of an active connection due to several technical, administrative and other considerations that differ with different access technologies used in the deployed wireless networks - i.e. some glitch in the service may be noticed during a handoff. Moreover, the term seamless is subjective and decided by the user. A real-time service user would accept delays below 50 milliseconds for seamless transfer of his services, whereas a non real-time user would tolerate several seconds of delay. Comparatively, a non-real-time user who has several active connections to the Internet (which for example he started while working in his office), would prefer to have these connections active as he travels to his next destination; for him delay is not an important factor. Seamless service support can thus have different significance for different users.

The examples cited so far indicate a transparency from the application perspective, i.e. real-time, non-real-time and an Internet connection. There are users who would prefer a lower service quality if available at a lower price - a very good example here would be students who are a vast majority of the roaming user population. For such users, the service changeover may not be as smooth as the users who prefer high quality of service and are willing to pay for it. However, both have opted for seamless handover, but their preferences for "seamless service" and hence requirements differ considerably. To achieve seamless or near seamless handoff, several processes at several layers of the protocols stack are required to cooperate in a holistic manner, which is a major technical challenge. In this chapter, we explain several handoff considerations and processes by highlighting some significant work conducted in this topic area.

Several decades ago, the term "seamless roaming" had no significance. This was the time when there were two major communications systems - the telecommunications networks, and the data networks namely the Internet. At that time, there was no necessity to support any type of roaming across these networks for the simple reason that user mobility and wireless networks had not proliferated much into the 'communications user' community. With the phenomenal advances in hardware and software technologies, network technologies and networks also experienced a huge growth. Needless to say that day to day e-business transactions, e-commerce, distance education, web based databases and commercial infiltration impacted the growth of communications networks, which was already being spurred by the growth in the underlying hardware and software technologies. With the Internet staging a phenomenal growth a couple of decades ago, combined with the proliferation of new and innovative wireless technologies and networks, and the increasing necessity for users to communicate and access personal data anywhere anytime, led to the critical need to bridge the two major networking technologies - the wireless and the wired. One noticed at this time, the attempts by all wireless networks to get tethered to the Internet to leverage its capabilities and extend Internet to the roaming users. Meanwhile the distinction between telecommunications and data communications started diminishing, resulting in a convergence in the services supported by the two major networks, leading to an all-IP communications paradigm. Traditional cellular networks, which were predominantly catering to voice services were faced with accepting a partnership with *wireless local area networks* (WLAN), which was predominantly for data communications. These integration efforts in the two major distinctly different wireless communications paradigm was the first to trigger interest in vertical handoff to facilitate integrated seamless services support across different wireless networks.

The continuous strive for better technologies both from a vendor's perspective to increase revenue and improve services, and from a researcher's perspective to overcome the challenges in wireless communications, has resulted in the emergence of numerous other wireless networks based on several different types of wireless technologies. Under the varied network and service requirement scenario, it is obvious why vertical handoff has gained importance and will probably continue to be a major interest area as newer technologies and services continue to proliferate the wireless networking market.

Seamless roaming or mobility is crucial to ubiquitous computing and requires network management operations to avoid service degradation. Handoff management is one of the major functions under *mobility management* that embraces several aspects of user mobility and mobility support procedures for wireless networks. Location management which includes paging and location update is the other major function. Handoff management includes wireless terminal handoff management considerations within one network called horizontal handoff and handoff management across different wireless networks which could be based on different wireless access technologies which is termed vertical handoff. The focus of this article is primarily on vertical handoff and associated mobility management considerations.

7.2 Advantages of Vertical Handoff

Though there are currently several deployed wireless networks, none of these can provide high bandwidth, low latency, low power consumption and wide area data service to large numbers of mobile users. However, when one considers each of the deployed wireless networks separately, they are capable of supporting a partial subset of such service features and by unifying their capabilities, we can achieve the highly desirable combined service capabilities mentioned above. For example, let us consider the two major wireless networks more widely deployed today i.e. the cellular networks and WLAN. The capabilities and underlying technologies in these networks are complementary. Bandwidth consuming applications and WLAN hot spots result in clustered traffic, which can be a major concern. Cellular networks that have overlapping coverage with WLANs can off-load some of the traffic from such applications to WLANs where available and conserve their costly resources while offering cheaper services to users by supporting them on the WLAN. This will help wireless Internet service providers to generate more revenue while using existing infrastructure. Network subscribers meanwhile will enjoy the best features of both technologies, universal coverage, larger bandwidth when required and lower costs for the seamlessly combined services on a single bill.

It is not uncommon today to find multimode terminals capable of working across WLAN and cellular networks. It is hence a fair inference that future mobile terminals will have multimode capabilities to maintain connectivity with different radio access technologies. Ubiquitous computing can become highly efficient if user mobility across several wireless networks is supported in a seamless manner as they will receive intelligent, context-aware services. The seamless movement could refer to personal mobility which allows a user to receive services at any terminal device, and terminal mobility which allows the device to receive services as it moves across different wireless networks.

Having given an introduction to vertical handoff and some of its advantages, some of the technical challenges in the implementation of efficient vertical handoff are discussed in the next section.

7.3 Technical Challenges to Vertical Handoff

The handoff of a session across heterogeneous wireless networks, if not executed properly, will result in the mobile user experiencing significant variations in quality due to the handoff delay. This delay can be caused by

- message exchanges,
- multiple database access,
- service negotiation and renegotiation,
- different service definitions and provisions due to different service providers,
- terminal mobility, and
- resource availability.

In extreme conditions the mobile user may experience a connection break. The vertical handoff process is hence critical and calls for consideration at several layers of the protocol stack from a technical perspective, complex involvement and service level agreements across wireless service providers from an administrative perspective and user's requirements and preferences from the subscriber perspective. We highlight several of these technical challenges in the following sub-sections.

7.3.1 Heterogeneity in Wireless Technologies

We start with the heterogeneity of existing wireless networks based on a heterogeneous set of wireless technologies, as this poses a major technical challenge. In Fig. 7.1, we cover a subset of the existing wireless technologies that are current major players and would continue to impact vertical handoff considerations in the future.

Of the technologies noted in Fig. 7.1, cellular networks and WLAN are the most popular ones from the user's point of view. WPANs are getting popular with the numerous personalized devices that have proliferated the market in the last few years. As the person moves, these devices may move as a group. Hence the considerations

Network	Standard	Data rate	Frequency band
Cellular networks	UMTS, 3G,	Up to 2 Mbps	1900-2025 MHZ
	4G	100 Mbit/s (high speeds) 1 Gbit/s (stationary conditions)	
WLAN	IEEE 802.11b	1-11Mbps	2.4GHz
	IEEE802.11n	100-540 Mbps	2.4 GHz, 5 GHz
Wireless Personal Area Networks (WPAN)	IEEE 802.15.3	11-55 Mbps	2.4 GHz
Zigbee	IEEE 802.15.4	20-250 Kbps	868 Mhz, 915 Mhz
Wireless Metropolitan Area Networks (WMAN)	IEEE 802.16.a	75 Mbps	2-11 Ghz
WiMAX	IEEE 802.16c	134 Mbps	10-66 GHz
Wireless Wide Area Networks (WWAN)	IEEE 802.20	2.25 – 18 Mbps	3.5 GHz
HIPERLAN/1 HIPERLAN/2 HIPERLAN/3 HIPERLAN/4	ETSI BRAN	23.5 Mbps 1-54 Mbps 25-100 mbps Up to 155 Mbps	5 GHz 5 GHz 40.5 -43.5 GHz 17 GHz

Fig. 7.1. Existing wireless technologies.

of such networks and WPAN devices in a seamless roaming study is very important. Zigbee has been getting popular more from an industry requirements perspective. WMAN and WiMAX are technologies that are competing for wide area coverage supported by cellular networks. Mesh networks are not listed because they primarily define a wireless network topology for backbone networks but are of interest both to the wireless and wired network operators. A contemporary of WLAN in several European countries is the HiperLAN, and hence to be included in the roaming studies.

7.3.2 Handoff at Layer 2 and Layer 3

Handoff of a user service across cells in cellular networks or across WLAN access points within one subnet can be completely executed at layer 2 of the protocol stack. Based on basic handoff decision criteria such as *received signal strength* (RSS), *signal to noise ratio* (SNR), and channel availability, it may be decided that the best base station in a cellular network or the best access point in WLAN is not the current one, and the mobile requires to be transferred to one of the neighboring base stations or access points. As this handoff is executed local to the wireless network or subnet, there is no necessity to involve layer 3 protocols. This type of handoff falls under the general category of horizontal handoff.

When the node changes wireless networks, its IP address (and/or its *mobile switching center* (MSC) if it is a cellular network node) may change. In such cases, the handoff cannot be executed completely at layer 2, as decisions have to be made at higher layers. With the all-IP paradigm and the advent of IP into cellular networks, inter-wireless networking and vertical handoffs experience a change in IP address, requiring the involvement of protocols at layer 3. Note that vertical handoff decisions also require the criteria used for horizontal handoff, namely RSS, SNR and channel availability, i.e. vertical handoff cannot be executed in isolation from the handoff process at layer 2. Vertical handoff thus faces some serious challenges as it becomes necessary for layer 2 and layer 3 handoff mechanisms to work with each other. Typically, layer 2 handoff should trigger layer 3 handoff and the timings of handoff initiation, execution and completion at the two layers has to be fine-tuned to reduce vertical handoff delays, which would otherwise have a major impact on the vertical handoff performance.

Horizontal handoff mechanisms are very mature in cellular networks. In WLANs, due to the limited coverage afforded by a single WLAN access point and the medium access mechanism which typically follows a random scheme, horizontal handoff is handled differently. When a user roams across several access points in the same subnet, *Inter Access Point Protocol* (IAPP) provides message exchanges between the access points, thus facilitating a smooth handoff of the mobile user across the access points. However a mobile user can cross several subnets or domains, which then requires notification and handoff initiation at layer 3. The delays in the upper layers can extend to several seconds, which can heavily impair the handoff performance in WLANs spanning different domains or subnets. This issue with WLANs has been intelligently circumvented by micro mobility management techniques. Micro mobility is able to efficiently execute handoff across subnets but within a domain. Before understanding micro mobility and its significance, it is important to understand macro mobility. These two mechanisms are discussed in the following section.

Several research solutions have opted for predictive handoff to reduce handoff latency. Quick handoff and or predictive handoff also require interaction between layer 2 and layer 3 protocols. Several such schemes can be found in [1]- [5]. These schemes target layer 2 and layer 3 interaction for WLANs or for cellular networks.

7.3.3 Macro and Micro Mobility

Macro mobility refers to mobility across administrative domains or inter-domain mobile movement, which primarily gets handled at layer 3 of the OSI model. *Mobile IP* is used for macromobility and is a concept which is fundamental to most vertical handoff solutions in the all-IP paradigm. The mobile IP concept was introduced when several wireless networks had proliferated the networking market and required access to the rich services offered by the Internet. Providing wireless Internet services to the stationery user is not a major problem, but nomadic or mobile users change their network or point of attachment. The major challenge was posed because of the way *Internet Protocol*, routed packets to destinations based on the destination network ID in the IP addresses. If the destination node is a mobile, its point of attachment will change as it moves across different wireless networks, thus resulting in change of IP addresses and the network ID, making service transparent to mobility impossible. To maintain transport-layer connections, the IP address must be the same for the duration of a transport layer session and any change in the IP address can cause the connection to be disrupted, resulting in data loss. Delivery of packets to the mobile node's current point of attachment requires an IP address for the mobile node in the new network.

Mobile IP solved the above problem by allowing the mobile node to use two IP addresses, the *home IP address* which is the permanent address of the mobile node and a *care-of address* that changes as the mobile node moves to different networks. Mobile IP proposed the use of a home agent in the home network, which would intercept the packets for the mobile node which is away from its home network. To subsequently forward the intercepted packets, the home agent should be aware of the mobile node's new IP address. The mobile node hence registers its newly acquired IP address (or care-of address) with the home agent using binding updates [6]. The home agent constructs a new IP packet with a header that contains the mobile node's care-of address as the destination IP address, encapsulates the original IP packet in the payload and forwards this IP packet to the mobile node in the foreign network. Such encapsulation is also called *tunneling*. This process is explained in Fig. 7.2.

Micro mobility was designed to avoid passing the handoff decision control to layer 3 as long as the mobile user did not cross domains. The approach is simple and is able to support intra-domain roaming from a layer 2 perspective. Mobile IP uses binding updates to inform the home network of a mobile node's new care-of-address, which can cause an increase in signaling load as well as delays, leading to a compromise in the quality of service. Micro mobility protocols reduce the need to send such messages through the scheme illustrated in Fig. 7.3 [7]. A hierarchy of base stations is organized under a foreign agent in the foreign network (in Fig. 7.3, it is shown as the mobility access point) such that when the mobile node changes base stations within the same domain, location information is propagated only through the domain's local routers and requires no binding update message to be sent to the home network. However, when a mobile node moves into a new domain, a traditional mobile IP handoff is necessary.

Categorization of Handoff Procedures

It should be clear by now that handoff is an event whereby a mobile device gets handed over from one transceiver station to another. Vertical and horizontal handoffs are different from the wireless access network perspective while micro and macro mobility are distinguished from an administrative domain perspective. It follows intuitively that one can categorize the handoff procedures into four different categories of roaming and handoff scenarios as indicated below.

- *Vertical macro mobility*: Different wireless access network under different administrative domains.
- *Horizontal macro mobility*: Same type of wireless access networks under different administrative domains.
- *Vertical micro mobility*: Different wireless access networks under same administrative domain.
- *Horizontal micro mobility*: Same type of wireless access networks under same administrative domain.

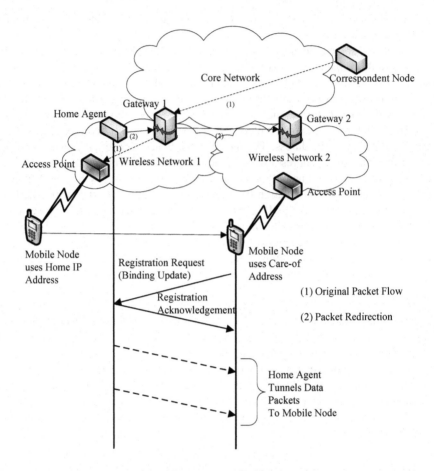

Fig. 7.2. Operation of Mobile IP.

7.4 Background - Evolution of Wireless Networks

It is important at this point to cover some aspects in the cellular network evolution which were driven by user preferences, their requirements for voice and data convergence, user mobility requirements and the influence of IP centric services and Internet penetration. The advent of vertical handoff can be noticed clearly as one follows the evolution of the cellular networks.

7.4.1 First Generation (1G) Networks

First Generation Wireless Networks (also known as 1G Networks) were primarily designed for low data rate voice and data communications [8]. Newer generations were developed for high speed data, video and multimedia support. Also, the emphasis on mobile content delivery increased with every new generation of wireless

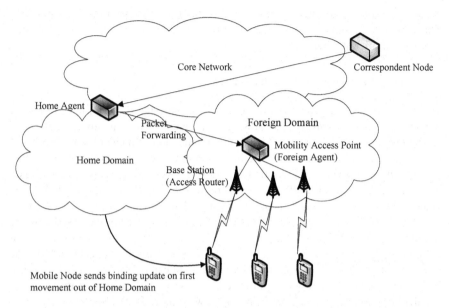

Fig. 7.3. Micro mobility.

networks. 1G Networks initially developed in Japan in 1979 were based on analog transmissions. Some of the key 1G standards are AMPS (Advanced Mobile Phone System), TACS (Total Access Communication System), JTACS (Japanese Total Access Communication System) and NMT (Nordic Mobile Telephone). AMPS (United States), TACS (United Kingdom) and JTACS (Japan) were designed to operate in the 800-900 MHz band while NMT (Denmark, Finland, Norway and Sweden) was originally developed to work around the 450 MHz band [9]. The range of frequencies and difference in operational aspects among the first generation cellular networks is worth noting. *Frequency shift keying* (FSK) was used for signaling and *frequency division multiplexing* (FDM) was used for spectrum allocation in these networks.

CDPD (Cellular Digital Packet Data) was introduced as an early wireless data networking technology. It was designed primarily to take advantage of the unused bandwidth of cellular networks and was deployed as an overlay on the analog systems that were already in existence. Hence this technology was attractive for the service providers as it would make use of the silent periods in conversations, the time when a call was undergoing handoff and so on. The greatest advantage that CDPD offered was the fact that it was designed as an extension of existing IP networks. The throughput achieved in CDPD ranged from 9.6 Kbps to 19.2 Kbps.

Horizontal handoff specifications were provided as part of the 1G standards. Based on the RSS by the *base transceiver station* (BTS) connected to a mobile, the BTS was able to determine whether the mobile is within the cell or moving away from the current cell. If the BTS concluded that the mobile is moving away from the cell, the surrounding BTSs were alerted and the BTS that received the strongest signal from the mobile picked up the call as the mobile crossed the cell boundary. This process of assigning the frequency of the neighboring cell for call continuation resulted in horizontal handoff.

7.4.2 Second Generation (2G) Networks

The advent of second generation or 2G networks introduced digital technology for wireless communications. By late 80's, cellular networks were gaining popularity at a very fast rate and the analog wireless communication framework was not able to handle the increasing demands for tetherless communications. In order to increase the capacity of 1G networks, the first 2G networks were introduced. While 1G networks were designed primarily for voice traffic support, 2G systems could support voice and limited data communications, such as *short message services* (SMS), paging, email, and fax services [8]. Spectrum efficiency was greatly increased with the introduction of digital technology. Compared to 1G systems, 2G systems used digital multiple access technology, such as *Time Division Multiple Access* (TDMA) and *Code Division Multiple Access* (CDMA). The architecture of 2G systems was similar to the 1G systems. However, in order to increase the capacity of the network, smaller cells were deployed and hence there were more BTSs. Several BTSs were now connected to a *Base Station Controller* (BSC) that was in turn connected to the MSC. The MSC handled most of the operational aspects of wireless connections. Roaming, billing and interconnections were handled by several databases such as *home location register (HLR), visitor location register (VLR), authentication center* and *equipment identity center* that were co-located with the MSC.

   Global System for Mobile Communications (GSM) was one of the most popular 2G standards. It was first introduced in 1992 and achieved unprecedented success all over the world. It was the fastest growing wireless communications technology of all times and according to a survey conducted in January 2006, about 670 commercial GSM networks were serving customers in 210 countries worldwide [9]. Depending on the implementation band, GSM operates in the 850 - 1900 MHz frequency range. It uses TDMA with *Frequency Division Duplex* (FDD) and does not depend on the underlying 1G analog systems. Some of the other 2G standards that emerged at this time were Digital-AMPS based on TDMA and IS-95 based on CDMA.

   In addition to changes in the MSC-BSC-BTS design, the mobile-assisted handoff mechanism was introduced in 2G networks. The mobiles were designed to sense the signals received from adjacent base stations and had the capability of initiating handoffs by exchanging control messages with the network.

7.4.3 2.5G Networks

2.5G networks were essentially a bridge between the 2G and 3G networks. They may also be considered an extension of 2G networks as they used circuit switching for voice and packet switching for data transmissions. Some of the technologies that fall under this category are *general packet radio service* (GPRS), *enhanced data rates for global evolution* (EDGE) and the *GSM EDGE radio access network* (GERAN).

GPRS is a radio technology for GSM networks characterized by packet-switching protocols and shorter setup time for *Internet service provider* (ISP) connections. Besides providing the convenience of getting connected from anywhere, IP-based network connectivity provides the mobile user with advanced-features such as colored Internet browsing, video streaming, multimedia messaging and location-based services. From the consumer's point of view, one of the main advantages of IP based packet-switching is the fact that billing is done on the amount of data transmitted and not on the channel hold time as was done for voice in circuit switched networks, which was the predominant technology used in early cellular systems.

EDGE builds on GPRS and was focused on a personal multimedia environment. With EDGE, GSM service providers could operate on existing GSM radio bands to offer wireless multimedia IP-based services and applications at speeds up to 384 Kbps with a bit-rate of 48 Kbps to 69.2 Kbps per timeslot. It also used TDMA (like GSM). Hence deployment of EDGE was fairly simple for GSM operators. GERAN is an extension of EDGE that was projected to have a data rate of up to 1920 Kbps and could support voice as well as real-time data services. It provided an interface for GSM to an all-IP core network of the third generation.

In essence, GPRS and 2.5G networks marked the beginning of the vertical handoff era. Since the focus had shifted to providing reliable data services, seamless mobility between heterogeneous networks (e.g. WLAN, which had started entering the wireless network market and GPRS) was a much desired feature and was addressed in the future generations.

7.4.4 Third Generation (3G) Networks

3G networks were proposed to eliminate some of the problems faced by 2G and 2.5G networks. Foreseeing the rise in demand for multimedia wireless services, 3G was the first step towards the broadband wireless realm. 3G networks are wide area cellular networks which evolved to incorporate high-speed Internet access and video telephony. Large scale implementation of 3G wireless was expected to launch in 2001, but it did not come into existence till several years later and 3G deployment continued not so successfully. Some of the challenges in 3G implementation included licensing of new spectrum for some countries, deployment of a new infrastructure, lack of 3G handsets, interoperability issues with 2G networks, high operating costs for providers resulting in higher cost for the consumers and lower than expected performance results.

The primary goal of 3G networks was to provide the end user with seamless connectivity and roaming, while availing multimedia data and voice services. Hence the user would get high-quality audio and video services along with global roaming that would enable him to connect to any of the available wireless networks anywhere. Fig. 7.4 illustrates the envisioned seamless roaming feature of 3G, where this feature is addressed through the umbrella coverage concept, which was getting popular then.

In order to achieve interoperability between different wireless networking technologies, vertical handoffs were now a required feature along with horizontal handoffs. However, one of the reasons that prevented wide-scale deployment of 3G net-

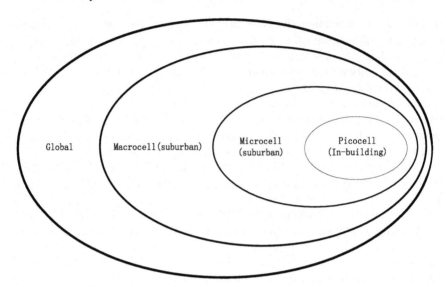

Fig. 7.4. Seamless roaming feature of 3G networks.

works was its too difficult to achieve goals and the requirement for separate stan-dardization entities all over the world to resolve the interoperability issues.

7.4.5 Fourth Generation (4G) Networks

Even though 3G is still evolving and many of its challenges are still not addressed, the next generation wireless networks i.e. 4G, also known as "wireless broadband" and beyond 3G or B3G is already being developed and is anticipated to be deployed in the 2010-2015 time frame [9].

4G networks are anticipated to not only address the problems and limitations of 3G, but also provide better QoS, more bandwidth and lower the cost for the service providers and the consumers. The main differences between 3G and 4G networks will be the higher data rates that can be expected in 4G i.e. up to 100 Mbps, bet-ter multimedia support and global roaming capability. 4G networks will be based completely on packet switching, unlike the combined circuit and packet switching feature of 3G. IPv6 is gradually being adopted in the telecommunication infrastruc-ture and 4G networks will be based on this upcoming technology. Each mobile node will have a *home* IPv6 address and a *care-of* IPv6 address based on its current lo-cation to harness the mobileIPv6 capabilities. When the mobile needs to be reached by any device on the Internet, it will send the traffic based on the home IP address of the mobile. A server on the home network of the mobile node will in turn tunnel the traffic to its care-of address. This server will also let the sending device know of care-of address so that the future communication link between the sender and the receiver can be setup directly. This mechanism is very different from the HLR/VLR concept for mobility support in the earlier generation networks.

Besides the location management mechanism being different for 4G networks, the handoff mechanism is also enhanced. Since 4G networks are envisioned to integrate different access technologies and networks, support for seamless roaming across these networks will require very accurate vertical handoff schemes and mechanisms to minimize the handoff call dropping rate. The challenge lies in developing a robust vertical handoff mechanism that incorporates features such as varying node mobility patterns, different cell radii like macro cells, microcells and picocells. Fast and seamless handoff is a big challenge for 4G networks that are designed to support real-time high speed multimedia applications and require small handoff delay coupled with high data-rate transmission.

The aim of 4G cellular networks is a framework for truly ubiquitous IP-based access by mobile users, with special emphasis on the ability to use a wide variety of wireless and wired access technologies to access the common information infrastructure. The 4G vision also embraces local-area access technologies, such as IEEE 802.11-based WLANs and Bluetooth-based WPANs. The development of mobile terminals with multiple physical or software-defined interfaces is expected to allow users to seamlessly switch between different access technologies which are expected to have overlapping areas of coverage and different cell sizes. Figure 4 is schematic of the multi-technology vision of 4G. Conventional wide-area cellular coverage will be available in all outdoor locations while a corporation based 802.11 WLAN access would provide service in public indoor locations such as parking lots with Bluetooth based access in every individual office. As mobile users commutes, their ongoing voice over IP calls will be seamlessly switched from the cellular to the WLAN and finally to WPAN access point. A *domain* can comprise of several such multiple access technologies and mobility management protocols should be capable of handling *vertical handoffs* across these heterogeneous technologies.

The revolutionary drivers for 4G include a push toward universal wireless access and ubiquitous computing through seamless personal and terminal mobility. The goals of 4G or B3G can be summarized to be based on the following five elements:

1. fully converged services,
2. ubiquitous mobile access,
3. diverse user devices,
4. autonomous users, and
5. software dependency.

Realization of the envisioned 4G requires unified efforts in the areas of standardization and development of the enabling technologies providing access network convergence in an evolutionary manner and designing system architectures that can enable access network convergence.

Fig. 7.6 illustrates the growth not only in the wireless technologies but also in the applications data rates, which were primarily spurred by data services.

There are a variety of issues related to handoff [10]. These issues can be divided into architectural issues and handoff decision issues. Architectural issues are those that are related to the methodology, control, hardware and software elements

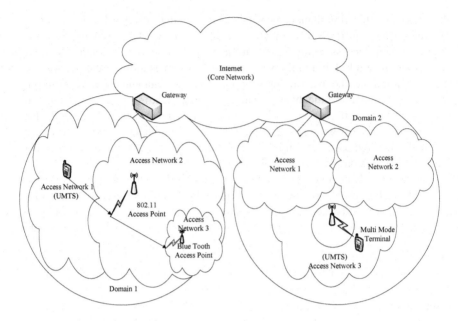

Fig. 7.5. Seamless vertical handoff in 4G networks.

involved in rerouting the connections. Issues related to the handoff decision are the type of algorithms and metrics used by the handoff decision schemes.

Having given the appropriate background, we now discuss the technical considerations related to vertical handoff.

7.5 Vertical Handoff Considerations

7.5.1 Handoff Process

In a handoff process, as the mobile terminal transitions from one transceiver to another, the channel (frequency, time, spreading code or combinations of them), associated with its current connection has to be changed while the call or session is in progress [11]. The whole handoff process can be thought of as consisting of three phases [12].

System discovery: In this phase, the system may periodically monitor for a better network to which the mobile terminal can be handed over. In some cases, the discovery process is initiated only when the current network conditions have deteriorated below a certain threshold. In either case, depending on the algorithms and the goals set for handoff, the monitoring and handoff considerations may include several different criteria.

Handoff decision: The handoff decision uses an algorithm that optimizes based on a selected set of criteria to decide when to handover and the best network to

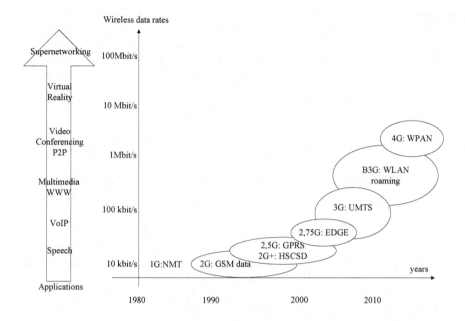

Fig. 7.6. Growth of wireless networks and applications.

handover to. The decision is very crucial and several different interesting solutions have been proposed to address this problem.

Handoff execution: This is the last phase in any handoff procedure where messages are exchanged between the two networks to reroute the user call to the new network. The handoff is executed based on a preplanned approach and has to take into consideration the implementation issues and should cause the least disruption to existing infrastructure.

7.5.2 Vertical Handoff Criteria

The above three processes are applicable both to horizontal handoff and vertical handoff. While in horizontal handoff, decision are based primarily on RSS, SNR and channel availability, in the case of vertical handoff, battery lifetime, congestion in the network, network coverage, velocity of mobile users, number of users in the network, and the number of users crossing the network among several others could be the deciding criteria. With increasing number of wireless networks and the pervasiveness of different user centric services, it has become important to consider soft parameters like user preferences and costs [11]. The decision criteria is further impacted because users and service providers have widely varying goals.

Timely and reliable transfer of a mobile user's connection(s) is important in vertical handoff. This requires transfer of the context of the link including the security associations, QoS guarantees, and any special processing operations. A context transfer protocol is necessary to allow exchange of state information regarding a mobile node's packet treatment. These include protocols for managing QoS guarantees, header compression, and authentication, authorization, and accounting (AAA).

According to 4G vertical handoff requirements as specified in RFC 2753 of the Internet Engineering Task Force, following are some of the criteria to be considered for handoff. A combination of these can be defined in a database from which the handoff algorithm picks the ones that it intends to use.

- *Different service types* require various combinations of reliability, latency and data transfer rates and it is very important that handoff decisions take the service type into consideration.
- *Monetary considerations*: Different networks implement different billing strategies, which will affect the users choice on the network he wishes to be handed over to.
- *Network conditions*: Network traffic, available bandwidth, latency in the network and congestion in the network have to be considered in making decisions on the most suitable network to handover a user to. This information could also be used for load balancing and relieving congestion in the network.
- *System performance*: SNR, Bit Error Rate (BER) and battery power are required to guarantee a certain system performance to the user.
- Mobile Terminal related: The velocity of the mobile, its mobility pattern, the mobility history, and its current location are also important handoff consideration factors.
- *User preference*: Some users may have special preference to systems with certain types of behavior.

Most of the handoff criteria that are mentioned above are highly correlated and cannot be addressed separately, thus making it important that the vertical handoff decisions optimize across multiple criteria. A multi criteria based handoff has a greater potential for achieving the desired balance among different systems. While simultaneous consideration of several significant aspects can enhance handoff procedure and satisfy system requirements, service provider goals and user preferences, the algorithm to optimize across several criteria gets complex.

The set of criteria for handoff between any two pairs of wireless networks may not be uniform as the procedure is governed by the goals targeted by the handoff algorithms, which in turn depend on the service providers and the underlying network technologies. Network heterogeneity compounds the problem as the criteria for handoff across different networks vary not only due to technical considerations, but also due to the administrative and billing policies in the network. Asymmetric handoff is common where the handoff while moving from one type of wireless network to another type is different when the user makes the movement in the reverse direction. For example, handoff considerations while moving from cellular to WLAN is not the same as when moving from WLAN to cellular. This is because the motivations for handoff from WLAN to cellular are very different from the motivation for handoff from cellular to WLAN. This disparity in handoff requirements has led to distinct consideration of the two processes for moving in and moving out of WLAN to a cellular coverage.

Though any vertical handoff consideration should look for roaming across heterogeneous wireless networks, during the early 2000s, there was a great drive to facilitate seamless roaming across cellular and WLAN. This was due to the fact that cellular networks had permeated considerably into the mobile user population and were offering low bandwidth high quality voice services with serious limitations in handling data traffic, while WLANs had the highly desirable features such as high data rates, lower costs and reduced coverage area, which complemented the cellular network capabilities. Hence one notices a flurry of research initiatives to integrate cellular and WLAN during this period, which continued as researchers and vendors wanted to improve the solutions proposed by their predecessors. Since then, several new wireless technologies and networks like Wireless Mesh Networks, WiMAX and Mobile Ad Hoc Networks have emerged but they have not achieved the popularity, success or wide deployment of either the cellular or WLANs. Hence the focus to efficiently support seamless roaming across these two highly proliferated network technologies continue. In the following discussions on handoff decision criteria and handoff algorithms, we primarily consider handoff between cellular and WLAN and vice versa.

7.5.3 Handoff Decision Criteria

The criteria which can be considered for efficient vertical handoff can be numerous, but it is neither feasible nor necessary to take all of them into cosideration for the decision making process. Several solutions have been proposed using only a subset of the criteria, based on some predefined goals. In most cases, there is a high dependency on the service providers and what they think is optimal to satisfy their customers. We highlight some of these criteria cosiderations and the rationale behind the selection of the criteria.

- In cellular networks, handoff can be between neighboring cells or between micro and macro cells. The criteria for handoff in both cases are considered separately

  - *Micro cell handoff*: RSS with threshold and hysteresis is an important criteria for handover [13]. RSS with threshold avoids handing over a mobile node too quickly to the neighboring base station. Handoff is initiated only if the signal strength of the current base station drops below a threshold value. Similarly RSS with hystersis avoids the ping pong effect at the border of the cell, whereby handover is initiated with a neighboring cell only if its signal strength is above a certain value.

  - *Macro-microcell handoff*: The overlay concept, which is so popular with current WLAN and cellular networks, existed even with cellular networks where a macro cell would provide an umbrella coverage to the microcells and the two could be using different radio links. In this case, velocity of the mobile was an important consideration. Slow users would be put under the micro cell coverage, while fast users were supported preferably under the macro cell coverage. This was addressed through a multi-tiered control approach [13].

- Bandwidth, latency, bit error rate and network costs could change with user mobility across networks and combining them with RSS, SNR and channel availability (commonly used criteria for horizontal handoffs) would be a good minimal criteria set for vertical handoffs. Bandwidth, bit error rate and latency have direct impact on user preferred QoS and hence require consideration [11].

- Bandwidth, velocity of mobile terminal and number of users in the target or after-handoff network is another combination which is based on the following rationale. Bandwidth is important because of the difference in bandwidth availability in WLAN and cellular networks. Velocity of the mobile and its mobility pattern is important as a fast moving mobile may quickly cross over the coverage range of a WLAN and it may not be appropriate to handover such a user from cellular to WLAN as this could result in quick successive handoffs and associated signaling overheads and delays [14]. Lastly, the number of users in the target network is important. Whether it is cellular networks which primarily uses some scheduled multiple access scheme or WLAN, which uses carrier sense random access scheme, increase in the number of users beyond a threshold value can severely impact the service quality in the network.

- Application QoS, call duration and a signaling cost function that takes into account network resources, signaling and processing overloads incurred in call rerouting (including authentication, authorization and context transfer) is another combination that may be considered.

- Maximizing the QoS experienced by each user, where the user services are optimized by deciding the best network for all sessions or deciding on a set of networks suited to the different sessions of the user, has also been considered as a possible decision criterion [15]. A network elimination process is used to cut down on the suitable networks to be considered for handover. A handoff cost function that accounts for the dynamic values inherent to vertical handoff is defined and is used to measure the benefit obtained by transferring either all sessions holistically to one network or different sessions to different networks.

- Another interesting handoff decision goal is the overall performance optimization of the access network in terms of battery life of mobile nodes and load balancing in the access networks. This attempt was a natural follow-on to the adoption of the Media Independent Handoff Function (MIHF) proposed in IEEE 802.21 for message exchanges. MIHF maintained the battery life information of the mobile nodes and also the load on the access points [16].

- Several initiatives have focused on restricting the number of handoffs to a minimal while considering the integrated network utilization or wireless resource utilization and QoS of the active applications in terms of bandwidth and latency.

- Another interesting approach is extending WLAN utilization as long as possible to lower costs and provide better data rates for data applications [17]. The dwell time in WLANs is an important factor to consider if we want to extend WLAN utilization, especially when moving out of the WLAN into cellular. This approach however is not suitable for real time applications.

7.5.4 Handoff Decision Algorithms

In the following sections, we list some of the proposed algorithms categorized by the approach used by each of them. There are several interesting handoff decision algorithms which have been investigated and we discuss a selected subset of such algorithms.

Knowledge-Based Systems

Pattern recognition can simplify the decision process when multiple criteria are to be considered and knowledge-based systems are considered highly suitable for this purpose. Fuzzy logic and neural networks are suitable due to their non-linearity and generalization capabilities that aid in pattern classification. Neural networks have training overheads and require training data and pre-training. However, once trained, multi criteria algorithms can be used to optimize them, even when there are conflicting criteria. Fuzzy logic allows simultaneous evaluation of several handoff criteria, thus increasing the precision of the handoff decisions through simple mathematical operations. However it suffers due to the limitation on complexity in the algorithm and calculations as the number of criteria is increased. Several handoff criteria like RSS, SNR, cost of handoff, BER, latency and data transmission rate can be combined into pattern vectors [11]. A mobile node maintains several such pattern vectors depending on the pilot signals that it picks up from the access points or base stations. The six dimensional pattern vectors are then fed into a fuzzifier which determines the best access point or point of attachment to which the mobile terminal should get connected to. The output of the fuzzifier is a membership value. If the membership value falls below a threshold for the current access point, the mobile terminal will then initiate handoff to the access point with the highest membership value.

A predictive algorithm which uses neural networks to make predictions on the number of users in the after-handoff network can be an important handoff criteria [14]. This predicted value is one of the inputs along with bandwidth required and the velocity of mobile. These inputs are fed into a fuzzy multi-criteria inference system which makes the decision on the most suitable network for handoff. The change in the number of users in the after-handoff network is nonlinear. Hence the nonlinear dynamic characteristics are mapped into an Elman neural network (ENN). To speed up convergence of ENN, the authors in [14] proposed a modified ENN (MENN). The speed of the mobile was used to determine the rate at which the user may move in and out of the coverage of a WLAN, which is used to make the handoff decision.

Markov Models

A *Markov Decision Process* (MDP) can be used effectively for handoff decisions [12]. A link reward function for each application was defined based on the application QoS. A signaling cost function which defines the signaling and processing overheads during vertical handoff was considered and it depends on the complexity of rerouting of the call, which in turn depends on the type of networks, the service providers and their billing policies. In the MDP, epochs are defined when the mobile terminal makes decision on the actions it has to take, which in turn depend on the current status of the different access points which are maintained in the MDP states. The MDP states carry information on network ID, bandwidth and delay in the co-located networks. A decision rule was used to take the action. Also, the reward fuction was incorporated into the MDP, which was defined as the difference between link reward function and the signaling cost function. The goal is to maximize the reward function.

The network selection for vertical handoff can be modeled by an Analytic Hierarchy Process (AHP) and the Grey Relational Analysis (GRA) [18] or can be decided using a utility based strategy [19]. A multi-layer framework for vertical handoff will have distinct advantages [20]. Vertical handoff decision can be evaluated via a handoff cost function and a handoff threshold function, which can be adapted to changes in the network environment dynamically [21].

Optimization

Several handoff criteria can be defined and an optimization across all these criteria to select the best target network could be attempted [15]. The use of a cost function to determine the best target network for handoff based on user and network valued metrics like monetary costs, network conditions, mobile terminal conditions, offered service and user preferences with a goal to maximize the QoS to the user may be adopted. A network elimination process to reduce delays and processing power in handoff calculations is useful. The problem was formulated as an optimization problem and two approaches were investigated. The first case applies to a user that has several sessions. In this case, optimization is performed collectively across all his sessions and all his sessions are handed over to the same network. In the second case, optimization is done on a service basis, so depending on the service, the different sessions get handed over to different networks.

In [16], the authors considered a joint optimization, where they start with optimizing the battery life time in the network subject to not exceeding the load on each access point. The authors select these metrics for their optimized decisions because these metrics are available at the MIHF modules as specified in 802.21. However optimizing the network overall battery life time is not fair among all users, hence they introduce the max-min fairness requirements with the constraint of maximizing the overall battery life. To take into consideration the load on the access points, a load based cost function is defined and the load distribution is optimized across all access points subject to not exceeding the load in each access point. A joint optimization is then performed across battery lifetime and fairness in load at the access points by using a cost function and weighting these two metrics. In [22], a policy-enabled handoff decision algorithm was proposed along with a cost function that considers several of the factors mentioned above.

7.5.5 Handoff Execution Strategies

There are three strategies for executing handoff: mobile-controlled handoff (MCHO), network-controlled handoff (NCHO), and mobile-assisted handoff (MAHO). MCHO is used in IEEE 802.11 WLANs, where the mobile node continuously monitors the signal of access points and initiates the handoff procedure when some handoff criteria are met. In contrast, MAHO has been widely adopted mostly in wide area wireless networks where a mobile node measures the signals of surrounding base stations and then decides whether or not to initiate the handoff procedures with the network.

MCHO is advantageous because of the low complexity in network equipment. However, it can result in high latency and loss of large numbers of packets during inter-subnet handoff, because the network does not have the knowledge about the mobile node's movements till the mobile announces its presence in the new subnet. Rerouting can be initiated only after the detection of the mobile in the new network. Micro mobility solutions discussed earlier address the inter-subnet roaming problem. However the inter-domain roaming will still incur high delays.

Handoff decision in MAHO is made by the network, which can aid in global optimization and coordination among mobile nodes. However, for handoff between heterogeneous wireless access network technologies, as only the mobile nodes have the knowledge about the kind of interfaces they are equipped with, the network dependency on the mobile node is high. Also the network may not have knowledge about the future network to which the mobile is moving into, unless some predictive schemes are employed. Therefore, MCHO and some assistance from the networks is considered more suitable for vertical handoff.

7.5.6 Handoff Implementation Considerations

While the above discussions have addressed handoff decision criteria and algorithms that can be used to optimize the handoff process, implementation aspects are equally important, which should address mechanisms for message exchanges and easy facilitation of solution implementation. In this section, we address some such issues. Some guideline questions that will help in addressing the implementation aspects are as follows:

1. What are the typical messages to be exchanged?
2. Which are the network entities that are most suited to handle these messages?
3. What is the order in which the messages are to be exchanged and the functions to be executed?

Besides the decision criteria and optimization across several such criteria, it is equally important that the actual process of handoff be initiated at the right time and completed as expected within some predefined time limits. The implementation should be easily achieved with least changes to the existing network entities and mobility processes. Needless to say that the implementation should consider the use of non-stale data and processing delays are critical in this aspect. Very often, under such conditions, it becomes necessary to go for predictive schemes which are also very popular in several schemes.

The IETF Framework

A framework for policy based admission control was proposed in [23], which advocates a policy based networking architecture for vertical handoff decision algorithms. The *Request for comments* (RFC) in [23] defines a framework to transform a set of policy rules to network device configuration. The two main architectural elements for policy are the policy enforcement point (PEP) and the policy decision point (PDP). PEP is a component that runs on the policy aware network node, which is represented by the access point in a network controlled handoff or a mobile assisted handoff and is represented by the mobile terminal in a mobile controlled handoff. This is the point at which the policies are enforced. Policy decisions are made at the PDP based on the policies extracted from the policy database.

Media-Independent Handoff Functions (MIHF)

A simple implementation strategy was proposed in [14], which uses existing protocols and functions for exchange of handoff messages and which can be used across multiple access technologies as well as multiple operators. A common language as provided by the MIHF [24] can be used for this purpose. Vertical handoff decisions are to be implemented at the multiple vertical handoff decision controllers (VHDC) located in the access networks. The MIHF provides the decision inputs to the VHDC. The MIHF provides standard message exchanges between access points to share information on current link conditions, traffic load, network capacity etc. The link layer triggers (LLT) available at MIHF are used for triggering the VHDC operation. The LLT can be triggered due to the RSS at an access point dropping below threshold, or the RSS of one or more of the neighboring access points going above the threshold. Depending on whether the trigger is due to either criteria, several tailor-made decisions can be adopted.

A Framework Approach

In [21], the authors proposed a framework for vertical handoff execution across several different wireless access networks. The main focus in this work was to make the solution easily implementable. Although no particular standards are adopted in this proposal, the authors focused on leveraging off the existing mobility schemes that would cause least disruption to the existing schemes, while targeting adaptability of the solution to any future new wireless networks and technologies. They define a global mobility protocol, which consists of several components protocols within the framework. These components reside at several nodes in the access networks and the core network, thus extending the solution framework across all these networks. This also results in a distributed approach that reduces single point dependencies and failures. The mobility control module called the *hierarchical intersystem mobility agent* (HIMA) for the mobile user, as he roams across several wireless networks can be located at a suitable hierarchy, depending on the mobility span of the user. The components of the global mobility protocol are shown in Fig. 7.7 along with the local mobility protocols in the wireless networks. The arrows show the interaction and any control and information flow between the component protocols. Using this approach, a smooth handoff interaction between layer 3 and layer 2 functions can be achieved. The dotted blocks show the provision for extending the framework to any number of hierarchical levels. Even though HIMA is suggested to be co-located at some control points in the core network and is supposed to have minimal database processing, the cost overheads of extending it to more than two hierarchical levels may not be always justified. Hence, the number of levels of hierarchical control using HIMA is at the discretion of the network and service provider.

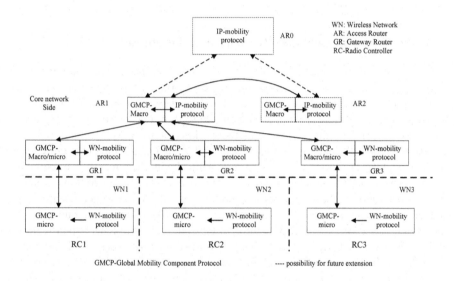

Fig. 7.7. Distributed global mobility protocol - a framework.

7.6 Architectures for Integration of Heterogeneous Wireless Networks

A network design typically starts with a study of an architecture suited to the application for which the network is targeted. While designing the network architecture, considerations such as the type of user population, their desired service support, the data rate requirements, the available capabilities and cost constraints are to be taken into account. Similarly, while designing or developing a solution to a problem that spans several networks, it is essential to start with an architecture that is able to address the problem under consideration in the best possible manner. In this case, a seamless handoff of mobile users with active connections across several heterogeneous wireless networks is desired. The infrastructure of the underlying networks, their technologies, several different service providers with distinct policies and goals and last but not the least, user preferences and requirements have to be considered. Most of the time such architectural studies are influenced by the goals of the investigating agencies and different architectures emerge based on what the investigators think is optimal, though all efforts are directed towards a common goal. An inefficient architecture for location and mobility management can lead to inefficiencies in the signaling scalability and handoff latency. This may not be acceptable in an infrastructure that must provide global mobility support to potentially billions of mobile nodes with some stringent performance bounds. These performance bounds associated with real-time multimedia traffic, which is getting to be the predominant traffic type in current networks.

7.6.1 Tightly-Coupled and Loosely-Coupled Architectures

Before we present any specific architecture that addresses the vertical handoff problem, we would like to mention about the two categories of heterogeneous wireless networks integration architectures, namely, tighlty and loosely coupled architectures. These architectures were investigated under European Telecommunications Standards Institute (ETSI) that targeted internetworking between WLAN and cellular networks in 2001 [9]; one of the major focus under this integration architecture was vertical handoff. Though the categories were defined with the integration of cellular and WLANs in mind, the concepts can easily be extended to integration of other wireless networks. In the tightly coupled architecture, the WLAN operates as a virtual radio access network (RAN) in the cellular system. To connect to the Internet, the data traffic in the WLAN must pass through the cellular network. The main feature of the tightly coupled architecture is that handoff and security issues in WLAN were totally controlled by the cellular handoff and security systems. The main drawback was the difficulty in supporting the WLAN interface in the cellular nodes namely the serving GPRS support node, and the performance overhead of the cellular system, which was incurred for data transfers between WLAN and Internet.

In a loosely coupled architecture, the WLAN was to be deployed as an access network that complemented the cellular networks. In other words, access to Internet via the WLAN did not have to go via the cellular networks. The loosely coupled architecture provides a flexible and independent environment for both wireless technologies, but required Mobile IP to support mobility across the two access networks. More details on these two architectures can be found in [25]. Of the two architectures, the loosely coupled one is the more used architecture because of the independence coupled with a capability for successful coexistence.

7.6.2 Macro Mobility Architectures

MobileIPv6

Earlier in this chapter, we discussed the Mobile IP concepts. This discussion was based more on Mobile IPv4. With the advent of Mobile IPv6 and most current vertical handoff solutions trying to define Mobile IPv6 as an inherent part of the architecture, it is appropriate to review at this point the difference in mobility support in IPv6 [6]. Mobile IPv6 retains the ideas of a home network, home agent, and encapsulation to deliver packets from the home network to the mobile node's current network or point of attachment. Discovery of a care-of address is still required, but a mobile node can configure its care-of address by using stateless address autoconfiguration and neighbor discovery. Thus, foreign agents are not required in Mobile IPv6. IPv6-within-IPv6 tunneling is part of the specifications. Route optimization is achieved through a mobile node sending binding updates to the correspondent node, whereby a remote correspondent node can deliver packets directly to the mobile node in its foreign network. However, all IPv6 nodes are expected to have the correspondent node capability, which makes it easy from an implementation perspective. Binding updates can be handled through IPv6 destination option capability. IPv6 correspondent nodes can use IPv6 routing headers, similar to IPv4 source routing option instead of tunneling.

Having discussed some fundamentals in architectural considerations that can aid in seamless vertical handoff execution, we now present some of the efforts which are based on these concepts. Of the early efforts in designing architectures for integrating heterogeneous wireless networks, two major projects that handled roaming on a wide scale spanning a number of different wireless networks, are worth mentioning. The first one is the project by the European Commission Research Program (ECRP) on expanding the "Wireless Universe" [26] and the second one is BARWAN (Bay Area Research Wireless Networks) project [27] undertaken at University of Berkeley.

European Commission Research Program

ECRP had a number of initiatives which targeted a wider scope of requirements via different projects, to address a number of related issues. MATRICE[1] (all acronyms are expanded in this footnote) focused on layers 1 to 3 for support of IP and handover algorithms for indoor/outdoor transition from 3G to WLAN. The Terrestrial Wireless Systems and Networks (TWSN) project used the key aspects of QoS evaluations and management for multimedia radio environments. BRAIN/MIND are two sub projects within TWSN that advocated an open architecture for convergence of fixed Internet and UMTS/GSM/GPRS to provide end-to-end IP.

BARWAN

In contrast to the above initiative which consisted of several sub-projects, the BARWAN project was a megascope test-bed implementation to study and investigate the concept of umbrella coverage afforded by different types of wireless networks using an overlay network approach. The project unified several heterogeneous networks of

---

[1]MATRICE - MC-CDMA Transmission techniques for Integrated Broadband Cellular Systems, BRAIN (Broadband Radio Access for IP Based Networks), MIND (Mobile IP based Network Development).

varying coverage into one logical network to provide the best of the services of all networks. The wireless networks in the logical network ranged from satellite networks with regional coverage, cellular networks with wide area coverage down to campus and in-building coverage. The project handled several issues such as improving TCP efficiency, service discovery and seamless handoff within the scope of the umbrella coverage. The goal was to move the complexity to the infrastructure, to enable the use of small lightweight inexpensive mobile devices with high capabilities.

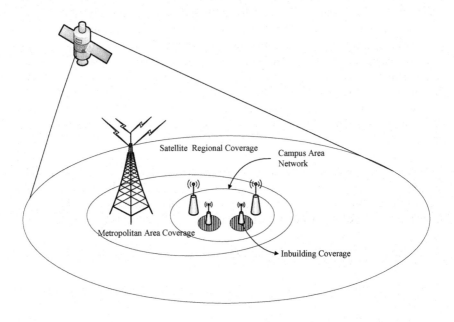

Fig. 7.8. Barwan project umbrella coverage concept.

Fig. 7.8 is the schematic of the architecture proposed under project BARWAN, which shows how the project leveraged existing network infrastructure to build an overlaid collection of networks. In Fig. 7.8, we see a network in a building which is under the umbrella coverage of a campus network that is covered by a metropolitan area network, which in turn is covered by a satellite network that has very wide geographic coverage but has latency issues. Lower levels in the hierarchy of the proposed overlay architecture comprise of high bandwidth wireless cells that cover a relatively small area while higher levels in the hierarchy provide a lower bandwidth per unit area connection over a large geographic area.

The BARWAN project was a test-bed and so they used existing networking technologies without much modifications. An aspect worth noting is the use of any powerful infrastructure that was highly available, cost effective and sufficiently scalable. The BARWAN project handoff scheme minimized handoff latency for individual users, while keeping bandwidth and power overheads low. As mentioned earlier they addressed vertical handoff with two considerations; upward handoff and downward handoff. Downward handoff was considered less critical as the mobile device could stay connected to the upper overlay network till handed off. Handoff was initiated by the mobile terminal, which would monitor the base stations from which it received beacons. The upward vertical handoff was initiated when the mobile terminal did not receive beacons from its current lower layer networks. A predictive scheme was adopted where the mobile terminal determined two base stations, one which would forward the packets and one which would buffer packet in anticipation of a handoff, thereby achieving the seamlessness required for vertical handoff.

Cellular /WLAN Integration Architectures

Early 2000s saw a flurry of research initiatives to integrate cellular and WLANs [28]-[32]. This was the time when cellular networks were getting digital and started supporting data services and WLANs were getting popular with their high data rate support and Internet connections. At this time several research and standardization efforts were directed towards this integration study. 3GPP introduced a standard architecture [14] to enable 3GPP system operators to provide public WLAN access as an integral component of their total service offered to their cellular subscribers. In [17], an architecture for integrating UMTS and 802.11 WLAN was presented, which allows mobile nodes to maintain data connection through WLAN and voice connection through UMTS in parallel.

A Conceptual Architecture

A completely IP centric approach to provide efficient seamless roaming service to end users while roaming between a WLAN and WWAN was discussed in [28]. The architecture of this approach is illustrated in Fig. 7.9. The architecture is based on the following basic functional modules:

- A connection manager (CM) that is able to intelligently detect the conditions on different types of networks and their availability. It also handles the handoff from WWAN to WLAN which requires layer 2 and layer 1 sensing in the WLAN, while the handoff from WLAN to WWAN is initiated as soon as there is failure to detect a WLAN.
- A virtual connectivity manager (VC) maintains connection continuity when handoff occurs, by utilizing the information from the CM. It uses a local connection translation which is a map between the original connection information and the current connection information and a subscription/notification service.
- A roaming decision maker and a context database provide the interconnection between the CM and VC. The context database facilitates context aware roaming, which the roaming decision maker takes into account.

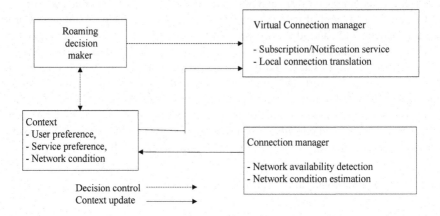

Fig. 7.9. Conceptual framework for roaming between WLAN and WWAN.

7.6.3 Micro Mobility Architectures

CellularIP and HAWAII

Cellular IP [7] and HAWAII [33] were two popular introductory micro-mobility schemes. HAWAII used a domain-based approach for supporting mobility with specialized path setup schemes with host-based forwarding entries installed in specific routers to support intra domain micro mobility. However, it assumed that the base stations had IP routing functionality and incurred a higher processing overhead on the power-limited mobile nodes. Cellular IP on the other hand maintained a distributed cache for location management and routing purposes. A distributed routing cache maintained the position of all active mobiles and dynamically refreshed the routing state in response to the handoff. When an active mobile approached a new base station, it would transmit a route-update packet from the old to the new base station.

Hierarchical Mobile IPv6 (HMIPv6)

HMIPv6 is a hierarchical architecture for micro mobility based on Mobile IPv6 [34], with a two level architecture; the root being at a *mobility anchor point* (MAP) and the second level supported by access routers. Each mobile node in HMIPv6 requires two care of addresses (CoAs): a regional CoA and an on-link CoA. When the mobile node moves to a foreign network, it sends two binding updates one to the home agent and/or the correspondent node and the other which is like a local binding update to a MAP, which is the root of the hierarchical architecture. When the mobile node moves to another subnet within the same domain, it then has to only send updates of its on-link CoA which is the local binding update to the MAP. Implementing HMIPv6 in a simple mobile node with low processing capabilities may be inefficient.

Hierarchical Mobile IPv6 (HiMIPv6)

Fig. 7.10 shows the architecture adopted in HiMIPv6, which is based on concepts form HMIPv6 but introduces more hierarchy to reduce the latency in handling binding updates to home networks and correspondent nodes. The routers in the proposed hierarchy can be root *foreign mobility agent* (FMA), which provides the care of address for the mobile node while it is in its footprint; they can be leaf FMAs, which are the access routers to which the mobile node is attached or they could be an interior FMA. Logically any of these routers can perform the functions of the switching FMA, which acts as an intermediate agent that intercepts the binding updates to the home network and the correspondent nodes to forward them subsequently and meanwhile acting as a proxy home agent it locally completes the registration process for the mobile node. To speed up the handoff process, the leaf FMAs use the capabilities of IAPP and belong to a multicast IPv6 group. On detecting the mobile movement, they take action promptly as the mobile node leaves one leaf FMA to join another.

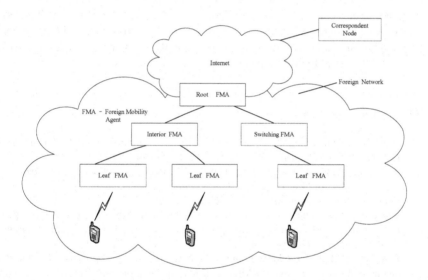

Fig. 7.10. Hierarchical Mobile IPv6.

Intra-Domain Mobility Management Protocol

The *intra-domain mobility management protocol* (IDMP) [32] for managing node mobility within a specific domain was introduced to define any additional mobility-related feature at the IP-layer, and not rely upon or assume the existence of specific features from the underlying link layers. Like most micro mobility proposals, IDMP envisions that multiple IP-subnets are aggregated into a single domain; as long as the mobile node moves within a single domain, all its mobility-related signaling remains localized to specialized nodes within that domain. Since the mobile changes domains fairly infrequently, this localization reduces both the global signaling load and the update latency considerably. IDMP is a two-level generalization of the mobile IP architecture, through a special node called the *mobility agent* (MA) which provides packet redirection. IDMP assumes an all-IP access network where IP-layer functionality extends all the way to the nodes at the wireless edge of the access domain, i.e., the base station and it is assumed that the *subnet agent* (SA), a specialized IDMP node that provides subnet-specific support to the mobile, is co-located with the base station.

Conclusion

The goal of this chapter was to make the reader aware of the importance and significance of seamless roaming across several different wireless networks. Vertical handoff is a very important process for achieving the seamless roaming feature and continues to be a challenging problem, for the simple reason that new and innovative wireless technologies and networks are still evolving. This chapter highlights some of the important technical challenges underlying this very important feature, which will be fundamental to all future networking endeavors.

References

1. H. Lach, C. Janneteau, and A. Petrescu, "Network mobility in beyond-3G systems," *IEEE Communications Magazine*, vol. 41, no. 7, pp. 52-57, July 2003.
2. S. Pack and Y. Choi, "Performance analysis of hierarchical MobileIPv6 in IP-based cellular networks," in *Proc. of 14th IEEE International Symposium on Personal, Indoor and Mobile Radio Communications*, pp. 2818-2822, 2003.
3. I. Okajima, N. Umeda, and Y. Yamao, "Architecture and MobileIPv6 extensions supporting mobile networks in mobile communications," in *Proc. of IEEE Vehicular Technology Conference (VTC)-Fall*, vol. 4, pp. 2533-2537, 7-11 Oct. 2001.
4. K. Omae, M. Inoue, I. Okajima, and N. Umeda, "Handoff performance of mobile host and mobile router employing HMIP extension," in *Proc. of IEEE Wireless Communications and Networking Conference (WCNC)*, New Orleans, Louisiana, USA, pp. 1218-1223, 2003.
5. P. Xue-hai, Z. Hong-ke, H. Jiu-Chuan, and Z. Si-dong, "Modeling in hierarchical MobileIPv6 and intelligent mobility management scheme," in *Proc. of 14th IEEE International Symposium on Personal, Indoor and Mobile Radio Communications (PIMRC)*, pp. 2823-2827, 2003.
6. C. E. Perkins, "IP Mobility Support for IPv4," *Internet Engineering Task Force* RFC3344, 2002.
7. H. C. Chao and C. Y. Huang, "Micro-mobility mechanism for smooth handoffs in an integrated ad-hoc and cellular IPv6 network under high-speed movement," *IEEE Transactions of Vehicular Technology*, pp. 1576-1593, 2003.
8. L. Goleniewski, *Telecommunications Essentials*. Addison Wesley, 2nd Ed., 2006.

9. "2G-4G networks: Evolution of technologies, standards, and deployment," in *Encyclopedia of Multimedia Technology and Networking*, Ideas Group Publisher, pp. 964-973, May 2005.

10. K. Pahlavan, P. Krishnamurthy, A. Hatami, and M. Yliantillai, "Handoff in hybrid mobile data networks," *IEEE Personal Communications Magazine*, vol. 7, no. 2, pp. 34-47, Apr. 2000.

11. S. Milena and P. Mahonen, "Algorithmic approaches for vertical handoff in heterogeneous wireless environment," in *Proc. of IEEE WCNC'07*, pp. 3783-3788, Hong Kong, China, Mar. 2007.

12. E. S.-Navarro, V. W. S. Wong, and Y. Lin, "A vertical handoff decision algorithm for heterogeneous wireless networks, " in *Proc. of IEEE WCNC'07*, pp. 3201-3206, Hong Kong, China, Mar. 2007.

13. G. P. Pollini, "Trends in Handover design," *IEEE Communications Magazine*, pp. 82-90, March 1996.

14. Q. Guo, J. Zhu, and X. Xu, "An adaptive multi-criteria vertical handoff decision algorithm for radio heterogeneous network," in *Proc. of IEEE International Conference on Communications (ICC)*, vol. 4, pp. 2769-2773, May 2005.

15. F. Zhu and J. Mcnair, "Optimization for vertical handoff decision algorithms," in *Proc. of IEEE WCNC'04*, pp. 867-872, 2004.

16. S. Lee et al., "Vertical handoff decision algorithm providing optimized performance in heterogeneous wireless networks," in *Proc. of IEEE Globecom'07*, pp. 5164- 5169, 2007.

17. A. H. Zahran and B. Liang, "Performance evaluation framework for vertical handoff algorithms in heterogeneous networks," in *Proc. of IEEE ICC'05*, vol. 4, pp. 173-178, May 2005.

18. Q. Song and A. Jamalipour, "A network selection mechanism for next generation networks," in *Proc. of IEEE ICC'05*, Seoul, Korea, May 2005.

19. O. Ormond, J. Murphy, and G. Muntean, "Utility-based intelligent network selection in beyond 3G systems," in *Proc. of IEEE ICC'06*, Istanbul, Turkey, June 2006.

20. A. Sur and D. Sicker, "Multi layer rules based framework for vertical handoff," in *Proc. of BROADNETS'05*, Boston, MA, Oct. 2005.

21. N. Shenoy and R. Montalvo, "A framework for seamless roaming across cellular and wireless local area networks," *IEEE Wireless Communications*, Special issue on "Towards Seamless Internetworking of Wirless LAN and Cellular Networks", vol. 12, no. 3, pp. 50-57, June 2005.

22. H. J. Wang , R. H. Katz , and J. Giese, "Policy-enabled handoffs across heterogeneous wireless networks," in *Proc. of Second IEEE Workshop on Mobile Computer Systems and Applications*, p.51, Feb. 25-26, 1999.

23. R. Yavatkar, "A framework for policy based admission control," *RFC2753 www.ietf.org*.

24. IEEE 802.21 Media Independent Handover Working Group, http://www.ieee802.org/21/.

25. W. K. Lai and J. C. Chu, "Improving handoff performance in wireless overlay networks by switching between two-layer IPv6 and one-layer IPv6 addressing," *IEEE Journal on Selected Areas in Communications*, vol. 23, no. 11, pp. 2129-2137, Nov. 2005.

26. S. Fabrizio, S. S. Joao, and F. Jose, "Expanding the wireless universe: EU research on the move," *IEEE Communications Magazine*, pp. 132-140, Oct. 2002.

27. E. A. Brewer et al., "A network architecture for heterogeneous mobile computing," *IEEE Personal Communications*, vol. 5, no. 5, pp. 8-24, Oct. 1998.

28. Q. Zhang, C. Guo, Z. Guo, and W. Zhu, "Efficient mobility management for vertical handoff between WWAN and WLAN," *IEEE Communications Magazine*, vol. 41, no. 11, pp. 102-109, Nov. 2003.

29. C. E. Perkins, "Mobile networking through Mobile IP," *IEEE Internet Computing*, pp. 58-69,1998.

30. C. W. Lee, L. M. Chen, M. C. Cheng, and Y. S. Sun, "A framework for handoff in wireless overlay networks based on Mobile IPv6," *IEEE Journal on Selected Areas in Communications*, vol. 23, no. 11, pp. 2118-2128, Nov. 2005.

31. J. Mcnair and F. Zhu, "Vertical handoffs in fourth generation multinetwork environments," *IEEE Wireless Communications*, pp. 8-15, June 2004.

32. A. Misra, S. Das, A. Dutta, A. McAuley, and S. K. Das, "IDMP-based fast handoffs and paging in IP-based 4G mobile networks," *IEEE Communications Magazine*, pp.138-145, 2002.

33. R. Ramjee, K. Varadhan, L. Salgarelli, S. Thuel, S. Y. Wang, and T. La Porta, "HAWAII: A domain-based approach for supporting mobility in wide-area wireless networks," in *Proc. of IEEE Int. Conf. Netw. Protocols*, pp. 396-410, June 1999.

34. "INRIA Hierarchical Mobile IPv6 Proposal (HMIPv6)," `http://www.inrialpes.fr/planete/people/bellier/hmip.html`.

35. A. T. Campbell et al., "Comparison of IP micromobility protocols", *IEEE Wireless Communications*, pp. 72-82, 2002.

36. M. Ylianttila, *Vertical handoff and mobility - system architecture and transition analysis*, Oulu University Press, 2005.

37. D. Johnson, C. E. Perkins, and J. Arkko, "Mobility support in IPv6", *Internet Engineering Task Force* RFC3775, 2004.

38. W. Zhang, "Handover decision using fuzzy MADM in heterogeneous networks," in *Proc. of IEEE WCNC'04*, Atlanta, GA, March 2004.

39. N. Shenoy, "A framework for seamless roaming across heterogeneous next generation wireless networks," *ACM Journal of Wireless Networks (WINET)*, Nov. 2005.

40. E. S.-Navarro and V. Wong, "Comparison between vertical handoff decision algorithms for heterogeneous wireless networks," in *Proc. of IEEE VTC'06-Spring*, Melbourne, Australia, May 2006.

41. A. Hassawa, N. Nasser, and H. Hassanein, "Tramcar: A context-aware cross-layer architecture for next generation heterogeneous wireless networks," in *Proc. of IEEE ICC'06*, Istanbul, Turkey, June 2006.

42. I. F. Akyildiz, J. McNair, J. S. M. Ho, H. Uzunalioglu, and W. Wang, "Mobility management in next generation wireless systems," *Proceedings of IEEE*, vol. 87, no. 8, pp. 1347-1384, August 1999.

43. M. Stemm and R. Katz, "Vertical handoffs in wireless overlay networks," *ACM Mobil Networking (MONET)*, Special Issue on Mobile Networking in the Internet, pp. 335-350, Summer 1998.

44. W. Chen and Y. Shu, "Active application oriented vertical handoff in next generation wireless networks," in *Proc. of IEEE WCNC'05*, New Orleans, LA, March 2005.

45. W. Zhang, "Handover decision using fuzzy MADM in heterogeneous networks," in *Proc. of IEEE WCNC'04*, Atlanta, GA, March 2004.

46. J. W. Jung, R. Mudumbai, D. Montgomery, and H. K. Kahng, "Performance evaluation of two layered mobility management using mobile IP and session initiation protocol," in *Proc. of IEEE Global Telecommunications Conference*, San Francisco, California, USA, pp. 1190-1194.

47. L. Ma, F. Yu, V. C. M. Leung, and T. Randhawa, "A new method to support UMTS/WLAN vertical handover using SCTP," *IEEE Wireless Communications*, pp. 44-51, 2004.

48. C. Politis, K. A. Chew, N. Akhtar, M. Georgiades, R. Tafazolli, and T. Dagiuklas, "Hybrid multilayer mobility management with AAA context transfer capabilities for all-IP networks," *IEEE Wireless Communications*, pp. 76-88, 2004.

49. N. Shenoy, "Global Roaming in Next Generation Wireless Networks," in *Proc. of CITC3*, Rochester, New York, Sept. 19-21, 2002.

50. K. Ahmavaara, H. Haverinen, and R. Pichna, "Internetworking architecture between 3GPP and WLAN systems," *IEEE Communications Magazine*, vol. 41, no. 11, pp. 74-82, Nov. 2003.

51. M. M. Buddhikot, G. Chandranmenon, S. Han, Y. Lee, S. Miller, and L. Salgarelli, "Design and implementation of a WLAN/CDMA2000 interworking architecture," *IEEE Communications Magazine*, vol. 41, no. 11, pp. 90-101, Nov. 2003.

52. W. Zhuang, Y.-S. Gan, K.-J. Loh, and K.-C. Chua, "Policy-based QoS management architecture in an integrated UMTS and WLAN environment," *IEEE Communications Magazine*, vol. 41, no. 11, pp. 118-125, Nov. 2003.

53. A. Doufexi, E. Tameh, A. Nix, S. Armour, and A. Molina, "Hotspot wireless LANs to enhance the performance of 3G and beyond cellular networks," *IEEE Communications Magazine*, vol. 41, no. 7, pp. 58-65, July 2003.

54. W. Wang and I. F. Akyildiz, "A new signalling protocol for intersystem roaming in next generation wireless systems," *IEEE Journal on Selected Areas in Communications*, vol. 19, no. 10, pp. 2040-2052, Oct. 2001.

55. M. T.-Moreno, X. P.-Costa, and S. S.-Ribes, "A performance study of fast handovers for MobileIPv6," in *Proc. of 28th Annual IEEE International Conference on Local Computer Networks*, pp. 89-98, 20-24 Oct. 2003.

56. N. Montavont and T. Noel, "Handover management for mobile nodes in IPv6 networks," *IEEE Communications Magazine*, pp. 38-43, Aug. 2002.

57. W. J. Floroiu, R. Ruppelt, D. Sisalem, and J. V. Stephanopoli, "Seamless handover in terrestrial radio access networks: A case study," *IEEE Communications Magazine*, vol. 41, no. 11, pp. 110-117, Nov. 2003.

58. L. Lamont, M. Wang, L. Villasenor, T. Randhawa, and S. Hardy, "Integrating WLANs and MANETs to the IPv6 based internet," in *Proc. of IEEE International Conference on Communications (ICC'03)*, vol. 2 , pp. 1090-1095, 2003.

59. J. Jun and M. L. Sichitiu, "The nominal capacity of wireless mesh networks," *IEEE Wireless Communications*, vol. 10, no. 5, pp. 8-14, Oct. 2003.

60. S. I. Maniatis, E. G. Nugikolouzou, and I. S. Venieris, "QoS issues in the converged 3G wireless and wired networks," *IEEE Communications Magazine*, pp. 44-53, Aug. 2002.

61. 3GPP, "Group Services and System Aspects: 3GPP Systems to Wireless Local Area Network (WLAN) Interworking; System Description (Release 6)", TS 23.234 v6.0.0.

62. M. Jaseemuddin, "An architecture for integrating UMTS and 802.11 WLAN," in *Proc. of Eight IEEE International Symposium on Computers and Communications*, pp. 716-723, 2003.

63. H. Harri, K. Pehkonen, M. T. Neimi, and A. T. Leino, "WCDMA and WLAN for 3G and beyond," *IEEE Wireless Communications*, pp. 14-18, Apr. 2002.

64. A. K. Salkintzis, C. Fors, and R. Pazhyannur, "WLAN GPRS integration for next generation mobile data networks," *IEEE Wireless Communications*, pp. 112-124, Oct. 2002.

65. W. Wenye and I. F. Akyildiz, "A new signalling protocol for intersystem roaming in next generation wireless systems," *IEEE Journal on Selected areas in communications*, vol. 19, no. 10, pp. 2040-2052, Oct. 2001.

8

Network Selection for Heterogeneous Wireless Access Networks

Wei Song and Weihua Zhuang

Department of Electrical and Computer Engineering, University of Waterloo, Canada
{wsong, wzhuang}@bbcr.uwaterloo.ca

8.1 Introduction

Motivated by the ever-increasing demand for wireless communications, the past decade has witnessed rapid evolution and successful deployment of wireless networks. It is widely accepted that next-generation wireless networks will be heterogeneous in nature with multiple wireless access technologies. While the heterogeneity poses new challenges to achieve interoperability among different wireless networks, their complementary characteristics can be exploited with the interworking to enhance service provisioning.

The popular cellular networks and wireless local area networks (WLANs) are two most promising technologies. Originally designed for voice service over wide areas, cellular networks have evolved to the third generation (3G) and are further augmented with multimedia service support. For example, the commercialized universal mobile telecommunication system (UMTS) supports a data rate up to 2 Mbps with improved capacity. However, the deployment cost remains high due to expensive radio spectrum and implementation complexity. On the other hand, WLANs have also achieved great success and provide higher data rates at a much lower cost. For example, the most popular WLAN standard IEEE 802.11b operates at the 2.4 GHz license-exempt industrial, scientific, and medical (ISM) frequency band. It supports a data rate up to 11 Mbps. However, designed as a wireless extension to the wired Ethernet, a WLAN can only cover a small geographic area.

We can see that the two types of networks present complementary strengths in terms of mobility support, data rate, and implementation cost. Cellular/WLAN interworking can provide mobile users with both ubiquitous connectivity and high-rate data service in hotspots. An essential interworking issue is network selection for a heterogeneous cellular/WLAN integrated network. With the widely entrenched infrastructure, cellular networks provide almost ubiquitous connectivity and support user mobility levels from fast highway vehicles to stationary users in an indoor environment. In contrast, WLANs are usually deployed disjointly in hotspot local areas, where the traffic intensity is typically much higher than surrounding areas. Thus, the

E. Hossain (ed.), *Heterogeneous Wireless Access Networks*,
DOI: 10.1007/978-0-387-09777-0_8, © Springer Science+Business Media, LLC 2008

cellular/WLAN interworking results in a hierarchical overlay structure. Both cellular access and WLAN access are available to mobiles within WLAN-covered areas. Given incoming service requests, a proper target network should be selected between the overlay cell and WLAN for purposes such as quality-of-service (QoS) enhancement, congestion relief, and cost reduction.

Most previous works in this area address only single service [1] and cannot effectively exploit the interworking gain. Further, network selection can be complicated by the heterogeneous QoS provisioning capability of underlying networks. While fine-grained QoS is enabled in the cellular network with the centralized control and reservation-based resource allocation, QoS provisioning of WLANs is rather limited due to contention-based random access. Nevertheless, not much research attention has been directed to exploiting the complementary QoS provisioning in network selection for multi-service support. Finally, the most popular speed-sensitive selection strategy for hierarchical cellular networks [2] is not applicable to heterogeneous cellular/WLAN integrated networks. This is because the indoor deployment of most WLANs results in very low mobility in WLANs. The user residence time within a WLAN is heavy-tailed [3] and user mobility within a large cell becomes location-dependent. The unique user mobility characteristics also affect network selection and are neglected in most previous works [4].

In this chapter, we investigate the network selection issue for the heterogeneous wireless access environment induced by cellular/WLAN integrated networks. Effective network selection strategies are proposed and analyzed, taking into account location-dependent user mobility, multi-service traffic characteristics, and heterogeneous QoS support. First, a service-differentiated network selection strategy is studied for conversational voice service and interactive data service. Second, we explore a randomized implementation to enable distributed control. Third, the heavy-tailedness of data traffic is further exploited with a size-based network selection strategy. Also, network reselection via dynamic vertical handoff is considered for interworking enhancement. Effective analytical approaches are developed to evaluate system performance. Important insights are gained about the impact of user mobility and multi-service traffic characteristics on network selection.

8.2 System Model

In this study, we consider a heterogeneous wireless network integrating a 3G cellular network and WLANs. The 3G cellular network provides ubiquitous connectivity over wide areas but occupies a relatively small bandwidth. Fine-grained QoS is enabled with the centralized infrastructure and code division multiple access (CDMA). On the other hand, WLANs are deployed disjointly in hotspot local areas, following physical and medium access control (MAC) specifications similar to those of IEEE 802.11 standards. We assume that there is one overlay WLAN in a cell and the WLAN is overlaid with one and only one cell. The cell and its overlay WLAN are referred to as a *cell/WLAN cluster*. The mobile devices are dual-mode and equipped with network interfaces to both the cellular network and WLAN. Thus, both cellular access and WLAN access are available to dual-mode mobiles within the WLAN-covered areas, which are referred to as *double-coverage areas*. In contrast, the areas with only cellular coverage are referred to as *cellular-only areas*.

8.2.1 Location-Dependent User Mobility Model

It is known that most WLANs are deployed in indoor environments like cafés and offices. Users within these areas are mostly stationary or only maintain a pedestrian-level mobility. Thus, it is not reasonable to apply a homogeneous mobility model for mobiles within the coverage of a large cell. To statistically characterize user mobility, user residence time should vary with the location, which is either within the cellular-only area or double-coverage area.

As shown in [3], the indoor deployment and low user mobility result in a heavy-tailed user residence time within a WLAN. However, performance analysis tends to be extremely difficult with heavy-tailed distributions involved in the system model. To render tractable analysis, it was proposed in [5] to fit a large class of heavy-tailed distributions with hyper-exponential distributions. To reduce the number of parameters, the user residence time within a WLAN, denoted by T_r^w, is modeled with an approximate two-stage hyper-exponential distribution [6] with mean $(\eta^w)^{-1}$ and probability density function (PDF)

$$f_{T_r^w}(t) = \frac{a}{a+1} \cdot \frac{1}{\frac{1}{a} \cdot \frac{1}{\eta^w}} e^{-a\eta^w t} + \frac{1}{a+1} \cdot \frac{1}{a \cdot \frac{1}{\eta^w}} e^{-\frac{\eta^w}{a}t}, \qquad a \geq 1, \quad t > 0. \quad (8.1)$$

This model well captures the so-called "*mice-elephants*" phenomenon [7] of heavy-tailedness. It implies that most users stay within a WLAN for a quite short time while a small fraction of users have an extremely long residence time. According to (8.1), a large fraction $\frac{a}{a+1}$ of the users stay within the WLAN for a mean time $\frac{1}{a} \cdot \frac{1}{\eta^w}$, while the other $\frac{1}{a+1}$ of the users have a mean residence time of $a \cdot \frac{1}{\eta^w}$. A larger value of a indicates T_r^w of higher variability.

On the other hand, the user residence time in the area of a cell with only cellular access, denoted by T_r^c, is assumed to be exponentially distributed with parameter η^c. Users moving out of the cellular-only area enter neighboring cells with a probability p^{cc} and enter the coverage of the overlay WLAN in the target cell with a probability $p^{cw} = 1 - p^{cc}$. Therefore, the residence time of users admitted in the cell follows more complicated phase-type distributions as shown in Fig. 8.1. Let T_{r1}^c and T_{r2}^c denote the cell residence time of a call from the cellular-only area and that of a call from the double-coverage area, respectively. The moment generating functions (MGFs) of T_{r1}^c and T_{r2}^c are derived based on Fig. 8.1 as

$$\Phi_1(s) = \sum_{i=1}^{\infty} (p^{cw})^{i-1} p^{cc} \frac{\eta^c}{\eta^c - s} [\psi(s)]^{i-1} \qquad (8.2)$$

$$\Phi_2(s) = \sum_{i=1}^{\infty} (p^{cw})^{i-1} p^{cc} [\psi(s)]^i \qquad (8.3)$$

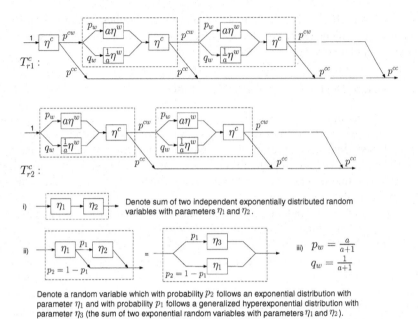

i) Denote sum of two independent exponentially distributed random variables with parameters η_1 and η_2.

ii) $p_2 = 1 - p_1$ = $p_2 = 1 - p_1$

iii) $p_w = \dfrac{a}{a+1}$

$q_w = \dfrac{1}{a+1}$

Denote a random variable which with probability p_2 follows an exponential distribution with parameter η_1 and with probability p_1 follows a generalized hyperexponential distribution with parameter η_3 (the sum of two exponential random variables with parameters η_1 and η_2).

Fig. 8.1. Modeling of user mobility within a cell/WLAN cluster.

where $\psi(\cdot)$ is the MGF of $T_r^c + T_r^w$, given by

$$\psi(s) = \mathrm{E}\!\left[e^{s(T_r^c + T_r^w)}\right] = \frac{\eta^c}{\eta^c - s}\left[\frac{a}{a+1} \cdot \frac{a\eta^w}{a\eta^w - s} + \frac{1}{a+1} \cdot \frac{\frac{1}{a}\eta^w}{\frac{1}{a}\eta^w - s}\right]. \quad (8.4)$$

8.2.2 Multi-Service Traffic Model

As an essential requirement for future wireless networks, multi-service support is an important motivation for cellular/WLAN interworking. In this study, we consider both conversational voice service and interactive data service.

Conversational Voice Service

The conversational class is meant for real-time services and characterized by a two-way communication pattern. A constant bandwidth is usually required to satisfy very demanding QoS constraints. A typical example is voice telephony, which has been extensively studied and widely deployed. Voice calls are assumed to arrive as a Poisson process, having an exponentially distributed duration T_v with mean in the order of several minutes.

Interactive Data Service

The interactive class comprises non-real time services with a request-response pattern, such as Web browsing, voice messaging, and file transfer. A main QoS criterion is the transfer delay (response time) to measure the responsiveness, e.g., how fast a Web page is successfully downloaded and appears after it has been requested. Although the transfer delay should be bounded to maintain fluent interactions, the delay requirement is much less stringent than that of conversational services [8]. A transfer delay of $2 - 4$ seconds per page is the proposed bound for Web browsing and a desirable target is 0.5 seconds.

As an essential call-level traffic characteristic, the heavy-tailedness of data file size has been extensively studied in the literature. There are many models to capture the statistics of real file size with well-known heavy-tailed distributions such as Pareto and Weibull distributions. In [9], the data file size L_d is modeled by a Weibull distribution, whose PDF is given by

$$f_{L_d}(x) = \frac{\alpha_d}{\beta_d} \left(\frac{x}{\beta_d} \right)^{\alpha_d - 1} e^{-(x/\beta_d)^{\alpha_d}}, \qquad 0 < \alpha_d \leq 1, \quad \beta_d > 0, \quad x > 0 \quad (8.5)$$

where α_d is the shape parameter and β_d is the scale parameter. The PDF of the Weibull distribution is denoted by $W_b(x, \alpha_d, \beta_d)$ for simplicity. The mean of L_d is given by $\mathrm{E}[L_d] \triangleq \overline{L}_d = \beta_d \, \Gamma(1 + \frac{1}{\alpha_d})$, where $\Gamma(\cdot)$ is the Gamma function. The exponential distribution is actually a special case of the Weibull distribution with $\alpha_d = 1$, while the Weibull distribution is heavy-tailed if $0 < \alpha_d < 1$. To determine the degree of heavy-tailedness, we use *Weibull factor* introduced in [10], which actually equals the shape parameter α_d for a Weibull distribution. The smaller the α_d value, the heavier the tail in a given Weibull distribution.

To avoid the complexity of directly applying heavy-tailed distributions in theoretical analysis, the data file size L_d can also be approximated by a two-stage hyper-exponential distribution [6], whose PDF is defined as

$$f_{L_d}(x) = \frac{b}{b+1} \cdot \frac{1}{\frac{1}{b} \cdot \overline{L}_d} e^{-\frac{b}{\overline{L}_d} x} + \frac{1}{b+1} \cdot \frac{1}{b \cdot \overline{L}_d} e^{-\frac{1}{b} \cdot \frac{1}{\overline{L}_d} x}, \qquad b \geq 1, \quad x > 0 \quad (8.6)$$

where the parameters b and $\overline{L}_d$ can be obtained by the first and second moments fitting. In particular, b can characterize the *"mice-elephants"* feature. A larger value of b corresponds to a data call size with a higher variability. Furthermore, since the hyper-exponential distribution consists of a linear mixture of exponentials, the analytical study involving (8.6) can be extended to higher-order hyper-exponential distributions with more exponential components, which more accurately approach the original heavy-tailed distribution.

8.2.3 Network Capacity Model

To ensure QoS satisfaction to admitted traffic, the traffic load in the cell or WLAN should be properly limited within the corresponding capacity region.

CDMA-based Cell Capacity

Consider a CDMA cellular system with integrated voice and data services. The maximum numbers of simultaneously admitted voice and data users should be limited to bound the interference level and satisfy user QoS requirements for the ratio of bit energy to noise and interference power spectral density ($\frac{E_b}{N_0}$). Specifically, the capacity can be evaluated using a cell load factor [11]

$$\eta_{DL} = \sum_{i=1}^{n_v^c} \frac{\frac{\rho + f_{DL}}{W_c}}{\left(\frac{E_b}{N_0}\right)_v \alpha_v R_{b,v}^c} + \rho + \sum_{i=1}^{n_d^c} \frac{\frac{\rho + f_{DL}}{W_c}}{\left(\frac{E_b}{N_0}\right)_d R_{b,d}^c} + \rho \qquad (8.7)$$

where n_v^c and n_d^c are the numbers of voice and data users, respectively, ρ is the orthogonality factor, f_{DL} is the ratio between intercell interference and total intracell power measured at the user receiver, W_c is the total cell bandwidth, α_v is the activity factor of voice users, $R_{b,v}^c$ and $R_{b,d}^c$ are the bit rates of voice and data users, respectively, and $\left(\frac{E_b}{N_0}\right)_v$ and $\left(\frac{E_b}{N_0}\right)_d$ are the $\frac{E_b}{N_0}$ requirements of voice and data users, respectively. Then, the power limitation of the base station is equivalent to bounding the cell load factor by

$$\eta_{max} = 1 - \frac{P_p + P_N X_n}{P_{T,max}} \qquad (8.8)$$

where

$$X_n = \sum_{i=1}^{n_v^c} \frac{\frac{L_{p,i}}{W_c}}{\left(\frac{E_b}{N_0}\right)_v \alpha_v R_{b,v}^c} + \rho + \sum_{i=1}^{n_d^c} \frac{\frac{L_{p,i}}{W_c}}{\left(\frac{E_b}{N_0}\right)_d R_{b,d}^c} + \rho \qquad (8.9)$$

with $L_{p,i}$ being the path loss for the ith user, P_p the power devoted to common control channels, P_N the background noise power, and $P_{T,max}$ the maximum transmitted power of the base station.

WLAN Capacity

In a contention-based WLAN similar to IEEE 802.11b, data transmission can follow the request to send (RTS)-clear to send (CTS)-DATA-ACK handshaking for channel access, while the basic access method is better for voice flows due to the small payload size of voice packets. With the analytical approach proposed in [12], we can derive the WLAN capacity region to satisfy the network stability constraints, i.e., the packet service rate is larger than the arrival rate. The WLAN capacity region is defined as a feasible set of (n_v^w, n_d^w) vectors, which limit the maximum numbers of voice and data flows[1] that can be simultaneously admitted in the WLAN, respectively.

Moreover, we can obtain the corresponding packet service rate for each data flow, denoted by $\xi_d^w(n_v^w, n_d^w)$. As observed from the capacity analysis, when more service calls contend for access, the achievable WLAN throughput is severely degraded due to a larger contention overhead. As a result, the number of service calls admitted in

---

[1]Each voice call in the WLAN has two voice flows from and to the mobile, while each data call has a one-way data flow to the mobile.

the WLAN are rather limited to maintain a small collision probability and satisfy the delay requirement of voice service. Meanwhile, a high throughput is achievable for each data call.

8.3 Service-Differentiated Network Selection

As discussed in Section 8.1, for incoming calls to the double-coverage area, a network selection/reselection decision needs to be made between the overlay cell and WLAN. The target network decides whether to accept or reject the incoming call according to its admission control policy. Due to heterogeneous underlying technologies, it is necessary to jointly consider factors such as traffic characteristics, user mobility, and QoS support capability in network selection/reselection.

8.3.1 Network Selection Strategy with Service Differentiation

For cellular networks, the centralized infrastructure enables dedicated resource allocation and provides hard QoS guarantee. The large cell size and ubiquitous coverage can reduce handoff frequency and the impact of handoff latency on delay-sensitive real-time services. In contrast, as WLANs suffer from a large contention overhead, only a very limited number of voice calls can be admitted. The achievable throughput may also be severely jeopardized. Further, the small and disjoint coverage of WLANs has an adverse effect on voice calls as frequent vertical handoff may be involved. On the other hand, interactive data calls can accept elastic bandwidth. A larger bandwidth leads to a faster departure from the system. The large WLAN bandwidth can be efficiently utilized by elastic data traffic to improve the multiplexing gain. A data call is very likely to complete within the WLAN and relieved from the cell.

In view of the above observation, we develop the following network selection/reselection strategy with service differentiation. A new voice call originating in the double-coverage area first selects the cell to request admission and overflows to the WLAN if rejected by the cell according to its admission control policy. A similar strategy is applied to new data calls in the double-coverage area except that data calls first try the WLAN for admission. When a mobile moves from the cellular-only area into the WLAN coverage, its associated voice calls are not handed over from the cell to the WLAN. No network reselection is applied to avoid QoS degradation induced by vertical handoff and the inefficient real-time service support of the WLAN. In contrast, ongoing data calls served by the cell will attempt to hand over to the WLAN.

8.3.2 Performance Analysis of Service-Differentiated Strategy

As shown in Section 8.3.1, the final selection of target network also depends on the admission regions of the cell and WLAN. In this work, simple admission control policies are considered in evaluating the performance of the service-differentiated network selection strategy.

Admission Control Policies

In the cellular network, the centralized control enables reservation-based resource sharing and effective service differentiation. A restricted access mechanism [13] is used to share the cell bandwidth between voice and data services. Voice service is offered preemptive priority over data service and only occupies up to a minimum amount of bandwidth to meet its QoS requirements. As data traffic can adapt to elastic bandwidth, all the bandwidth unused by voice traffic is shared equally by active data calls. Further, a limited fractional guard channel policy shown in Fig. 8.2 is considered to provide new and handoff traffic in the cellular-only area a priority over new traffic in the double-coverage area. This is because a rejected call in the double-coverage area can overflow to the other network to request admission, whereas a call in the cellular-only area is cleared from the system if rejected by the cell. Here, N_v^c is the maximum number of voice calls allowed in the cell, while the guard bandwidths G_{v1}^c and G_{v2}^c are fractional because call blocking/dropping probabilities are very sensitive to the amount of reserved bandwidth. Correspondingly, data admission region of the cell is given by $(N_d^c, G_{d1}^c, G_{d2}^c)$.

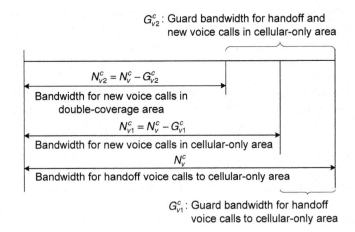

Fig. 8.2. Limited fractional guard channel policy for voice calls in the cell.

On the other hand, the WLAN capacity region can be derived as a feasible set of (n_v^w, n_d^w) vectors, which are the maximum numbers of voice and data calls that can be simultaneously admitted in the WLAN. It is observed that the data packet service rate $\xi_d^w(n_v^w, n_d^w)$ decreases dramatically with a larger n_v^w. Hence, a best operating point can be selected to limit the maximum numbers of voice and data calls allowed in the WLAN by N_v^w and N_d^w, respectively. As vertical handoff from the cell to the WLAN is optional, two other admission parameters G_v^w and G_d^w are introduced to denote the WLAN bandwidth reserved for new traffic. For the service-differentiated selection strategy, no voice calls are handed over to the WLAN while vertical handoff proceeds for a data call if there is spare WLAN capacity. This case can be modeled by setting $G_d^w = 0$ and $G_v^w = N_v^w$.

Given the service-differentiated selection strategy and admission control policies, the performance evaluation is quite complex because multiple dimensions are involved with the coupling between the cell and the WLAN, resource sharing between voice and data services, and differentiation of new and handoff traffic in different areas. The following approach employs proper decomposition and statistical averaging techniques to simplify the analysis.

QoS Evaluation for Voice Service

Let λ_{v1} and λ_{v2} denote the mean arrival rates of new voice calls in the cellular-only area and the double-coverage area, respectively. According to the service-differentiated selection strategy, the mean rate of new voice calls offered to the WLAN is $\lambda_{nv}^w = \lambda_{v2} B_{v2}^c$. The channel holding time of voice calls in the WLAN is $T_v^w = \min(T_v, T_r^w)$, whose PDF can be derived from (8.1) as

$$f_{T_r^w}(t) = \frac{a}{a+1} E_x(a\eta^w + \mu_v) + \frac{1}{a+1} E_x(\eta^w/a + \mu_v) \tag{8.10}$$

where $\mu_v = \mathrm{E}^{-1}[T_v]$ and $E_x(\cdot)$ is the PDF function of an exponential distribution, defined as

$$E_x(\lambda) = \lambda e^{-\lambda t}, \qquad \lambda > 0, \quad t > 0. \tag{8.11}$$

The mean channel holding time is then

$$\mathrm{E}[T_v^w] = \frac{a}{a+1} A_I(a\eta^w, \mu_v) + \frac{1}{a+1} A_I(\eta^w/a, \mu_v) \triangleq (\mu_v^w)^{-1} \tag{8.12}$$

where $A_I(\cdot)$ is defined as

$$A_I(\nu_1, \nu_2) = \frac{1}{\nu_1 + \nu_2}, \qquad \nu_1 > 0, \quad \nu_2 > 0. \tag{8.13}$$

For tractability, both new and handoff call arrivals to the WLAN are assumed to be Poisson. Then, voice calls in the WLAN can be modeled by an $M/G/K/K$ queueing system. As the steady-state probabilities of an $M/G/K/K$ queue are insensitive to the service time distribution, the probability of having k_v^w voice calls in the WLAN is obtained as

$$\pi_v^w(k_v^w) = \pi_v^w(0) \prod_{i=1}^{k_v^w} \frac{\lambda_v^w(i)}{i \cdot \mu_v^w}, \qquad k_v^w = 1, ..., N_v^w \tag{8.14}$$

where $\lambda_v^w(i) = \lambda_{nv}^w + \lambda_{hv}^{cw}$ if $i \leq N_v^w - G_v^w$, $\lambda_v^w(i) = \lambda_{nv}^w$ if $N_v^w - G_v^w + 1 \leq i \leq N_v^w$, and λ_{hv}^{cw} is the mean rate of handoff voice calls from the overlay cell to the WLAN. The voice call blocking and rejection probabilities of the WLAN (B_v^w and D_v^w, respectively) can then be obtained from (8.14).

Due to the limited fractional guard channel policy and varying mobility within the cell, QoS evaluation for the cell is more complicate. We model voice calls in the cell with a two-dimensional Markov process with a state transition rate diagram given in Fig. 8.3. The state (k_{v1}^c, k_{v2}^c) denotes the numbers of existing voice calls in the cellular-only area and the double-coverage area, respectively, where $0 \leq k_{v1}^c + k_{v2}^c \leq N_v^c$. For clarity of presentation, the diagram is divided into several areas and only example transitions are shown in each area between a tagged state and its neighboring states.

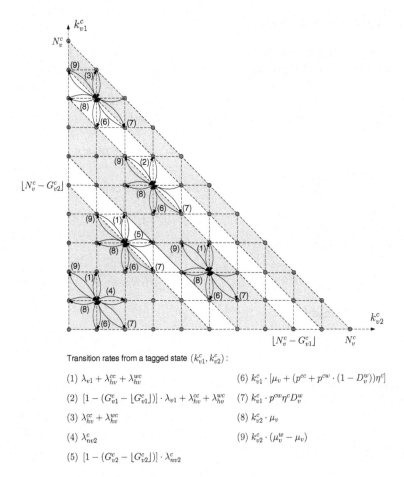

Transition rates from a tagged state (k_{v1}^c, k_{v2}^c):

(1) $\lambda_{v1} + \lambda_{hv}^{cc} + \lambda_{hv}^{wc}$

(2) $[1 - (G_{v1}^c - \lfloor G_{v1}^c \rfloor)] \cdot \lambda_{v1} + \lambda_{hv}^{cc} + \lambda_{hv}^{wc}$

(3) $\lambda_{hv}^{cc} + \lambda_{hv}^{wc}$

(4) λ_{nv2}^c

(5) $[1 - (G_{v2}^c - \lfloor G_{v2}^c \rfloor)] \cdot \lambda_{nv2}^c$

(6) $k_{v1}^c \cdot [\mu_v + (p^{cc} + p^{cw} \cdot (1 - D_v^w))\eta^c]$

(7) $k_{v1}^c \cdot p^{cw}\eta^c D_v^w$

(8) $k_{v2}^c \cdot \mu_v$

(9) $k_{v2}^c \cdot (\mu_v^w - \mu_v)$

Fig. 8.3. State transition rate diagram for voice calls in the cell.

Under the Poisson assumption for call arrivals, the mean arrival rate of voice calls in the cellular-only area is $\lambda_{v1} + \lambda_{hv}^{cc} + \lambda_{hv}^{wc}$, where λ_{hv}^{cc} and λ_{hv}^{wc} denote the mean rates of handoff voice calls from neighboring cells and the overlay WLAN, respectively. Let λ_{nv2}^c denote the mean arrival rate of new voice calls in the double-coverage area.

For the service-differentiated selection strategy, $\lambda_{nv2}^c = \lambda_{v2}$, i.e., all new voice calls in the double-coverage area first try the cell to get admitted. According to the limited fractional guard channel policy, the transition rate from (k_{v1}^c, k_{v2}^c) to $(k_{v1}^c + 1, k_{v2}^c)$ is given by (1), (2), and (3) shown in Fig. 8.3. Similarly, the transition rate from (k_{v1}^c, k_{v2}^c) to $(k_{v1}^c, k_{v2}^c + 1)$ is then (4) and (5) given in Fig. 8.3.

On the other hand, a voice call in the cellular-only area may depart from the cell due to call completion or handoff to a neighboring cell. As both voice call duration and user residence time in the cellular-only area are exponentially distributed, the cell may transit from state (k_{v1}^c, k_{v2}^c) $(k_{v1}^c \geq 1)$ to $(k_{v1}^c - 1, k_{v2}^c)$ if a voice call ends, hands over to a neighboring cell, or gets admitted to the overlay WLAN. Thus, the transition rate from (k_{v1}^c, k_{v2}^c) to $(k_{v1}^c - 1, k_{v2}^c)$ is approximated by

$$(k_{v1}^c, k_{v2}^c) \rightarrow (k_{v1}^c - 1, k_{v2}^c): \qquad (6)\ k_{v1}^c \cdot [\mu_v + (p^{cc} + p^{cw} \cdot (1 - D_v^w))\eta^c], \text{ if } k_{v1}^c \geq 1.$$

In contrast, the transition from (k_{v1}^c, k_{v2}^c) to $(k_{v1}^c - 1, k_{v2}^c + 1)$ is induced by user movement into the WLAN area. The transition rate is approximated by

$$(k_{v1}^c, k_{v2}^c) \rightarrow (k_{v1}^c - 1, k_{v2}^c + 1): \qquad (7)\ k_{v1}^c \cdot p^{cw} \eta^c D_v^w, \text{ if } k_{v1}^c \geq 1. \qquad (8.15)$$

Moreover, a call departure in the double-coverage area may incur transitions from (k_{v1}^c, k_{v2}^c) to $(k_{v1}^c, k_{v2}^c - 1)$ or to $(k_{v1}^c + 1, k_{v2}^c - 1)$ with rates

$$\begin{aligned}
(k_{v1}^c, k_{v2}^c) \rightarrow (k_{v1}^c, k_{v2}^c - 1): & \qquad (8)\ k_{v2}^c \cdot \mu_v, \text{ if } k_{v2}^c \geq 1 \\
(k_{v1}^c, k_{v2}^c) \rightarrow (k_{v1}^c + 1, k_{v2}^c - 1): & \qquad (9)\ k_{v2}^c \cdot (\mu_v^w - \mu_v), \text{ if } k_{v2}^c \geq 1
\end{aligned} \qquad (8.16)$$

where μ_v^w is given by (8.12), which is the total departure rate of a voice call leaving the double-coverage area due to either call completion or user movement. While voice call completion results in the transition from (k_{v1}^c, k_{v2}^c) to $(k_{v1}^c, k_{v2}^c - 1)$ with rate $k_{v2}^c \cdot \mu_v$, the transition rate from (k_{v1}^c, k_{v2}^c) to $(k_{v1}^c + 1, k_{v2}^c - 1)$ is approximated by $k_{v2}^c \cdot (\mu_v^w - \mu_v)$.

By solving the balance equations of this Markov process, we can obtain the steady-state probability of (k_{v1}^c, k_{v2}^c), denoted by $p_v^c(k_{v1}^c, k_{v2}^c)$. Then, the probability of having k_v^c voice calls in the cell is given by

$$\pi_v^c(k_v^c) = \sum_{i=0}^{k_v^c} p_v^c(i, k_v^c - i), \qquad k_v^c = 0, 1, ..., N_v^c. \qquad (8.17)$$

The voice call handoff dropping probability (D_v^c) and blocking probabilities of the cell for new voice calls in the cellular-only area and the double-coverage area (B_{v1}^c and B_{v2}^c, respectively) can then be obtained from (8.17).

As seen from the preceding analysis, the call blocking/dropping probabilities are directly dependent on handoff traffic load. With the inter-dependence between handoff traffic and steady-state probabilities, the QoS metrics need to be evaluated recursively. With an equilibrium assumption for the system, the mean rate of incoming handoff voice calls from neighboring cells λ_{hv}^{cc} is equal to that of outgoing handoff voice calls, which can be obtained by

$$\lambda_{hv}^{cc} = \sum_{i=0}^{N_v^c} \sum_{j=0}^{N_v^c - i} i \cdot p^{cc} \eta^c \cdot p_v^c(i, j). \qquad (8.18)$$

Similarly, the mean rate of potential handoff voice calls into the WLAN coverage is given by $\lambda_{hv}^{cw} = \sum_{i=0}^{N_v^c} \sum_{j=0}^{N_v^c - i} i \cdot p^{cw} \eta^c \cdot p_v^c(i,j)$. On the other hand, a voice call admitted to the WLAN may complete within the WLAN, or it may need to hand over to the cell if it is not finished when the mobile moves out of the WLAN coverage. The mean arrival rate of handoff voice calls from the WLAN to the cell is $\lambda_{hv}^{wc} = [(1 - B_v^w)\lambda_{nv}^w + (1 - D_v^w)\lambda_{hv}^{cw}] \cdot H_v^{wc}$, where the handoff probability H_v^{wc} is given by $H_v^{wc} = \Pr[T_v > T_r^w] = \frac{a}{a+1} \cdot \frac{a\eta^w}{a\eta^w + \mu_v} + \frac{1}{a+1} \cdot \frac{\frac{1}{a}\eta^w}{\frac{1}{a}\eta^w + \mu_v}$.

QoS Evaluation for Data Service

For interactive data services such as Web browsing, the mean data response time is bounded within seconds to guarantee fluent interaction. In contrast to voice calls with a duration of minutes, data calls arrive and depart in a much smaller time scale. In an extreme case that there are no voice call arrivals or departures during a data call duration, the QoS evaluation for data service can be decomposed from voice service. In particular, this limiting behavior for a Markov process is referred to as *nearly complete decomposability* [14]. Thus, we can first analyze the data performance conditioned on the number of voice calls in the system and then obtain the QoS approximation by averaging over the steady-state probabilities of voice calls.

Let (k_v^w, k_d^w) denote the current state of the WLAN with k_v^w voice calls and k_d^w data calls. Assume data call arrivals to the WLAN are a Poisson process with a mean rate λ_d^w. Then, the new and handoff data calls to the WLAN can be viewed as two virtual service classes with Poisson arrival rates $\frac{b}{b+1}\lambda_d^w$ and $\frac{1}{b+1}\lambda_d^w$, respectively, and exponentially distributed service requirements with mean $\frac{1}{b} \cdot \overline{L}_d$ and $b \cdot \overline{L}_d$, respectively [15]. As shown in [16], when the WLAN works in the proper operating range, the packet collision probability is quite small and each packet sees an approximately constant service rate. Hence, the two classes' departure rates due to call completion and handoff out of the WLAN can be derived from (8.1) and (8.6) as

$$\mu_{d1}^w(k_v^w, k_d^w) = \left[\frac{a}{a+1} A_I(a\eta^w, b\nu_d^w(k_v^w, k_d^w)) + \frac{1}{a+1} A_I(\eta^w/a, b\nu_d^w(k_v^w, k_d^w))\right]^{-1}$$

$$\mu_{d2}^w(k_v^w, k_d^w) = \left[\frac{a}{a+1} A_I a\eta^w, \nu_d^w(k_v^w, k_d^w)/b) + \frac{1}{a+1} A_I(\eta^w/a, \nu_d^w(k_v^w, k_d^w)/b)\right]^{-1}$$

where $\nu_d^w(k_v^w, k_d^w) = \frac{\xi_d^w(k_v^w, k_d^w)}{\overline{L}_d}$, $\xi_d^w(\cdot)$ is the average data packet service rate for each data call derived from WLAN capacity analysis, and $A_I(\cdot)$ is defined in (8.13). The offered data traffic load at state (k_v^w, k_d^w) is then

$$\rho_d^w(k_v^w, k_d^w) = \frac{\frac{b}{b+1} \cdot \lambda_d^w(k_d^w)}{k_d^w \cdot \mu_{d1}^w(k_v^w, k_d^w)} + \frac{\frac{1}{b+1} \cdot \lambda_d^w(k_d^w)}{k_d^w \cdot \mu_{d2}^w(k_v^w, k_d^w)} \qquad (8.19)$$

where $\lambda_d^w(k_d^w) = \lambda_{nd}^w + \lambda_{hd}^{cw}$ for $k_d^w \leq N_d^w - G_d^w$, and $\lambda_d^w(k_d^w) = \lambda_{nd}^w$ when $N_d^w - G_d^w + 1 \leq k_d^w \leq N_d^w$, with λ_{nd}^w and λ_{hd}^{cw} being the mean arrival rates of new and handoff data calls to the WLAN, respectively. For the service-differentiated strategy under study, $\lambda_{nd}^w = \lambda_{d2}$ and $G_d^w = 0$. Therefore, the probability of having k_d^w data calls in the WLAN is

$$\pi_d^w(k_d^w) = \sum_{i=0}^{N_v^w} \pi_v^w(i) \prod_{j=1}^{k_d^w} \pi_d^w(0)\rho_d^w(i,j), \qquad k_d^w = 0, 1, ..., N_d^w \qquad (8.20)$$

where $\pi_v^w(\cdot)$ is the steady-state probabilities of voice calls in the WLAN given by (8.14). Thus, data call blocking/rejection probabilities of the WLAN (B_d^w and D_d^w, respectively) can be obtained from (8.20). According to the Little's law, the mean response time of data calls carried by the WLAN is given by

$$\overline{T}_d^w = \sum_{i=0}^{N_v^w} \pi_v^w(i) \sum_{j=1}^{N_d^w} \frac{j \cdot \prod_{l=1}^{j} \pi_d^w(0)\rho_d^w(i,l)}{\lambda_{nd}^w \cdot (1 - B_d^w) + \lambda_{hd}^{cw} \cdot (1 - D_d^w)}. \qquad (8.21)$$

Next, data calls in the cell are modeled by a two-dimensional Markov process similar to that in Fig. 8.3. Consider a tagged state (k_{d1}^c, k_{d2}^c) with k_{d1}^c and k_{d2}^c denoting the numbers of data calls in the cellular-only area and the double-coverage area of the cell, respectively. Due to call arrivals in the cellular-only area, the transition rate from (k_{d1}^c, k_{d2}^c) to $(k_{d1}^c + 1, k_{d2}^c)$ is

$$(k_{d1}^c, k_{d2}^c) \rightarrow (k_{d1}^c + 1, k_{d2}^c): \qquad (8.22)$$
(1) $\lambda_{d1} + \lambda_{hd}^{cc} + \lambda_{hd}^{wc}$, if $k_{d1}^c + k_{d2}^c \le \lfloor N_d^c - G_{d1}^c \rfloor - 1$
(2) $[1 - (G_{d1}^c - \lfloor G_{d1}^c \rfloor)]\lambda_{d1} + \lambda_{hd}^{cc} + \lambda_{hd}^{wc}$, if $k_{d1}^c + k_{d2}^c = \lfloor N_d^c - G_{d1}^c \rfloor$
(3) $\lambda_{hd}^{cc} + \lambda_{hd}^{wc}$, if $\lceil N_d^c - G_{d1}^c \rceil \le k_{d1}^c + k_{d2}^c \le N_d^c - 1$

where λ_{hd}^{cc} and λ_{hd}^{wc} are the mean arrival rates of handoff data calls from neighboring cells and the overlay WLAN, respectively. Incurred by call arrivals to the cell in the double-coverage area, the state transition from (k_{d1}^c, k_{d2}^c) to $(k_{d1}^c, k_{d2}^c + 1)$ has the following mean rate

$$(k_{d1}^c, k_{d2}^c) \rightarrow (k_{d1}^c, k_{d2}^c + 1): \qquad (8.23)$$
(4) λ_{nd2}^c, if $k_{d1}^c + k_{d2}^c \le \lfloor N_d^c - G_{d2}^c \rfloor - 1$
(5) $[1 - (G_{d2}^c - \lfloor G_{d2}^c \rfloor)]\lambda_{nd2}^c$, if $k_{d1}^c + k_{d2}^c = \lfloor N_d^c - G_{d2}^c \rfloor$

where $\lambda_{nd2}^c = \lambda_{d2} B_d^w$ is the mean arrival rate of new data calls rejected by the WLAN and overflowing to the cell in the double-coverage area.

The state transitions due to departure events are more complex with the hyper-exponentially distributed data call size and WLAN residence time. First, state (k_{d1}^c, k_{d2}^c) can transit to $(k_{d1}^c - 1, k_{d2}^c)$ when a data call completes within the cell, hands over to a neighboring cell, or hands over to the overlay WLAN with sufficient spare capacity to admit the call. Given that a data call finds sufficient spare WLAN capacity with a probability $(1 - D_d^w)$, the conditional user residence time in the cellular-only area, denoted by $\tilde{T}_r^c$, can be modeled by an exponential distribution with parameter $\tilde{\eta}^c = [p^{cc} + (1 - D_d^w)p^{cw}]\eta^c$. On the other hand, given k_v^c voice calls and k_d^c data calls carried by the cell, the time that a data call stays with the cell before it completes, denoted by $\tilde{T}_d^c(k_v^c, k_d^c)$, is approximately hyper-exponential with a PDF

$$f_{\tilde{T}_d^c(k_v^c,k_d^c)}(t) = \frac{b}{b+1}E_x(b\nu_d^c(k_v^c,k_d^c)) + \frac{1}{b+1}E_x(\nu_d^c(k_v^c,k_d^c)/b) \qquad (8.24)$$

where $E_x(\cdot)$ is defined in (8.11) and $\nu_d^c(k_v^c,k_d^c) = \frac{R_{b,d}^c}{\overline{L}_d}$ with $R_{b,d}^c$ being the data service rate of the cell in (8.7). Given the state transition from (k_{d1}^c,k_{d2}^c) to (k_{d1}^c-1,k_{d2}^c), the conditional channel holding time of data calls in the cell, denoted by $T_{d1}^c(k_v^c,k_d^c)$, is $\min[\tilde{T}_r^c,\tilde{T}_d^c(k_v^c,k_d^c)]$ with mean

$$\begin{aligned}E[T_{d1}^c(k_v^c,k_d^c)] &= \frac{b}{b+1}A_I(\tilde{\eta}_c,b\nu_d^c(k_v^c,k_d^c)) + \frac{1}{b+1}A_I(\tilde{\eta}_c,\nu_d^c(k_v^c,k_d^c)/b)\\ &\triangleq [\phi_{d1}^c(k_v^c,k_d^c)]^{-1}.\end{aligned} \qquad (8.25)$$

Thus, having k_v^c voice calls in the cell, the state transition rate from (k_{d1}^c,k_{d2}^c) to (k_{d1}^c-1,k_{d2}^c) is

$$(k_{d1}^c,k_{d2}^c) \to (k_{d1}^c-1,k_{d2}^c): \qquad (6)\ k_{d1}^c \cdot \phi_{d1}^c(k_v^c,k_{d1}^c+k_{d2}^c),\ \text{if } k_{d1}^c \geq 1. \qquad (8.26)$$

The transition from (k_{d1}^c,k_{d2}^c) to (k_{d1}^c-1,k_{d2}^c+1) indicates that a data call attempts to hand over to the WLAN but is rejected. So it remains served by the cell but moves into the double-coverage area. Thus, the transition rate is

$$(k_{d1}^c,k_{d2}^c) \to (k_{d1}^c-1,k_{d2}^c+1): \qquad (7)\ k_{d1}^c \cdot p^{cw}\eta^c D_d^w,\ \text{if } k_{d1}^c \geq 1. \qquad (8.27)$$

Given data call completion within the double-coverage area, the cell state transits from (k_{d1}^c,k_{d2}^c) to (k_{d1}^c,k_{d2}^c-1) with a rate

$$(k_{d1}^c,k_{d2}^c) \to (k_{d1}^c,k_{d2}^c-1): \qquad (8)\ k_{d2}^c \cdot \mu_d^c(k_v^c,k_{d1}^c+k_{d2}^c),\ \text{if } k_{d2}^c \geq 1 \qquad (8.28)$$

where $\mu_d^c(k_v^c,k_{d1}^c+k_{d2}^c) = E^{-1}[\tilde{T}_d^c(k_v^c,k_{d1}^c+k_{d2}^c)]$. When the cell carries k_v^c voice calls and k_d^c data calls, for a data call in the double-coverage area and served by the cell, the channel holding time before it completes or moves out of this area with unfinished service, denoted by $T_{d2}^c(k_v^c,k_d^c) = \min[T_r^w,\tilde{T}_d^c(k_v^c,k_d^c)]$, follows a hyperexponential distribution with PDF

$$\begin{aligned}f_{T_{d2}^c(k_v^c,k_d^c)}(t) &= \frac{a}{a+1}\frac{b}{b+1}E_x(a\eta^w+b\nu_d^c(k_v^c,k_d^c)) + \frac{a}{a+1}\frac{1}{b+1}E_x(a\eta^w+\nu_d^c(k_v^c,k_d^c)/b)\\ &+ \frac{1}{a+1}\frac{b}{b+1}E_x(\eta^w/a+b\nu_d^c(k_v^c,k_d^c)) + \frac{1}{a+1}\frac{1}{b+1}E_x(\eta^w/a+\nu_d^c(k_v^c,k_d^c)/b).\end{aligned} \qquad (8.29)$$

Then, the total transition rate from (k_{d1}^c,k_{d2}^c) to (k_{d1}^c,k_{d2}^c-1) or (k_{d1}^c+1,k_{d2}^c-1) is $k_{d2}^c \cdot \phi_{d2}^c(k_v^c,k_{d1}^c+k_{d2}^c)$, where

$$\phi_{d2}^c(k_v^c,k_{d1}^c+k_{d2}^c) = E^{-1}[T_{d2}^c(k_v^c,k_{d1}^c+k_{d2}^c)]. \qquad (8.30)$$

To take advantage of data traffic elasticity, data calls in the cell are served under the processor sharing (PS) discipline, i.e., all active data calls equally share the bandwidth unused by ongoing voice calls. With admission control in place, the system is similar to an $M/G/1/K - PS$ queue, whose steady-state probabilities are insensitive to service requirement [17]. Although the time that the cell stays at state

(k_{d1}^c, k_{d2}^c) before transiting to $(k_{d1}^c + 1, k_{d2}^c - 1)$ is not exponentially distributed, we can approximate the transition rate by

$$(k_{d1}^c, k_{d2}^c) \rightarrow (k_{d1}^c + 1, k_{d2}^c - 1) :$$
$$(9)\ k_{d2}^c \cdot [\phi_{d2}^c(k_v^c, k_{d1}^c + k_{d2}^c) - \mu_d^c(k_v^c, k_{d1}^c + k_{d2}^c)], \text{ if } k_{d2}^c \geq 1. \tag{8.31}$$

By numerically solving the balance equations of the preceding Markov process, we can obtain the steady-state probability of (k_{d1}^c, k_{d2}^c) given k_v^c voice calls carried by the cell, denoted by $\tilde{p}_d^c(k_{d1}^c, k_{d2}^c | k_v^c)$. Then, the probability of having k_d^c data calls in the cell is given by

$$\pi_d^c(k_d^c) = \sum_{i=0}^{N_v^c} \pi_v^c(i) \sum_{j=0}^{k_d^c} \tilde{p}_d^c(j, k_d^c - j | i), \qquad k_d^c = 0, 1, ..., N_d^c \tag{8.32}$$

where $\pi_v^c(\cdot)$ is the steady-state probabilities of voice in the cell given by (8.17). The data call blocking and dropping probabilities of the cell (B_{d1}^c, B_{d2}^c, and D_d^c) can then be obtained from (8.32), while the mean data response time is derived from the Little's law as

$$\overline{T}_d^c = \sum_{i=0}^{N_v^c} \pi_v^c(i) \sum_{j=0}^{N_d^c} \sum_{k=0}^{N_d^c-j} \frac{(j+k) \cdot \tilde{p}_d^c(j, k|i)}{(1 - B_{d1}^c)\lambda_{d1} + (1 - B_{d2}^c)\lambda_{nd2}^c + (1 - D_d^c)(\lambda_{hd}^{cc} + \lambda_{hd}^{wc})}.$$

8.4 Randomized Network Selection for Distributed Implementation

In Section 8.3, we have introduced a service-differentiated network selection strategy for the cellular/WLAN integrated network. The rationale has been to properly distribute the multi-service traffic load to the integrated cell and WLAN so as to effectively exploit their complementary strengths. In this section, we further generalize this strategy with randomized selection to enable distributed implementation.

8.4.1 Decentralization with Randomized Selection

Due to network heterogeneity, it may be very challenging for a central controller to timely obtain updated information of loosely coupled systems (e.g., the numbers of ongoing calls in the cell and overlay WLANs) to make an optimal network selection for each admission request. Consequently, distributed control is more practical, although the selection decision may not be optimal in terms of maximizing resource utilization with QoS guarantee. A simple distributed strategy with randomized network selection [18] is studied in the following. An incoming voice (data) call in the double-coverage area selects the cell with a probability θ_v^c (θ_d^c), while it requests admission to the WLAN with a probability $\theta_v^w = 1 - \theta_v^c$ ($\theta_d^w = 1 - \theta_d^c$). The selection parameters θ_v^c and θ_d^c (or θ_v^w and θ_d^w) are determined for a given traffic load and broadcast to the associated mobiles. Then, a mobile can make a decision on its own according to these parameters and send the admission request to the corresponding target network. With the simplicity, the randomized selection strategy can be implemented in a distributed manner. The cellular network and the integrated WLANs only need to exchange information and update the above selection parameters with traffic variation.

8.4.2 Determination of Selection Parameters Based on MGFs

For the randomized network selection strategy, the parameters θ_v^w and θ_d^w (or corresponding θ_v^c and θ_d^c) should be properly determined to distribute the voice and data traffic load to the overlay cell and WLAN. First, the voice traffic load should be measured and estimated, as voice calls fluctuate in a larger time scale. Given the voice traffic load, the selection parameters can then be determined to maximize the acceptable data traffic load (λ_d) with QoS satisfaction [18]. That is,

$$\max_{(\theta_v^w, \theta_d^w)} \lambda_d, \quad s.t.$$

$$\theta_v^w B_v^w + \theta_v^c B_{v2}^c \leq Q_{PB}, \quad B_{v1}^c \leq Q_{PB}, \quad D_v^c \leq Q_{PD}$$

$$\theta_d^w B_d^w + \theta_d^c B_{d2}^c \leq Q_{PB}, \quad B_{d1}^c \leq Q_{PB}, \quad D_d^c \leq Q_{PD}, \quad \overline{T}_d^c \leq Q_T, \quad \overline{T}_d^w \leq Q_T$$

where $\theta_v^w B_v^w + \theta_v^c B_{v2}^c$ is the blocking probability in the double-coverage area for new voice calls, $\theta_d^w B_d^w + \theta_d^c B_{d2}^c$ is that for new data calls, Q_{PB}, Q_{PD}, and Q_T are the upper bounds for call blocking and dropping probabilities and mean data response time, respectively. The analytical model proposed in Section 8.3.2 is also applicable to the QoS evaluation for this case. The mean arrival rates of new voice and data calls to the cell from the double-coverage area are respectively

$$\lambda_{nv2}^c = \theta_v^c \cdot \lambda_{v2}, \qquad \lambda_{nd2}^c = \theta_d^c \cdot \lambda_{d2}. \tag{8.33}$$

Similarly, the mean arrival rates of new voice and data calls to the WLAN from the double-coverage area are respectively

$$\lambda_{nv}^w = \theta_v^w \cdot \lambda_{v2}, \qquad \lambda_{nd}^w = \theta_d^w \cdot \lambda_{d2}. \tag{8.34}$$

Nonetheless, as the QoS evaluation for the cell is based on two-dimensional Markov processes, the computation complexity of solving large-scale balance equations increases with the size of state space. In this work, we simplify the computation by means of moment generating functions (MGFs).

In general, suppose X and Y are two independent random variables with X being exponentially distributed with parameter λ. Then,

$$\Pr[X > Y] = \int_0^\infty f_Y(y) \int_y^\infty \lambda e^{-\lambda x} dx \, dy = \Psi_Y(-\lambda) \tag{8.35}$$

where $f_Y(\cdot)$ and $\Psi_Y(\cdot)$ are the PDF and MGF of Y, respectively. Letting $Z = \min(X, Y)$, the mean of Z is

$$E[Z] = E[X] - \int_0^\infty f_Y(y) \frac{1}{\lambda} e^{-\lambda y} dy = \frac{1}{\lambda} - \frac{1}{\lambda} \Psi_Y(-\lambda). \tag{8.36}$$

Due to the location-dependent mobility within a cell, calls in the cellular-only area and the double-coverage area differ in channel holding time. Depending on the WLAN state, the average channel holding time of voice calls in the cellular-only area can be derived from (8.2) and (8.36) as

$$E[\min(T_v, T_{r1}^c)] = \frac{1}{\mu_v} - \frac{1}{\mu_v}\Phi_1(-\mu_v) \triangleq \frac{1}{\mu_{v1}^c} \qquad (8.37)$$

when there is not sufficient spare capacity in the WLAN for a voice call; and it is $1/(\mu_v + \eta^c)$ when the incoming voice call can be admitted to the WLAN. Similarly, for voice calls in the double-coverage area, the average channel holding time is $\frac{1}{\mu_v} - \frac{1}{\mu_v}\psi(-\mu_v)$ if there is room for one more voice call in the WLAN or otherwise

$$E[\min(T_v, T_{r2}^c)] = \frac{1}{\mu_v} - \frac{1}{\mu_v}\Phi_2(-\mu_v) \triangleq \frac{1}{\mu_{v2}^c} \qquad (8.38)$$

where $\Phi_2(\cdot)$ is given by (8.3). To simplify analysis, we take an average for the mean service rates of voice calls in the cellular-only area and the double-coverage area, which are respectively given by

$$\tilde{\mu}_{v1}^c = D_v^w \mu_{v1}^c + (1 - D_v^w)(\mu_v + \eta^c) \qquad (8.39)$$

$$\tilde{\mu}_{v2}^c = D_v^w \mu_{v2}^c + (1 - D_v^w)\frac{\mu_v}{1 - \psi(-\mu_v)}. \qquad (8.40)$$

Since voice traffic admitted to the cell from the cellular-only area and the double-coverage area has different average channel holding time approximated by $(\tilde{\mu}_{v1}^c)^{-1}$ and $(\tilde{\mu}_{v2}^c)^{-1}$, respectively, the cell can be viewed as a multi-service loss system. A recursive method was proposed in [19] to approximate the steady-state distribution, which is shown to be accurate for a wide range of traffic intensities and when the service rates (such as $\tilde{\mu}_{v1}^c$ and $\tilde{\mu}_{v2}^c$) do not greatly differ from each other. Moreover, call blocking probabilities are *almost* insensitive to service time distributions. Hence, the probability of having k_v^c voice calls in the cell can be obtained as

$$\pi_v^c(k_v^c) = \pi_v^c(0) \prod_{i=1}^{k_v^c} \left[\frac{\lambda_{v1}^c(i)}{i \cdot \tilde{\mu}_{v1}^c} + \frac{\lambda_{v2}^c(i)}{i \cdot \tilde{\mu}_{v2}^c} \right], \qquad k_v^c = 1, ..., N_v^c \qquad (8.41)$$

where

$$\lambda_{v1}^c(i) = \begin{cases} \lambda_{v1} + \lambda_{hv}^{cc} + \lambda_{hv}^{wc}, & i \le \lfloor N_v^c - G_{v1}^c \rfloor \\ [1 - (G_{v1}^c - \lfloor G_{v1}^c \rfloor)]\lambda_{v1} + \lambda_{hv}^{cc} + \lambda_{hv}^{wc}, & i = \lceil N_v^c - G_{v1}^c \rceil \\ \lambda_{hv}^{cc} + \lambda_{hv}^{wc}, & \lceil N_v^c - G_{v1}^c \rceil + 1 \le i \le N_v^c \end{cases}$$

$$\lambda_{v2}^c(i) = \begin{cases} \lambda_{nv2}^c, & i \le \lfloor N_v^c - G_{v2}^c \rfloor \\ [1 - (G_{v1}^c - \lfloor G_{v1}^c \rfloor)]\lambda_{nv2}^c, & i = \lceil N_v^c - G_{v2}^c \rceil. \end{cases} \qquad (8.42)$$

The voice call blocking and dropping probabilities of the cell can then be obtained from $\pi_v^c(k_v^c)$, $k_v^c = 0, 1, ..., N_v^c$.

Given the inter-dependence between the QoS metrics and handoff traffic, the mean arrival rates of handoff calls out of the cell can be derived recursively. According to (8.35), the handoff probability of voice calls in the cellular-only area to neighboring cells, denoted by H_v^{cc}, can be obtained as

$$H_v^{cc} = p^{cc} \cdot \Pr[T_v > T_{r1}^c] = p^{cc}\Phi_1(-\mu_v). \tag{8.43}$$

Similarly, the handoff probability of voice calls in the cellular-only area to the overlay WLAN, denoted by H_v^{cw}, is given by $H_v^{cw} = p^{cw}\Phi_1(-\mu_v)$. Then, the handoff traffic between neighboring cells and that between the cell and the overlay WLAN can be obtained by solving the following equations

$$\lambda_{hv}^{cc} = H_v^{cc}\left[\lambda_{v1} \cdot (1 - B_{v1}^c) + (\lambda_{hv}^{wc} + \lambda_{hv}^{cc})(1 - D_v^c) + \lambda_{nv2}^c \cdot (1 - B_{v2}^c)H_v^{wc}\right]$$

$$\lambda_{hv}^{cw} = H_v^{cw}\left[\lambda_{v1} \cdot (1 - B_{v1}^c) + (\lambda_{hv}^{wc} + \lambda_{hv}^{cc})(1 - D_v^c) + \lambda_{nv2}^c \cdot (1 - B_{v2}^c)H_v^{wc}\right]$$

$$\lambda_{hv}^{wc} = H_v^{wc}\left[\lambda_{nv}^w \cdot (1 - B_v^w) + \lambda_{hv}^{cw} \cdot (1 - D_v^w)\right]. \tag{8.44}$$

The QoS metrics of data calls in the cell can be evaluated under the assumption of nearly complete decomposition of data traffic from voice. Given j voice calls and k data calls carried by the cell, the cell operates like a symmetric queue [20] for data calls with

$$\phi(k) = k \cdot \tilde{\mu}_d^c(j, k), \qquad \gamma(l, k) = \delta(l, k) = \frac{1}{k} \tag{8.45}$$

$$l = 1, 2, ..., k, \qquad k = 1, 2, ..., N_d^c$$

where $\phi(k)$ ($\phi(k) > 0$ if $k > 0$) is the total service rate when there are k customers (data calls) in the queue in positions $l = 1, 2, ..., k$; $\tilde{\mu}_d^c(\cdot)$ is the service rate dedicated to each customer; $\gamma(l, k)$ is the fraction of the service rate directed to the customer in position l ($\sum_{l=1}^k \gamma(l, k) = 1$); $\delta(l, k+1) = \gamma(l, k+1)$ (symmetric condition) is the probability that an arriving customer moves into position l. A data call carried by the cell may depart due to call completion or a handoff to another cell or WLAN. Since the remaining bandwidth unused by current voice calls is shared equally by existing data calls in a PS manner, the departure is independent of the queuing position of the data call and a fair share of the total service rate is still dedicated to each data call. The service rate $\tilde{\mu}_d^c(\cdot)$ in (8.45) needs to be extended as follows.

Similar to the QoS evaluation for data traffic in Section 8.3.2, data calls admitted in the cell are differentiated into two virtual classes with exponentially distributed service requirements with mean $\frac{1}{b} \cdot \overline{L}_d$ and $b \cdot \overline{L}_d$, respectively [15]. Then, data service in the cell is modeled by a symmetric queue serving multiple classes. Given j voice calls and k data calls in the cell, as in (8.39) and (8.40), the service rates of the two virtual classes of data calls in the cellular-only area can be approximated by

$$\tilde{\mu}_{d1}^{c1}(j, k) = D_d^w \mu_{d1}^{c1}(j, k) + (1 - D_d^w)\left[b \cdot \nu_d^c(j, k) + \eta^c\right] \tag{8.46}$$

$$\tilde{\mu}_{d2}^{c1}(j, k) = D_d^w \mu_{d2}^{c1}(j, k) + (1 - D_d^w)\left[\nu_d^c(j, k)/b + \eta^c\right] \tag{8.47}$$

where $\nu_d^c(j,k) = \frac{R_{b,d}^c}{L_d}$ with $R_{b,d}^c$ given by (8.7) and

$$\mu_{d1}^{c1}(j,k) = \frac{b \cdot \nu_d^c(j,k)}{1 - \Phi_1(-b \cdot \nu_d^c(j,k))}, \qquad \mu_{d2}^{c1}(j,k) = \frac{\nu_d^c(j,k)/b}{1 - \Phi_1(-\nu_d^c(j,k)/b)}.$$

Similarly, the service rates of the two virtual classes of data calls admitted to the cell from the double-coverage area can be obtained as

$$\tilde{\mu}_{d1}^{c2}(j,k) = D_d^w \mu_{d1}^{c2}(j,k) + (1 - D_d^w)\frac{b \cdot \nu_d^c(j,k)}{1 - \psi(-b \cdot \nu_d^c(j,k))} \tag{8.48}$$

$$\tilde{\mu}_{d2}^{c2}(j,k) = D_d^w \mu_{d2}^{c2}(j,k) + (1 - D_d^w)\frac{\nu_d^c(j,k)/b}{1 - \psi(-\nu_d^c(j,k)/b)} \tag{8.49}$$

where

$$\mu_{d1}^{c2}(j,k) = \frac{b \cdot \nu_d^c(j,k)}{1 - \Phi_2(-b \cdot \nu_d^c(j,k))}, \qquad \mu_{d2}^{c2}(j,k) = \frac{\nu_d^c(j,k)/b}{1 - \Phi_2(\nu_d^c(j,k)/b)}. \tag{8.50}$$

For symmetric queues such as processor-sharing queues and multi-server queues without waiting room (i.e., loss systems), a product-form stationary queue occupancy distribution exists for arbitrarily distributed service requirements [20]. Hence, given k_v^c voice calls in the cell, the equilibrium distribution of the symmetric queue for data traffic in the cell is given by

$$\tilde{\pi}_d^c(k_d^c|k_v^c) = \tilde{\pi}_d^c(0|k_v^c) \prod_{i=1}^{k_d^c} \left[\frac{\frac{b}{b+1}\lambda_{d1}^c(i)}{i \cdot \tilde{\mu}_{d1}^{c1}(k_v^c,i)} + \frac{\frac{1}{b+1}\lambda_{d1}^c(i)}{i \cdot \tilde{\mu}_{d2}^{c1}(k_v^c,i)} \right.$$
$$\left. + \frac{\frac{b}{b+1}\lambda_{d2}^c(i)}{i \cdot \tilde{\mu}_{d1}^{c2}(k_v^c,i)} + \frac{\frac{1}{b+1}\lambda_{d2}^c(i)}{i \cdot \tilde{\mu}_{d2}^{c2}(k_v^c,i)} \right] \tag{8.51}$$

where $\lambda_{d1}^c(\cdot)$ and $\lambda_{d2}^c(\cdot)$ are the mean arrival rates of data calls from the cellular-only area and the double-coverage area, respectively, given by

$$\lambda_{d1}^c(i) = \begin{cases} \lambda_{d1} + \lambda_{hd}^{cc} + \lambda_{hd}^{wc}, & i \le \lfloor N_d^c - G_{d1}^c \rfloor \\ [1 - (G_{d1}^c - \lfloor G_{d1}^c \rfloor)]\lambda_{d1} + \lambda_{hd}^{cc} + \lambda_{hd}^{wc}, & i = \lceil N_d^c - G_{d1}^c \rceil \\ \lambda_{hd}^{cc} + \lambda_{hd}^{wc}, & \lceil N_d^c - G_{d1}^c \rceil + 1 \le i \le N_d^c \end{cases}$$

$$\lambda_{d2}^c(i) = \begin{cases} \lambda_{nd2}^c, & i \le \lfloor N_d^c - G_{d2}^c \rfloor \\ [1 - (G_{d1}^c - \lfloor G_{d1}^c \rfloor)]\lambda_{nd2}^c, & i = \lceil N_d^c - G_{d2}^c \rceil. \end{cases} \tag{8.52}$$

Let $\pi_d^c(\cdot)$ denote the steady-state probability of data calls in the cell. Then, $\pi_d^c(k_d^c) = \sum_{i=0}^{N_v^c} \pi_v^c(i)\tilde{\pi}_d^c(k_d^c|i)$, $k_d^c = 0, 1, ..., N_d^c$. The data call blocking and dropping probabilities and mean data response time can be obtained from π_d^c as in Section 8.3.2.

8.5 Size-Based Network Selection and Reselection via Vertical Handoff

As discussed in Section 8.3 and Section 8.4, the overall resource utilization of the cellular/WLAN integrated network can be maximized by means of initial network selection during call admission and reselection via vertical handoff crossing WLAN borders. In this section, we study another network selection strategy based on a data call size threshold. Further, we consider network reselection via dynamic vertical handoff within the overlay area triggered by network states instead of user mobility.

8.5.1 Network Selection and Scheduling for Heavy-Tailed Data Calls

As proposed in the service-differentiated strategy, an incoming voice call preferably selects the cell as its target serving network and overflows to the WLAN only if there is not sufficient spare capacity in the cell. Here, we further consider network reselection for voice calls via dynamic vertical handoff from the WLAN to the cell, which can be performed whenever the cell has spare capacity to accommodate more voice calls. As such, voice calls are more concentrated in the cell and provisioned fine QoS guarantee. The bandwidth unused by voice traffic in the two systems can then be pooled to serve data calls. By exploiting the cellular/WLAN interworking and vertical handoff, we can maximize the multiplexing gain.

Because voice calls are preferably distributed to the cell via initial network selection and reselection via dynamic vertical handoff, the average cell bandwidth available to data traffic is relatively low when the voice traffic load is high. It is necessary and feasible to serve data calls in the cell with a more efficient service discipline. In this work, we consider the shortest remaining processing time (SRPT) discipline [21], in which only one call with the least remaining data to transmit is scheduled first and receives service at an instant. In contrast, under the PS, each on-going call shares an equal quantum of service. The SRPT is observed to significantly outperform the PS and be optimal in terms of minimizing the mean response time. It may be suspected that the improvement of SRPT over the PS comes at the expense of a longer response time for calls with a larger data size. However, it is proved in [22] that the unfairness of SRPT diminishes with heavy-tailed call size.

Consider some specific elastic data applications such as Web browsing and file transfer. The Web documents or data files are usually pre-stored in a Web server or file server. It is possible to know some meta information of the file (such as content type and size) *a priori* from session signaling [23]. Hence, we can exploit the data call size in network selection. In particular, a data call selects the cell if the call size is not greater than a threshold Φ_d and the cell bandwidth available to data traffic is at least R_d^c. Otherwise, the data call selects the WLAN as its target network. By properly determining the size threshold (to be discussed in Section 8.5.2), we can improve the overall resource utilization without degrading the user QoS experience.

8.5.2 Performance Analysis of Size-Based Strategy

In this section, we analytically evaluate the QoS metrics such as voice/data call blocking probabilities and mean data response time, based on which we can determine the data call size threshold (Φ_d).

QoS Evaluation for Voice and Data Services

With the contention-based access of WLAN, data calls in the WLAN share the available bandwidth in a PS manner. Under the PS, the mean response time is insensitive to the call size distribution if the overall service capacity is fixed. Although the insensitivity is generally lost due to varying WLAN capacity [24], it can also be retained with admission control in place for a high load condition, where proper resource allocation and load control are critical to prevent QoS violation. In a light load case, the call blocking probability is usually sufficiently low and all admitted calls are provided satisfactory QoS. Hence, we assume that the QoS of data traffic in the WLAN is insensitive to the heavy-tailed call size distribution.

Assume that voice and data call arrivals to the double-coverage area are independent Poisson processes with mean rates denoted by λ_v and λ_d, respectively. Since data calls select the overlay cell and WLAN based on data call size, the data call arrivals to the cell and WLAN are still Poisson processes with mean rates denoted by λ_d^c and λ_d^w, respectively. Given the insensitivity assumption for data service in the WLAN, we can model the integrated cell/WLAN cluster with a three-dimensional Markov process, in which the state (i, j, k) denotes the numbers of voice and data calls in the WLAN (i and j, respectively) and the number of voice calls in the cell (k). The steady-state probability is denoted by $\pi(i, j, k)$. Based on the bandwidth occupancy of voice traffic in the cell and the overall data call size distribution given in (8.5), the mean data call arrival rate to the cell can be derived as

$$\lambda_d^c = \lambda_d \cdot \delta_d^c \cdot \chi_d^c \qquad (8.53)$$

$$\delta_d^c = \int_0^{\Phi_d} f_{L_d}(x)\, dx, \quad \chi_d^c = \sum_{(i,j)} \sum_{k:\, C_d^c(k)\, \geq\, R_d^c} \pi(i, j, k)$$

where δ_d^c is the fraction of data calls with a size not greater than Φ_d, $(1 - \chi_d^c)$ is the probability that such a data call is blocked by the cell due to congestion, and $C_d^c(k)$ is the maximum cell capacity available to data traffic when there are k voice calls in progress. Similarly, the mean data call arrival rate to the WLAN is then $\lambda_d^w = \lambda_d \cdot [\delta_d^c \cdot (1 - \chi_d^c) + (1 - \delta_d^c)] = \lambda_d \cdot (1 - \delta_d^c \cdot \chi_d^c)$.

The corresponding state transition rates of the three-dimensional Markov process are given by

$$
\begin{aligned}
(i,j,k) &\to (i,j,k+1): \lambda_v, & i \leq N_v^w,\ j \leq N_d^w(i),\ k \leq N_v^c - 1 \\
(i,j,k) &\to (i,j,k-1): k \cdot \mu_v, & i = 0,\ j \leq N_d^w(i),\ 1 \leq k \leq N_v^c \\
(i,j,k) &\to (i+1,j,k): \lambda_v, & i \leq N_v^w - 1,\ j \leq N_d^w(i+1),\ k = N_v^c \\
(i,j,k) &\to (i-1,j,k): (i+k) \cdot \mu_v, & 1 \leq i \leq N_v^w,\ j \leq N_d^w(i),\ k = N_v^c \\
(i,j,k) &\to (i,j+1,k): \lambda_d^w, & i \leq N_v^w,\ 0 \leq j \leq N_d^w(i) - 1,\ k \leq N_v^c \\
(i,j,k) &\to (i,j-1,k): j \cdot \xi_d^w(i,j)/g_d^w, & i \leq N_v^w,\ 1 \leq j \leq N_d^w(i),\ k \leq N_v^c
\end{aligned}
$$
$$(8.54)$$

where N_v^c and N_v^w are the maximum numbers of voice calls admitted in the cell and the WLAN, respectively, $N_d^w(i)$ is the maximum number of data calls allowed in the

WLAN with i voice calls in progress[2], $\xi_d^w(i,j)$ is the mean service rate provided to each data call with i voice calls and j data calls in the WLAN, and g_d^w is the mean size of data calls flowing to the WLAN. Note that the transition rate from state (i,j,k) to state $(i-1,j,k)$ consists of two components. One is due to the completion of the i voice calls in the WLAN with a mean rate of $i \cdot \mu_v$, and the other is due to the completion of the k voice calls in the cell with a mean rate $k \cdot \mu_v$. When one of the k voice calls in the cell completes and makes room for a new voice call, one of the i voice calls in the WLAN can be handed over to the cell. According to the selection strategy and data call size distribution given in (8.5), g_d^w can be derived as

$$g_d^w = \frac{(1-\chi_d^c)\displaystyle\int_0^{\Phi_d} x f_{L_d}(x)\,dx + \int_{\Phi_d}^{\infty} x f_{L_d}(x)\,dx}{\delta_d^c \cdot (1-\chi_d^c) + (1-\delta_d^c)}. \tag{8.55}$$

The first term in the numerator of (8.55) corresponds to data calls of a size not greater than Φ_d, which are blocked by the cell due to congestion with a probability $(1-\chi_d^c)$ and overflow to the WLAN. The second term in the numerator accounts for the data calls that have a size larger than Φ_d and select the WLAN to request admission. The denominator is a normalization constant for the size distribution of data calls flowing to the WLAN.

Due to the interdependence between i and k as shown in the state transition rates of (8.54), the balance equations of the Markov process is very sparse and the steady-state probabilities $\pi(i,j,k)$ can be obtained easily. Thus, the voice call blocking probability B_v is given by

$$B_v = \sum_{\substack{(i,j):\, i \le N_v^w \\ j > N_d^w(i+1)}} \pi(i,j,N_v^c). \tag{8.56}$$

That is, an incoming voice call is blocked if there are N_v^c voice calls in the cell and not sufficient spare capacity is available for one more voice call, and if the WLAN is also congested with i voice calls and j data calls, which means that, with the j data calls already in progress, the admission of one more voice call in the WLAN will result in delay violation to the admitted i voice calls.

As observed in previous analysis on PS queues, when overload occurs, the mean response time under the PS increases dramatically with the offered load and the number of admissible calls (N_d), while the call blocking probability converges and cannot be reduced by increasing N_d. In contrast, in an underload case, the call blocking probability is sufficiently small with a reasonably large value of N_d and the mean response time is almost independent of N_d. Similar phenomenon is observed for the SRPT discipline. Hence, the QoS of data calls can be assured by maintaining an underload condition for data traffic in the cell, so that the call blocking probability

---

[2]N_v^c, N_v^w, and $N_d^w(i)$ are obtained from the admission regions of the cell and the WLAN, i.e., the feasible sets of vectors (n_v^c, n_d^c) and (n_v^w, n_d^w), respectively. Here, $N_v^c = \max(n_v^c)$, $N_v^w = \max(n_v^w)$, and $N_d^w(i) = \max(n_d^w)$, given $n_v^w = i$.

is bounded and a sufficiently high throughput is maintained for admitted calls [25]. This can be achieved by properly determining the size threshold Φ_d. Then, the data call blocking probability B_d can be obtained as

$$B_d = \left[\delta_d^c \cdot (1 - \chi_d^c) + (1 - \delta_d^c)\right] B_d^w = (1 - \delta_d^c \cdot \chi_d^c) \cdot B_d^w \qquad (8.57)$$

where B_d^w is the data call blocking probability of the WLAN and is given by

$$B_d^w = \sum_{\substack{(i,j):\, i \le N_v^w \\ j+1 > N_d^w(i)}} \sum_{k=0}^{N_v^c} \pi(i, j, k). \qquad (8.58)$$

That is, the admission of a new data call should not degrade the WLAN capacity so much that the bandwidth requirement of ongoing voice calls cannot be satisfied. From the Little's law, the mean response time of data calls served in the WLAN can be obtained as

$$\overline{T}_d^w = \frac{1}{\lambda_d^w \cdot (1 - B_d^w)} \sum_{\substack{(i,j):\, i \le N_v^w \\ j \le N_d^w(i)}} \sum_{k=0}^{N_v^c} j \cdot \pi(i, j, k). \qquad (8.59)$$

On the other hand, the mean response time of data calls admitted to the cell can be obtained approximately from the $M/G/1 - SRPT$ queue. This is because data call arrivals to the cell is still a Poisson process with a mean rate λ_d^c given in (8.53). The data call blocking probability is negligibly small if an underload condition is guaranteed by the threshold Φ_d. The average bandwidth allocated to data calls is

$$\overline{C}_d^c = \sum_{\substack{(i,j):\, i \le N_v^w \\ j \le N_d^w(i)}} \sum_{k=0}^{N_v^c} C_d^c(k) \cdot \pi(i, j, k). \qquad (8.60)$$

Based on the formulas in [21], the mean response time is approximated by

$$\overline{T}_d^c = \int_0^{\Phi_d} \frac{1}{\delta_d^c} f_{L_d}(x)\, \Gamma_d^c(x)\, dx \qquad (8.61)$$

where $\frac{1}{\delta_d^c} f_{L_d}(x)$ $(0 < x \le \Phi_d)$ is the PDF of the data call size in the cell, and $\Gamma_d^c(x)$ is the conditional response time for a data call of size x, given by

$$\Gamma_d^c(x) = \int_0^y \frac{dt}{1 - \rho_d^c(t)} + \frac{\lambda_d^c \left[\int_0^y t^2 g_{L_d}(t)\, dt + y^2(1 - G_{L_d}(y))\right]}{2\left[1 - \rho_d^c(y)\right]^2} \qquad (8.62)$$

$$\rho_d^c(y) = \lambda_d^c \int_0^y t \cdot g_{L_d}(t)\, dt, \quad y = \frac{x}{\overline{C}_d^c} \qquad (8.63)$$

$$g_{L_d}(t) = \frac{1}{\delta_d^c} W_b(t, \alpha_d, \beta_d/\overline{C}_d^c), \quad 0 < t \le \Phi_d/\overline{C}_d^c. \qquad (8.64)$$

Here, $g_{L_d}(\cdot)$ denotes the PDF of a bounded Weibull distribution and $G_{L_d}(\cdot)$ the corresponding cumulative distribution function. In contrast to the data call size distribution $W_b(x, \alpha_d, \beta_d)$ given in (8.5), the scale parameter β_d is proportionally modified with $\overline{C}_d^c$ to switch the unit from data call size to service time. Taking into account the size-based selection for data traffic, the overall mean response time of data calls can be evaluated by

$$\overline{T}_d = \frac{\delta_d^c \chi_d^c \cdot \overline{T}_d^c + \left[\delta_d^c \cdot (1 - \chi_d^c) + (1 - \delta_d^c)\right](1 - B_d^w) \cdot \overline{T}_d^w}{\delta_d^c \chi_d^c + \left[\delta_d^c \cdot (1 - \chi_d^c) + (1 - \delta_d^c)\right](1 - B_d^w)}. \tag{8.65}$$

Determination of Data Call Size Threshold

Based on the observations in Section 8.5.1, there are some important principles to follow in determining the data size threshold Φ_d. First, an underload condition should be ensured for data traffic in the cell. That is, the data load factor in the worst case, denoted by $\hat{\rho}_d^c$, is less than 1. Similar to (8.63), $\hat{\rho}_d^c$ can be obtained as

$$\hat{\rho}_d^c = \lambda_d^c \int_0^{\Phi_d/R_d^c} t \cdot \frac{1}{\delta_d^c} W_b(t, \alpha_d, \beta_d/R_d^c) \, dt \tag{8.66}$$

where R_d^c is the minimum cell bandwidth available to data traffic, and $\frac{1}{\delta_d^c} W_b(t, \alpha_d, \beta_d/R_d^c)$, $0 < t \le \Phi_d/R_d^c$, denotes the PDF of a bounded Weibull distribution with shape parameter α_d and scale parameter β_d/R_d^c. Moreover, data calls with a smaller size usually expect a shorter response time than those with a larger size. As data calls in the cell have a smaller size than most of those in the WLAN, our second principle is to guarantee that $\overline{T}_d^c \le \overline{T}_d^w$. The mean response time $\overline{T}_d^w$ and $\overline{T}_d^c$ are given by (8.59) and (8.61), respectively.

Last, when determining the size threshold Φ_d, we should make a good trade-off between user-perceived QoS such as mean data response time and QoS in terms of call blocking probabilities. An appropriate threshold Φ_d can be determined to satisfy the following condition:

$$B_d(\Phi_d) < B_d(\Phi_d^*) \Rightarrow \overline{T}_d(\Phi_d) > \overline{T}_d(\Phi_d^*), \quad \forall \, \Phi_d \ne \Phi_d^*. \tag{8.67}$$

That is, the size threshold Φ_d should be chosen so that the mean response time $\overline{T}_d$ is minimized without increasing the data call blocking probability B_d. As such, the resource utilization is improved without degrading the QoS performance. As $\overline{T}_d$ and B_d can be evaluated analytically with the model given in Section 8.5.2, Φ_d can be determined with various search techniques such as the golden section search.

8.6 Numerical Results and Discussions

In this section, we first validate the QoS analytical approaches given in Sections 8.3.2, 8.4.2, and 8.5.2. Then, we investigate the dependence of user-perceived QoS and resource utilization on network selection parameters such as selection probabilities and

Table 8.1. System parameters.

Parameter	Value	Parameter	Value
$(\eta^c)^{-1}$	10 min	$(\eta^w)^{-1}$	14 min
p^{cc}	0.76	p^{cw}	0.24
W_c	3.84 Mchips/s	C^w	11 Mbit/s
$(\mu_v)^{-1}$	140 s	$R_{b,v}^c$	12.2 kbit/s
λ_{v1}	0.12 calls/s	λ_{v2}	0.18 calls/s
Q_{PB}	0.01	Q_{PD}	0.001
Q_T	4.0 s	L_d	64 kbytes
ρ	0.4	f_{DL}	0.55
α_v	0.43	$P_{T,max}$	43 dB
P_p	33 dB	P_N	-106 dB
$\left(\frac{E_b}{N_0}\right)_v$	4.60 dB	$\left(\frac{E_b}{N_0}\right)_d$	4.65 dB

data call size threshold. Further, we study the impact of mobility and traffic characteristics on selection parameters and resource utilization. Last, the performance of the size-based strategy studied in Section 8.5 is compared with those of the service-differentiated strategy and randomized strategy, discussed in Section 8.3 and Section 8.4, respectively. Given in Table 8.1 are the system parameters for the numerical analysis.

8.6.1 Accuracy Validation of QoS Evaluation Approaches

In Section 8.3.2, we propose a QoS evaluation approach based on two-dimensional Markov processes. The analytical model is further simplified by means of MGFs in Section 8.4.2. Here, we conduct computer simulation to verify the analysis accuracy. As an example, Fig. 8.4 compares the simulation results of voice call blocking/dropping probabilities and corresponding analytical results based on Markov processes. It is observed that the analytical results agree well with the simulation results. Fig. 8.5 illustrates the results of mean data response time in the cell ($\overline{T}_d^c$). It can be seen that the gap increases slightly with the data call variability parameter b. The error is induced by the assumption of nearly complete decomposability to decouple the analysis for data calls from voice traffic. Also, the insensitivity of mean data response time to data call size distribution is generally lost with the varying cell capacity available to data traffic. As a result, $\overline{T}_d^c$ is overestimated with a higher variability of data call size [24], i.e., a larger value of b. Nonetheless, the analysis is still quite accurate especially when data calls arrive and depart in a much smaller time scale than voice calls.

Figs. 8.6-8.7 compare the analytical results of the two approaches given in Section 8.3.2 and Section 8.4.2. It can be seen that the analytical results match well and are tightly bounded by the corresponding requirements. Moreover, as shown in Fig. 8.7, the data call QoS significantly improves with the data call variability parameter b. When there is a higher data call variability, not only is more traffic load

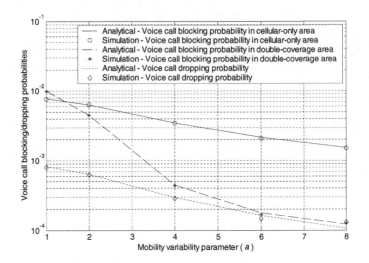

Fig. 8.4. Simulation results of voice call blocking and dropping probabilities and corresponding analytical results based on Markov processes.

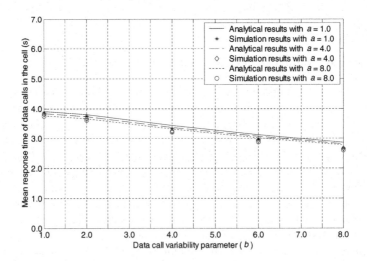

Fig. 8.5. Simulation results of mean response time of data calls in the cell ($\overline{T}_d^c$) and corresponding analytical results based on Markov processes.

relived from the cell by the WLAN, but also most data calls in the cell have a shorter channel holding time and depart from the system faster.

For the size-based selection strategy studied in Section 8.5, Fig. 8.8 illustrates the simulation results of voice/data call QoS and corresponding analytical results

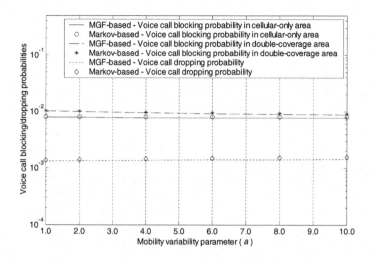

Fig. 8.6. Analytical results of voice call blocking and dropping probabilities with Markov-based and MGF-based approaches.

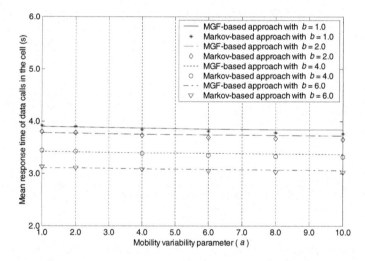

Fig. 8.7. Analytical results of mean response time of data calls in the cell $(\overline{T}^c_d)$ with Markov-based and MGF-based approaches.

based on the evaluation approach given in Section 8.5.2. For investigation simplicity, we fix the voice traffic load by taking $\lambda_v = 0.45$ (calls/s). It can be seen that the analytical results agree with the simulation results for a heavy-tailed data call size $(0 < W_{L_d} < 1)$, except that B_v is slightly overestimated when $W_{L_d} \leq 0.3$. This is

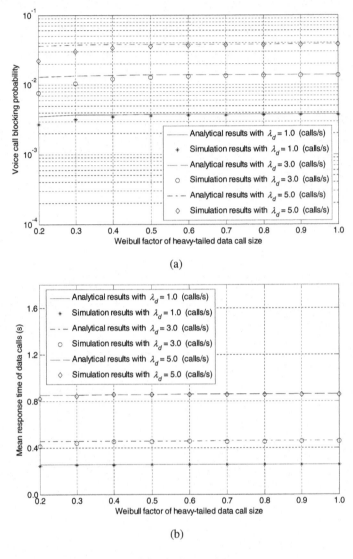

Fig. 8.8. Analytical and simulation results of voice and data QoS vs. Weibull factor (W_{L_d}) for a heavy-tailed data call size. (a) Voice call blocking probability (B_v). (b) Mean data response time ($\overline{T}_d$).

due to the increase of heavy-tailedness and a greater variability of data call size with a small W_{L_d}. Although the insensitivity assumption used in Section 8.5.2 is impaired due to the varying WLAN capacity, it is expected to retain when the call blocking probabilities are sufficiently small. For example, as seen in Fig. 8.8, with a relatively light traffic load and a smaller B_d, the gap between the analytical and simulation results when $W_{L_d} \leq 0.3$ is much smaller. As the system is usually designed to ensure call blocking probabilities in the order of 10^{-3} - 10^{-2}, the analytical model in Section 8.5.2 is valid for numerical analysis.

8.6.2 Dependence of Utilization on Selection Parameters

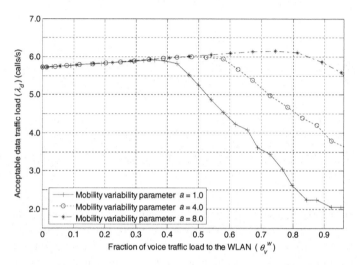

Fig. 8.9. Acceptable data traffic load (λ_d) vs. selection parameter θ_v^w with different mobility variability parameters (a).

Fig. 8.9 shows the dependence of the acceptable data traffic load (λ_d) on the selection parameter θ_v^w, which is the probability that an incoming voice call in the double-coverage area selects the WLAN as its target network. It can be seen that there exist optimal values of θ_v^w that maximize the acceptable λ_d. Here, the voice traffic load is fixed for investigation simplicity. Hence, a maximum λ_d indicates a maximum resource utilization. This results from the load sharing of voice and data traffic within the overlay cell and WLAN. On one hand, when more voice traffic in the double-coverage area is directed to the WLAN with a larger θ_v^w, more cell bandwidth is available for data calls. The congestion of the cell and in turn the whole system can be effectively relieved. On the other hand, when θ_v^w is larger than a threshold, the total acceptable traffic load starts to decrease. This is because the WLAN is very inefficient in supporting voice traffic and the WLAN achievable throughput is significantly reduced when overload occurs.

Fig. 8.10 shows the variation of the acceptable data traffic load (λ_d) with θ_d^w, i.e., the probability that a data call in the double-coverage area selects the WLAN, which is correlated with θ_v^w. With a smaller θ_v^w to carry a less voice traffic load in the WLAN, θ_d^w can be larger to admit more data calls and provide enough bandwidth for each admitted call. When θ_v^w is sufficiently large and θ_d^w is even smaller, the WLAN

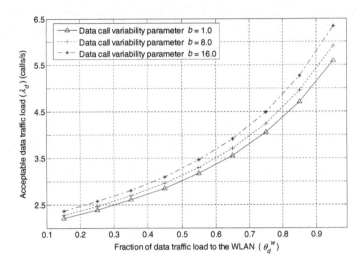

Fig. 8.10. Acceptable data traffic load (λ_d) vs. selection parameter θ_d^w with different data call variability parameters (b).

cannot carry a large portion of the data traffic load in the double-coverage area. As a result, the bottleneck effect of the cell becomes evident. Thus, as shown in Fig. 8.10, the acceptable λ_d decreases with a smaller θ_d^w. The effectiveness of the WLAN as a complement to the cell may be greatly jeopardized with a small θ_d^w and a very large θ_v^w. To maximize the utilization, θ_v^w should be large enough to balance voice traffic load from the cell and also small enough to avoid an inefficient utilization of the WLAN for voice support.

For the size-based selection strategy discussed in Section 8.5, the data call size threshold (Φ_d) should be properly determined to improve user-perceived QoS and resource utilization. Fig. 8.11 illustrates the impact of Φ_d on voice and data QoS. It is observed that B_v and $\overline{T}_d$ slightly decrease with Φ_d when Φ_d is sufficiently small to meet the underload condition. The larger the value of Φ_d, the more the data calls of a small size select the cell and can efficiently utilize the cell bandwidth under the SRPT. On the other hand, after a certain threshold (say, $\Phi_d = 102.4$ kbits), B_v begins to decrease faster with Φ_d. When Φ_d is even larger (e.g., $\Phi_d \geq 640.0$ kbits), $\overline{T}_d$ increases exponentially with Φ_d. This is due to the congestion of the cell. With a small cell bandwidth and high occupancy by voice calls, the data call performance is degraded substantially if the cell is overloaded.

8.6.3 Impact of User Mobility and Traffic Characteristics

The curves in Fig. 8.9 are obtained with different mobility variability parameters (a). It is observed that the acceptable data traffic load (λ_d) is larger with a larger value of a. That is, a higher utilization is achievable when the variability of user mobility in the double-coverage area is higher. As illustrated in Fig. 8.9, when $a = 1.0, 4.0$, and

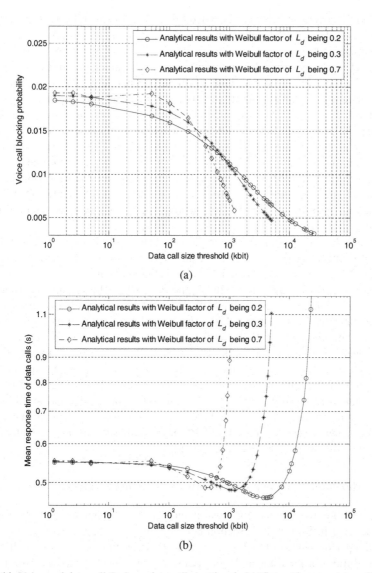

Fig. 8.11. Voice and data call QoS vs. data call size threshold (Φ_d) with mean data call arrival rate $\lambda_d = 3.6$ (calls/s) and different heavy-tailedness of data call size. (a) Voice call blocking probability (B_v). (b) Mean data response time ($\overline{T}_d$).

8.0, the highest utilization is achieved with $\theta_v^w = 0.36, 0.48$, and 0.74, respectively. A larger parameter a indicates that more users staying within the WLAN for a shorter time. Then, more voice calls may have a smaller channel holding time and occupy the WLAN bandwidth for a less time. Given the maximum number of voice calls allowed in the WLAN, when the parameter a is larger, a larger fraction of voice calls in the double-coverage area can be carried by the WLAN and relieved from the cell. Therefore, the data call throughput in the cell is higher and more traffic is acceptable with QoS satisfaction.

The data call variability also affects the selection parameters and resource utilization. As shown in Fig. 8.10, more data traffic is acceptable with a larger value of b, which indicates a higher variability of data call size. For a fixed mean data call size, a larger value of b indicates that more data calls have a smaller size and less have an extremely larger size. Hence, more data calls have a shorter channel holding time and can be carried by the WLAN with a high throughput. Also, more data calls are likely to complete within the WLAN and do not need to hand over to the cell when users move out of the WLAN coverage. Therefore, the WLAN bandwidth is more effectively utilized to reduce the traffic load to the cell of small bandwidth. With the MGF-based approach, we can take into account the variability of user mobility and data traffic in determining the selection parameters.

In Fig. 8.11, we can observe the impact of the heavy-tailedness of data call size. Let the Weibull factor $W_{L_d} = \alpha_d$ denote the degree of heavy-tailedness. The smaller the value of W_{L_d}, the heavier the tail of the distribution of data call size. It can be seen in Fig. 8.11(a) that B_v decreases with Φ_d more slowly if W_{L_d} is smaller. When the data size threshold Φ_d is relatively small, B_v even increases with a larger W_{L_d}. On the other hand, as shown in Fig. 8.11(b), the mean data response time $\overline{T}_d$ first slowly decreases with Φ_d until a sufficiently large Φ_d leads to an explosive increase of $\overline{T}_d$ due to system overload. For a smaller W_{L_d} (say, 0.2), $\overline{T}_d$ decreases more slowly and can achieve an even smaller lower bound. This is due to the mice-elephants property of heavy-tailed distributions. A smaller W_{L_d} (i.e., a higher level of heavy-tailedness) implies that there is a larger fraction of even shorter data calls and that less data calls have a much larger size. Given the same size threshold Φ_d, more data calls can then be efficiently served under the SRPT in the cell. As a result, a smaller $\overline{T}_d$ is achievable with an appropriate size threshold.

8.6.4 Performance Comparison of Network Selection Strategies

In the following, we compare the performance of the three network selection strategies discussed in Section 8.3, Section 8.4, and Section 8.5, respectively. Fig. 8.12 shows the performance of the three selection strategies in terms of voice call blocking probability (B_v) and mean data response time $(\overline{T}_d)$. The size-based strategy is observed to significantly outperform the other two. Actually, the performance improvement lies in the fact that the size-based strategy effectively exploits the heavy-tailedness of data call size in initial network selection. Moreover, network reselection via dynamic vertical handoff can further enhance performance by maximizing the multiplexing gain. Nonetheless, the size-based strategy requires that the data call size be known *a priori* via session signaling. The signaling and control overhead for dynamic vertical handoff may increase the implementation complexity.

Moreover, it is observed in Fig. 8.12 that an even larger performance gain is achievable with the size-based strategy for $\overline{T}_d$ when W_{L_d} is smaller, i.e., the data call size is distributed with a heavier tail. For example, when W_{L_d} decreases from

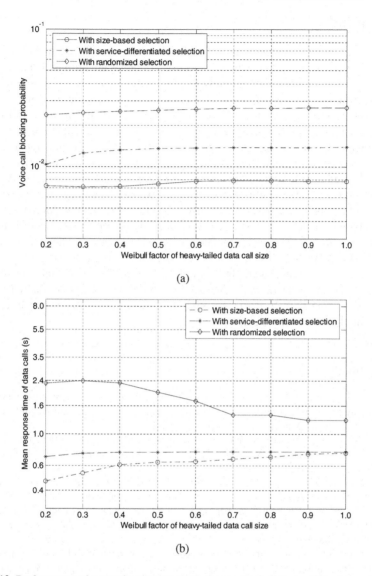

Fig. 8.12. Performance of network selection strategies vs. Weibull factor (W_{L_d}) for a heavy-tailed data call size and mean data call arrival rate $\lambda_d = 3.6$ (calls/s). (a) Voice call blocking probability (B_v). (b) Mean data response time ($\overline{T}_d$).

0.8 to 0.2, the reduction of $\overline{T}_d$ with respect to the randomized strategy increases from 49.6% to 79.7%. In comparison with the service-differentiated strategy, the size-based strategy reduces $\overline{T}_d$ by 7.7% when $W_{L_d} = 0.8$ and by 32.8% when $W_{L_d} = 0.2$. The reduction of $\overline{T}_d$ with W_{L_d} is due to the much higher call blocking probabilities, which restrict the total admissible traffic load to share the bandwidth.

Conclusion

In this chapter, we investigate the network selection issue for heterogeneous wireless overlay networks. Three network selection strategies are proposed and analyzed for the cellular/WLAN integrated network. The selection strategies can significantly improve resource utilization and QoS satisfaction by exploiting user mobility and multi-service traffic characteristics, such as delay-sensitivity of voice traffic and heavy-tailedness of elastic data traffic. The interworking effectiveness is further enhanced with network reselection via vertical handoff between the integrated cell and WLAN. The network reselection can be triggered by user mobility or network states and performed at WLAN border crossing or within the WLAN coverage. As such, the multi-service traffic load is properly shared between the integrated systems and the overall resources are efficiently utilized for QoS provisioning. Moreover, effective analytical models are developed to evaluate QoS performance and determine selection parameters. The analytical approaches take into account the unique characteristics of cell/WLAN cluster, such as complementary network strengths, location-dependent user mobility, and highly variable user residence time within the WLAN. It is observed that user mobility and traffic variabilities have a substantial impact on resource utilization. In particular, high data traffic variability is constructive to improving the interworking gain.

References

1. T. E. Klein and S.-J. Han, "Assignment strategies for mobile data users in hierarchical overlay networks: performance of optimal and adaptive strategies," *IEEE J. Select. Areas Commun.*, vol. 22, no. 5, pp. 849-861, June 2004.
2. W. Shan, "Performance evaluation of a hierarchical cellular system with mobile velocity-based bidirectional call-overflow scheme," *IEEE Trans. Parallel Distrib. Syst.*, vol. 14, no. 1, pp. 72-83, Jan. 2003.
3. S. Thajchayapong and J. Peha, "Mobility patterns in microcellular wireless networks," *IEEE Trans. Mobile Comput.*, vol. 5, no. 1, pp. 52-63, Jan. 2006.
4. F. Yu and V. Krishnamurthy, "Optimal joint session admission control in integrated WLAN and CDMA cellular network," *IEEE Trans. Mobile Comput.*, vol. 6, no. 1, pp. 126-139, Jan. 2007.
5. A. Feldmann and W. Whitt, "Fitting mixtures of exponentials to long-tail distributions to analyze network performance models," *Perform. Eval.*, vol. 31, no. 3-4, pp. 245-279, Jan. 1998.
6. O. J. Boxma, A. F. Gabor, R. N. Queija, and H.-P. Tan, "Performance analysis of admission control for integrated services with minimum rate guarantees," in *Proc. 2nd Conf. on Next Generation Internet Design and Engineering (NGI)*, Apr. 2006, pp. 41-47.
7. N. Benameur, S. B. Fredj, F. Delcoigne, S. Oueslati-Boulahia, and J. W. Roberts, "Integrated admission control for streaming and elastic traffic," in *Proc. 2nd Int'l Wksp. on Quality of Future Internet Services*, Sept. 2001, pp. 69-81.
8. 3GPP, "Services and service capabilities," 3GPP TS 22.105 V8.4.0, June 2007.
9. K. M. Rezaul and A. Pakštas, "Web traffic analysis based on EDF statistics," in *Proc. 7th Annual PostGraduate Symposium on the Convergence of Telecommunications, Networking and Broadcasting (PGNet)*, June 2006.

10. R. Daris and L. Torelli, "Some indices for heavy-tailed distributions," in *Proc. 31st Int'l ASTIN Colloquium*, June 2000, pp. 45-54.

11. J. Pérez-Romero, O. Sallent, R. Agusti, and M. A. Diaz-Guerra, *Radio Resource Management Strategies in UMTS*. New York: Wiley, 2005.

12. W. Song, H. Jiang, and W. Zhuang, "Performance analysis of the WLAN-first scheme in cellular/WLAN interworking," *IEEE Trans. Wireless Commun.*, vol. 6, no. 5, pp. 1932-1952, May 2007.

13. M. Naghshineh and A. S. Acampora, "QoS provisioning in micro-cellular networks supporting multiple classes of traffic," *Wireless Networks*, vol. 2, no. 3, pp. 195-203, Aug. 1996.

14. R. N. Queija, "Processor-sharing models for integrated-services networks," Ph.D. dissertation, Eindhoven University of Technology, Jan. 2000.

15. M. Marsan, "Performance analysis of hierarchical cellular networks with generally distributed call holding times and dwell times," *IEEE Trans. Wireless Commun.*, vol. 3, no. 1, pp. 248-257, Jan. 2004.

16. H. Zhai, X. Chen, and Y. Fang, "How well can the IEEE 802.11 wireless LAN support quality of service?" *IEEE Trans. Wireless Commun.*, vol. 4, no. 6, pp. 3084-3094, Nov. 2005.

17. F. Delcoigne, A. Proutière, and G. Régnié, "Modeling integration of streaming and data traffic," *Perform. Eval.*, vol. 55, no. 3-4, pp. 185-209, Feb. 2004.

18. W. Song, Y. Cheng, and W. Zhuang, "Improving voice and data services in cellular/WLAN integrated network by admission control," *IEEE Trans. Wireless Commun.*, vol. 6, no. 11, pp. 4025-4037, Nov. 2007.

19. P. Tran-Gia and F. Hübner, "An analysis of trunk reservation and grade of service balancing mechanisms in multiservice broadband networks," in *Proc. IFIP Workshop TC6*, 1993, pp. 83-97.

20. F. P. Kelly, *Reversibility and Stochastic Networks*. New York: Wiley, 1979.

21. L. E. Schrage and L. W. Miller, "The queue M/G/1 with the shortest remaining processing time discipline," *Operations Research*, vol. 14, no. 4, pp. 670-684, Jul.-Aug. 1966.

22. N. Bansal and M. Harchol-Balter, "Analysis of SRPT scheduling: Investigating unfairness," *ACM SIGMETRICS Performance Evaluation Review*, vol. 29, no. 1, pp. 279-290, June 2001.

23. M. Garcia-Martin, M. Isomaki, G. Camarillo, and S. Loreto, "A session description protocol (SDP) offer/answer mechanism to enable file transfer," Internet draft, June 2007.

24. R. Litjens and R. J. Boucherie, "Elastic calls in an integrated services network: The greater the call size variability the better the QoS," *Perform. Eval.*, vol. 52, no. 4, pp. 193-220, May 2003.

25. S. B. Fredj, S. Oueslati-Boulahia, and J. W. Roberts, "Measurement-based admission control for elastic traffic," in *Proc. 17th Int'l. Teletraffic Congress*, Dec. 2001, pp. 161-172.

9

Modeling and Performance Analysis of Voice Admission Control in Next Generation Heterogeneous Mobile Networks

Racha Ben Ali and Samuel Pierre

Computer Engineering Department, Ecole Polytechnique Montreal, Canada
{racha.benali, samuel.pierre}@polymtl.ca

9.1 Introduction

Because no single mobile wireless technology can ever provide both high system capacity and cost effective global service coverage for more demanding mobile users, existent and future heterogeneous mobile networks have to be efficiently integrated for this purpose profiting to an increasing number of multi-mode mobile stations. This integration efficiency in what we call a next generation heterogeneous mobile network (NGHMN) is achieved primarily using efficient wireless bandwidth management mechanisms while guaranteeing quality of service. For this purpose, numerical models are used to accurately estimate both new call blocking probability and handoff call dropping probability in the integrated NGHMN. In the remaining of this chapter, we start by presenting the different NGHMN characteristics that can be used to design accurate models estimating their performance in managing the heterogeneous wireless bandwidth. Then we provide models and algorithms in a newly proposed generic analytical framework for accurately evaluating call level quality of service (QoS) parameters of any kind of NGHMN implementing vertical handoff and different network selection strategies. Using this analytical framework, we estimate and compare the handoff rates and the blocking probabilities under a standard exponential model, a general Gamma model and an accurate hyper-exponential model of cell residence times in UMTS cells and WLAN cells.

9.2 Characteristics of NGMHN and Resource Management

9.2.1 Wireless Resources: Hierarchy and Heterogeneity

Generally, the wireless bandwidth resources are spread over a given geographical coverage, which is defined in the planning phase of the mobile network, in different ways depending on several parameters such as the density of the different regions, the geographical mobility needs and also to meet the service and users demands.

E. Hossain (ed.), *Heterogeneous Wireless Access Networks*,
DOI: 10.1007/978-0-387-09777-0_9, © Springer Science+Business Media, LLC 2008

In order to improve the capacity of the cellular network and accommodate a growing number of mobile users, the cell size has to be reduced in dense geographical areas. In fact, due to relatively scarce radio spectrum resources, frequency reuse in neighboring homogenous cells have to maximized and therefore the best way to increase cell's capacity in dense areas is to reduce the cell size in order to decrease the interferences with neighboring cells using the same frequency. For instance, in networks based on the code division multiple access (CDMA) method, adjacent cells share the same frequency and therefore the interference-based soft capacity of a cell can be increased by limiting the cell's transmission power which consequently decreases the cell's size. For several other wireless technologies, such as IEEE 802.11a and IEEE 802.11b/g it is possible to provide very high wireless bandwidth in local area coverage composed of small cells usually called pico-cells.

However, using small cells requires not only a higher number of expensive base stations compared to using larger cells for the same coverage, but also induces a high signaling cost which consumes excessive resources due to the high handoff rate between small cells especially for highly mobile environment. Therefore, in order to circumvent these problems, hierarchical cellular coverage is designed using overlaying heterogeneous cells with different sizes, frequencies or/and access technologies on several layers. Opposed to low demand urban areas which can be largely covered with large cells with low capacity, dense areas are usually covered with several hierarchical layers of heterogeneous cells and a special layer selection and assignment strategy is performed to different mobile users depending on their speed and the access technologies available to them. In fact, in order to limit the signaling cost of a high handoff rate in these dense areas, slow users are assigned to small cells whereas fast users are assigned to overlaying large cells.

This hierarchical architecture generally deployed in homogenous cellular networks can be extended to heterogeneous mobile networks with the availability of more and more multi-mode mobile stations able to connect to heterogenous wireless networks using multiple heterogeneous interfaces. In fact, since there is no wireless access technology that can only by itself provide both high capacity and large coverage to access a large number of services, various heterogeneous mobile networks have to be integrated to profit to an increasing number of multi-mode mobile stations. As a result, a NGMHN is based on the integration of various types of wireless access networks especially in indoor, local and urban dense areas where the most of these networks are available. Among these heterogeneous wireless access networks, the following are significant:

- Wireless Personal Area Networks (WPAN): e.g. pico-nets using bluetooth standard;
- Wireless Local Area Networks (WLAN): e.g. hotspots using IEEE 802.11 a/b/g standard;
- Wireless Metropolitan Area Networks (WMAN): e.g. large cells using either fixed or mobile WiMAX IEEE 802.11d/e standard;
- Wireless Wide Area Networks (WWAN): e.g. cellular networks in different generations and under different standards (for example, 2G GSM standard; 2.5G

GPRS and EDGE standards; 3G cdma2000, UMTS, HSDPA standards; 4G LTE standard, etc.).

Table 9.1. Characteristics of some heterogeneous wireless access technologies.

Wireless access type	Standards	Multiple access	Cell average size
Satellite	VSAT	FDMA, TDMA	2900-25000m
2G	GSM, GPRS	TDMA	600-5000m
3G	UMTS, HSDPA	(W)CDMA	250-4000m
WLAN	IEEE 802.11b,g,a	CSMA/CA	30-150m
WiMAX	IEEE 802.16d,e	(MIMO-)OFDMA	10000-50000m
4G	HSOPA, LTE	MIMO-OFDMA	5000-30000m

In order to efficiently integrate these heterogeneous networks in NGHMN, existent wireless bandwidth management mechanisms, used usually to efficiently allocate the scarce wireless resources while guaranteeing users' QoS, have to interoperate using a distributed admission control mechanism or to be re-adapted and integrated in a centralized admission control mechanism as we will see further in this chapter.

9.2.2 Interconnection Architecture and Radio Resource Management Distribution

A mobile network is generally planned based on an architecture that decomposes it in one wireless access network and one transport core network. The access network is composed of nodes that implement data link protocols for wireless and wired access, whereas the core network is composed of nodes that implement network layer protocols that are usually IP routing nodes. Interconnecting several mobile networks using different heterogeneous wireless technologies can be performed at different levels of the interconnection architecture. As illustrated in Fig. 9.1, if the interconnection is performed at the level of nodes at the access network, this is referred to as a tightly coupled scheme. If the interconnection is performed at the level of nodes interconnecting the access network to the core network, this is referred to as a loosely coupled scheme. If the interconnection is performed at the level of nodes interconnecting the core network to an external IP network such as an IP backbone, this is referred to as a very loosely coupled scheme.

In a tightly coupled scheme, wireless bandwidth allocation has good efficiency since it usually uses a centralized call admission control (CAC) in a common radio resource manager hosted in the radio network controller. In fact, this efficiency is due to the common radio network controller located as near as possible to the wireless resources, where admission control decisions can be done in real-time, thus reflecting the very recent state of cell's resource occupation. Besides, a common resource manager has all the specific and proprietary details of each radio access technology which additionally improves the bandwidth efficiency.

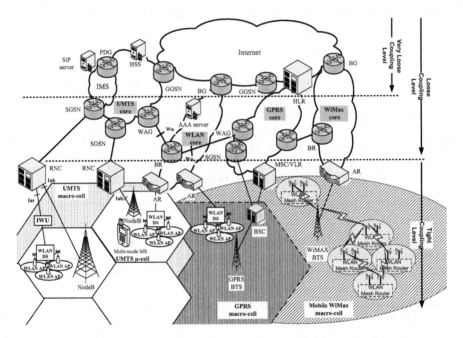

Fig. 9.1. NGHMN interconnection architecture: Levels for different coupling schemes.

It is known that the loosely coupled scheme provides much more flexibility than a tightly coupled scheme. In fact, the loose coupling scheme, opposed to the tight one, allows the integration of several heterogeneous mobile networks under different administrative authorities or belonging to different mobile operators since it uses IETF based open protocols that easily interconnect equipments of different mobile operators at the Internet protocol (IP) layer. However, ensuring resource efficiency by using a centralized CAC under a loose coupling scheme is very challenging since the nearest common node in which the CAC can be placed in the coupling architecture is a lot far away from the wireless resources. Consequently, centralized CAC decisions generally do not reflect the most recent state of wireless bandwidth occupation due to the long path that signaling messages traverse to reach a radio resource controller located in a common node shared by several heterogeneous wireless networks loosely inter-connected. Therefore, a distributed CAC can be more suitable for this kind of loosely coupled architectures than a centralized CAC. In fact, each heterogeneous base station is able to broadcast to each multi-mode mobile station an admission factor which instantly varies depending on the instantaneous load of cell's wireless bandwidth. Using these instantaneous admission factors coming from different wireless networks, the multi-mode mobile station, acting as a bridge relay between loosely coupled wireless networks, decides on which wireless network to send its CAC request. CAC decisions will be taken in the radio resource controller of each heterogeneous network selected by the mobile station in a distributed fashion. Even this distributed CAC does not provide the same high level of efficiency as a centralized common CAC in tightly coupled architectures, it is much more efficient than a centralized common CAC in a loosely coupled architecture. Furthermore, IETF policy based framework can be used to easily deploy this distributed CAC across heterogeneous wireless networks. Centralized common CAC in tight coupling architectures and distributed policy-based CAC in loose coupling architectures in 3G/WLAN integrated network, for example, are illustrated in Fig. 9.2.

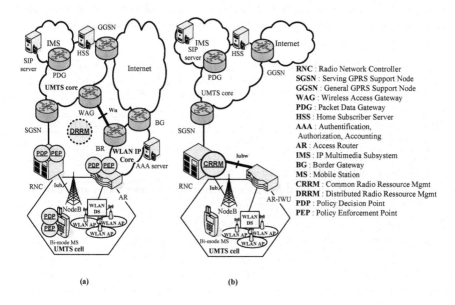

Fig. 9.2. (a) Centralized CAC in tightly coupled and (b) a distributed CAC in a loosely coupled 3G/WLAN integrated network.

9.2.3 Resource Access Priority of Handoff Call Classes

In order to be able to communicate with a mobile wireless network, each mobile station has to be within range of at least one base station serving a cell of a given type. The type of a cell can depend on the wireless access network type, size of the base station, and transmission and reception power of each base station which typically defines the dimension and the shape of the cell. Generally, in new generation networks especially those using CDMA and OFDMA access technologies, adjacent cells belonging to the same network overlap, and consequently, a mobile station is within the coverage of more than one base station for the majority of time. However, cells belonging to different networks are overlaid within each other. Depending on several factors, a mobile station has to decide when and to which cell initiate the handoff process in order to change its point of attachment or add, in the case of a soft handoff, another point of attachment to the available networks. There are several categorization factors based on which it is possible to classify several types of handoffs. Consequently, we can have different types of handoff calls, knowing that a handoff call is an active call hosted by a mobile station during a handoff process. Due to cell's limited wireless bandwidth, different types of handoff calls have to ensure a predefined share or partition of this bandwidth which is more or less important depending on the priority of each handoff call class in guaranteeing the QoS.

Table 9.2. Wireless bandwidth access priority of handoff call classes.

Classification factor	Access priority to bandwidth	
	high	low
Network types involved	Horizontal HO call	Vertical HO call
*Each network gives priority to its own calls		
Number of connections involved	Hard HO call	Soft HO call
*Opposed to Hard HO, Soft HO can be delayed		
Administrative domains involved	Intra-administrative HO Call	Inter-administrative HO call
*Each administrative domain gives priority to its calls		
Necessity of handoff	Obligatory HO Call	Voluntary HO call
*Opposed to voluntary HO, obligatory HO maintains call's continuity		

9.3 General Assumptions for the NGHMN System Model

9.3.1 Service Classes

One of the important motivations of integrating several heterogeneous wireless mobile networks in one single NGMHN is to be able to provide users with a combination of various communication services that a single homogenous mobile network cannot provide while ensuring ubiquity and affordable cost. Generally, the different communication services are classified depending on their delay sensitivity due to their use of the same wireless medium constantly affected by radio interferences. For instance, the delay sensitivity classification criterion has been used by the third generation partnership project (3GPP) to define the four classes of service defined in the UMTS standard. Since other types of mobile networks can use different service classification approaches, a QoS mapping mechanism as the one proposed by [1] have to be ensured between different types of networks integrated in a single NGMHN. This mechanism ensures end-to-end QoS consistency since each heterogeneous wireless network may have its own definition of service classes.

In order to take into account the multi-service aspect of a NGMHN in our modeling of wireless bandwidth management for multiple classes of service, we assume that the bandwidth is shared using the restricted access mechanism initially introduced in wired networks by [2]. This assumption saves us from the development of very complex multiple classes' analytical models that have been already studied in the literature [3, 4]. Thus, we assume that we know at all time the approximate capacity of a given cell in terms of maximal number of calls that can be accommodated without violating their packet-level QoS constraints. Some analytical models providing the capacity of certain types of heterogeneous cells will be briefly presented in this chapter.

9.3.2 Packet-Level and Call-Level QoS

Packet-level QoS is defined using a combination of several metrics such as packet loss, delay and jitter. It is the responsibility of each heterogeneous radio access network to ensure these packet-level QoS for each call depending on its class and the access technique used. In our modeling, we assume that the packet-level QoS is guaranteed stochastically by allocating enough effective bandwidth [5] to the call in order to ensure a packet loss not exceeding a given threshold and/or a delay/jitter violation not exceeding a given probability. A call admission control mechanism for an integrated NGMHN will focus more on call-level QoS provisioning, i.e. ensuring call blocking and dropping probabilities. Therefore, in this chapter we only focus on modeling and evaluating the call-level QoS of NGHMN.

9.4 Mobility Models Depending on Localization

In general, mobility can be modeled using either analytical or simulation models. Simulation is able to provide realistic mobility models since it uses a large amount of details to build the trajectory of the mobile user by periodically tracking its location in small time steps. However such models are generally intractable when considering large network coverage topologies with large number of users. Therefore, analytical mobility models based on random variables and stochastic processes, used to model cell residence time, are more suitable in this case. In this Section, we develop a new analytical mobility model in a heterogeneous coverage to study the impact of general and/or accurate cell residence time distributions on admission control performance in terms of blocking and dropping probabilities. For this purpose, we propose a location dependant mobility model within the larger cell of a typical NGHMN integrating WLAN, 3G and 2.5G wireless access networks.

9.4.1 Mobility Model in a WLAN Multiple Coverage Area

The wireless local area networks (WLAN) under the IEEE 802.11 standard are usually deployed in indoor environments (buildings, airports, shopping malls, etc.), and consequently the user mobility in this type of coverage is relatively low. We note CRT_w the random variable representing the cell residence time (CRT), i.e. the time period spent by a mobile station within a WLAN cell. Usually, we assume a network selection strategy that gives resource allocation preference to the low-cost WLAN rather than the generally highly loaded 3G. Due to the low mobility in WLAN environments (indoor environments in general) most users stay within this region for a short time, while few users stay for a very long time. It was shown in [6] that heavy-tailed Pareto distributions provide the best fit to captured realistic WLAN residence times. Besides, it was shown that usual exponential distributions are no more valid for modeling WLAN residence time. Therefore, in order to capture the Pareto effect without losing the Markov property needed in queuing analysis, we use a hyper-exponential distribution with a mean rate η_w and a mobility variability parameter α_w to model the WLAN residence time. Thus, the probability density function (pdf) of this residence time and the corresponding Laplace transforms can be expressed as follows:

$$f_{CRT_w}(t) = \frac{\alpha_w}{1 + \alpha_w}\alpha_w\eta_w e^{-\alpha_w\eta_w t} + \frac{1}{1 + \alpha_w}\frac{\eta_w}{\alpha_w}e^{-\frac{\eta_w}{\alpha_w}t} \qquad (9.1)$$

$$F_{CRT_w}(s) = \frac{(\alpha_w^2 - \alpha_w + 1)s + \alpha_w\eta_w}{(\alpha_w s + \eta_w)(s + \alpha_w\eta_w)}. \qquad (9.2)$$

9.4.2 Mobility Model in a 3G/2G Cellular-Only Coverage Area

Opposed to WLAN, generally deployed in local and indoor environments characterized by low mobility, 3G networks, under the UMTS standard for example, are deployed in urban outdoor environments where mobile users are highly mobile (pedestrians and vehicular users). As we explained in a previous section that 3G networks are planned using small cells in dense areas in order to limit the interferences that decreases their effective capacity, the cell residence time in a 3G micro-cells have to be modeled differently to 2G macro-cells. In fact, it was shown by [7] that when the CRT has a low variance, generally induced by small sized cells, the exponential distribution is no more valid in modeling this CRT. Therefore, in order to capture this low variance in the 3G CRT, other distributions such as Gamma, Erlang and hyper-Erlang are proposed. In our case, we use a simple Erlang distribution with a mean parameter η_u and a shape parameter m to model the CRT in 3G micro-cells. Thus, the probability density function (pdf) of this residence time can be expressed as follows:

$$f_{CRT_u}(t) = \frac{(m\eta_u)^m t^{m-1}}{(m-1)!}e^{-m\eta_u t}. \qquad (9.3)$$

9.4.3 Generalized Mobility Model in a NGMHN

It has not been proven that a hyper-exponential distribution can be universal in capturing the precise values of any CRT in any WLAN cell visited under various user mobility profiles. Similarly for an Erlang distribution capturing the CRT in 3G micro-cells. In our generalized model, we use a Gamma distribution, which is universal in capturing the CRT in any kind of heterogeneous cell at the price of an eventual loss of model accuracy. Consequently, further in this chapter, we will compare NGHMN performance results under general and accurate distributions for mobility environments in heterogeneous cells. Besides its generalization characteristic, the Gamma distribution with its two parameters of mean and variance, have the interesting feature that allows us to observe the effect of the CRT variance on NGHMN performance while keeping the mean parameter unchanged. However, in order to provide a good accuracy in blocking probabilities in a typical NGHMN integrating WLAN, UMTS and GPRS using Markov chains, we will rather use hyper-exponential, Erlang, and exponential distribution for modeling the CRT in respectively WLAN, UMTS and GPRS cells. In fact, model accuracy is enhanced not only because these distributions are more accurate but also because they keep the Markovian property [8] needed for our queuing analysis due to their rational Laplace transforms.

9.5 Voice Capacity Model

Depending on the access technology used, there are several analytical methods in the literature that bound the maximal number of voice calls, i.e. the number of channels, that a cell can accommodate with the required packet-level QoS.

9.5.1 3G Cell Voice Capacity Model

We assume that the 3G network is based on the popular UMTS standard which uses the wide band code-division multiple access (W-CDMA) cellular system. Assuming different types of calls, the W-CDMA soft capacity is modeled using the very popular load expression in [9]. Using the expression in [9] that bounds the interference, we can estimate the 3G cell voice capacity in term of maximal number of voice calls which is a function of the actual number of calls admitted in each other class. In order to simplify our analysis we assume that voice calls are preemptively prioritized over other types of calls. Therefore, the voice capacity region can be easily evaluated independently from other types of calls.

9.5.2 WLAN Cell Voice Capacity Model

Even though the DCF-based QoS differentiation mechanism introduced in the recent IEEE 802.11e standard [10] can provide transmission priority to voice traffic it cannot guarantee its delay bound since it cannot preempt ongoing transmissions of other low-priority traffic using the distributed operation mode. However, despite the quite lower resource efficiency of the PCF operation mode when no voice traffic is present due to the unnecessary MS polling, this centralized access method can provide strict delay guarantees for voice calls, as opposed to the DCF. The algorithm proposed in [11] provides an analytical expression giving the number of nodes that a PCF-based AP can accommodate in order to satisfy a given delay constraint at each of the nodes. It was shown that an AP configured with Contention Free Period repetition interval $CFPri$ of 25 ms can ensure the strict QoS of 18 voice calls using the G.723.1 codec. In our work, it is possible to limit the maximal number of voice calls to 15 so that a minimal Contention Period (CP) will be guaranteed for data traffic using the DCF as an example of a restricted access resource sharing scheme [2]. In fact, if the MS has no active voice sessions it will be removed from the polling list and a longer CP will be reserved for data traffic.

9.6 Admission Control Policy

Usually when we can derive the voice capacity c_c for the 3G cell [9] (resp. c_w for the PCF-based WLAN cell [11]) in terms of maximal number of voice calls supported without violating the packet-level QoS, a simple admission control algorithm can be designed. However, since it is commonly considered that the dropping of an ongoing call has much more negative impact on users' perception than the blocking of a newly initiated call, we have to prioritize handoff calls over new ones in accessing the wireless bandwidth resources. An acceptance probability function $\beta_k(i,j)$ (resp. $\beta'_k(i,j)$) similar to the one used in [12] is designed for admitting newly initiated voice calls (resp. VHO calls) in the loosely coupled 3G/WLAN network. Where k is either w for WLAN or c for cellular 3G. We assume that since in a loosely coupled 3G/WLAN network the 3G and the WLAN bandwidth resources belong usually to two distinct administrative domains, VHO calls which are controlled and triggered at the MS level are processed as new calls. Therefore, the priority to access the voice capacity for VHO calls and for new calls are equal to $(\beta'_k(i,j) = \beta_k(i,j))$. This acceptance function is tightly related to the the limited fractional guard channel (LFGC) [12] which defines the non-integer number of channels g_k exclusively reserved for HHO calls. Up to $c_k - g_k$ channels can be occupied by new calls and up to $c_k - g'_k$ can be occupied by VHO calls. Note that $g'_k = g_k$ since $\beta'_k(i,j) = \beta_k(i,j)$. In fact, $\beta_k(i,j)$ is equal to 1 if $i + j < \lfloor c_k - g_k \rfloor$, $\lceil g_k \rceil - g_k$ if $i + j = \lfloor c_k - g_k \rfloor$, and 0 if $i + j > \lfloor c_k - g_k \rfloor$. g_k have to be chosen for the optimal solution of the CAC problem presented later.

9.7 Call Traffic Model and Channel Holding Times

Considering a NGMHN integrating N several heterogeneous networks on N layers, we make the following assumptions:

- New calls arrival to a cell at layer k follows a Poisson process with mean rate λ_k which is a commonly used assumption in cellular networks.
- The call holding time represents the call duration which generally depends on user behavior and habits during periods of the day for example. We assume that

the call holding time follows a commonly used exponential distribution with a mean rate μ.

- Horizontal handoff calls arrive following a Poisson process with a mean rate ν_k. This approximation of a handoff traffic using a Poisson process was verified by [13].

- Horizontal handoff calls, that are failed either due to CAC rejection or out of service of the current layer coverage, can be recovered using obligatory vertical handoff calls overflowing to overlaying layers.

- The channel holding time (CHT) represents the service time of a call by one given cell. It was shown in [14] that for exponentially distributed call holding time and a Poisson model for the new call arrival process, the CHT of new calls and the CHT of handoff calls are exponentially distributed if and only if the CRT is exponentially distributed. In Fig. 9.3, the mean CHT of new calls and the mean CHT of handoff calls are defined, respectively, as follows:

$$\frac{1}{\mu_k^{new}} = min\{t_c, r_1\} \tag{9.4}$$

$$\frac{1}{\mu_k^{ho}} = min\{r_m, t_m\}. \tag{9.5}$$

Given the hypothesis of an exponential call holding time and a Poisson process for the new call arrival process, then Theorem 2 in [14] allows us to develop the following expressions (in (9.6) and (9.7), where F_{CRT_k} is the Laplace transform of the pdf of the CRT in cell at layer k of the NGHMN) for the mean CHT of new calls and the mean CHT of handoff calls which are equals if and only if the CRT is exponentially distributed as shown in [7]. However, this is not the case for our mobility model since we consider non-exponential distributions for modeling the CRT in WLAN and UMTS cells. Particularly, we use the Gamma distribution for a general CRT model in any kind of cell and a hyper-exponential distribution for an accurate CRT model in WLAN cells.

$$\frac{1}{\mu_k^{new}} = -(F_{CHT}^{new,k})'(0) = \frac{1}{\mu} - \frac{\eta_k}{\mu^2}(1 - F_{CRT_k}(\mu)) \tag{9.6}$$

$$\frac{1}{\mu_k^{ho}} = -(F_{CHT}^{ho,k})'(0) = \frac{(1 - F_{CRT_k}(\mu))}{\mu}. \tag{9.7}$$

9.8 Network Selection Strategies

Since the NGHMN is composed of several heterogeneous cellular layers hierarchically overlaid, a multi-mode mobile station has to adopt a strategy to select the most suitable network accessible through one of these layers. Several network selection strategies (NSS) can be defined depending either on obligatory critical criteria in order to guarantee both packet and call-level QoS or on optional criteria reflecting user

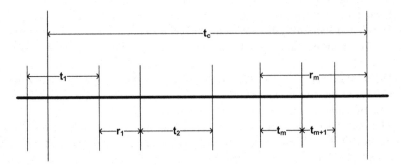

Fig. 9.3. Temporal diagram for CRTs $(t_1, t_2, \cdots, t_{m+1})$, CHTs $(r_1, t_2, \cdots, r_m)$, and call duration t_c.

and/or operator preferences such as cost, security, load balancing, etc. In order to keep the general characteristic of our models, we initially consider any kind of NSS. However, in order to explicitly calculate some performance metrics such as dropping probabilities in integrated networks, we have to choose a commonly used NSS for handoff calls. The overflow scheme usually used in hierarchical homogenous networks is adopted for handoff calls in the NGHMN heterogeneous layers which ensure call's continuity by allowing obligatory handoffs only. This NSS have the advantage to minimize the number of vertical handoffs which, opposed to horizontal handoffs, can degrade packet-level QoS due to the high latency of vertical handoff especially in loosely coupled systems. It has the advantage of being easy to model and tackle analytically. This NSS assumes that call traffic blocked at a lower layer, due to CAC rejections or out of coverage in that layer, overflow to cells at higher layers using obligatory upward vertical handoffs. Besides this NSS, other generic NSS allowing voluntary vertical handoffs, are modeled by (9.18) and (9.19) expressing the rates of different types of handoffs. In these equations, $\alpha_{i,j}$ denotes the fraction of voluntary vertical handoff calls.

9.9 Analysis of Blocking Probabilities

Under an exponential CRT it is easy to develop closed form expressions for steady state probabilities, since we can model the wireless capacity occupation using one-dimensional finite Markov chain without distinguishing between new calls and handoff calls. However, [7] showed that the exponential CRT and CHT hypothesis is no more valid in new generation mobile networks such as 3G and WLAN networks. In this case, the blocking probability analysis and estimation may present some undesirable inaccuracies. Therefore, if we do not know the accurate distributions that we have to use to model the CRT and the CHT in a NGMHN, it is preferable to use a general distribution such as the Gamma distribution. In fact, authors in [15] used this Gamma distribution to model the CRT and the CHT in a homogenous cellular network composed of only one layer. However, our models consider N heterogeneous layers in a NGHMN. In each layer k, we distinguish between the handoff probability of a new call $P_h^{new,k} = \eta_k/\mu F_k(\mu)$ and the handoff probability of a handoff call $P_h^{ho,k} = F_k(\mu)$, where $F_k(\mu)$ is the Laplace transform of the pdf of the CRT following a Gamma distribution and γ_k is the shape parameter of the Gamma function. The Gamma function is defined as follows: $F_k(\mu) = \left(\frac{\gamma_k \eta_k}{\mu + \gamma_k \eta_k}\right)^{\gamma_k}$. Note that choosing an integer γ_k helps us specialize a CRT following an Erlang distribution suitable for 3G/UMTS micro-cells and $\gamma_k = 1$ helps us specialize a CRT following an exponential distribution for 2.5G/GPRS macro-cells.

Since in our NGHMN model we use a generalized CRT following a Gamma distribution, the Markovian property may not be verified and in this case we would rather use this distribution for approximation only. Hyper-exponential, Erlang, and exponential distributions, respectively, for WLAN CRT, UMTS CRT and GPRS CRT, are used in a more accurate model since it preserves the Markovian property as stated in [8].

In any case, we model the wireless capacity occupation using a bi-dimensional Markov chain to distinguish between new calls and handoff calls not having the same CHT since we consider non-exponentially distributed CRTs. Balance equations can be easily deduced from Fig. 9.4 and are explicitly given in Algorithm 1.

Since it is extremely difficult to find closed form expressions for steady-state probabilities from these balance equations, we use the Gauss-Seidel iterative method in Algorithm 1 to estimate these probabilities iteratively. For this purpose, we initialize the system parameters and we start the Gauss Seidel algorithm with "good quality but not enough accurate" state probabilities values given by the closed form expressions from the one-dimensional Markov model. Then we iterate on the balance equations until the convergence is reached with an error margin ϵ. Having the estimated steady-state probabilities, the blocking probabilities are easy to compute using closed form formulas.

Assuming a fractional guard capacity (FGC) [12] CAC policy which as general policy for the limited fractional guard capacity (LFGC) policy, we express the blocking probability of a new call from a cell at layer k, depending on the new call acceptance probability function β_k as follows:

$$P_{b_k} = P_{b_k}(g_k) = \sum_{i+j=0}^{c_k} \beta_k(i+j)P_{i,j,k}. \tag{9.8}$$

Admission of a horizontal handoff call in a cell at layer k is rejected if the current number of calls occupying the cell reached the call's capacity of the cell, i.e. the maximum possible number of calls c_k. However, a handoff call can be rejected from the current layer if there is an out of coverage, i.e. no layer k cell exists where the mobile station is moving. In both the cases, this is referred to as a failed horizontal handoff call at layer k having the following probability:

$$P_{f_k} = P_{f_k}(g_k) = (1 - P_{cov_k}) \sum_{i+j=c_k} P_{i,j,k}(g_k) + P_{cov_k} \tag{9.9}$$

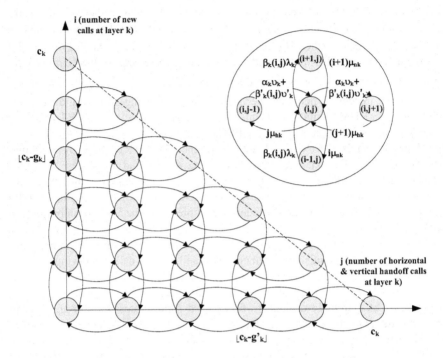

Fig. 9.4. Bi-dimensional Markov chain for voice occupation model in cell k (WLAN or 3G).

where P_{cov_k} is the out of coverage probability at layer k.

If we distinguish between horizontal handoff calls and vertical handoff calls such as in a loosely coupled NGHMN, then we note P_{fv_k} the probability of failing a vertical handoff call. Under the FGC CAC policy, a vertical handoff call requesting the admission to a cell at layer k of the NGHMN is accepted depending on an acceptance probability function β'_k. Therefore, the probability of failing a vertical handoff call at layer k can be expressed as follows:

$$P_{fv_k} = P_{fv_k}(g'_k) = \sum_{i+j=0}^{c_k} \beta'_k(i+j)P_{i,j,k}. \qquad (9.10)$$

9.10 Dropping Probabilities in NGHMN

We express the dropping probability of a call at layer k as the probability of having at least one failed horizontal handoff at layer k. This is the same expression as in homogenous cellular systems [16] as follows:

$$P_{d_k} = \sum_{H_k=0}^{\infty} (P_{h_k})^{H_k}(1-P_{f_k})^{H_k-1}P_{f_k} = \frac{P_{h_k}P_{f_k}}{1-P_{h_k}(1-P_{f_k})} \qquad (9.11)$$

where H_k is the number of succeeded horizontal handoffs before failing the one that drops the call from layer k only. In the following, we derive a more general expression for the dropping probability in a typical NGHMN integrating the three commonly used heterogeneous networks: WLAN (layer $k = w$), UMTS (layer $k = u$) and GPRS (layer $k = g$). We assume a NGHMN using the overflow NSS so that only obligatory vertical handoff are considered. We denote by P_c the probability that a call not initially blocked is terminated normally without any failed handoff. P_c is the sum of the probability of the following events:

- The call terminate normally in a WLAN cell,
- The call terminate normally in a WLAN cell following several successful WLAN-WLAN horizontal handoffs,
- The call terminate normally in a UMTS cell following several successful WLAN-WLAN, WLAN-UMTS, UMTS-UMTS, handoffs,
- The call terminate normally in a GPRS cell following several successful WLAN-WLAN, WLAN-UMTS, UMTS-UMTS, UMTS-GPRS, GPRS-GPRS.

Therefore,

$$
\begin{aligned}
P_c = {} & (1 - P_{h_w}) \\
& + (1 - P_{h_w}) \sum_{i=1}^{\infty} P_{h_w}^i (1 - P_{f_w})^i \\
& + (1 - P_{h_u}) \sum_{i=1}^{\infty} \sum_{j=1}^{i} P_{h_w}^j (1 - P_{f_w})^{j-1} P_{f_w} (1 - P_{f v_u}) P_{h_u}^{i-j} (1 - P_{f_u})^{i-j} \\
& + (1 - P_{h_g}) \times \\
& \sum_{i=1}^{\infty} \sum_{j=1}^{i} \sum_{k=1}^{j} P_{h_w}^k (1 - P_{f_w})^{k-1} P_{f_w} P_{h_u}^{j-k} (1 - P_{f_u})^{j-k} P_{f_u} (1 - P_{f v_g}) P_{h_g}^{i-j} (1 - P_{f_g})^{i-j}.
\end{aligned}
\tag{9.12}
$$

Since we assume that in a loosely coupled NGHMN, there is no distinction between horizontal handoff calls and vertical handoff calls, i.e. both types of calls are treated similarly regarding the access priority to cell's capacity. Therefore, $P_{f_k} = P_{f v_k}$ since $\beta_k(.) = \beta'_k(.)$. Thus

$$
\begin{aligned}
P_c = {} & (1 - P_{h_w}) \frac{1}{1 - P_{h_w}(1 - P_{f_w})} \\
& + \frac{(1 - P_{h_u}) P_{f_w} P_{h_w}(1 - P_{f_u})}{[1 - P_{h_w}(1 - P_{f_w})][1 - P_{h_u}(1 - P_{f_u})]} \\
& + \frac{(1 - P_{h_g}) P_{f_w} P_{h_w} P_{f_u}(1 - P_{f_g})}{[1 - P_{h_w}(1 - P_{f_w})][1 - P_{h_u}(1 - P_{f_u})][1 - P_{h_g}(1 - P_{f_g})]}.
\end{aligned}
\tag{9.13}
$$

$$P_d = 1 - P_c$$

$$= \frac{P_{h_w} P_{f_w} P_{f_u} P_{f_g}}{[1 - P_{h_w}(1 - P_{f_w})][1 - P_{h_u}(1 - P_{f_u})][1 - P_{h_g}(1 - P_{f_g})]}$$

$$= P_{d_w} \frac{P_{d_u}}{P_{h_u}} \frac{P_{d_g}}{P_{h_g}}. \tag{9.14}$$

By generalizing our development to any NGHMN integrating N heterogeneous wireless mobile networks in N layers , we can easily deduce the global dropping probability in any NGHMN as follows:

$$P_d = P_{h_N} \prod_{k=1}^{N} \frac{P_{f_k}}{[1 - P_{h_k}(1 - P_{f_k})]}$$

$$= P_{d_N} \prod_{k=1}^{N-1} \frac{P_{d_k}}{P_{h_k}}. \tag{9.15}$$

9.11 Handoff Rate and Blocking Probability Estimation Algorithms

The handoff call departure rate from a given cell at layer k is equal to the sum of all the call traffic rates arriving to the cell and are willing to leave it through handoff. Two terms contribute to this handoff call departure rate denoted by ν_k^{dep}. The first term is the arrival rate of new calls λ_k , depending on the chosen NSS for new calls expressed in $F^{new}(.)$, with an acceptance probability of $1 - P_{b_k}$ and a handoff probability of $P_{h_k}^n$. The second term is the handoff call arrival rate which depends on the chosen NSS for horizontal handoff calls expressed in $F^{hho}(.)$ and for vertical handoff calls expressed in $F^{vho}(.)$. Thus

$$\nu_k^{dep} = \lambda_k(1 - P_{b_k})P_h^{new,k} + \nu_k^{arr}(1 - P_{f_k})P_h^{ho,k} + \nu'^{arr}_k(1 - P_{fv_k})P_h^{ho,k} \tag{9.16}$$

where: $\nu_k^{arr} = F_k^{hho}(\nu_1, \nu_2, \cdots, \nu_N)$ and $\nu'^{arr}_k = F_k^{vho}(\nu_1, \nu_2, \cdots, \nu_N)$. Note that the functions $F_k^{hho}(.)$ and $F_k^{vho}(.)$ are explicitly given in (9.18) and (9.19), respectively, which express the handoff rates depending on the chosen NSS.

Assuming that time intervals in which call and mobility traffic mean parameters are constant, our Markovian system is homogenous during a time interval long enough to neglect the transient state. Thus, at steady state, we have the flow conservation, i.e. the mean call arrival rate to a given cell at layer k is equal to the mean handoff call departure rate, i.e., $\nu_k^{dep} = \nu_k^{arr} = \nu_k$. Therefore,

$$\nu_k = F_k^{new}(\Lambda)(1 - P_{b_k})P_h^{new,k} + F_k^{hho}(\nu_1, \nu_2, \cdots, \nu_N)(1 - P_{f_k})P_h^{ho,k}$$

$$+ F_k^{vho}(\nu_1, \nu_2, \cdots, \nu_N)(1 - P_{fv_k})P_h^{ho,k}. \tag{9.17}$$

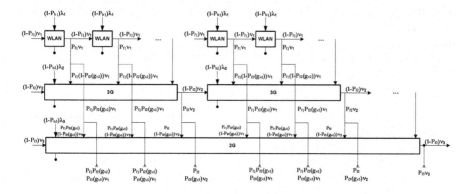

Fig. 9.5. Call traffic in a typical NGHMN.

$$F_1^{hho}(\nu_1, \nu_2, \nu_3) = \nu_1$$
$$F_1^{vho}(\nu_1, \nu_2, \nu_3) = (1/m_1)\alpha_{2,1}\nu_2 + (1/m_1)(1/m_2)\alpha_{3,1}\nu_3$$
$$F_2^{hho}(\nu_1, \nu_2, \nu_3) = \alpha_{2,2}\nu_2 + \alpha_{2,1}\nu_2 P_{f_1'}$$
$$F_2^{vho}(\nu_1, \nu_2, \nu_3) = m_1 P_{f_1}\nu_1 + (1/m_2)\alpha_{3,2}\nu_3 \qquad (9.18)$$
$$F_3^{hho}(\nu_1, \nu_2, \nu_3) = \alpha_{3,3}\nu_3 + \alpha_{3,2}\nu_3 P_{f_2'} + \alpha_{3,1}\nu_3 P_{f_1'}$$
$$F_3^{vho}(\nu_1, \nu_2, \nu_3) = m_1 m_2 P_{f_1} P_{f_2}\nu_1 + m_2 P_{f_2}\nu_2 + \nu_3$$

$$\alpha_{3,1} + \alpha_{3,2} + \alpha_{3,3} = 1$$
$$\alpha_{2,1} + \alpha_{2,2} = 1 \qquad (9.19)$$
$$\alpha_{1,1} = 1.$$

9.12 Numerical Results and Interpretations

In order to study the performance provided by the models and algorithms that we propose for NGHMN, we consider a typical NGHMN integrating the most commonly available wireless access networks: GPRS, UMTS, and WLAN.

9.12.1 NGHMN Parameters

We evaluate the performance of a typical NGHMN in which the heterogeneous wireless coverage area is composed of 7 UMTS micro-cells that overlay 7 WLAN pico-cells each and all the cells are overlaid by one GPRS macro-cell. For the mobility models, we use commonly used parameters values illustrated in Table 9.3. Particularly, realistic values of the CRT in WLAN cells are presented in [6]. In addition,

Algorithm 1 Estimation of blocking probabilities under the generalized model for any NGHMN

Require: Traffic parameters for each layer k: λ_k, μ, η_k et γ_k
Require: NGHMN parameters: $c_k, \alpha_{k,k}, \beta_k(.), \beta'_k(.), m_k, S_k$
Ensure: $P_{b_k}, P_{f_k}, P_{f'_k}$ et P_{d_k} for each layer k
1: Call Algorithm 1
2: Initialize: $\nu_k \leftarrow 0, k = 1, \cdots, N$
3: **while** all handoff rates $\nu_k, k = 1, \cdots, N$ have not converged to their respective steady state values with error margin ϵ_ν **do**
4:     **for** each layer k from 1 to N **do**
5:         // Estimate probabilities P_{b_k} et P_{f_k} using the Gauss-Seidel algorithm
6:         **while** $(\max_i \max_j |P_k(i,j) - P_k^{anc}(i,j)| > \epsilon_P)$ **do**
7:             $P_k^{anc}(i,j) \leftarrow P_k(i,j)$ for all $(i,j) \in [0, \cdots, c_k]^2$
8:             $P_{b_k} \leftarrow 0$ et $P_{f_k} \leftarrow 0$
9:             **for** i=0 to c_k **do**
10:                 **for** j=0 to c_k **do**
11:                     **if** $0 \le i + j < c_k$ **then**
12:                         $P_k(i,j) \leftarrow (\beta_k(i+j-1)\lambda_k P_k(i-1,j) + (i+1)\mu_{nk}P_k(i+1,j) + (\alpha_{k,k}\nu_k + \beta'_k(i+j-1)\nu'_k)P_k(i,j-1) + (j+1)\mu_{hk}P_k(i,j+1))/(\beta_k(i+j)\lambda_k + \alpha_{k,k}\nu_k + \beta'_k(i+j)\nu'_k + i\mu_{nk} + j\mu_{hk})$
13:                     **else if** $i + j \ge c_k - g_k$ **then**
14:                         $P_{b_k} \leftarrow P_{b_k} + (1 - \beta_k(i+j))P_k(i,j)$
15:                         $P_{f'_k} \leftarrow P_{b_k} + (1 - \beta'_k(i+j))P_k(i,j)$
16:                     **else if** $i + j = c_k$ **then**
17:                         $P_k(i,j) \leftarrow (\beta_k(i+j-1)\lambda_k P_k(i-1,j) + (\alpha_{k,k}\nu_k + \beta'_k(i+j-1)\nu'_k)P_k(i,j-1))/(i\mu_{nk} + j\mu_{hk})$
18:                         $P_{b_k} \leftarrow P_{b_k} + (1 - \beta_k(i+j))P_k(i,j)$
19:                         $P_{f'_k} \leftarrow P_{b_k} + (1 - \beta'_k(i+j))P_k(i,j)$
20:                         $P_{f_k} \leftarrow P_{f_k} + P_k(i,j)$
21:                     **end if**
22:                 **end for**
23:             **end for**
24:             $P_{f_k} \leftarrow (1 - P_{cov_k})P_{f_k} + P_{cov_k}$
25:             Compute P_{d_k} depending on the chosen NSS
26:         **end while**
27:         // Estimate the arrival rate of new calls at layer k
28:         $\lambda_k \leftarrow F_k^{new}(\lambda_1, \lambda_2, \cdots, \lambda_N)$
29:         // Estimate the arrival rate of handoff calls leaving cells at layer k
30:         $\nu_k \leftarrow \lambda_k(1 - P_{b_k})P_h^{new,k} + F_k^{hho}(\nu_1, \nu_2, \cdots, \nu_N)(1 - P_{f_k})P_h^{ho,k} + F_k^{vho}(\nu_1, \nu_2, \cdots, \nu_N)(1 - P_{f'_k})P_h^{ho,k}$
31:     **end for**
32: **end while**

---

the value most commonly used in similar studies for the mean call holding time is 4 min. For the mean call arrival rate λ_k, we use the same value 3 calls/min for all heterogeneous layers to reflect the density of the traffic in different WLAN, UMTS, and GPRS coverage areas.

We assume that the cells capacities in terms of the maximal number of voice calls that can be supported following the assumed restricted access resource sharing scheme are 10, 13, and 15, respectively, for GPRS, UMTS, and WLAN cells.

Following preliminary numerical tests of NGHMN performance, we choose the integer guard band values 4, 3, and 0, respectively, for GPRS, UMTS and WLAN cells since they reflect the intensity of the handoff rate (both horizontal and vertical) in each layer. Note that the absence of a guard band capacity for WLAN cells will not degrade handoff call dropping probability due to the low handoff and overflow traffic at the WLAN layer. Besides, there is no obligation to modify the IEEE 802.11 standard to distinguish between admission control requests coming from new calls from those coming from handoff calls.

Table 9.3. CRT parameters values characterizing heterogeneous cells.

Layer k	E[CRT]=$1/\eta_k$	CRT std dev. σ_k	Distributions
GPRS	moderate (3 mn)	moderate (3 mn)	exp. , Gamma(1)
UMTS	low (1 mn)	low (0.33 mn)	exp. , Gamma(9), Erlang
WLAN	high (50 mn)	high (1200 mn)	exp. , Gamma(0.0017), hyper-exp.

9.12.2 Handoff Rate Convergence

Results in Figs. 9.6-9.7 show that the fixed point algorithm used for estimating hand-off rates at steady state converge to different values for the different GPRS, UMTS and WLAN layers, reflecting the mobility and the traffic intensities in these layers. We notice that at the UMTS layer the horizontal handoff rate converges quite slowly to the highest rate, between 9 and 9.5 calls/min (Fig. 9.6(b)). This is in line with UMTS cells characteristics presented in Table 9.3, i.e., cells of small sizes in which mobile station have a relatively high mobility. However, in WLAN cells, the horizontal handoff rate converges to the lowest rate between 0.15 and 0.25 calls/min (Fig. 9.7). This is in line with WLAN cell characteristics presented in Table 9.3, i.e., cells of small sizes in which mobile station have a relatively low mobility.

Fig. 9.8 shows that the highest vertical handoff rate is the one for traffic over-flowing from the UMTS layer (including those overflowing from WLAN layer) to the GPRS layer ($\nu_{UMTS/GPRS}^{vert} = 0.75$ calls/mn). This value, which is relatively low compared to highest horizontal handoff rate, shows that each network gives priority to horizontal handoff over vertical ones. Fig. 9.8 also shows that during the transient period of the fixed point algorithm, if the horizontal handoff rate increases, the corresponding vertical handoff rate decreases and vice-versa. This confirms the effect of blocking probabilities on computing effective horizontal and vertical handoff rates which is not generally considered in most of the models presented in the literature. Our model is indeed more accurate since rather than using any given value of the handoff rate that may be inconsistent with other system parameters, we use this fixed point algorithm to estimate the real values of the horizontal and vertical handoff rates depending on other traffic parameters. Note that this algorithm is restarted for each new time interval triggered when a call or a mobility traffic mean parameter changes or whenever a guard band capacity is adjusted by an optimal admission control algorithm.

Handoff rate estimation using different mobility models based on CRT distributions in different heterogeneous cells, illustrated in Fig. ??, shows that the handoff rate between WLAN cells, although it is the lowest in intensity, is the most affected by CRT distributions. In fact, an exponential distribution usually used in the literature over-estimate WLAN handoff rate and a Gamma distribution used for a generalized model, like the one proposed in this chapter, under-estimates the WLAN handoff rate. The hyper-exponential distribution is the one that captures the best the CRT in WLAN cells as was shown in [6]. For UMTS cells, an hyper-exponential distribution provides UMTS handoff rates quite near the ones provided by a Gamma distribution. However, exponential distribution under-estimates UMTS handoff rate and provides a fixed point function that converges slower than those under other distributions. Finally, given the validity of the exponential distribution for modeling CRT in GPRS macro-cells, the other distributions clearly provide the same GPRS handoff rates.

9.12.3 Impact of CRT Distribution on NGHMN CAC Blocking Performance

We study the CAC blocking performance of the typical NGHMN integrating WLAN, UMTS, and GPRS under different CRT distribution and depending on WLAN network load. We present the performance results in Fig. 9.9, which shows that a Gamma distributed CRT in UMTS cells provides different blocking probabilities than the ones provided by an exponentially distributed CRT. This confirms the fact that the standard exponential model is no more valid for UMTS mobile networks. In addition, there is no difference between blocking probabilities provided by an hyper-exponentially distributed CRT and a Gamma distributed CRT, since both distributions take in account two parameters.

Given the low handoff rate between WLAN cells due to a mean CRT relatively high compared to the mean holding time, we consider a scenario in which WLAN does not maintain any guard band capacity (Fig. 9.10(a)) and another scenario in which it maintains a very low guard band capacity of a single voice call (Fig. 9.10(b)). Fig. 9.10 shows that the difference in the blocking probabilities between those given under the hyper-exponentially distributed CRT and those under the Gamma distributed CRT, is sensibly accentuated and becomes more and more important when the WLAN CRT variability decreases from 10^0 to 10^{-3}.

In Fig. 9.10, we present blocking probabilities for three network selection strategies (NSS). The first NSS, noted NSS A, refers to the overflow strategy that permit only horizontal and vertical obligatory handoff calls ensuring voice call's continuity. The second NSS, noted NSS B, refers to the NSS that permit few voluntary vertical

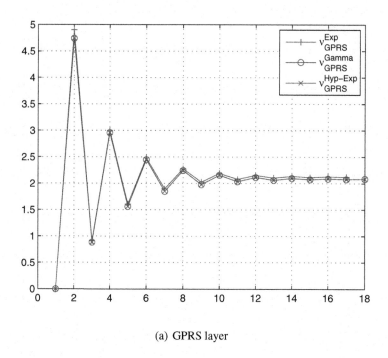

(a) GPRS layer

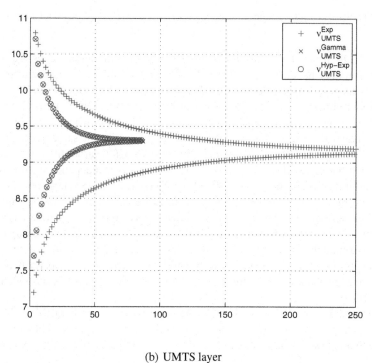

(b) UMTS layer

Fig. 9.6. Handoff rate convergence for GPRS and UMTS cells under different CRT models.

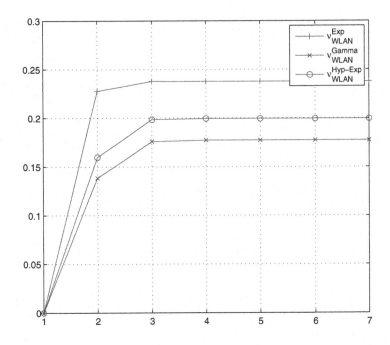

Fig. 9.7. Handoff rate convergence for WLAN cells under different CRT models.

handoff calls particularly only a fraction $\alpha = 5\%$ of all handoff calls are permitted as voluntary handoff calls. The third NSS, noted NSS C, is similar to previous NSS B except it does not permit WLAN horizontal handoffs since each WLAN cell is considered as a "hotspot" isolated from its neighbors. NSS C is used when the WLAN does not have the capability to maintain voice QoS when performing a fast WLAN handoff. Results show that NSS B and NSS C give the same new call blocking probabilities for WLAN cells under all WLAN CRT variabilities. Besides, we notice that the new call blocking probability for NSS A is independent from WLAN CRT variability when there is no WLAN guard band capacity. These results can be explained by the insensitivity property of blocking probability to the variability of CRT distribution when the call arrival follows a Poisson process. With a guard band capacity, the handoff calls generated from new calls having a higher blocking may be less Poissonian than without a guard band capacity, and therefore the traffic arriving to a WLAN cell is no more Poissonian. Thus, the insensitivity property is no more valid since the blocking probability depends on the CRT variability as is evident from Fig. 9.10(b).

Besides, from Fig. 9.10, we notice that an hyper-exponentially distributed WLAN CRT, which is more accurate for this kind of cells, provides a new call blocking probability that is lower than the one provided by a Gamma distributed WLAN CRT, and

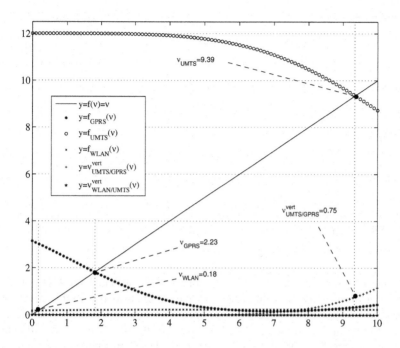

Fig. 9.8. Convergence of fixed point functions for different handoff rates in an NGHMN integrating WLAN, UMTS, and GPRS.

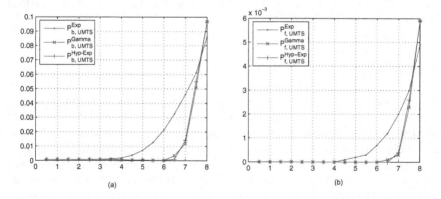

Fig. 9.9. Impact of the CRT distribution on UMTS blocking and handoff failure probabilities depending on new call arrival rate.

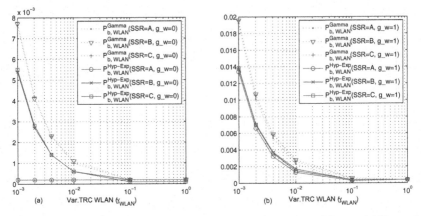

Fig. 9.10. WLAN blocking probabilities depending on the WLAN mobility variability under different NSS with (a) $g_w = 0$, and (b) $g_w = 1$.

much higher than the one provided by an exponentially distributed WLAN CRT, illustrated as a special case with variability parameter equals to 10^0. In this case, the difference between the blocking probabilities given by these distributions is accentuated when the variability parameter decreases from 10^0 to 10^{-3}. Thus, if we estimate the performance results under the generalized model based on a Gamma distributed CRT, we can increase the guard capacity while keeping the same real performance results that are given by the more precise hyper-exponential model. Therefore, under different CRT variabilities, blocking probabilities under an exponentially distributed CRT are over-estimated for UMTS cells and largely under-estimated (even invalid) for WLAN cells. In addition, these blocking probabilities are over-estimated under a Gamma distributed WLAN CRT.

Conclusion

We have presented new generalized models and algorithms for evaluating voice admission control blocking probabilities in a next generation heterogeneous mobiles network (NGHMN) integrating different kinds of wireless access networks. Numerical results have shown that the commonly used exponential distribution in cell residence time (CRT) largely under-estimates WLAN and UMTS blocking probabilities compared to a general model based on a Gamma distributed CRT. A more accurate model would be based on a hyper-exponentially distributed CRT, both tacking in account the CRT variability, which characterizes heterogeneous cells and makes the exponential model obsolete in new generation networks such as WLAN and UMTS. Using accurate and/or generalized models for NGHMNs provide more realistic performance results that may help adjusting more efficiently the guard band capacities for guaranteeing call-level QoS through voice admission control.

References

1. R. Ben Ali, S. Pierre, and Y. Lemieux, "UMTS to IP backbone QoS mapping for voice and video-telephony services," *IEEE Network*, vol. 18, no. 2, pp. 26-32, Mar-Apr. 2005.
2. M. Naghshineh and A.S. Acampora, "QoS provisioning in micro-cellular networks supporting multiple classes of traffic," *Wireless Networks*, vol. 2, no. 3, pp. 195-203, 1996.
3. S. Malomsoky, S. Racz, and S. Nadas, "Connection admission control in UMTS radio access networks," *Computer Communications*, vol. 26, no. 17, pp. 2011-2023, Nov. 2003.
4. N. Nasser and H. Hassanein, "Optimal multi-class guard channel admission policy under hard handoff constraints," in *Proc. of 3rd ACS/IEEE International Conference on Computer Systems and Applications*, Jan. 2005, p. 53.
5. F. P. Kelly, "Notes on effective bandwidth," *Stochastic Networks: Theory and Applications*, vol. 4, pp. 141-168, 1996.
6. S. Thajchayapong and J. M. Peha, "Mobility patterns in microcellular wireless network," *IEEE Transations on Mobile Computing*, vol. 5, no. 1, pp. 52-63, Jan. 2006.
7. Y. Fang and I. Chlamtac, "Teletraffic analysis and mobility modeling of PCS networks," *IEEE Transations on Communications*, vol. 47, no. 7, pp. 1062-1072, July 1999.
8. Y. Fang, "Modeling and performance analysis for wireless mobile networks: A new analytical approach," *IEEE/ACM Transactions on Networking*, vol. 13, no. 5, pp. 989-1002, 2005.
9. H. Holma and A. Toskala, *WCDMA for UMTS: Radio Access for Third Generation Mobile Communications*, John Wiley and Sons, 2004.
10. IEEE 802.11e-Working-Group, "IEEE Standard 802.11e wireless LAN medium access control and physical layer specifications, medium access control quality of service enhancements," *IEEE Standards*, 2005.
11. B. Sikdar, "An analytic model for the delay in IEEE 802.11 PCF MAC based wireless networks," *IEEE Transactions on Wireless Communications*, vol. 6, no. 4, 2007.
12. R. Ramjee, D. Towsley, and R. Nagarajan, "On optimal call admission control in cellular network," *Wireless Networks*, vol. 3, no. 1, pp. 29-41, 1997.
13. E. Chlebus and W. Ludwin, "Is handoff traffic really Poissonian?," in *Proc. of Fourth IEEE International Conference on Universal Personal Communications*, pp. 348-353, 1995.
14. Y. C. Fang and Y. B. Lin, "Channel occupancy times and handoff rate for mobile computing and PCS networks," *IEEE Transactions on Computers*, vol. 47, no. 6, pp. 679-692, 1998.
15. H. Zeng and I. Chlamtac, "Adaptive guard channel allocation and blocking probability estimation in pcs networks," *Computer Networks*, vol. 43, no. 2, pp. 163-176, 2003.
16. D. Hong and S. Rappaport, "Traffic model and performance analysis for cellular mobile radio telephone systems with prioritized and nonprioritized handoff procedures," *IEEE Transactions on Vehicular Technology*, vol. 35, no. 3, pp. 77-92, Aug. 1986.

10

Energy Saving Aspects for Mobile Device Exploiting Heterogeneous Wireless Networks

Gian Paolo Perrucci, Frank H. P. Fitzek, and Morten V. Petersen

Electronic Systems, Aalborg University, Denmark
{gpp,ff,mvpe}@es.aau.dk

10.1 Introduction and Motivation

Mobile phones have been undergoing a breathtaking evolution over the last decade starting from simple mobile phones with only voice services towards the transition of smart phones offering Internet access, localization information and even more. It seems that there are simply no limitations for mobile devices getting smaller, offering higher data rates, brighter displays. Unfortunately this assumption is not correct. The limiting factor is known, namely the energy and power consumption of mobile devices being battery driven. As the complexity within the mobile device is increasing dramatically due to new services such as GPS modules, digital photo cameras, mp3 players and others, the improvement of battery capacity is quite moderate. The increase in complexity of the mobile device is related to the fact that mobile handset vendors need new services to market their products and therefore exploit all the computational power of the given hardware, which is following Moore's law. Even the wireless air interfaces are getting more and more complex starting from simple TDMA systems towards the planned OFDMA/MIMO systems for the 4G wireless communication systems. The increased complexity has two key impacts. First, the enormous power consumption will lead to heating problems. Thus, the mobile device cannot cope with the heat using passive cooling anymore and active cooling would be necessary. To illustrate the problem of heating, Figs. 10.1 and 10.2 show the temperature for a Nokia N95 in the offline mode and for WLAN IEEE802.11b/g downloading, respectively. The temperature difference is approximately 8 degrees Celsius. The situation could be even worse if the mobile device would be covered by holding it in the hand.

Secondly, the increased energy consumption would lead to lower stand-by times. Those stand-by times are limiting the time a customer can use its mobile device, which in turn makes the customer as well as the network operator unsatisfied. The customer cannot use any service and the network operator does not make any revenue. In other words, the degree on mobility for the customer depends on mobile device with low recharge cycles. That energy consumption is not only a problem of the future can be derived by Apple's announcement that battery life was the main

E. Hossain (ed.), *Heterogeneous Wireless Access Networks*,
DOI: 10.1007/978-0-387-09777-0_10, © Springer Science+Business Media, LLC 2008

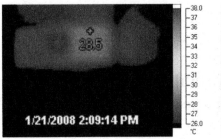 

Fig. 10.1. Thermal plot for N95 device using WLAN in offline mode $(28, 5C)$.

Fig. 10.2. Thermal plot for N95 device using WLAN IEEE 802.11b/g $(36, 3C)$.

reason for the delay behind a faster 3G version of their iPhone. As given in [2], Steve Jobs said:

The 3G chipsets that are available to semiconductors work reasonably well except for power. They are real power hogs. So as you know, the handset battery life used to be 5-6 hours for GSM, but when we got to 3G they got cut in half. Most 3G phones have battery lives of 2-3 hours.

As 4G mobile phones will be even more complex than 3G phones, the question arises how the battery should cope with this new challenge. The increase in battery consumption by mobile phones is the reason that public battery chargers at airport and hotels are getting more diffused. Those public chargers have several cable plugs for different mobile phones to charge the device, but the user needs to pay several dollars for the service – energy has a price!

Therefore, the main focus in this chapter is the energy saving potential for mobile devices exploiting heterogeneous wireless network combinations. Besides the cellular air interfaces, mobile devices are equipped with several local area network technologies such as Bluetooth or WLAN 802.11. The different heterogeneous wireless technologies lead to energy savings if they are used properly. By the example of cooperative wireless networks, energy saving potentials are derived for different heterogeneous wireless technology combination.

In contrast to cellular systems, where the mobile devices are only connected to the base station (see Fig. 10.3), in cooperative wireless networks, the mobile device, in addition to the cellular communication, will establish short range links to neighboring mobile devices within its proximity (see Fig. 10.4). In prior work [3, 5] it was shown that the newly formed cooperative cluster, also referred to as wireless grid, can offer each participating mobile device a better performance in terms of data rate, delay, robustness, security, and energy consumption in contrast to any stand alone device. The improved data rate, delay, and robustness come obviously by the accumulated cellular links with its inherent diversity. The decreased energy consumption was introduced in [4] for the first time and is extended in this book chapter for multiple combinations of heterogeneous wireless networks. In a nutshell, as long as the energy per bit ratio is better in the short range connection than the cellular link, the cooperation will improve the energy consumption. This is the case for all available

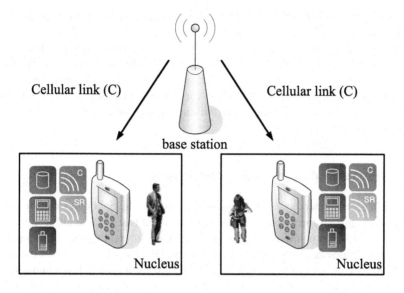

Fig. 10.3. State of the art architecture.

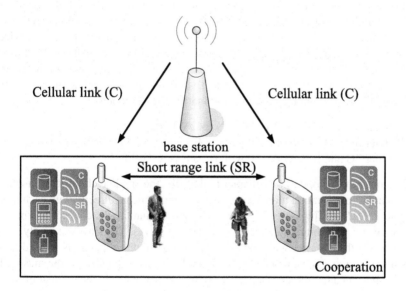

Fig. 10.4. Cooperative architecture.

and future wireless technologies as the path loss is much smaller in short range communication (around 10 m) than cellular communication (several 100 m).

In this chapter the energy consumption of all embedded wireless communication technologies that can be found on any commercially available smart phone is presented, namely cellular networks, IEEE802.11 and Bluetooth. Knowing about the energy consumption of the individual wireless technologies, adaptive schemes to switch among those technologies are presented using the example of cooperation among mobile devices.

10.2 General Notations and Assumptions

Three different architectures using cooperation among the mobile devices are given in Fig. 10.5. For all three architectures, it is assumed that the overall access point is *seeding* sub-streams into the mobile devices. Two different application scenarios are taken into consideration, namely *streaming* and *file download*. In case of streaming, the mobile devices of one given cooperation cluster listen on different multicast channels of the access point. A mobile device that is not cooperating with other mobile devices needs to receive all sub-streams by itself. Cooperating device will receive only a subset of those channels, in the extreme only one. An example of the streaming service could be a live stream of a sport event. In the case of file download, the access point is setting up several dedicated channels for each mobile device and therefore parallel download from the network is possible. The cooperating devices will receive the partial information over the cellular link and then combine those information over the short range link. An example could be a download of gaming maps or file download through BitTorrent. To build up the different architectures, the following three different technology combinations are assumed:

- Cellular 3G and IEEE802.11 WLAN (3G/WLAN),
- Cellular 3G and Bluetoothv2.0 without Broadcast (3G/BTwoBr), and
- Cellular 3G and Bluetoothv2.0 with Broadcast (3G/BTwBr).

It is out of the scope of this chapter to describe the single technologies in detail, but it is noted here that all mobile devices can communicate directly with each other using IEEE802.11. This is not the case for Bluetooth that has the concept of a master connected to a maximum number of seven active slaves. The slaves can only communicate via their master with each other and more data exchange is needed here. How much data is exchanged depends on the broadcast capability of the master. In case the master cannot broadcast packets to its slaves, obviously the number of transmission is larger than for the case the master is able to broadcast. Nevertheless, in both cases more packets are transmitted than in the 3G/WLAN case.

In Table 10.2 the mathematical notations used in this chapter are given. J is the number of mobile devices within a cooperative cluster served by one access point. As each mobile device is using two wireless technologies for cooperation, the overall energy consumption in this case, namely E_{Coop}, is composed of $E_{overlay}$ and $E_{short\ range}$. To answer the question whether cooperation is beneficial in terms of

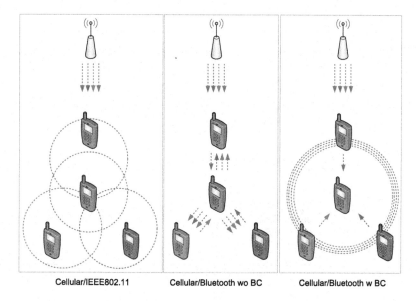

Cellular/IEEE802.11   Cellular/Bluetooth wo BC   Cellular/Bluetooth w BC

Fig. 10.5. Three different architectures for cooperative wireless networking.

energy consumption, E_{Coop} has to be compared with E_{noCoop}, which is the energy for a stand-alone mobile device with only one active wireless technology (only the cellular one and no short range). The energy saving S is given in (10.2).

$$E_{Coop} = E_{overlay} + E_{short\ range}. \tag{10.1}$$

$$S = 1 - \frac{E_{Coop}}{E_{noCoop}}. \tag{10.2}$$

In Fig. 10.6 the general notation for frames, slots, and mini-slots is given. A frame is the time period the centralized access point is conveying J slots. One slot out of J can be used to convey one of the multicast channels introduced beforehand. A frame has the nominal length of 1 and as a result of that, a slot has the length $1/J$. The transmission on the short range can be realized at a rate Z times larger than on the cellular link. If a mobile device is *forwarding* one slot, a time period of $1/JZ$ is needed on the short range link and referred to as mini-slot. Forwarding all mini slots will take $1/Z$.

As can be seen from Fig. 10.6, the energy consumption E_{Coop} is composed of transmitting (tx), receiving (rx), and idle (i) phases.

$$E_{Coop} = \underbrace{E_{c,rx} + E_{c,i}}_{overlay\ contribution} + \underbrace{E_{sr,tx} + E_{sr,rx} + E_{sr,i}}_{short\ range\ contribution}. \tag{10.3}$$

The individual energy values depend on the power level of that phase and the related time the mobile device is in that phase. The power values and the value of Z

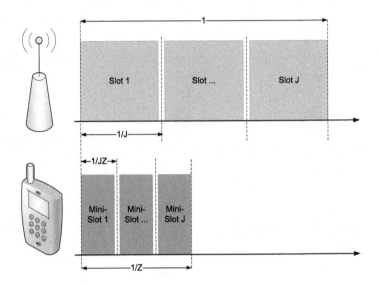

Fig. 10.6. General notation for frames, slots, and mini-slots.

are given by the technology used, while J depends on the scenario. The power levels can be measured and are presented in Section 10.3. The time values $t_{c,rx}$ and others are calculated in later sections.

$$E_{Coop} = \underbrace{t_{c,rx} \cdot P_{c,rx} + t_{c,i} \cdot P_{c,i}}_{cellular\ contribution} + \underbrace{t_{sr,tx} \cdot P_{sr,tx} + t_{sr,rx} \cdot P_{sr,rx} + t_{sr,i} \cdot P_{sr,i}}_{short\ range\ contribution}.$$

(10.4)

10.3 Energy Measurements

As described in the last section a detailed investigation on the energy consumption for the individual actions and achievable data rates is needed. Therefore measurement results of the data rates and energy consumption for the cellular network and the short range network on the mobile device are presented shortly. A more detailed description of the measurement setup and results can be found in [6].

For the WLAN IEEE802.11 measurements, one N95 is used for sending broadcast packets of 1000 byte length towards ten other N95s distributed equidistantly between 3 m and 30 m. Several tests were conducted, where one test was sending five times 5.000 packets in a row as fast as possible with 10 sec pauses in between. In the active phase the sending and receiving power where measured and in the pause intervals the idle power was measured. For the energy measurements the energy profiler of Nokia were used [1]. While doing the WLAN measurements, the cellular connection was offline. The results of this measurement campaign is given in Table 10.2.

Table 10.1. Notations.

J	number of cooperative mobile devices
Z	ratio of short range data rate and cellular data rate
S	energy savings for cooperation compared with no cooperation
D	delay to provide a given service
α	auxiliary variable
$P_{c,rx}$	power level for receiving state using cellular
$P_{c,i}$	power level for idle state using cellular
$P_{sr,tx}$	power level for sending state using short range
$P_{sr,rx}$	power level for receiving state using short range
$P_{sr,i}$	power level for idle state using short range
$t_{c,rx}$	receiving time for the cellular link
$t_{c,i}$	idle time for the cellular link
$t_{sr,tx}$	sending time for the short range link
$t_{sr,rx}$	receiving time for the short range link
$t_{sr,i}$	idle time for the short range link
$E_{c,rx}$	energy level for receiving state using cellular
$E_{c,i}$	energy level for idle state using cellular
$E_{sr,tx}$	energy level for sending state using short range
$E_{sr,rx}$	energy level for receiving state using short range
$E_{sr,i}$	energy level for idle state using short range
E_{Coop}	overall energy level for cooperation
E_{noCoop}	overall energy level for no cooperation

The power levels for sending is larger than those of the receiving state. The receiving power levels for different distances between sender and receiver differ, which may be a reason that the mobile device with more distance to the sender is not receiving all packets and less energy is used in the signal processing part. The data rate is high with over 5 Mbps.

The Bluetooth measurements were similar to those of the WLAN, but only between two N95 devices. The energy levels will not differ from unicast to broadcast transmission as only the recepient address will change. While doing the Bluetooth measurements, the cellular connection was offline. Table 10.3 gives the results for the Bluetooth measurements. The power levels are now quite low compared to WLAN,

Table 10.2. Power levels and data rate for WLAN broadcast - 1000 byte.

state	power value [W]	data rate [Mbps]
sending	1.629	5.623
receiving @ 3 m	1.375	5.379
receiving @ 30 m	1.213	5.115
idle @ 3m	0.979	−
idle @ 30m	0.952	−

but the data rate is also smaller. Beneficial for the cooperation, as will be explained later on, is the low power level in idle mode.

Table 10.3. Power levels and data rate for Bluetooth - 224 bytes/DM5.

state	power value [W]	data rate [Mbps]
sending	0.422	0.631
receiving	0.367	0.631
idle	0.161	−

The measurement setup for the cellular link was composed of an Internet server transmitting UDP packets towards a Nokia N95 device over a 3G operator in Aalborg, Denmark. The results for the energy consumption for the receiving and the idle state are given in Table 10.4. As expected the mobile device will consume more energy in the receiving mode than in the idle mode. The receiving power level is even lower than the one used in WLAN. But as the data rate for the cellular link with 0.193 Mbps is significantly lower than that of WLAN, the energy per bit ratio is worse for the cellular link.

Table 10.4. Power levels and data rate for cellular - 100 byte.

state	power value [W]	data rate [Mbps]
receiving	1.314	0.193
idle	0.661	−

10.4 Cellular and IEEE802.11

In this section the energy consumption for cellular technology such as 3G for the overlay network and 802.11 technology for the short range exchange is calculated.

10.4.1 Streaming

For the streaming scenario it is assumed that the sub streams of the overlay network are conveyed in multicast channels. Those channels are sent out sequentially within one frame. A stand alone device would need to receive all substreams over the cellular link, while cooperative devices will share this task by receiving a subset of multicast channels and exchange the missing pieces over the short range link. As given in Fig. 10.7 one cooperating mobile device is receiving one slot on the cellular link and will stay the rest of the frame idle. The received slot will be offered over the short range link in $1/(JZ)$ time. The $J - 1$ missing slots will be received over the short range link as well in $(J - 1)/(JZ)$ time. When the local exchange is over also the short range link is idle until the next frame starts. The cooperative energy consumption E_{Coop}^{stream} is given in (10.5).

$$E_{Coop}^{stream} = \underbrace{\frac{1}{J}\ P_{c,rx}}_{t_{c,rx}} + \underbrace{(1 - \frac{1}{J})\ P_{c,i}}_{t_{c,i}} + \underbrace{\frac{1}{J \cdot Z}\ P_{sr,tx}}_{t_{sr,tx}} + \underbrace{\frac{J - 1}{J \cdot Z}\ P_{sr,rx}}_{t_{sr,rx}} + \underbrace{(1 - \frac{1}{Z})\ P_{sr,i}}_{t_{sr,i}}.$$

$$(10.5)$$

10.4.2 File Download

For the file download scenario it is assumed that all cooperating mobile devices have a dedicated cellular link receiving partial information that will be combined over the short range link. If $1/J \geq 1/Z$, as given in Fig. 10.8, there will be no idle time at all on the cellular link but some idle time on the short range link. Fig. 10.8 shows the case when the number of cooperating entities is small and the ratio Z is high. Equation (10.6) gives the energy consumption in the cooperative case for $1/J \geq 1/Z$.

$$E_{Coop}^{down} = \underbrace{\frac{1}{J}\ P_{c,rx}}_{t_{c,rx}} + \underbrace{0\ P_{c,i}}_{t_{c,i}} + \underbrace{\frac{1}{J \cdot Z}\ P_{sr,tx}}_{t_{sr,tx}} + \underbrace{\frac{J - 1}{J \cdot Z}\ P_{sr,rx}}_{t_{sr,rx}} + \underbrace{(\frac{1}{J} - \frac{1}{Z})\ P_{sr,i}}_{t_{sr,i}}.$$

$$(10.6)$$

In Fig. 10.9 the case for $1/J < 1/Z$ is shown. Now there is no idle time on the short range link, but some remaining idle time on the cellular link. The energy consumption E_{Coop}^{down} for the cooperative case is given by (10.7) if $1/J < 1/Z$.

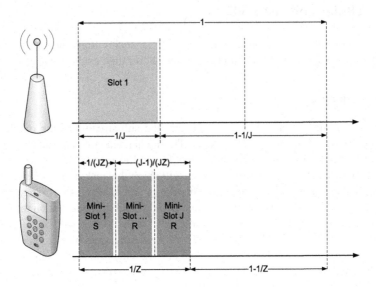

Fig. 10.7. Streaming scenario for cooperative wireless network using cellular and IEEE802.11 technology.

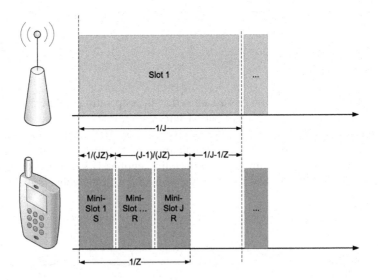

Fig. 10.8. File download scenario for cooperative wireless network using cellular and IEEE802.11 technology for $1/J \geq 1/Z$.

$$E_{Coop}^{down} = \underbrace{\frac{1}{J}}_{t_{c,rx}} P_{c,rx} + \underbrace{(\frac{1}{Z} - \frac{1}{J})}_{t_{c,i}} P_{c,i} + \underbrace{\frac{1}{J \cdot Z}}_{t_{sr,tx}} P_{sr,tx} + \underbrace{\frac{J-1}{J \cdot Z}}_{t_{sr,rx}} P_{sr,rx} + \underbrace{0}_{t_{sr,i}} P_{sr,i}.$$

$$(10.7)$$

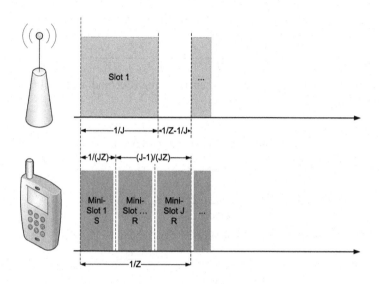

Fig. 10.9. File download scenario for cooperative wireless network using cellular and IEEE802.11 technology for $1/J < 1/Z$.

10.5 Cellular and Bluetooth without Broadcast

As has been mentioned before, the energy consumption in Bluetooth for the master and the slave differs. Therefore, a separate investigation on the master and the slaves is presented before deriving the mean value of it.

10.5.1 Streaming

As for the WLAN scenario, all mobile devices are receiving one slot over the cellular network, which is used for exchange over the short range technology. The master will then receive $J - 1$ packets from its $J - 1$ slaves as given in Fig. 10.10 using the example of three cooperating mobile devices. The master will send its own packet to $J - 1$ slaves, while it will relay $J - 2$ packets from other slaves to each individual slave. This will end up in $(J - 1)^2$ transmissions for the master. A slave will send only one packet and receive $J - 1$ packets from its master. The rest of the time the slave is idle. By subtracting the time needed to transmit and to receive from one frame time, the idle time is received. To calculate the t_{sr} values, the values of the master and slaves are weighted to build the mean value.

$$t_{sr,tx} = \frac{\frac{(J-1)^2}{JZ} + (J-1) \cdot \frac{1}{JZ}}{J} = \frac{J-1}{JZ}. \tag{10.8}$$

The same is done for the idle time for the short range communication.

$$t_{sr,i} = \frac{0 + (J-1) \cdot \frac{JZ-1}{Z}}{J} = \frac{JZ - 2J + 2}{JZ}. \tag{10.9}$$

All time values for the streaming scenario using Bluetooth without broadcast are given in Table 10.5.

Table 10.5. Streaming 3G/BTwoBr.

Cellular receive	$t_{c,rx}$		$1/J$
Cellular idle	$t_{c,i}$		$1 - 1/J$
Master	send		$\frac{(J-1)^2}{JZ}$
Master	receive		$\frac{J-1}{JZ}$
Master	idle		$\frac{J(Z-J+1)}{JZ}$
Slave	send		$\frac{1}{JZ}$
Slave	receive		$\frac{J-1}{JZ}$
Slave	idle		$\frac{JZ-J}{JZ}$
Mean	send	$t_{sr,tx}$	$\frac{J-1}{JZ}$
Mean	receive	$t_{sr,rx}$	$\frac{J-1}{JZ}$
Mean	idle	$t_{sr,i}$	$\frac{JZ-2J+2}{JZ}$

As long as $(J-1)/Z$ is smaller than one or in other words, as along as the short range exchange fits into one frame, the energy consumption can be calculated by

$$E_{Coop}^{stream} = \underbrace{\frac{1}{J}}_{t_{c,rx}} P_{c,rx} + \underbrace{(1 - \frac{1}{J})}_{t_{c,i}} P_{c,i} +$$
$$\underbrace{\frac{J-1}{J \cdot Z}}_{t_{sr,tx}} P_{sr,tx} + \underbrace{\frac{J-1}{J \cdot Z}}_{t_{sr,rx}} P_{sr,rx} + \underbrace{(\frac{JZ - 2J + 2}{J \cdot Z})}_{t_{sr,i}} P_{sr,i}. \tag{10.10}$$

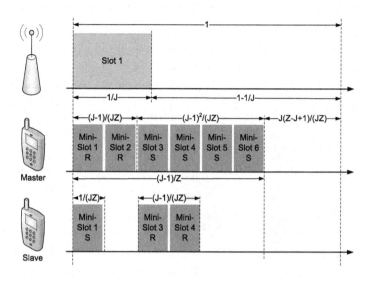

Fig. 10.10. Streaming scenario for cooperative wireless network using cellular and Bluetooth technology without broadcast functionality.

10.5.2 File Download

In case $1/J \geq (J - 1)/Z$, there is no idle time on the cellular communication link ($t_{c,i} = 0$). The master and slave will act in the same way as in the streaming scenario, but the idle time are shorter now. The individual idle time can be calculated by subtracting the time for sending and receiving from $1/J$. To calculate $t_{sr,i}$, the mean of the values of the master and slaves is taken (see (10.11)). For $1/J \geq (J - 1)/Z$ cooperative energy consumption E_{Coop}^{down} is given in (10.12).

$$t_{sr,i} = \frac{\frac{Z-J^2+J}{JZ} + (J - 1) \cdot \frac{Z-J}{JZ}}{J} = \frac{Z - 2J + 2}{JZ}. \qquad (10.11)$$

$$E_{Coop}^{down} = \underbrace{\frac{1}{J}}_{t_{c,rx}} P_{c,rx} + \underbrace{(0)}_{t_{c,i}} P_{c,i} +$$
$$\underbrace{\frac{J - 1}{J \cdot Z}}_{t_{sr,tx}} P_{sr,tx} + \underbrace{\frac{J - 1}{J \cdot Z}}_{t_{sr,rx}} P_{sr,rx} + \underbrace{(\frac{Z - 2J + 2}{J \cdot Z})}_{t_{sr,i}} P_{sr,i}. \qquad (10.12)$$

In case the short range communication is not fast enough to exchange the data within the cooperative cluster ($1/J < (J - 1)/Z$), the Bluetooth master is creating idle time on the cellular technology and extending those of the slaves. The idle time of the cellular link equals $1/J$ subtracted from the time the Bluetooth master

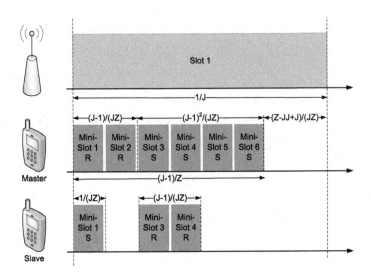

Fig. 10.11. File download scenario for cooperative wireless network using cellular and Bluetooth technology without broadcast functionality for $1/J \geq (J-1)/Z$.

needs to send and receive packets. All time values for the different cases are given in Table 10.6.

$$t_{sr,i} = \frac{\frac{Z-J^2+J}{JZ} + (J-1) \cdot \frac{J-2}{JZ}}{J} = \frac{(J-1)(J-2)}{JZ}. \qquad (10.13)$$

$$E_{Down}^{down} = \underbrace{\frac{1}{J}}_{t_{c,rx}} P_{c,rx} + \underbrace{(\frac{J-1}{Z} - 1/J)}_{t_{c,i}} P_{c,i} + $$
$$\underbrace{\frac{J-1}{J \cdot Z}}_{t_{sr,tx}} P_{sr,tx} + \underbrace{\frac{J-1}{J \cdot Z}}_{t_{sr,rx}} P_{sr,rx} + \underbrace{(\frac{(J-1)(J-2)}{J \cdot Z})}_{t_{sr,i}} P_{sr,i}. \qquad (10.14)$$

10.6 Cellular and Bluetooth with Broadcast

In contrast to the previous section, in this section it is assumed that the master in the Bluetooth piconet is able to broadcast packets to the slaves which is part of the Bluetooth standard but not implemented or accessible on all Bluetooth chip sets.

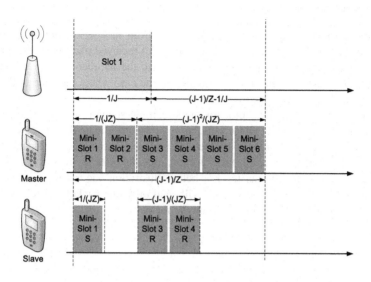

Fig. 10.12. File download scenario for cooperative wireless network using Cellular and Bluetooth technology without broadcast functionality for $1/J < (J-1)/Z$.

Table 10.6. File download 3G/BTwoBr.

			$1/J \geq (J-1)/Z$	$1/J < (J-1)/Z$
Cellular	receive	$t_{c,rx}$		$1/J$
	idle	$t_{c,i}$	0	$\frac{J-1}{Z} - 1/J$
Master	send		$\frac{(J-1)^2}{JZ}$	
	receive		$\frac{J-1}{JZ}$	
	idle		$\frac{Z-J^2+J}{JZ}$	0
Slave	send		$\frac{1}{JZ}$	
	receive		$\frac{J-1}{JZ}$	
	idle		$\frac{Z-J}{JZ}$	$\frac{J-2}{Z}$
Mean	send	$t_{sr,tx}$	$\frac{J-1}{JZ}$	
	receive	$t_{sr,rx}$	$\frac{J-1}{JZ}$	
	idle	$t_{sr,i}$	$\frac{Z-2J+2}{JZ}$	$\frac{(J-1)(J-2)}{JZ}$

10.6.1 Streaming

Enabling the master to broadcast packets will result in a significant lower amount of packets that will be transmitted. The number of received packets by the master is not changed compared to the previous case. Also the number of received and sent packets is the same for the slaves compared to the case where the Bluetooth master was not able to broadcast the messages.

In Fig. 10.13 the activity plot for the streaming scenario is given. One mobile device is receiving one slot on the cellular and remains idle on this specific air interface for the rest of the frame. The Bluetooth master will receive $J - 1$ packets from its $J - 1$ slaves and broadcast those packets together with its own packet resulting in J transmissions for the master. For better illustration one master and two slaves are assumed in Fig. 10.13. Any slave will send only one packet and receive $J - 1$ packets over the broadcast channel from the master.

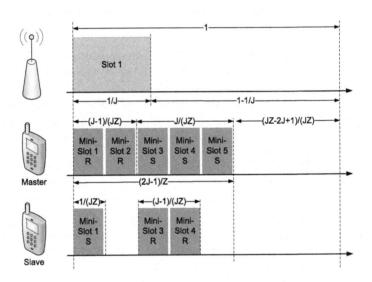

Fig. 10.13. Streaming scenario for cooperative wireless network using cellular and Bluetooth technology with broadcast functionality.

To calculate the idle time of the master, the time the master is receiving and sending is subtracted from the frame time.

$$t_{sr,i,master} = 1 - \frac{J}{JZ} - \frac{J-1}{JZ} = \frac{JZ - 2J + 1}{JZ}. \qquad (10.15)$$

But for $J = 2$, the idle time for the master needs to be rewritten as given in (10.16).

$$t_{sr,i,master,J=2} = 1 - \frac{1}{JZ} - \frac{J-1}{JZ} = \frac{J(Z-1)}{JZ}. \qquad (10.16)$$

In Table 10.7 a general description of the idle time is given by using variable α. For $J = 2$, α equals $1/J$ and 1 otherwise. This variable is used throughout this chapter. The same adjustment needs to be done for the mean sending time $t_{sr,tx}$. A slave will send only one packet and receive $J - 1$ packets from its master. The rest of the time the slave is idle, such that $t_{sr,i,slave}$ equals

$$t_{sr,i,slave} = 1 - \frac{1}{JZ} - \frac{J-1}{JZ} = \frac{J(Z-1)}{JZ}. \qquad (10.17)$$

Using $t_{sr,i,master}$ and $t_{sr,i,slave}$, the mean value $t_{sr,i}$ can be calculated. In Table 10.7 the values for the individual timing are given.

Table 10.7. Streaming 3G/BTwBr.

Cellular receive $t_{c,rx}$		$1/J$
Cellular idle $\quad t_{c,i}$		$1 - 1/J$
Master	send	$\frac{\alpha J}{JZ}$
Master	receive	$\frac{J-1}{JZ}$
Master	idle	$\frac{JZ-J-\alpha J+1}{JZ}$
Slave	send	$\frac{1}{JZ}$
Slave	receive	$\frac{J-1}{JZ}$
Slave	idle	$\frac{J(Z-1)}{JZ}$
Mean	send $\quad t_{sr,tx}$	$\frac{J+\alpha J-1}{J^2 Z}$
Mean	receive $t_{sr,rx}$	$\frac{J-1}{JZ}$
Mean	idle $\quad t_{sr,i}$	$\frac{J^2(Z-1)-\alpha J+1}{J^2 Z}$

The energy spent for the *streaming* scenario is given in (10.18).

$$E_{Coop}^{stream} = \underbrace{\frac{1}{J}}_{t_{c,rx}} P_{c,rx} + \underbrace{\left(1 - \frac{1}{J}\right)}_{t_{c,i}} P_{c,i} +$$

$$\underbrace{\frac{J+\alpha J-1}{J^2 \cdot Z}}_{t_{sr,tx}} P_{sr,tx} + \underbrace{\frac{J-1}{J \cdot Z}}_{t_{sr,rx}} P_{sr,rx} + \underbrace{\left(\frac{J^2(Z-1)-\alpha J+1}{J^2 \cdot Z}\right)}_{t_{sr,i}} P_{sr,i}.$$

$$(10.18)$$

10.6.2 File Download

The *file download* scenario is similar to the previous cases, except that only that the master is sending less packets than before.

In Fig. 10.14 the case that the entire short range exchange can be done within one slot is depicted. For illustration purpose, three cooperating entities are assumed without loosing generality. The master will receive $J - 1$ packets from the slaves and send J packet by himself using the broadcast capability. A slave on the other side receives $J - 1$ packets and sends one packet towards the master. As the short range communication is fast enough, the master has still some idle time (given in Table 10.8). The idle time of the slaves is even larger, resulting in the mean idle time for all mobile devices given by

$$t_{sr,i} = \frac{\frac{Z-J-\alpha J+1}{JZ} + (J-1) \cdot \frac{Z-J}{JZ}}{J} = \frac{(JZ + 1 - \alpha J - J^2)}{J^2 \cdot Z}. \tag{10.19}$$

The energy spent in this case is

$$E_{Coop}^{down} = \underbrace{\frac{1}{J}}_{t_{c,rx}} P_{c,rx} + \underbrace{(0)}_{t_{c,i}} P_{c,i} +$$
$$\underbrace{\frac{J + \alpha J - 1}{J^2 \cdot Z}}_{t_{sr,tx}} P_{sr,tx} + \underbrace{\frac{J - 1}{J \cdot Z}}_{t_{sr,rx}} P_{sr,rx} + \underbrace{(\frac{JZ + 1 - \alpha J - J^2}{J^2 \cdot Z}}_{t_{sr,i}}) P_{sr,i}.$$
$$\tag{10.20}$$

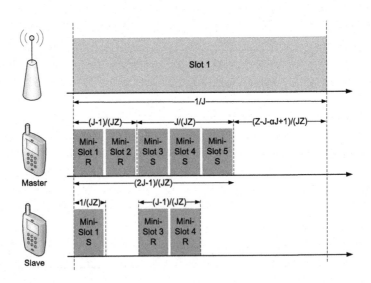

Fig. 10.14. File download scenario for cooperative wireless network using cellular and Bluetooth technology with broadcast functionality (for $1/J \geq (2J - 1)/(JZ)$).

If the short range exchange needs more time than the reception of one slot over the cellular link, the master has no idle time anymore, but is sending and receiving mini slots as given in Fig. 10.15. On the other side the cellular link has now idle time that needs to be taken under consideration such that the overall energy consumption is as follows:

$$
E_{Down}^{down} = \underbrace{\frac{1}{J}\ P_{c,rx}}_{t_{c,rx}} + \underbrace{(\frac{2J-1}{JZ} - 1/J)\ P_{c,i}}_{t_{c,i}} +
$$

$$
\underbrace{\frac{J+\alpha J-1}{J^2 \cdot Z}\ P_{sr,tx}}_{t_{sr,tx}} + \underbrace{\frac{J-1}{J\cdot Z}\ P_{sr,rx}}_{t_{sr,rx}} + \underbrace{(\frac{(J-1)^2}{J^2 \cdot Z})\ P_{sr,i}}_{t_{sr,i}}. \qquad (10.21)
$$

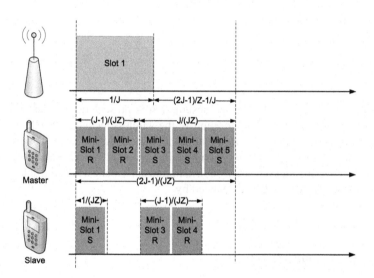

Fig. 10.15. File download scenario for cooperative wireless network using cellular and Bluetooth technology with broadcast functionality (for $1/J < (2J-1)/(JZ)$).

In Table 10.8 the individual time values for the two cases $1/J \geq (2J-1)/(JZ)$ and $1/J < (2J-1)/(JZ)$ are given. The related energy calculation is given in (10.20) and (10.21), respectively.

10.7 Results

Now the energy measurements of Section 10.3 and the analytical derivations of the energy consumption are combined to show potential benefits of cooperative wireless networking in terms of energy saving and delay (latter one only for the file download

Table 10.8. File download 3G/BTwBr.

		$1/J > (2J-1)/(JZ)$	$1/J < (2J-1)/(JZ)$
Cellular	receive $t_{c,rx}$		$1/J$
	idle $t_{c,i}$	0	$\frac{2J-1}{Z} - 1/J$
Master	send		$\frac{\alpha J}{JZ}$
	receive		$\frac{J-1}{JZ}$
	idle	$\frac{(Z-J-\alpha\cdot J+1)}{JZ}$	0
Slave	send		$\frac{1}{JZ}$
	receive		$\frac{J-1}{JZ}$
	idle	$\frac{Z-J}{JZ}$	$\frac{J-1}{JZ}$
Mean	send $t_{sr,tx}$		$\frac{J+\alpha J-1}{J^2 Z}$
	receive $t_{sr,rx}$		$\frac{J-1}{JZ}$
	idle $t_{sr,i}$	$\frac{JZ+1-\alpha J-J^2}{J^2 Z}$	$\frac{(J-1)^2}{J^2 Z}$

scenario). Also, the overall energy consumption is presented to understand where most of the energy is consumed to improve the wireless technology. The result section is splitted up for the *streaming* and the *file download* scenario. The results are given for three different technology combinations, namely 3G/WLAN, 3G/BTwoBR, and 3G/BTwBR. Besides the energy consumption and savings, also the delay is given in this section. The delay D is calculated by

$$D = max(t_{sr,tx} + t_{sr,rx} + t_{sr,i}, t_{c,rx} + t_{c,i}). \tag{10.22}$$

10.7.1 File Download Results

First the file download results are presented. In Fig. 10.16 the overall energy consumption is given for the three different cases. In Fig. 10.17 the detailed energy consumption for the three different technology combinations are given. For each value of J, six columns are given. The first two are for the 3G/BTwBr case, while the left one is for the cellular energy consumption and the right one for the short range. The left column has always two parts, namely the receiving and the idle part. The right column has three parts with receiving, idle, and transmitting (from the bottom to the top). Column three and four are representing the 3G/BTwoBr case and the last columns are for the 3G/WLAN case. For a given case, the overall energy consumption is the sum of both columns.

The 3G/WLAN case is using always less energy with each additional cooperating mobile device. For $J = 2$ the energy consumption is only a little bit smaller than the non-cooperative case. As the non-cooperative case has only one air interface powered up, in case of any cooperation the second air interface needs to be switched on. On top of that, for $J = 2$, each received packet has to be responded with a transmission of a packet on a short range link. This ratio becomes better for larger number of J.

In case Bluetooth is used instead of WLAN, the energy consumption for $J = 2$ becomes significantly better. For larger number of J, the Bluetooth with broadcast capabilities is using only a little bit less with each cooperating device. Bluetooth without broadcast capabilities is using more energy for a larger number of J. For the settings presented in the chapter, the energy consumption becomes even larger than the non cooperative case for $J > 5$. Now, the main energy consumption is taking place in the idle mode of the cellular air interface.

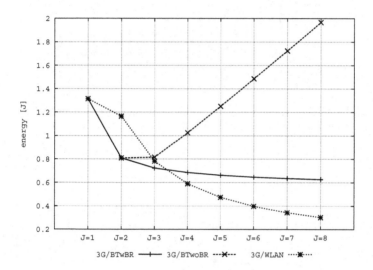

Fig. 10.16. Overall energy consumption of three different cooperative technology combinations versus number of cooperating mobile devices for the *file download* scenario.

In Fig. 10.18 the energy saving potential versus number of cooperating mobile devices for the *file download* scenario is given. As a reference, the *ideal* energy saving curve is used, where it is assumed that no energy at all is consumed for the short range communication. The ideal energy savings can be calculated by $1 - 1/J$. The energy saving plot is a direct result of the energy plot. It can be seen that for $J \geq 4$, WLAN is giving the best energy saving results. Bluetooth with broadcast capabilities is giving good results for smaller number of J, but is saturating very quickly. The situation becomes worse for Bluetooth without broadcast capabilities. An energy saving can only be reported for $J < 6$. But already for $J \geq 3$ the gain is becoming smaller than for $J = 2$. Note, due to the larger coverage in WLAN, a larger number of J can be expected for WLAN than for Bluetooth.

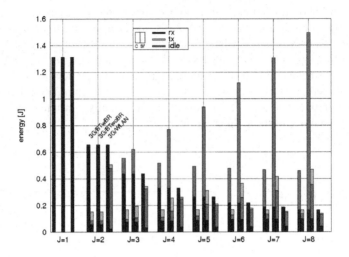

Fig. 10.17. Detailed energy consumption of three different cooperative technology combinations versus number of cooperating mobile devices for the *file download* scenario.

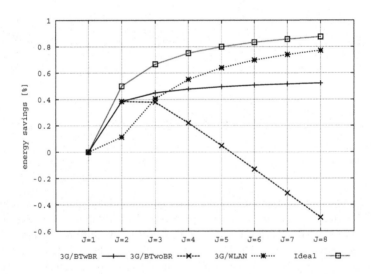

Fig. 10.18. Energy saving potential versus number of cooperating mobile devices for the *file download* scenario.

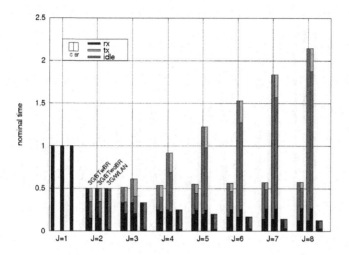

Fig. 10.19. Delay of three different cooperative technology combinations versus number of cooperating mobile devices for the *file download* scenario.

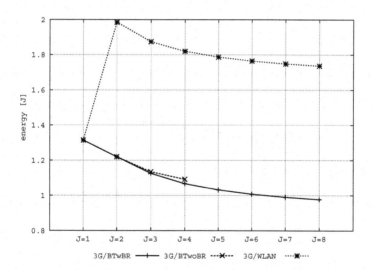

Fig. 10.20. Overall energy consumption of three different cooperative technology combinations versus number of cooperating mobile devices for the *streaming* scenario.

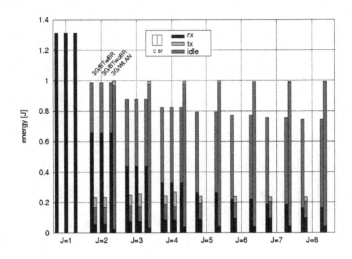

Fig. 10.21. Detailed energy consumption of three different cooperative technology combinations versus number of cooperating mobile devices for the *streaming* scenario.

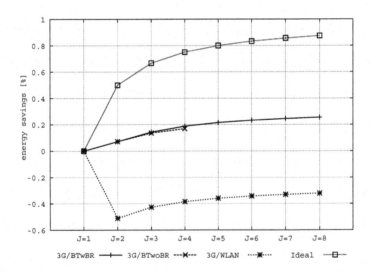

Fig. 10.22. Energy saving potential versus number of cooperating mobile devices for the *streaming* scenario.

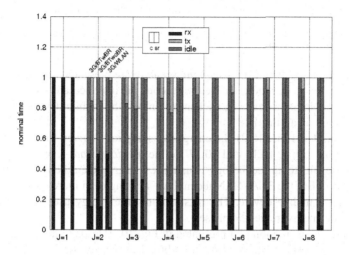

Fig. 10.23. Delay of three different cooperative technology combinations versus number of cooperating mobile devices for the *streaming* scenario.

Similar to the energy plot, the delay is presented in Fig. 10.19. Here the each of the columns is representing the delay (they do not have to be added as in the energy case). Both columns are given to understand how long a certain air interface is staying in which mode.

The 3G/WLAN scenario is reducing the delay with each cooperating device. The delay in this case is $1/J$ as long as the data rate of the WLAN technology is significantly higher than the cellular network. For Bluetooth the data rate are smaller and due to their communication architecture the traffic they have to support is much larger. For Bluetooth with broadcast capabilities, the delay is more or less stable over J resulting in a delay reduction to 50% for the specific values used in this example. The Bluetooth without broadcast capabilities is suffering by the large amount of data that has to be sent by the master. The master needs so much time that the cellular network interface is going into idle mode waiting for the short range to have finished the exchange. Also the Bluetooth slaves are most of the time in idle mode, such that most of the devices are spending their time, and as a result of that also their energy, in the idle mode. The only exception is the Bluetooth master, that is always receiving or transmitting on the short range without being idle on this particular air interface.

To sum up, for the *download file* scenario benefits for the energy consumption and the download delay can be expected for the 3G/WLAN and 3G/BTwBr. For 3G/BTwoBr a gain can only be achieved for a smaller number of cooperating mobile devices. Due to the large coverage of WLAN over Bluetooth, a larger number of cooperating devices can be assumed for that specific wireless technology. Note that, the performance of the cooperative wireless networking can be improved significantly by lower power consumption in the idle mode. Other parameters such as data rate and power levels for sending and receiving have less impact.

10.7.2 Streaming Results

In this section the results for the *streaming* scenario are presented. In contrast to the *file download* scenario, where the heterogeneous air interface are switched off as soon as the disjoint data has been exchanged among the mobile devices, here both air interfaces are on all the time.

In Fig. 10.20 and Fig. 10.21 the energy consumption is given for the *streaming* scenario similar to Fig. 10.16 and 10.17 for the *file download*, respectively. Obviously only the Bluetooth technology is using less energy than the non-cooperative mobile devices. But the 3G/WLAN technology is using even more energy than the non-cooperative mobile device. The reason is not the energy used in the activity phases, but the energy in the idle phases.

The energy consumption plot presented in Fig. 10.20 leads to the energy saving plot in Fig. 10.22. Here, Bluetooth without broadcast capabilities can only support four mobile devices as the data rate is not large enough to exchange the information over the short range fast enough.

In case the idle power of the WLAN technology could be reduced to zero, the 3G/WLAN combination would lead to an energy saving of 40% for $J = 8$. If further the idle time of the cellular would be zero (as for DVB-H), the energy saving would increase to 83%, which is only 4% below the *ideal case*.

The delay plot in Fig. 10.23 is giving the time values for the individual phases. The overall delay as such is obviously not improving as for the *file download*.

Conclusion

In this chapter a new way to combine heterogeneous wireless technologies, referred to as cooperative wireless networking, has been advocated. The main focus here has been on the energy consumption and delay behavior for such an architecture. By means of measurements on commercially available mobile phones, the energy values and the data rates for different heterogeneous plots have presented. A large part of the chapter has dealt with the analytical derivation of the energy consumption of a cooperative wireless network compared to non-cooperative state of the art ones. Potential gains for cooperative wireless networking have been presented by using the measurement results.

For two different scenarios it has been shown that cooperative wireless networking with existing heterogeneous wireless technologies, such as 3G, IEEE 802.11, and Bluetooth, can offer large energy savings and reduced delay for the *file download* scenario. For the second scenario, the *streaming* scenario, the gains have been observed to be small in case of Bluetooth or even non-existent in case of WLAN. The reason lies in the scenario as such and the large power consumption in the idle phase in the cellular and the short range communication technology. As cellular systems already offer power consumption values near to zero for DVB-H, solutions have to be found for the short range technology which simply uses too much energy in the idle mode.

Acknowledgment

We would like to thank Nokia for providing technical support as well as mobile phones to carry out the measurement campaign. Special thanks to Mika Kuulusa, Gerard Bosch, Harri Pennanen, Nina Tammelin, and Per Moeller from Nokia. This work was partially financed by the X3MP project granted by Danish Ministry of Science, Technology and Innovation.

References

1. Nokia - energy profiler.
   http://www.forum.nokia.com/main/resources/development_process/power_man
2. Steve job - press release for the iphone.
   `http://www.macnn.com/articles/07/09/18/jobs.uk.cell.`
   `carrier.qa/`.
3. F. H. P. Fitzek and M. Katz, editors. *Cognitive Wireless Networks: Concepts, Methodologies and Visions Inspiring the Age of Enlightenment of Wireless Communications*. ISBN 978-1-4020-5978-0. Springer, July 2007.
4. F. H. P. Fitzek, P. Kyritsi, and M. Katz, "Cooperation in Wireless Networks – Power Consumption and Spectrum Usage Paradigms in Cooperative Wireless Networks," chapter 11, pp. 365-386. Springer, 2006.
5. F. H. P. Fitzek and F. Reichert, editors. Mobile Phone Programming and its Application to Wireless Networking. Number 10.1007/978-1-4020-5969-8 in ISBN 978-1-4020-5968-1. Springer, June 2007.
6. M. V. Petersen, G. P. Perrucci, and F. H. P. Fitzek, "Energy and link measurements for mobile phones using ieee802.11b/g," in *Proc. of 4th International Workshop on Wireless Network Measurements (WiNMEE 2008)* - in conjunction with WiOpt 2008, Berlin, Germany, March 2008.

11

Qos-Aware Routing in Heterogeneous Wireless Access Networks

Yumin Wu[1], Kun Yang[1], and Hsiao-Hwa Chen[2]

[1] Department of Computing and Electronic Systems, University of Essex, U.K.
{ywud,kunyang}@essex.ac.uk
[2] Department of Engineering Science, National Cheng Kun University, Taiwan
hshwchen@ieee.org

11.1 Introduction

Current wireless access networks are designed to work independently, without cooperating with each other. The upcoming 4G network which is normally referred to as HWAN (Heterogeneous Wireless Access Network) brings us some new challenges of utilizing various wireless technologies and making them work together. These heterogeneous network structures can be classified as "integrated networks" or "inter-working networks" in the literature according to their own network functions. In integrated networks, such as iCAR [1] (Integrated Cellular and Ad Hoc Overlaying Systems), interfaces from different networks are tightly coupled at the Radio Access Network (RAN) or Core Network (CN) level [2]. Inter-working networks, such as the EuQoS (End-to-end Quality of Service Support over Heterogeneous Networks) system [3] and the structures mentioned in [4], are normally constructed by adding gateways between different networks, and the exchange of information between different networks is through these gateways. The various wireless access technologies and the various cooperating methods in heterogeneous wireless networks lead to research issues which are waiting for the solutions. Routing is probably one of the most important components of HWANs to make the proposed HWAN systems work. Quality of service (QoS) as a part of routing in heterogeneous wireless networks plays a critical role on realizing the design purposes of proposed heterogeneous system, such as balancing traffic and guaranteeing high-data-rate (HDR) service.

In conventional wireless networks such as an 802.11-based ad hoc network, routing protocols are generally divided into two groups, proactive routing protocols and reactive routing protocols [5]. In proactive routing protocols, every node gathers the network topology information by periodically sending routing packets, and builds up or updates its routing table according to the gathered information, such as DSDV [6]. In reactive routing protocols, route discovery process is only launched as a node cannot find a route to its proposed destination from its routing table, such as DSR [7]. QoS routing, which is also called as constraint-based routing, is designed to determine paths that satisfy QoS constraints [8].

E. Hossain (ed.), *Heterogeneous Wireless Access Networks*,
DOI: 10.1007/978-0-387-09777-0_11, © Springer Science+Business Media, LLC 2008

In conventional routing process, some general QoS effects such as delay and bandwidth are considered to fulfill users' application requirements [9]. Multiple QoS constraints involved in a routing process will lead to a NP (Non-deterministic Polynomial time) -complete problem. Therefore, a routing process with multiple QoS constraints, which can give an accurate route decision, may not exist in practical environment. As a result, most QoS-aware routing protocols are designed specially focusing on one QoS constraint.

Most of the heterogeneous network architectures are deployed to further utilize network resources and to solve the problems existing in present homogeneous networks. These various and specific design objectives of HWANs encourage people to design special routing protocols to firstly satisfy the original design purposes of the heterogeneous networks. For example, UCAN [10] (Unified Cellular and Ad Hoc Network) was proposed to utilize the out-of-band bandwidth from ad hoc interface to guarantee the HDR service in UMTS network. In other words, the routing protocols designed for UCAN aim at setting up relaying routes to support HDR transmission. Thus, the relaying routes in UCAN should be able to bear HDR service from the source to the destination. In other network scenario like MADF [11] (Mobile-Assisted Data Forwarding), the author tried to balance the traffic between adjacent cellular cells through ad hoc relaying routes. Then, the relaying routes in MADF play a role on discovering a route from a congested cell to a non-congested cell for a blocked calling service. As for other inter-working networks like EuQoS, the whole network is running by adding gateways between separate wireless networks. Hence, the routing protocols proposed for EuQoS should focus on connecting routes from separate wireless networks together through those gateways. Thus, the first target of QoS-aware routing is to realize the original design objectives of HWANs.

In addition, setting up QoS-aware routes in HWANs not only introduces the difficulties on fulfilling users' application requirements, but also brings the troubles on the cooperation and harmonization of different radio technologies (RTs). For example, Wireless LANs [12] include IEEE 802.11a (up to 54 Mbps in the 5 GHz band), IEEE 802.11b (up to 11 Mbps in the 2.4 GHz band) and IEEE 802.11g (up to 54 Mbps in the 2.4 GHz band). They offer users high data rate service but with limited transmission range up to 300 m. UMTS (Universal Mobile Telecommunication System) can cover a much larger area up to 20 km, but only offer a lower data transmission rate compared with the IEEE 802.11 technology. Other RTs also have their own specific data transmission rates and operation frequencies. Therefore, in order to design a successful QoS-aware routing protocol, the coordination of diverse RTs is also a critical issue in HWANs.

The organization of this chapter is as follows. In Section 11.2, we present a taxonomy of QoS-aware routing protocols in HWANs, which is based on the design purposes of these routing protocols and how the network routing messages are transmitted. In Section 11.3, we describe the general procedure of QoS-aware routing in HWANs, and some QoS-aware routing issues (such as Source Selection (SS), Route Selection (RS) and Destination Selection (DS)) are also pointed out in this section. In Section 11.4, we detail the operation of SS, and some algorithms proposed for SS. Section 11.5 gives some solutions for DS. In Section 11.6, some algorithms of RS are discussed to show the general operation of RS in HWANs. The chapter ends with a summary of QoS-aware routing issues in HWANs in Conclusion.

11.2 Overview

Although other taxonomies of routing protocols were proposed in homogeneous networks, we present two taxonomies that are based on how QoS-aware routing protocols make HWANs work and support HWANs with more functions. Such taxonomies provide the most natural framework within which to discuss QoS-aware routing in HWANs. The conventional taxonomies in homogeneous networks mainly focus on some special features involved in route establishment, such as flooding or directed flooding in terms of the broadcast method of route discover. The taxonomies in this chapter focus on the HWAN QoS-aware characteristics.

The first taxonomy of QoS-aware routing in HWANs is based on the network structure of proposed HWANs. As mentioned above, HWANs can be either inter-working heterogeneous networks and intra-working heterogeneous networks. Therefore, QoS-aware routing in HWANs can be generally classified into two big categories:

- inter-working HWAN QoS-aware routing and
- intra-working HWAN QoS-aware routing.

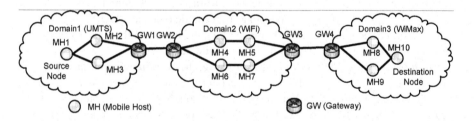

Fig. 11.1. Inter-working HWAN QoS-aware routing.

In *inter-working HWAN QoS-aware routing*, routes are constructed through various heterogeneous network domains, and routes in each separate domain are actually homogeneous, as shown in Fig. 11.1. In the EuQoS system [3], network domains include the most common access networks such as xDSL, UMTS, WiFi, and LAN. The main design objective of EuQoS system is to support customers with general QoS-based services through multiple heterogeneous domains. Each heterogeneous network domain in EuQoS is discussed in general without special specifications, and connected with each other through gateways. Thus, the QoS-based routes in EuQoS are designed through general air-interfaces and general access technologies. The QoS-aware routing protocols proposed in EuQoS is called QoSR [3] (QoS Routing), which is mainly designed to solve the inter-domain QoS routing issues and to establish QoS constraint paths to satisfy customers' service requirements. QoSR in the EuQoS system also includes some novel routing components to enhance the inter-domain QoS routing, such as p-SLSs (peering Service Level Specifications), EQ-BGP (Enhanced QoS Border Gateway Protocol) and TERO (Traffic Engineering and Resource Optimization).

The inter-domain routing through QoS-class planes [13] as another inter-working HWAN QoS-aware routing protocol focuses on providing qualitative QoS guarantees across multiple heterogeneous network domains, and further investigates the corporation between Q-BGP [14] (QoS-enhanced Border Gateway Protocol) and QoS-based route selection policies to allow several inter-domain QoS requirements to co-exist. Hence, one main issue in the inter-working HWAN QoS-aware routing is to extend the conventional BGP [15] (Border Gateway Protocol) with more QoS support for interconnecting adjacent heterogeneous network domains so as to adapt to the inter-domain heterogeneous routing protocols. For example, Domain1 in Fig. 11.1 is a UMTS network, and Domain2 in Fig. 11.1 is a WiFi network. In order to offer a connection between Domain1 and Domain2, GW1 (Gateway1) and GW2 should be developed with more QoS extensions so as to connect with each other. Another main issue in the inter-working HWAN QoS-aware routing is to design suitable QoS-based route selection policies for each heterogeneous network domain in order to build QoS-ware routes through multiple heterogeneous network domains, because the deployment and wireless techniques used in HWAN domains are different from each other. For example, multiple routes could exist in each domain in Fig. 11.1. Therefore, the route selection policies in Domain1, Domain2 and Domain3 should be specially designed to adapt to the network features of each domain. As a result, in order to successfully offer a route from a source node (such as MH1 in Fig. 11.1) to a destination node (such as MH10 in Fig. 11.1), the issues on the inter-working and inter-connection between HWAN domains should be solved.

In *intra-working HWAN QoS-aware routing*, the working air interface for each hop is selected from a list of available air interfaces to construct the whole route, and the available wireless techniques are tightly coupled with each other, as shown in Fig. 11.2. In A-GSM [16] (Ad Hoc GSM) network, the system aims at expanding the transmission range of a GSM network with ad hoc overlay. In GSM network, a MH (Mobile Host) which is located in the dead spot [17] is not able to be connected to a GSM BS (Base Station). However, by communicating through an ad hoc relaying route, this MH can get the connection to a GSM BS. The relaying routes in A-GSM are constructed by both ad hoc interface and cellular Interface, and the air interfaces through ad hoc interface and cellular interface are tightly coupled with each other. For example, if MH2 in Fig. 11.2 located in the dead spot is not able to communicate with any BS, MH2 in A-GSM can be connected to BS1 through a relaying route (such as MH2-MH1-BS1 in Fig. 11.2).

In Opportunity Driven Multiple Access (ODMA) [18], the system is designed to tightly couple UMTS network with ad hoc network so as to guarantee high-bit-rate service through the whole UMTS network. An UMTS cell in ODMA is divided into 2 regions, namely Region H (High-bit-rate Region) where MHs can get high-bit-rate service and Region L (Low-bit-rate Region) where MHs can only get low-bit-rate

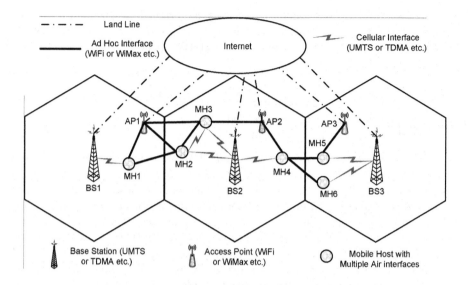

Fig. 11.2. Intra-working HWAN QoS-aware routing.

service due to hostile losses. ODMA can support MHs located within Region L with a high-bit-rate ad hoc relaying route to get the high-bit-rate service from Region H. For example, if MH4 Fig. 11.2 is located in Region L, and MH6 is located in Region H, then MH2 in ODMA can get high-bit-rate service through an ad hoc relaying route (such as MH4-MH6-BS2 in Fig. 11.2). Hence, in an intra-working HWAN system, the air interface selection policy in each hop plays a critical role in QoS-aware routing strategies. Like the route (MH2-MH3-AP2 in Fig. 11.2), MH2 can communicate with MH3 through both ad hoc interface and cellular interface. To choose a proper working air interface for this hop, bandwidth capacity, data transmission rate and other QoS-aware metrics could be involved in the air interface selection. In addition, the combination of different network nodes such as (MH, AP and BS in Fig. 11.2) is also very important in intra-working HWAN QoS-aware routing. For example, in order to connect MH1 with BS2 in Fig. 11.2, MH, AP and BS are involved in the route establishment. If the proposed HWAN only accept s the route composed of MH and BS, then MH1 only have two routes to BS2 (namely MH1-MH2-BS2 and MH1-MH2-MH3-BS2). However, if the HWAN system allows more types of nodes involved in the route establishment (such as AP1), MH1 can have much more optional routes to BS2, as shown in Fig. 11.2. Moreover, the combination of heterogeneous network nodes also needs to consider other QoS issues like interference and bandwidth utilization.

The second taxonomy of QoS-aware routing in HWANs is based on the design objectives of each routing strategies. Each HWAN is actually a combination of different wireless networks, and designed to solve present network issues such as network congestion and to take advantages of various wireless techniques. One main QoS-aware issue in a HWAN routing is to satisfy the original design purposes of this proposed HWAN system. According to the specific design objective of each HWAN system, we divide the routing protocols in each HWAN network into the following four main categories:

- balancing traffic between cellular cells,
- extending network transmission range,
- guaranteeing high-bit-rate service, and
- offering inter-domain QoS support.

To *balance traffic between cellular cells*, HWAN architectures like iCAR and MADF use ad hoc relaying routes to divert overloaded traffic from congested cells to non-congested cells. Then, the overall traffic of the whole network system is balanced through these ad hoc relaying routes. Another routing protocol named as ARCA [19] (an Adaptive Routing Protocol for Converged Ad hoc and Cellular Networks) was proposed to setup ad hoc relaying routes in a so-called CACN (Converged Ad hoc and Cellular Network) system. The main target of ARCA is to support the CACN system with relaying routes so as to realize the balance of cellular traffic. Thus, the CACN system can drop the call blocking probability of conventional cellular networks and solve the hot spots issues.

To *extend network transmission range*, HWAN routing protocols were designed to build heterogeneous routes to connect to the nodes uncovered by the original homogeneous networks. DARP [20] (Dynamic Adaptive Routing Protocol) in HWN (Heterogeneous Wireless Network) was proposed to setup combined ad hoc and cellular routes to extend the transmission range of ad hoc networks. In the proposed HWN system, cellular network and ad hoc network are tightly coupled. DARP uses cellular BSs in a one-hop ad hoc mode, which cooperate with normal ad hoc nodes to setup routes. Because the one hop route of cellular BSs has a longer transmission range than that of normal ad hoc nodes, the transmission range of ad hoc network can be extended through the combination of normal ad hoc nodes and cellular BSs.

To *guarantee high-bit-rate service*, routing protocols are normally used to build up high-bit-rate routes so as to support high-bit-rate service to MHs which is located in Region L in conventional UMTS network. In UCAN, the HWAN architecture is constructed by both UMTS network and ad hoc network. A MH with poor channel quality (a destination client) can firstly connect to a MH with high downlink data-rate [21] (a proxy client) through a relaying path (constructed by relay clients). Then, this destination client utilizes the high downlink data-rate channel from the proxy client to communicate with the UMTS BS. Three routing protocols, which are referred to as Greedy, On-demand and DST, were proposed in UCAN to build up the relaying routes.

To *offer inter-domain QoSsupport*, HWAN routes should fulfill the source MHs QoS requirements both within a single network domain and cross multiple HWAN domains. The inter-domain routing protocols in both [3] and [13] were proposed to support the accomplishment of various HWAN domain QoS requirements and to make the routes in different domain connect with each other. Finally, the system can build up a route from the source to the destination with a full QoS support. Additionally, QoS-aware HWAN routing protocols can be also used to setup routes to help a HWAN system to fulfill several QoS requirements. For example, the routing protocols proposed in SOPRANO [22] aim at establishing relaying routes to increase the network capacity, to extend cellular cell coverage, and to provide a broad connectivity.

11.3 General QoS-Aware Routing Procedure in HWANs

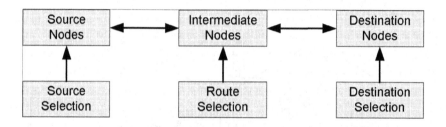

Fig. 11.3. General HWAN QoS-aware route structure.

In this chapter, we describe the QoS-aware routing in HWANs as a general concept, which spans several platforms and techniques. As for a route, nodes can be generally divided into three categories, namely, source nodes, intermediate nodes, and destination nodes as shown in Fig. 11.3. In a QoS-aware routing protocol, source nodes are normally used to initiate the route discovery, and to bring up the QoS requirements for the route discovery such as hop number limitation and transmission bit rate requirement. Intermediate nodes are used to build up routes from source to destination. In a single hop route, there could be no intermediate nodes. In a multiple hop route, each hop and each intermediate node should fulfill the QoS requirements. In other words, the main function of intermediate nodes is to offer successful connections between source and destination. A destination node is the final destination of a route discovery. The destination node returns a route reply (RREP) to the source node to inform the source node the availability of the route. Moreover, a destination node should also fulfill the source's QoS requirements.

In traditional ad hoc wireless network routing, the source and the destination are all pre-decided before the route discovery. Thus, during a route discovery process in ad hoc wireless network, only intermediate nodes need to be discovered. Hence, QoS routing in ad hoc wireless network mainly focuses on the route selection part to satisfy the QoS requirements of route discovery. However, QoS-aware routing issues in HWAN systems are much more complicated than those in homogeneous network.

As shown in Fig. 11.3, QoS-aware issues in source nodes, which we call as *Source Selection*, could lead to dynamic bandwidth allocation of the system and the starting point decision of route discovery. In terms of intermediate nodes in HWAN, *Route Selection* involves not only hop number limitation or bandwidth requirement, but also the cooperation among heterogeneous nodes and air-interfaces. As for destination nodes, we give a name as *Destination Selection* for the general issues involved in HWAN system. Actually, the destination selection in HWAN includes the single destination selection and multiple destination selection. In single destination selection, only one destination is needed for one route discovery. However, in multiple destination selection, several destinations can co-exist in one route discovery process. Thus, the route discovery process should select one final destination from a list of available destinations. Hence, destination selection in HWAN involves heterogeneous network selection and homogeneous destination selection.

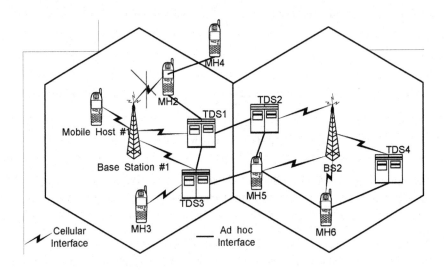

Fig. 11.4. The operation of CACN system.

The following is an example of HWAN system, which is named as CACN (Converged Ad hoc and Cellular Network). Fig. 11.4 gives the general operation of a CACN system, and Fig. 11.5 presents the cooperation among Source Selection, Route Selection, and Destination Selection, and how they successfully establish routes in CACN and make the system work properly.

As shown in Fig. 11.4, a CACN system has two air interfaces used for the communication between nodes: C (Cellular) interface that operates at a cellular network frequency (in-band); A (Ad-hoc) interface that operates at an ad-hoc network frequency (out-of-band, e.g. IEEE 802.11). The *Base Station* (BS) is same as the present cellular network base stations with C-interface. A base station uses its C-interface to communicate with mobile handsets in a wireless mode. However, the communication between base stations is actually performed in wired mode. *Traffic Diversion Stations* (TDSs) equipped with both A-interfaces and C-interfaces are deployed with managed mobility [23]. In a TDS, C-interfaces consuming in-band bandwidth are used for communicating with a BS or a MH with a C-interface. A-interfaces which operate in the unlicensed ISM (Industry, Science, and Medical) band and consumes out-of-band bandwidth, is used for communicating between TDSs or with MHs with an A-interface. *Mobile Hosts or Mobile Handsets* (MHs) in a CACN system are diverse mobile devices (such as mobile phone, laptop and PDA) are equipped with both A-interface and C-interface, or either of them.

Fig. 11.4 is a simplified two-cell model, which illustrates the general operation of a CACN system. In Fig. 11.4, if MH3 is trying to make a call through BS1 which is congested, this call will be dropped in conventional cellular networks. In a CACN system, this blocked call will be diverted to a non-congested cell through a relaying route (like MH3-TDS3-TDS1-TDS2-BS2 in Fig. 11.4). In another case, if MH1 (which is not covered by any TDS) is also trying to make a call through congested BS1, this call will not be able to be directly diverted to a non-congested cell. To successfully accept this call, a pseudo source (like MH2 in Fig. 11.4) is chosen to divert its on-going call to a non-congested cell through a relaying route, and then release its occupied bandwidth for the use of the original source (namely MH1 in Fig. 11.4). The selection of pseudo source (like MH2 in Fig. 11.4) can lead to the issue of Source Selection, and the selection of destination BSs (like BS2 in Fig. 11.4, or any other available non-congested BSs) brings up the issue of Destination Selection. Moreover, the establishment of relaying routes in Fig. 11.4 results in the Route Selection. Only a proper cooperation of these three parts can successfully setup routes and make the CACN system work.

Fig. 11.5 illustrates the cooperation procedure of Source Selection, Route Selection, and Destination Selection in CACN. It starts from the point where a MH initiates a voice call or data transmission request whereas no sufficient bandwidth in its home cell is available to accommodate this request. We denote the request initiating MH (also called source MH) as MH_s. We call the cell where MH_s is located as home cell and denote it by C_h. If there is a TDS of sufficient bandwidth within the transmission range of MH_s, then MH_s uses it directly as the next hop for relaying data. Otherwise, a *Source Selection* is triggered to find a proper MH (denoted as MH_j as illustrated in Fig. 11.5) within cell C_h so as to release this MH's channel to MH_s. In any case, a MH's data needs to be transmitted via relay as such a relaying route needs to be discovered. Unlike conventional ad hoc routing algorithms where the destination of a route is known, MH_s, either it being the genuine one that is making the request or the pseudo one MH_j that is selected by Source Selection, does not know its route destination. All it needs is a route to a cell that could accommodate its data transmission regardless of the cell's location. In this regard, Destination Selection is a necessity. Once a destination or a list of destinations is selected, a route discovery procedure is activated to find all routes satisfying the destination and bandwidth requirements. If multiple routes are found, a Route Selection is needed to select the most proper route to relay MH_s's data. Obviously, if Source Selection gets involved in the beginning, the pseudo source MH (MH_j) will have to release the channel it uses to the genuine source MH (MH_s) before invoking the data transmission to transmit data from MH_s using route r_k.

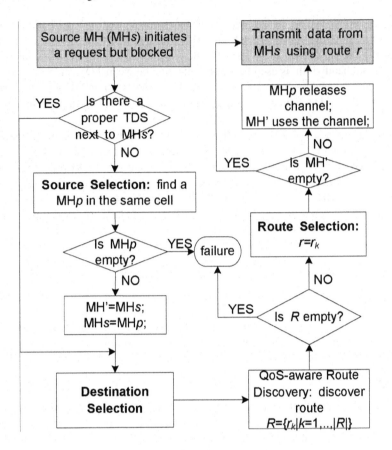

Fig. 11.5. Cooperation among source selection, route selection, and destination selection in CACN.

As mentioned above, to successfully setup QoS-aware routes in a HWAN system, a routing protocol should solve the issues of Source Selection, Route Selection, and Destination Selection. Additionally, proper cooperation of the three parts should also be considered to enable the establishment of QoS-aware routes in an effective and efficient manner.

11.4 Source Selection

In this section, we present some issues on source selection in HWAN, and some solutions with regards to the proposed HWAN structures are given for the source selection. Additionally, the solutions for source selection are mainly related to the QoS-aware routing process. Therefore, we also call the solution as SSP [24] (Source Selection Procedure), which works as a dynamic bandwidth allocation algorithm.

Different from other bandwidth allocation in wireless networks such as DBA (Dynamic Bandwidth Allocation) proposed for WCDMA systems in [25], SSP is designed for the QoS-aware routing in HWAN. In homogeneous network routing, the source node is known to the route discovery procedure. However, as the example shown above, several available sources could co-exist for one route discovery procedure. Thus, SSP is needed for the decision of just one certain source for one route discovery procedure. The selection of a certain source node may relate to the reallocation of bandwidth, and the dynamic bandwidth allocation in SSP is realized by using the bandwidth from other networks or other cells through relaying routes. The example HWAN system is also CACN, the same as the one shown in Section 11.3. In the following, we first present the general procedure of bandwidth allocation in SSP in CACN. Then, three SSP algorithms are given to detail the process of source selection, namely, how to choose a certain source node for a route discovery.

11.4.1 Bandwidth Reallocation in Source Selection

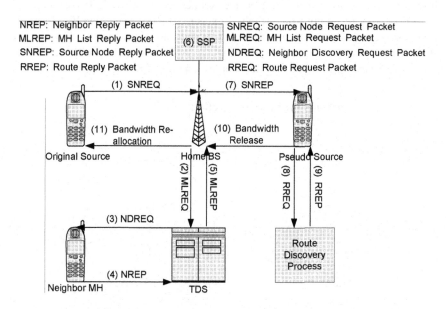

Fig. 11.6. General process of bandwidth reallocation in source selection.

In some HWAN architectures such as iCAR, MADF, and CACN, an MH can release its occupied bandwidth for the use of another MH, and its ongoing traffic is diverted to other cells or other networks through relaying routes. Here, we use CACN as an example to show how bandwidth reallocation is performed in such a HWAN system.

In a CACN system, a pseudo source is chosen by the home BS to release its occupied bandwidth without interrupting the present communication of the pseudo source. Fig. 11.6 shows the main steps of bandwidth reallocation in a certain cell.

If an MH in a congested cell is trying to make a call, it sends a Source Node Request Packet (SNREQ) to the home BS (see step 1 in Fig. 11.6). After receiving a SNREQ, the home BS broadcasts a MH List Request Packet (MLREQ) to all TDSs within the home cell (see step 2 in Fig. 11.6). Once a TDS receives a MLREQ, it broadcasts a Neighbor Discovery Request Packet (NDREQ) to all MHs within its coverage (see step 3 in Fig. 11.6), which respond Neighbor Reply Packets (NREP) to the TDS (see step 4 in Fig. 11.6). Then, the TDSs return a list of MHs to the home BS by sending a MH List Reply Packet (MLREP) (see step 5 in Fig. 11.6), which also contains the bandwidth status of the TDSs. After receiving the MLREPs from all the TDSs upper bounded to a timeout, the home BS applies a rational SSP algorithm to analyze the information included in MLREPs so as to choose a proper source node (see step 6 in Fig. 11.6). Following this, the home BS sends a Source Node Reply Packet (SNREP) to the decided pseudo source if the source node is not the original source (see step 7 in Fig. 11.6) so that the pseudo source starts a route discovery process by broadcasting Route Request Packets (RREQ) (see step 8 in Fig. 11.6). After receiving Route Reply Packets (RREP) (see step 9 in Fig. 11.6), the pseudo source releases its occupied bandwidth and starts diverting its ongoing traffic through a relaying route (see step 10 in Fig. 11.6). Finally, the home BS reallocates the released bandwidth to the original source (see step 11 in Fig. 11.6).

11.4.2 Three Algorithms for Source Selection

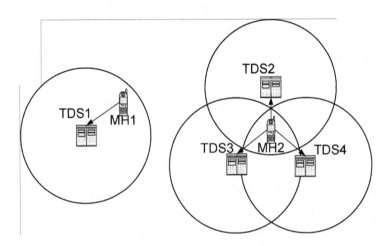

Fig. 11.7. An exmple of deployment of network nodes in CACN.

As mentioned above, SSP algorithms are designed to run in BSs. The objective of SSP is to choose a source node as the starting point of route discovery process in HWAN. Moreover, SSP should also consider the network situation to improve the overall performance of the system. In other words, if the proposed HWAN architecture aims at dropping call blocking rate, source nodes chosen by the home BS should have the most possibility of discovering a relaying route, and a large number of source nodes will not partially block TDSs (namely improve the overall performance of the system). Additionally, the operation of SSP also depends on the information gathered by the HWAN system (such as the information included in MLREP packets in CACN). The following is an example of MLREP packet mainly containing three fields: <TDS ID, MH ID List, Bandwidth Info>. In the MLREP packet, "TDS ID" refers to the address of a TDS in the home BS. "MH ID List" includes a list of the addresses of MH's within the transmission range of the TDS. "Bandwidth Info" is the available free bandwidth (or channel) in the TDS.

Table 11.1. Available pseudo sources.

MH ID: MHi	Home TDS ID: $TDSij$	Bandwidth Info: Bij
$MH1$	$TDS1$	B_{TDS1}
$MH2$	$TDS2$	B_{TDS2}
$MH2$	$TDS3$	B_{TDS3}
$MH2$	$TDS4$	B_{TDS3}

After receiving MLREP packets, the home BS gets a big list of MHs. In fact, a source node should satisfy the requirements that a source node is an MH within the transmission range of the home cell (relevant information can be obtained by checking the register information of MHs), and a source node should be covered by at least one TDS. Hence, the home BS chooses all the MHs, which fulfill the basic requirements, to build up a table of available source nodes. Then, the home BS chooses a final source node to divert its traffic according to a source selection algorithm. As shown in Table 11.1, MH_i is the address of an available source node. TDS_{ij} is the address of a reachable TDS of MH_i. B_{ij} contains the available bandwidth information of TDS_{ij}. Also, Table 11.1 gives some examples according to Fig. 11.7. In order to select a final source node, three SSPs are proposed according to different design purposes.

SSP1 chooses source nodes with the most number of reachable TDSs, because these nodes have the most possibility of successfully discovering a relay route (details are given below). Considering the Request Rejection Rate (RRR), the RRR is the probability of a TDS failing to find a relaying route after broadcasting RREQs. Then, we could also choose a source node with the lowest RRR. Assuming that the RRR of TDSij is $P_{ij}(P_{ij} \leq 1)$, then the RRR of MH1 is $P_{MH1} = P_{11}$ (P_{11} denotes the RRR of TDS1 in Fig. 11.7). Also, the RRR of MH2 is $P_{MH1} = P_{21}P_{22}P_{23}$ (P_{21}, P_{22} and P_{23} denote the RRR's of TDS2, TDS3 and TDS4 in Fig. 11.7, respectively). If $P_{11} = P_{21} = P_{22} = P_{23}$, then $P_{MH1} \geq P_{MH2}$). In general, if the number of reachable TDS's of MHi is Ni, the RRR of MHi is

$$P_{MHi} = \prod_{j=1}^{Ni} P_{ij}. \tag{11.1}$$

Thus, the source node chosen by SSP1 is the MH with minimum RRR, namely $\min(P_{MH1}, \cdots, P_{MHn})$ (assuming the number of available MHs in Table 11.1 is n). To simplify the notation, we assume that the RRR of each TDS is identical at P. Then, the RRR of MHi is

$$P_{MHi} = \prod_{j=1}^{Ni} P_{ij} = P^{Ni}. \tag{11.2}$$

Thus, SSP1 just chooses a source node with the maximum number of reachable TDSs (namely $\max(N_1, ..., N_n)$), because $P \leq 1$. As shown in Fig. 11.7, MH1 can only broadcast RREQ to TDS1, but MH2 can broadcast RREQs to TDS2, TDS3 and TDS4. MH2 stands a better chance to find a relaying route. One problem of the selection is that SSP1 could only choose MH's within a specific cross area with most reachable TDSs. As shown in Fig. 11.7, SSP1 could only choose source nodes within the cross area covered by TDS2, TDS3 and TDS4, but not choose source nodes covered by TDS1. The result is that the bandwidth of TDS's covering the specific cross area is firstly consumed. Then, these TDSs could become congested but other TDSs still have lots of free bandwidth unused. Therefore, we design other SSPs to solve this problem.

SSP2 tries to balance the bandwidth consumption amongst TDSs when choosing source nodes. Using the information in Table 11.1, the home BS can get the bandwidth status of all reachable TDSs of available pseudo sources. Bij refers to the free bandwidth of the reachable $TDSij$ of MHi. Then, the average bandwidth of all reachable TDS's of an available source node is:

$$B_{MHi} = \frac{\sum_{j=1}^{Ni} Bij}{Ni}. \tag{11.3}$$

To achieve the balance of TDS bandwidth consumption, SSP2 chooses pseudo sources with maximum B_{MHi}, namely, $\max(B_{MH1}, ..., B_{MHn})$. As a result, source nodes will consume the bandwidth of TDSs in average, and avoid partially congesting TDSs. However, such a selection of source nodes may lead to a relatively higher RRR, because SSP2 is designed not to find a source node with a minimum RRR but to choose a source node with a maximum average bandwidth of reachable TDS's. Then, the instantaneous RRR in SSP2 is determined by the RRR of current source nodes with a random number of reachable TDSs. Thus, compared with SSP1, SSP2 can avoid partially blocking TDSs by balancing the consumed bandwidth of TDSs, but the RRR in SSP2 could be higher due to the unpredictable number of reachable TDSs of source nodes.

SSP3 takes both the number and the average free bandwidth of reachable TDSs into consideration. Therefore, SSP3 can achieve a balance of the bandwidth consumption of TDSs without highly increasing the RRR of the system during the selection of source nodes. In SSP3, node MHi has two weights for selecting pseudo sources, namely, (P_{MHi}, B_{MHi}) as mentioned in SSP1 and SSP2. The combined weight calculated by SSP3 is $Wi = B_{MHi}(1 - PMHi)$, where $1 - PMHi$ refers to the possibility of successfully discovering a relaying route. Also, the weight (Wi) can be named as the average free bandwidth of reachable TDSs of MHi in probability, compared with B_{MHi}. According to formula (2) and (3) shown above, the combined weight is

$$W_i = \frac{\sum_{j=1}^{Ni} Bij}{Ni}[1 - P^{Ni}]. \tag{11.4}$$

SSP3 chooses pseudo sources with the maximum Wi, namely $\max(Wi, ..., Wn)$. Because P in the above formula is the average RRR of TDSs, the calculation of P depends on the statistical results calculated by BSs or TDSs. To reduce the amount of calculation in BSs or TDSs, the combined weight of a MHi can be simplified as following. For the MHs given in Table 11.1, both the number of reachable TDSs (Ni) and the average free bandwidth of reachable TDSs (B_{MHi}) are sorted in order. Then, instead of (P_{MHi}, B_{MHi}), the weight of MHi in SSP3 is (N_T, N_B) (N_T refers to the order of MHi in terms of the number of reachable TDSs, and N_B refers to the order of MHi in terms of the average free bandwidth of reachable TDSs). Then, the combined weight of MHi in SSP3 is simplified as

$$W_i = N_T N_B. \tag{11.5}$$

SSP3 could just choose pseudo sources with the simplified minimum Wi.

We evaluate the three SSPs in terms of the average request rejection rate of the overall network and the signaling overhead in a specific BS and some TDSs, by plugging in reasonable values of parameters in above formulas. The request rejection rate is the percentage of MHs which fail to divert its blocked traffic when requesting a diversion service. The results show that SSP1 has a higher request rejection rate, compared with SSP2 and SSP3. However, SSP1 adds less extra signaling overhead to the network. Compared with SSP2, SSP3 shows a slightly lower request rejection rate.

The SSP algorithms given above are specially designed for integrated cellular networks like iCAR and CACN, but the general design ideas of these algorithms can be extended to other types of HWAN system by considering the design objectives of these HWAN architectures.

11.5 Destination Selection

Unlike conventional homogeneous networks where each source node in a route discovery process has a certain destination, destinations in HWAN system may not been known by the source for route discovery. Thus, one critical issue on how to select or

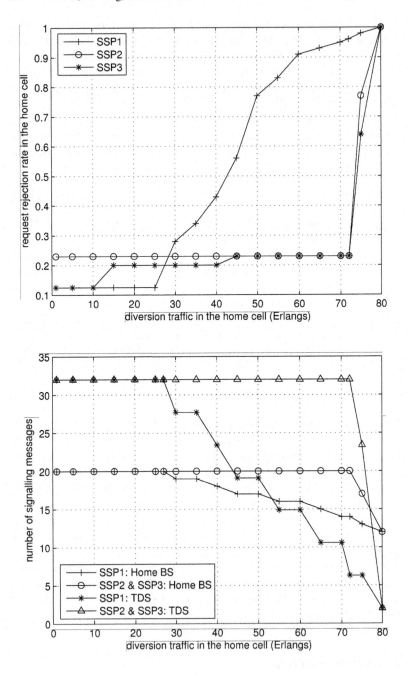

Fig. 11.8. Performance analysis of source selection algorithms.

decide destination nodes occurs in HWAN. For example, iCAR system allows users to divert their blocked traffic from congested cell to non-congested cell. The non-congested BS could be a list of available BSs, and a mobile user needs to decide which BS is going to be the destination accepting its diverted traffic. Additionally, when a user is requiring a service (such as a video stream) from a HWAN system, several service providers could have the same service available for the user. Then, to discover a route, the final selected service provider (namely the destination) needs to be selected. During the whole process of a QoS-aware route discovery in HWAN, the selection of destination nodes is quite important for improving the service provided to mobile users, and various HWAN architectures could lead to different destination decision methods. In this section, we discuss the destination selection method in general, and use some HWAN systems as examples to describe how the destination selection method works for route discovery.

If the access network such as WLAN or UMTS is considered as a service provider, a mobile user could have multiple service providers co-existing in the same HWAN system. Then, the issues on destination selection could be related to another issue in HWAN, namely, network selection. Many researches and algorithms such as [26] and [27] have been done for the further investigation of utilizing diverse network resources. In this section, we consider the destination selection only as a part of a route discovery process in QoS-aware routing in HWAN, and discuss how the destination selection affects the performance of routing and services provided by a heterogeneous wireless system. Here, destination selection methods have be generally classified as the following two main categories:

- reactive destination selection and
- proactive destination selection.

The *reactive destination selection* is applied during a route discovery process. In other words, the reactive method is used to discover destinations after a MH starts route discovery. The *proactive destination selection* is designed to select destination nodes before route discovery. Then, the route discovery process in this proactive method is similar to that in a homogeneous network. Both reactive and proactive methods need to find destinations which fulfill the requirements from the source. The requirements could be a certain service like a video or audio stream, or a bandwidth requirement. Hence, the destination selection methods discussed in this chapter actually include the discovery of destination nodes and the decision of destination nodes.

11.5.1 Reactive Destination Selection

The reactive destination selection is launched with the route discovery. Thus, the destination nodes are discovered and decided by intermediate nodes according to the network information collected by intermediate nodes, as shown in Fig. 11.9.

With the reactive destination selection, the whole routing process follows the following steps:

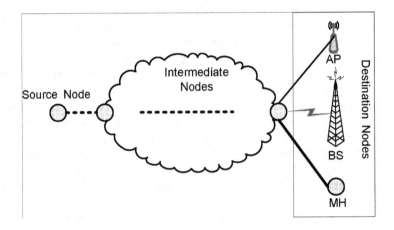

Fig. 11.9. Analysis of reactive destination selection.

- Step 1: A source node requires a service from the HWAN system, and starts route discovery.
- Step 2: Network information such as bandwidth and service content needs to be collected by intermediate nodes.
- Step 3: After receiving the QoS request from the source, intermediate nodes need to decide whether a destination has been discovered according the network information obtained, or just to rebroadcast the request to other intermediate nodes without finding a destination.
- Step 4: Once a destination node is discovered during the route discovered, a reply message will be sent back to inform the source about the presence of a certain destination.
- Step 5: The source needs to select a final destination from a list of discovered destinations, and a final route to this destination.
- Step 6: Once the destination and the route are decided, the source starts to receive its required service from the destination.

Considering some special issues involved in the above process, intermediate nodes in Step 2 need to decide a method to collect network information. Here, we have two options. One is to periodically collect network information. For example, the intermediate nodes can periodically collect the bandwidth information from BSs, as shown in Fig. 11.9. Then, if the source only wants to find a non-congested cell such as in iCAR and MADF, an intermediate node will notice the presence of the destination BS immediately after receiving the request from the source. Another way is to request the network information after receiving the QoS request from the source. If the source requires a movie from some service providers, an intermediate node can send a request to all service provides which it is connecting to, and to find out whether one of the service providers has the service content available.

In Step 5, the destination nodes discovered by intermediate nodes could be more than one. Then, the source should choose a final destination from all destinations discovered. Therefore, we can design some QoS-aware destination decision algorithms to find this final destination. For example, if the source is requiring some free bandwidth from a BS to divert its blocked traffic, it can choose a destination BS with the maximum amount of free bandwidth. Moreover, the destination decision algorithm can also cooperate with the route selection. If the source also wants to have the service provided with minimum latency, then the source could choose a destination node which has a route with the minimum hop number.

11.5.2 Proactive Destination Selection

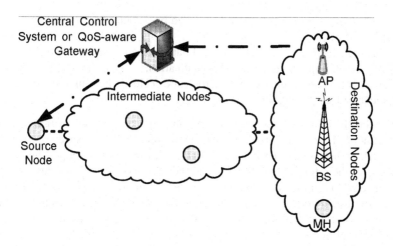

Fig. 11.10. Analysis of proactive destination selection.

In the proactive destination selection, all available destinations are decided before route discovery, and the list of destinations are given to the source for the following route discovery process. Thus, to find destination satisfying the QoS requirements of the source before route discovery, the HWAN system needs to maintain network information (like bandwidth) before starting the route discovery process. The pre-decided destinations can be simply implemented in normal QoS-aware routing process. As shown in Fig. 11.10, the destinations are decided by a central control system or a QoS-aware gateway, which can collect the required QoS information from prospective destinations nodes. The central control system which can maintain the network information (like the bandwidth information of BSs) exists in traditional cell networks such as a GSM network. The QoS-aware gateways are used to maintain the QoS-related network information cross multiple network domains. One example of

QoS-aware gateways is given in [28]. The destination list is sent back to the source by the central control or the gateway.

Compared with the routing with the reactive destination selection, the routing including the proactive destination selection has the following steps:

- Step 1: The central control system or QoS-aware gateway collects QoS information from all possible destination nodes in the HWAN system.
- Step 2: The source sends its service requirements to the central control or gateway.
- Step 3: After receiving the QoS request from the source, the destination decision module of the central control or gateway chooses one or a list of destinations for the route discovery, and send a reply back to the source.
- Step 4: According to the destinations discovered by the central control or gateway, the source starts the route discovery process to find available QoS-aware routes to the destination.
- Step 5: Once a destination node is reached by a route, the destination returns a reply to the source.
- Step 6: The source needs to select a final destination from a list of reachable destinations, and a final route to this destination.
- Step 7: Once the destination and the route are decided, the source starts to receive its required service from the destination.

Besides the issues mentioned in the routing with reactive destination selection, the routing with proactive destination selection presents some special research issues, such as how to extend normal gateway with more QoS support and how to make use of the information from the central control system. A good design of destination selection in the central control can help to improve the performance of the heterogeneous network system. In this section, we give an example from the CACN system mentioned above, which shows how to use the pre-decided destination list to reduce the routing overheads.

In the CACN system, the source in the route discovery discovers an ad hoc relaying route to divert its traffic to another non-congested cell. Since the central control can decide destinations for the source, a limited broadcasting can be applied to route discovery process, as shown in Fig. 11.11. Cells surrounding the source cell are divided into several circles according to the distance to the source cell. Circle-1 includes six cells of the $1st$ ring with the closest distance to the source cell. The nth ring includes 6n cells. Circle-n includes all cells from the $1st$ ring to the nth ring. Route Request (RREQ) packets are only broadcasted within one of these circles (namely the limited broadcast area). The limited broadcast area can be chosen according to the number of destinations required for a route discovery. More destinations included in RREQs lead to more destination BSs discovered, but also bring more routing overheads. However, in order to broadcast RREQs within the pre-decided circle, the central control has to give a full list of cells, in which RREQs can be broadcasted. Hence, with limited broadcast, once an intermediate receives a RREQ, it first checks whether it is within the cells listed in the RREQ. All intermediate nodes that locate out of the circle simply drop this RREQ.

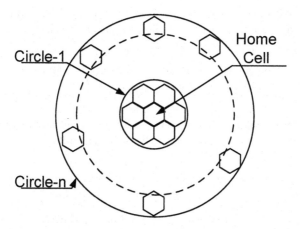

Fig. 11.11. Limited broadcast by using pre-decided destinations in CACN.

By using the proactive destination selection and location information, we can modify some present homogeneous routing protocols (such as LAR [29]) for the use of HWAN system. Then, by discovering destinations before starting the route discovery, the reuse of homogeneous routing protocols is fully possible in HWANs.

Considering the characteristics of both the reactive and the proactive destination selection, the implementation of both methods depends on the original network structures of HWANs. As a result, the reactive destination selection is suitable for a HWAN system without central control or QoS-aware gateway. And the routing with reactive destination selection may require the system bearing lots of routing overheads, compared with the routing with proactive destination selection. On the other hand, the proactive destination selection requires the HWAN system equipped with central control system or QoS-aware gateway. This special requirement prevents the application of the proactive method in other HWAN systems.

11.6 Route Selection

In this section, we analyze the process of the route establishment in intermediate nodes for general HWAN system, when the source and the destination have been decided by a route discovery process. Some general QoS-aware route requirements are discussed in this section. We also further discuss the QoS-aware route selection issues in both inter-working and intra-working HWAN systems.

In general QoS-aware route establishment in HWAN has a number of service requirements during data transmission. These service requirements can be classified as various QoS metrics, such as transmission delay, packet drop rate, bandwidth requirements, power consumption and throughput. These QoS metrics can be general divided into the following three categories.

In a route where $R = j \rightarrow k \rightarrow \cdots \rightarrow l \rightarrow m$ (j, k, l, m present the nodes through the route), the QoS metrics (d) of this route can be classified as follows:

- Adding Metrics: $d(R) = d(j, k) + \cdots + d(l, m)$,
- Multiplying Metrics: $d(R) = d(j, k) \times \cdots \times d(l, m)$, and
- Minimizing Metrics: $d(R) = \min(d(j, k), \ldots, d(l, m))$.

Here we have some examples for each metrics. The transmission delay, packet drop rate, and bandwidth of a route (R) are adding metrics, multiplying metrics and minimizing metrics, respectively, and can be defined as follows:

- $delay(R) = delay(j, k) + \cdots + delay(l, m)$,
- $drop(R) = 1 - (1 - drop(j, k)) \times \cdots \times (1 - d(l, m))$, and
- $bandwidth(R) = \min(bandwidth(j, k), \ldots, bandwidth(l, m))$.

However, the QoS-aware situations in each hop of a HWAN route are much more complicated than normal homogeneous networks. Because of the presence of multiple interfaces in each hop, the states of each hop in a metrics are more than one. For example, if the number of communication interfaces between j and k is n, then the QoS metrics between j and k could be from $d_1(j, k)$ to $d_n(j, k)$. Moreover, in different HWAN systems, the applications of multiple air interfaces are totally different from each other. The following sections discuss the route selection in terms of

- route selection in inter-working HWAN QoS-aware routing and
- route selection in intra-working HWAN QoS-aware routing.

11.6.1 Route Selection in Inter-working HWANs

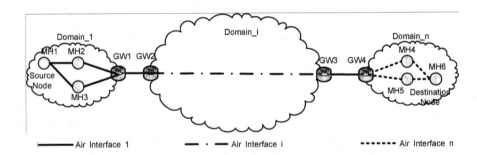

Fig. 11.12. Route selection analysis in inter-working HWAN.

The inter-working HWAN such as EuQoS is constructed by several homogeneous network domains, and these homogeneous network domains are connected with each other by gateways. As a result, the air interface used in each homogeneous network domain is identical. Only the gateways between each domain are used to connect the routes with different communication air interfaces or networks. As shown in Fig. 11.12, the HWAN system has n network domains, and the network domains are connected by gateways. Then, we have n homogeneous routes $(R_1, \ldots, R_n)$ to be connected with each other to setup one heterogeneous route, in order to fulfill the user's QoS requirements (such as MH1 in Fig. 11.12). The air interface used in each network domain could be different from each other.

As shown in Fig. 11.12, a route in $Domain_i$ is $R_i = j_i \to k_i \to \cdots \to l_i \to m_i$ (j_i, k_i, l_i, m_i present the nodes through the route in $Domain_i$). A mobile user's QoS requirement in $Domain_i$ can be defined as d_i. Then, we can calculate the QoS metrics in $Domain_i$ as follows:

- Adding Metrics: $d_i(R_i) = d_i(j, k) + \cdots + d_i(l, m)$,
- Multiplying Metrics: $d_i(R_i) = d_i(j, k) \times \cdots \times d_i(l, m)$, and
- Minimizing Metrics: $d_i(R_i) = \min(d_i(j, k), \ldots, d_i(l, m))$.

Therefore, the QoS-aware metrics through the whole heterogeneous route from $Domain_1$ to $Domain_n$ can be calculated as follows:

- Adding Metrics: $d(R) = \sum_{i=1}^{n} d_i(R_i)$,
- Multiplying Metrics: $d(R) = \prod_{i=1}^{n} d_i(R_i)$,
- Minimizing Metrics: $d(R) = \min(d_1(R_1), \ldots, d_n(R_n))$.

The transmission delay, packet drop rate, and bandwidth of the heterogeneous route (R) can be re-calculated as follows:

- $delay(R) = \sum_{i=1}^{n} delay_i(R_i)$,
- $drop(R) = 1 - \prod_{i=1}^{n}(1 - drop_i(R_i))$,
- $bandwidth(R) = \min(bandwidth_1(R_1), \ldots, bandwidth_n(R_n))$.

Thus, in inter-working HWAN systems, the gateways are designed to help communication between adjacent network domains. By collecting the QoS metrics information across multiple heterogeneous network domains, the route selection is performed according to the QoS requirements of the source. The above examples of QoS metrics mainly focus on single QoS constraint. Considering multiple QoS constraints, if the QoS-aware metrics in each network domain are different from each other, then we could not separate QoS metrics $(d_1, \ldots, d_n)$ from $Domain_1$ to $Domain_n$. Each QoS metrics in different domain can be decided independently. Then, the final route in an inter-working HWAN system is $R = R_1 \to \cdots \to R_n$.

11.6.2 Route Selection in Intra-working HWANs

In intra-working HWAN system such as iCAR and MADF, the number of communication air interface in each hop is more than one. In other words, communications through multiple air interfaces are highly coupled with each other. As shown in Fig. 11.13, in order to build a QoS-aware route in an intra-working HWAN network system, the working air interface should be decided at each hop to satisfy the user's QoS requirements. In other words, the working air interfaces in intra-working HWAN system are different from each other at each hop, unlike the inter-working

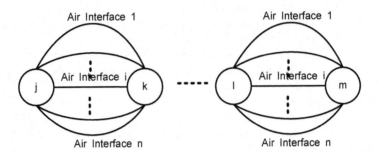

Fig. 11.13. Route selection analysis in intra-working HWAN.

HWAN system which allows the working air interface or network only switch with the change of network domain. In each hop of a HWAN route, the QoS metrics in air interface i can be defined as d_i. As a result, more than one air interface could co-exist for the decision of the communication interface. For example, if the source node (like node j in Fig. 11.13) has a bandwidth requirement through the heterogeneous route discovery process, more than one air interfaces which fulfill the bandwidth requirement could be discovered between two intermediate nodes (such as node j and k in Fig. 11.13). Then, to finally decide the communication interface between the two nodes, other QoS constraints may need to be involved in the decision process. For instance, we can choose an air interface with the maximum bandwidth or the minimum interference as the working interface.

Considering the adding metrics like transmission delay, we can have diverse combinations of working air interface to satisfy the QoS requirement of a user. Therefore, a minimizing metric can be added for the decision of the working air interface. Then, an air interface with minimum transmission delay is chosen at each hop to transmit data. Thus, the transmission delay of the whole heterogeneous route in Fig. 11.13 is as follows:

$$delay(R) = \min(delay_1(j, k), \ldots, delay_n(j, k)) + \ldots \\ + \min(delay_1(l, m), \ldots, delay_n(l, m))). \tag{11.6}$$

For the multiplying metrics such as packet drop rate, an air interface with minimum packet drop rate can also be chosen for the communication at each hop, so as to obtain an optimum route. Thus, the packet drop rate of the whole heterogeneous route in Fig. 11.13 is given by

$$drop(R) = 1 - (1 - \min(drop_1(j, k), \ldots, drop_n(j, k))) \times \ldots \\ \times (1 - \min(drop_1(l, m), \ldots, drop_n(l, m))). \tag{11.7}$$

In terms of minimization of metrics such as bandwidth, the performance of the QoS-aware HWAN route can be further improved by adding other QoS constraints, such as choosing the air interface with the maximum bandwidth. Thus, the bandwidth of the whole QoS-aware HWAN route in Fig. 11.13 is

$$bandwidth(R) = \min(\max(bandwidth_1(j,k), \ldots, bandwidth_n(j,k)),$$
$$\max(drop_1(l,m), \ldots, drop_n(l,m))). \tag{11.8}$$

Therefore, the route selection in an intra-working HWAN system is normally decided by more than one QoS constraints. In order to satisfy the original design purposes of a HWAN system, some special constraints could be involved in the route selection. For example, the relaying routes in iCAR system only allows a special type of node named as ARS to be involved in the route establishment, and the communication interface through intermediate nodes is only an ad hoc interface. Through the whole route in iCAR, only the hop from ARS to the destination base station accepts the communication in cellular interface. However, another HWAN system called SOPRANO allows the communication in cellular interface at both the hop from the source node to ARS and the hop from ARS to the destination base station, so as to add more flexibility to the route construction. Thus, we can add more QoS constraints in an intra-working HWAN system, not only to satisfy the QoS requirement of users but also to improve the overall performance of the system. In other words, we can aim at selecting the optimum QoS-aware route, instead of only selecting a route fulfilling the QoS requirement.

Conclusion

In this chapter we have discussed how QoS-aware routing strategies can be used in HWAN to improve the performance of the heterogeneous network system. QoS-aware routing algorithms have been described in general heterogeneous wireless access networks. Routing algorithms proposed in heterogeneous wireless networks have been classified according to their routing structures and design objectives. Considering the QoS aspects in heterogeneous wireless networks (such as bandwidth guarantee, transmission delay, and route discovery successful rate), the QoS-aware routing issues have been discussed in general heterogeneous wireless networks. We have also presented the general structure and procedure of heterogeneous QoS-aware routing, which are divided into three main parts, namely, source selection, route selection, and destination selection. Each part of the QoS-aware routing structures has been discussed separately to show how they affect the whole routing process and fulfill the QoS requirements. In addition, some example algorithms have been also presented to describe the working process of source selection, route selection, and destination selection.

Some open issues include how the in-band frequencies are used by TDSs and how the placement of TDSs affects the performance of the overall systems in integrated systems. The impact of other design motivations (such as expansion of cell transmission range) on SSPs can also be involved in integrated HWANs. Further investigations are required to design QoS-aware gateways in inter-working HWANs.

Acknowledgment

Some parts of this work were presented at IET Proceedings - Communications 2007. We thank the Institution of Engineering and Technology for supporting our works.

References

1. H. Wu, C. Qiao, S. De, and O. Tonguz, "Integrated cellular and ad hoc relaying systems: iCAR," *IEEE Journal on Selected Areas in Communications*, vol. 19, no. 10, pp. 2105-2113, Oct. 2001.
2. S. Uskela, "Key concepts for evolution towards beyond 3G networks," *IEEE Wireless Communications*, pp. 43-48, Feb. 2003.
3. X. Masip-Bruin, M. Yannuzzi, R. Serral-Gracia, J. Domingo-Pascual, J. Enriquez-Gaberias, M. A. Callejo, M. Diaz, F. Racaru, A. Beben, and W. Burakowski, "The EuQoS system: A solution for QoS routing in heterogeneous networks," *IEEE Communication Magazine*, pp. 96-103, Feb. 2007.
4. W. Song, H. Jiang, W. Zhuang, and X. Shen, "Resource management for QoS Support in WLAN/Cellular interworking," *IEEE Network - Special issue on 4G Network Technologies for Mobile Telecommunications*, vol. 19, no. 5, pp. 12-18, 2005.
5. S. Pierre, M. Barbeau, and E. Kranakis, "Ad-Hoc, Mobile, and Wireless Networks," in *Proc. of ADHOC-NOW 2003*, Montreal, Canada, pp. 293, Oct. 2003.
6. C. E. Perkins and P. Bhagwat, "Highly dynamic destination-sequenced distance-vector routing (DSDV) for mobile computers," in *Proc. of ACM SIGCOMM*, Oct. 1994.
7. D. B. Johnson, D. A. Maltz, and Y.-C. Hu, "The dynamic source routing protocol for mobile ad hoc networks (DSR)," draft-ietf-manet-dsr-08.txt, IETF MANET Working Group, INTERNET-DRAFT, 24 Feb. 2003.
8. P. Van Mieghem et al., "Quality of service routing," Ch. 2, *Quality of Future Internet Services: COST 263 Final Report*, Springer-Verlag, Oct. 2003.
9. B. X. Zhang and H. T. Mouftah, "QoS routing for wireless ad hoc networks: Problems, algorithms, and protocols," *IEEE Communications Magazine*, pp. 110-117, Oct. 2005.
10. H. Luo, R. Ramjee, P. Sinha, L. (E.) Li, and S. Lu, "UCAN: A unified cellular and ad-hoc network architechture," in *Proc. of ACM MobiCom'03*, Sept. 14-19, 2003.
11. X. Wu, S.-H. G. Chan, B. Mukherjee, and B. Bhargava, "MADF: Moblie-assisted data forwarding for wireless data networks," *KICS/IEEE Journal of Communications and Networks*, vol. 6, no. 3, pp. 216-225, Jan. 2002.
12. J. Schiller, *Mobile Communication*. Second Edition, Addison Wesley, pp. 269-297, 2003.
13. D. Griffin, J. Spencer, J. Griem, M. Boucadair, P. Morand, M. Howarth, N. Wang, G. Pavlou, A. Asgari, and P. Georgatsos, "Interdomain routing through QoS-class planes," *IEEE Communications Magazine*, vol. 45, no. 2, pp. 88-95, Feb. 2007.
14. M. Boucadair, "QoS-enhanced border gateway protocol," draft-boucadir-qos-bqp-spc-01.txt,IETF Internet Draft, July 2005.
15. Y. Rekhter and T. Li, "A border gateway protocol 4 (BGP-4)," IETF RFC 1771, March 1995.
16. G. N. Aggelou and R. Tafazolli, "On the relaying capacity of next-generation GSM cellular networks," *IEEE Personal Communications*, pp. 40-47, Feb. 2001.
17. ETSI, "WB-TDMA/CDMA: System description performance evaluation," Tech. Rep., European Telecommunications Standards Institute SMG2-24, Madrid, Spain, Dec. 1997.

18. "3rd generation partnership project: Opportunity driven multiple access," 3GPP, Tech. Rep. Tech. Spec. 3G TR 25.924 version 1.0.0 (1999-12), 1999.

19. Y. Wu, K. Yang, and H-H Chen, "ARCA: An adaptive routing protocol for converged ad hoc and cellular networks," *KICS/IEEE Journal of Communications and Networks (JCN)*, Special Issue on Broadband converged Network (BcN), vol. 8, no. 4, pp. 422-433, Dec. 2006.

20. E. H. -K. Wu and Y.-Z. Huang, "Dynamic adaptive routing for a heterogeneous wireless network," *Mobile Networks and Applications*, vol. 9, no. 3, pp. 219-233, June 2004.

21. H. Holma and A. Toskala, *WCDMA for UMTS*, New York: Wiley, 2000.

22. A. N. Zadeh, B. Jabbari, R. Pickholtz, and B. Bojcjic, "Self-organizing packet radio ad hoc networks with overlay (SOPRANO)," *IEEE Communications Magazine*, vol. 40, no. 6, pp. 149-157, 2002.

23. H. Wu, S. De, C. Qiao, E. Yanmaz, and O. Tonguz, "Managed mobility: A novel concept in integrated wireless systems," *IEEE Journal on Selected Areas in Communications*, pp. 537-539, 2004.

24. Y. Wu and K. Yang, "Source selection routing algorithms for integrated cellular networks," accepted for publication in *IET Proceedings - Communications*, 2007.

25. L. Xu, X. Shen, and J. W. Mark, "Dynamic bandwidth allocation with fair scheduling for WCDMA systems," *IEEE Wireless Communications*, vol. 9, no. 2, pp. 26-32, Apr. 2002.

26. F. Bari and V. C.M. Leung, "Automated network selection in a heterogeneous wireless network environment," *IEEE Network*, vol. 21, no. 1, pp. 34-40, Jan.-Feb. 2007.

27. Q. Song and A. Jamalipour, "Network selection in an integrated wireless LAN and UMTS environment using mathematical modeling and computing techniques," *IEEE Wireless Communications*, vol. 12, no. 3, pp. 42-48, June 2005.

28. Y. L. Morgan and T. Kunz, "A design framework for wireless MANET QoS gateway," in *Proc. of SNPD/SAWN'05*, pp. 420-427, May 2005.

29. Y.-B. Ko and N. H. Vaidya, "Location-aided routing in mobile ad hoc networks," *Wireless Networks*, vol. 6, pp. 307-321, July 2000.

12

Congestion Control in the Wired-Cum-Wireless Internet

Shan Chen[1], Xiaojun Hei[2], Junhua Zhu[1], and Brahim Bensaou[1]

[1] Department of Computer Science and Engineering
[2] Department of Electronic and Computer Engineering
The Hong Kong University of Science and Technology, Clear Water Bay, Kowloon, Hong Kong
{chenshan, csjhzhu, brahim}@cse.ust.hk, heixj@ece.ust.hk

12.1 Introduction

The Internet has significantly penetrated our daily activities since it was first released to the public in the mid 1980's. In contrast to other proprietary or standardized network platforms, the success of the Internet is due mainly to the great success achieved by the TCP/IP protocol stack in providing a robust yet simple network platform. TCP (Transmission Control Protocol) [1] is designed to provide a reliable, connection-oriented service over best-effort packet-switched networks while the underlying congestion control mechanism ensures that Internet flows share the network resources efficiently and fairly. Perhaps, the fact that TCP operates at end-systems was a determining factor in the success of the Internet, as this allowed the rapid design and deployment of the routers that made the Internet backbone. In contrast, other more promising network technologies with more evolved service architectures did not live up to their promises because of the complexity that they involved. For decades, TCP has proved to be resilient under many traffic scenarios and has been widely accepted as being a very effective and robust congestion control protocol; thus, it has been predominantly deployed on almost every device connected to the Internet. A number of continuous measurement studies show that most of the traffic carried in the Internet rides on top of TCP [2, 3].

As the Internet evolves, TCP also faces many new challenges posed by emerging network technologies. A large proportion of these challenges arise from the ever-growing popularity of wireless networks, driven by the users demands for untethered Internet access. Compared to their wired counterparts, wireless networks have the following major advantages:

- The deployment of wireless networks is very easy because tedious configuration, civil engineering infrastructure and placement planning procedures, which are usually needed in wired networks, can be to a large extent avoided.

E. Hossain (ed.), *Heterogeneous Wireless Access Networks*,
DOI: 10.1007/978-0-387-09777-0_12, © Springer Science+Business Media, LLC 2008

- Under certain circumstances wireless Internet access is more economical than the wired approach. For example, in remote villages with no wired network infrastructure, it is difficult to access the Internet. In this case employing a satellite link would be cheaper and more effective than deploying wires from the only available ISP center far away.
- Wireless networks can provide unconstrained network access, with possible user mobility. For example, in cellular networks users can stay connected even on a fast moving vehicle.

In a nutshell, the emergence of wireless networks, including satellite Internet links, wireless hot-spot (WLAN), mobile ad-hoc network, 3G cellular network, high-speed down-link packet access (HSDPA), and so on, has made the Internet a wired-cum-wireless communication platform: while the backbone of the network relies on wired, often optical links, access networks are relying more and more on wireless technologies.

Despite the advantages they bring about to Internet communication, wireless links suffer also from many disadvantages compared to their wired counterparts. Wireless communications take place in an open transmission medium which is subject to various environmental interferences and noises. In addition, the same wireless spectrum is often shared by many wireless nodes, resulting in multi-user interference and collisions. Due to various factors, such as, the interference, channel fading, hidden nodes, user mobility, and so on, non-congestion-related packet losses are very frequent in wireless networks. Furthermore, while wired network bandwidth evolved homogeneously across all the domains of the network[3], wireless networks bandwidth varies across a large spectrum and can be as narrow as several kbps in GSM cellular network, or as broad as hundreds of Mbps in wireless metropolitan area network (WMAN). This contributes significantly to increasing the bandwidth-delay product in the network. Finally, due to handoff and other factors, there could also be temporary disconnections in wireless networks, which affect the operation of the TCP congestion control mechanism. To effectively utilize network resources and achieve good fairness in such a heterogeneous wired-cum-wireless environment, new challenges are posed to the conventional TCP congestion control mechanisms.

In this chapter we discuss various TCP congestion control issues in the wired-cum-wireless Internet, review some of the challenges, and survey the related literature. We do not intend to present a comprehensive list of all the proposed schemes; instead, we provide a summary of major research efforts and highlight the representative approaches to address different issues. We categorize the state-of-the-art research progress with respect to the major issues faced by TCP congestion control in the wired-cum-wireless Internet. It is worth noting though that there is already a number of survey papers in the literature that deal with this subject [4–8]. Our work differs from the others in that it reviews the literature according to the issues addressed by the different proposed solutions and focuses on introducing what causes

---

[3]When the backbone bandwidth increased because of improved switching capacity to terabits per second, wired access networks bandwidth also improved with the invention of newer/better access methods, say Ethernet 100 Mbps, Gigabit Ethernet, xDSL, and so on).

the big challenges to congestion control in the wired-cum-wireless environment, and the possible approaches that can be applied to each of these new challenges. This can help the reader to understand the issues and solutions more comprehensively. And for readers who are interested in any one of these issues in particular, they can just focus on the proposed approaches to address it.

12.2 Wired-Cum-Wireless Internet: The Big Picture

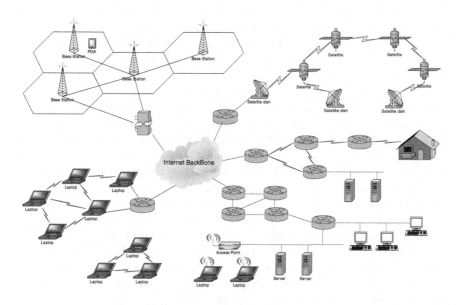

Fig. 12.1. A picture of the wired-cum-wireless Internet.

In only about 20 years, the Internet has grown into the most powerful knowledge and information repository, an extremely popular communication and entertainment media, and a platform on which practitioners have developed various networking applications. Driven by the need of accessing the Internet any time at any place, diverse access media and techniques have been developed. Due to the low deployment flexibility and the mobility potential they introduce, the wireless medium has attracted great attentions. Wireless telecommunication technologies, such as WiFi (IEEE 802.11) [9], WiMAX (IEEE 802.16) [10], 3G cellular networks [11], and even satellite access, are commercially very popular nowadays, and the demand for such access methods are forecast to grow in the near future. A generic big picture of what the wired-cum-wireless Internet will look like is shown in Fig. 12.1, while most of the backbone network is constructed on wired medium such as optical fiber with optical switches, access networks will be more and more based on wireless technologies.

12.2.1 WiFi (IEEE 802.11)

Among these wireless technologies, WiFi turns out to be the most successful and widely deployed. WiFi is based on the family of IEEE 802.11 standards [9] for wireless local area network (WLAN) data communications in the 2.4 GHz and 5 GHz public spectrum of the industrial, scientific and medical (ISM) radio bands. The IEEE 802.11 family includes over-the-air physical modulation techniques that use the same basic MAC protocol. IEEE 802.11 defines two medium access modes: the point coordination function (PCF) and the distributed coordination function (DCF). The PCF is a contention-free protocol that operates on top of the contention based DCF protocol. In the PCF the access point (AP) controls the access to the channel using a round robin scheduling. The DCF is contention-based protocol that uses carrier sensing multiple access with collision avoidance (CSMA/CA). Because of its superior effectiveness the DCF is much more commonly implemented in commodity WiFi APs.

In the DCF, stations compete for channel access to transmit data to the AP (uplink), while the AP also participates in this contention to transmit data to the stations (downlink). This contention-based MAC is often referred interchangeably as IEEE 802.11 MAC or CSMA/CA that tries to extend the Ethernet principles to the wireless medium. It uses binary exponential back-off to reduce collision probability among competing transmissions. After gaining the channel access, a frame-exchanging protocol needs to be followed to complete the transmission. Before transmitting a DATA frame, the sender and the receiver exchange optional Require-To-Send and Clear-To-Send frames (RTS/CTS) to reserve the channel thus avoiding unnecessary collision. After transmitting a DATA frame, the IEEE 802.11 sender waits for the receiver to send back an ACK frame to confirm that the DATA is successfully received. Otherwise the sender attempts to retransmit the DATA frame (after competing for the channel) for a fixed retransmission limit after which the DATA frame is simply dropped at the MAC layer if not successful.

Among the IEEE 802.11 family, IEEE 802.11b and IEEE 802.11g, which operate in the 2.4 GHz frequency band and provide a maximum data rate of 11 Mbps and 54 Mbps respectively, are the most popular today. Operating on the 5 GHz ISM band, IEEE 802.11a hasn't gained much popularity even though a maximum raw data rate of 54 Mbps is supported. This is due to the fact that the higher frequency leads to more difficulties for the signal to penetrate walls and other obstructions, thus limits the range of IEEE 802.11a networks. At the mean time, IEEE 802.11n, which is designed to provide a maximum data rate about 200 Mbps with improved signal range, is yet to be standardized.

A WiFi-enabled device, such as a laptop, game console, cell phone, MP3 player or PDA, can connect to the Internet within the range of an AP. The area covered by one or more interconnected access points is called a hotspot. Hotspots can cover as little as a single room with wireless-opaque walls or as much as many square miles covered by overlapping APs. Researchers also propose to form ad-hoc network to transmit data among stations using the DCF mode of IEEE 802.11, under the circumstances where AP-based infrastructure networks are not applicable, or in the situations when the coverage of APs is not large enough.

12.2.2 WiMAX (IEEE 802.16)

WiMAX is based on the IEEE 802.16 [10] standard for the deployment of wireless metropolitan area networks (WMAN). The WiMAX Forum describes WiMAX as "a standards-based technology enabling the delivery of last mile wireless broadband access as an alternative to cable and DSL." The original IEEE 802.16 uses the 10-66 GHz band with only a line-of-sight (LOS) capability. Later a specification on the use of the 2-11 GHz band was added. WiMax can provide high data rate over a long distance. The optimal data rate with WiMAX is about 70 Mbps when the range is small. WiMax can operate over a maximum range of 50 km by sacrificing the data rate. As a rule of thumb, the bit error rate (BER) increases with an increasing operation range. In practice, the data rate of 10 Mpbs at 10km could be reached with LOS. With this last mile wireless broadband access, the network deployment in urban areas can be simplified dramatically, in that a WiMAX router can simply be placed on the roof of each building to avoid tedious and costly civil engineering and cable layout work down the street. RoofNet is one example project [12].

12.2.3 Bluetooth

Bluetooth [13] is an industrial specification for wireless personal area networks (WPANs). Bluetooth provides a cordless way to connect and exchange information between devices such as mobile phones, laptops, printers, and digital cameras, and so on within a short distance. Bluetooth operates on the same 2.4 GHz ISM band as used by 802.11b/g. However, by adopting a frequency hopping spread spectrum (FHSS) multiplexing method, Bluetooth does not interfere with 802.11b/g in theory. Bluetooth and WiFi have slightly different applications in today's offices, homes, and on the move. While Bluetooth is meant as a cable replacement for a variety of applications within a personal area, thus the name wireless personal area networks (WPAN), WiFi is a cable replacement only for local area network access. WiFi provides higher data communication rate and covers greater distances, but requires more expensive hardware and higher power consumption. Bluetooth is often thought of as the wireless USB, whereas WiFi is the wireless Ethernet.

12.2.4 3G Family

The wireless wide area network (WWAN) infrastructure is nowadays more and more provided by the so-called 3G [11] wireless networks. Through improved spectral efficiency, in 3G networks the data rate available to a customer is higher than that provided by 2G cellular networks, thus Internet access and video telephony are becoming a reality on 3G cellular phones. With the advent of high-speed data service such as high speed downlink packet access (HSDPA) and high speed uplink packet access

(HSUPA), network access bandwidth on 3G cellular phones can reach megabits per second, making bandwidth-intensive applications and services, such as multimedia or video conferencing, more and more pervasive.

12.2.5 Satellite Networks

When the deployment of all the aforementioned networking technologies is not feasible, like on airplanes or sea-ships or in remote locations, one can still utilize satellite networks to access the Internet. There are commercial satellite ISPs, like HughesNet and WildBlue, that provide a high-speed Internet access with a download speed up to about 1.5 Mbps and an upload speed of several hundred Kbps [14]. Inter-networking with satellites began with the use of individual satellites in geosynchronous earth orbits (GEO). However, the requirements for lower propagation delays and propagation loss, in conjunction with the coverage of high latitude regions for personal communication services, have sparked the development of new satellite communication systems, namely, the low earth orbit (LEO), and medium earth orbit (MEO) satellite systems. In these systems, constellations of satellites form networks to relay packets for each other. Satellite links serve well where DSL and WiMAX do not apply and provide more agility for accessing the Internet.

12.2.6 Summary

Fig. 12.2 briefly summarizes two key link characteristics of popular wireless technologies, more specifically, the link rate vs. the signal range. The figure is only for references and does not mean to be very accurate and precise. For example, the figure doesn't include the fact that as the transmission distance increases, the achievable wireless link rate decreases, as usually more robust but slower physical modulation schemes are used to compensate for the signal to noise ratio (SNR) at the receiver. From Fig. 12.2 we can see that wireless technologies support a wide range of link rates and communication distances. Under different scenarios different wireless technologies can be adopted to provide satisfactory Internet access services.

The heterogenous Internet access technologies provide a very fertile ground for the development of the Internet into a global wired-cum-wireless service network with many new and exciting applications. However, the change of the underlying communication media from wired to wired-cum-wireless poses new challenges to the Internet design. In this survey we specifically focus on the consequent impact on the TCP congestion control. TCP congestion control was designed and later modified for wired networks. Since certain assumptions behind existing TCP's highly tuned algorithms do not hold any more for the emerging wireless networks, a revisit of the current TCP congestion control designs is required. Intensive research and development efforts have been directed into this area to meet the challenges of TCP in wireless networks.

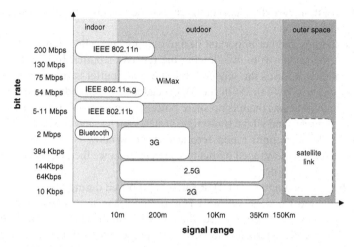

Fig. 12.2. Link characteristics of different wireless technologies.

12.3 Challenges in Congestion Control for Wired-Cum-Wireless Internet

In this section we review major challenges to the conventional congestion control mechanisms raised by the heterogenous wired-cum-wireless Internet environment. We first review briefly the traditional congestion control mechanism of TCP, then describe the different problems that arise due to the wireless component of the network.

12.3.1 Conventional TCP-Based Congestion Control

TCP was designed to provide congestion control on IP networks with the objective of achieving good efficiency while maintaining a relatively fair service for competing flows.

The congestion control algorithm in the conventional TCP operates in two phases: the "slow-start", a probing phase, where the sender tries to obtain an initial estimate of the bandwidth available to its connection, and the "congestion avoidance" phase which typically applies an additive-increase multiplicative-decrease (AIMD) rate adjustment algorithm to maintain the sender rate around the available connection bandwidth. Implemented as a sliding-window protocol, TCP congestion control maintains two state variables at the sender – congestion window ($cwnd$) and slow-start threshold ($ssthresh$) – to regulate the transmission rate: $cwnd$ controls the number of packets a connection can have in flight during a round trip time (RTT). Implicitly, $cwnd$ indicates the sending rate of the connection; $ssthresh$ sets the value of $cwnd$ at which TCP switches between the slow-start phase and and the congestion avoidance phases. In conventional TCP, packet losses are regarded as indications of network congestion. Usually a packet loss is detected when the sender receives three duplicate ACKs [4]

In TCP there is another important design component, called flow control, which is to prevent the sender from overflowing the receiver's buffer. In flow control, the receiver explicitly informs the sender about the spare buffer at the receiver through the "advertised window" fields in the TCP header. If a zero advertised window is received, which indicates there is no spare buffer at the receiver, the sender is forced into a *"persist mode"* and all the transmission states such as *cwnd*, *ssthresh*, and even RTO timer are frozen, until a non-zero advertised window is received. Such flow control mechanism is leveraged by many works to enhance the performance of TCP as will be introduced latter.

There have been a number of TCP variants proposed during the evolution of the protocol. In the most widely deployed TCP protocol, the so-called TCP Reno [15], when a new connection is established, the *cwnd* is initialized to one maximum segment size. The sender starts in the slow-start phase, where for each successfully received acknowledgement (ACK) packet, *cwnd* is increased by one full segment, thus *cwnd* doubles each RTT. The sender continues to increase *cwnd* exponentially in the slow-start phase, until *cwnd* reaches the *ssthresh*; then the sender enters the congestion avoidance phase. During this phase, the AIMD algorithm is invoked. In the congestion avoidance phase, *cwnd* is increased by a fraction of a segment proportional to the acknowledged data for each ACK, in the absence of packet losses (which corresponds to the AI part of AIMD). This *cwnd* increase adds up to one segment increase in each RTT. When packet losses are detected by triple duplicate ACKs, the congestion window is reduced by half (which corresponds to the MD part in AIMD), and *fast recovery* is invoked to recover the loss. Generally speaking, AI serves for bandwidth probing while MD applies a conservative measure meant to mitigate network congestion and to quickly prevent it from becoming more dramatic. The AIMD algorithm can be shown to possess a fairness property [16].

Despite the rapid evolution of the Internet, the core TCP/IP stack remains almost the same as 20 years ago. Until now, TCP Reno [15] and TCP NewReno [17] are still the dominant reliable transport protocols [18], on top of which the majority of Internet applications run.

12.3.2 Wireless Random Packet Losses

In almost all wireless environments, TCP congestion control faces the common challenge of high BERs and dynamic channel conditions. Unlike in wired communications where the communication medium is shielded from environment induced errors, wireless communications are carried out in the open air and are more sensitive to the surrounding environment interferences, such as, wind, weather, trees, walls and

---

[4]On arrival of each (or every the other) DATA packet from the TCP sender, the TCP receiver needs to reply with an ACK packet indicating the next packet expected from the sender. If duplicate ACKs are received by the sender, it means there is some packet missing at the receiver.

so on. When using the unlicensed radio spectrum, such as the ISM band, it is very likely that the channel is used simultaneously by by many other wireless devices or wireless nodes from other networks using the same frequency band. For example, 802.11b/g equipments may occasionally suffer interference from microwave ovens, cordless telephones, and Bluetooth devices. TCP was originally designed for a wired medium, which has a BER in the order of 10^{-6} to 10^{-8}, which translates into a packet loss rate of about 1% to 0.01% for 1500-byte packets. Due to channel fading, interference, and environmental noise, the BER is much higher in wireless environments; typically in the order of 10^{-3}, and sometimes can be as high as 10^{-1} [4].

With such a high packet loss rate, the conventional TCP congestion control mechanisms perform extremely poorly due to two major reasons: 1) lost packets need to be retransmitted; 2) with such a high packet loss rate, and without distinction between congestion losses and wireless random losses, the TCP congestion window is throttled unnecessarily, which leads to under-utilization of network bandwidth. This problem has been investigated extensively and many schemes have been proposed to improve congestion control in lossy wireless environments.

12.3.3 Mobility Support

One big advantage of wireless networks over their wired counterparts is that they permit hosts to move freely within a local or wide area network. However, the coverage area of a base station, in cellular networks or an access point, in WiFi hotspots, is usually limited. User mobility often leads to inevitable handoffs when users go over the coverage boundary between base stations. Depending on the technology used the ongoing communication might incur large RTTs or short durations of disconnection during the handoff. Similar disconnections may also occur when radio signals are blocked off by buildings and other similar obstacles, or simply mobile hosts move out of the coverage range of the base station temporarily. TCP treats sudden increased delays and losses caused by these short disconnection periods as congestion signals and triggers congestion back-offs. This leads to a reduced throughput and a loss of the connection state information. After the handoff procedure is completed, TCP needs to restart the bandwidth probing procedure to acquire its bandwidth share. The situation deteriorates if after the handoff the mobile node is allocated with a new IP address in a new network, as the ongoing TCP connection breaks up. Mobile IP [19] has been proposed for this issue to ensure continuous connectivity when the mobile hosts roams between different wireless networks; however, it is still not in widespread use nowadays.

To support mobility in wireless mobile networks, the user should be able to roam freely and enjoy network services without interruptions. Of course this requirement drives the underlying architecture to provide negligible handoff-related disconnection periods and delays and more properly to support Mobile IP. It is also desirable that the transport protocol is able to distinguish between handoff-related and congestion-related losses and delays; and resume data transmission state right after the handoff procedure.

12.3.4 High Bandwidth-Delay Product in Satellite Networks

In networks with high bandwidth-delay product, TCP congestion control faces an inefficiency issue. This inefficiency arises from the conservative nature of TCP's bandwidth probing mechanism with the AIMD algorithm. TCP injects successively one more packet in the network for each RTT, until congestion is detected. When the bandwidth-delay product is large, this probing procedure lasts for a long period before congestion occurs; this leads to severe under-utilization of link bandwidth. For example, when the RTT is about 500ms, increasing the $cwnd$ by 10 segments takes about 5 seconds.

In general, the bandwidth in wireless networks is less than that in the wired networks; however, some wireless networks, such as satellite networks, still exhibit high bandwidth-delay products because of the length of the wireless link. Satellite network delays are influenced by several factors, the main one being the orbit type. In most LEO satellite systems, one-way propagation delays are in the 20-25 ms range. These values increase to 110-150 ms in MEO and arises up to 250-280 ms for GEO satellites [20]. Nevertheless, since LEO systems use lower orbits, visibility windows are relatively small; hence, LEO satellite networks formed by constellations of satellites need to be formed to provide continuous coverage over large regions [21]. This leads to typically very large end-to-end delays; the actual delay varies depending on the number of satellite links spanned by the end-to-end path as well as orbital dynamics along the connection. Such high bandwidth-delay product in satellite networks requires modifications to the TCP congestion control mechanism in order to achieve high utilization on the expensive links. In this particular context, the exponential slow-start phase of TCP tends to be less effective in grabbing bandwidth, e.g., on a GEO satellite link it takes about 4 seconds for slow-start to reach a bandwidth of 1 Mbps [22].

12.3.5 Rate-Adaptive Wireless Links

The IEEE 802.11 physical layer specification provides several largely different data rates (differing in modulation schemes and coding rates) to be adopted dynamically according to channel conditions in an effort to achieve good signal-to-noise ratio (SNR) at the receiver. A rate-adaptation technique, in general, switches the link rate dynamically to match the wireless channel conditions, with the goal of selecting an optimal rate to maximize the throughput for a given channel condition. It is expected that rate-adaption with IEEE 802.11 enhances the data delivery ratio. However, the performance of TCP degrades dramatically when available bandwidth fluctuates. This problem is bound to become more important as other wireless standards, such as IEEE 802.16, adopt rate-adaptation as well.

12.3.6 Fairness and Efficiency in Mobile Ad-hoc Networks

Mobile ad-hoc networks pose a series of new challenges to the TCP congestion control mechanism. More serious than in single-hop wireless networks, these challenges include frequent route changes and failures, hidden and exposed stations, topology-related unequal channel access opportunity, and unpredictability of the end-to-end transmission delay.

In mobile ad-hoc networks, nodes may join and leave the network at wish during the network life time. This node characteristic leads to frequent route changes and failures. During route changes and failures, it is possible that some packets in transmission are lost and a temporary disconnection is experienced. As a result, TCP suffers from a degraded performance, similar to the cases of wireless random losses and handoff disconnections. It is worth noting that route changes and failures may also result in packet reordering, which in turn triggers the duplicate-ACK response. When such packet reordering is severe, TCP timeouts may trigger prematurely and reduce the throughput dramatically. Packet reordering may also occur when multi-path routing is adopted as proposed by many researchers. The unpredictable end-to-end transmission delay may also lead to false triggers of the retransmission timeout by the RTO timer at the TCP sender, thus reducing the link utilization dramatically.

Hidden and exposed stations can significantly decrease the efficiency and impact fairness in mobile ad-hoc networks. Hidden nodes are typically those nodes that are within interfering range of an intended receiver but out of the sensing range of the transmitter. In such a case, the receiver may not correctly receive the intended packet due to collisions from the hidden node. An exposed node is the one that is within the sensing range of the transmitter but out of the interfering range of the receiver. Though its transmission does not interfere with the receiver, it can not start transmission because it senses a busy channel; hence, spatial reuse cannot be achieved efficiently.

In a mobile ad-hoc network, the network topology, which is uncontrollable, also has a significant impact on a node's channel access opportunity. Typically nodes at the periphery of the network enjoy more access opportunities that those in the center. This is because a node placed in a high density area has to contend with more stations to gain access to the transmission medium. As a result, TCP flow with many hops are likely to cross such dense areas and thus suffer tremendously low throughput due to long RTTs and high packet dropping probability.

12.4 State-of-the-Art Congestion Control Enhancements for Wired-Cum-Wireless Internet

In Section 12.3, we have highlighted several major challenges on improving the performance of TCP in the wired-cum-wireless Internet. In this section we review the literature with the emphasis on how these research works address the fore-mentioned challenges.

12.4.1 Wireless Random Packet Losses

Inherently, almost all wireless channels suffer high BERs. To improve wireless network performance, technique such as forward error correction (FEC) and automatic repeat request (ARQ), are widely used. FEC techniques rely on adding redundant bits to the transmitted payload to enable the receiver to recover the original data even when there are some error bits in the received data. Due to its high computational and bandwidth overhead, FEC is not widely used in wireless networks. It is often invoked for protecting small amounts of control information such as packet headers. ARQ, commonly used in IEEE 802.11, is meant to provide reliability at the link layer over an unreliable channel. It rectifies physical layer introduced bit errors by retransmissions. Its aim is to provide upper-layer protocols with the same dependable communication channel as that in wired links. FEC and ARQ cannot thoroughly solve the problem of wireless random losses. There have been extensive research efforts to address this issue and improve congestion control in presence of random wireless losses. These schemes can be classified into three broad categories: proxy-based, end-to-end, and network-cooperation approaches.

Proxy-Based Approach

In the proxy-based approach, the end-to-end path is divided by a proxy into two segments: wired and wireless. Usually this proxy helps to make wireless random packet losses transparent to TCP congestion control. The advantage of this approach is that it leaves the TCP/IP implementations in wired networks unchanged. This allows the quick deployment of wireless links for Internet access without re-constructing the Internet architecture at large. Most proposals in this category assume a last-hop wireless scenario, such as wireless LAN and cellular networks. In non-last-hop wireless networks, the proxy-based approach encounters some difficulties.

I-TCP: Indirect TCP [23], proposed in 1995, is one of the earliest protocols in this category. More specifically I-TCP introduced a split-connection method to combat wireless random losses. The main idea is to split the connection between a fixed host (FH) in the wired network and a mobile host (MH) in the wireless network into two at the base station (BS)[5]. The BS serves as a relay proxy for the MH; therefore, one TCP connection is set up between the proxy; the other connection is established for the data transmission between the proxy and the MH. The latter connection is over a one-hop wireless link, and the traffic does not need to run on top of TCP. Instead, new protocols that tolerate random wireless losses can be adopted for optimized performance over this wireless link. A packet sent to the MH is first received by the proxy; the proxy is authorized to send back the corresponding ACK to the FH even before forwarding the packet to the MH. Now it is the responsibility of the proxy to make sure the packet is delivered to the MH promptly and properly. Thus, the proxy needs to buffer packets till they are successfully delivered to the MH.

The major advantage of the proxy-based approach is that it separates congestion control functionality on wireless links from that on wired networks. This provides the opportunity to use another lightweight transport protocol, which is robust to wireless random losses, on the one-hop wireless link. However, I-TCP violates the end-to-end semantics of TCP. When bandwidth asymmetry exists between the wired and wireless path, a huge buffer is required at the proxy to store and forward packets toward

---

[5]The BS could be an access point in WLAN or a cell tower in cellular networks.

the MH. In addition, the control overhead with I-TCP is considerable as the BS needs to maintain a significant amount of state information for each TCP connection. During hand-offs, long delays may occur when the state information is transferred from the old BS to the new BS.

snoop: The "snoop" protocol [24] is a proxy-based TCP-aware link layer enhancement to improve the performance of TCP in presence of wireless random losses. In snoop, packets are cached at the BS and are locally retransmitted across the wireless link when necessary. When a TCP connection is established between a FH and a MH, a snoop agent, residing on the BS, monitors the packets flowing over the wireless link as well as related ACKs, and maintains a cache of unacknowledged packet destined to the MH. When a packet loss is detected at the snoop agent (i.e., through capturing duplicate ACKs generated by the MH), the cached copy is used for local retransmission, and duplicate ACKs are suppressed instead of being forwarded to the FH. snoop also uses timeouts to locally retransmit packets if needed. These timeout intervals are set to be much less than the RTO value of the TCP connection. Therefore, if a local timeout expire, a local retransmission is triggered within the time span of a TCP RTO.

For data transmission from the MH to the FH, the BS keeps track of the packets that are lost on the wireless link. The BS then generates negative acknowledgments (NACKs) for those packets back to the MH. The MH uses these NACKs to selectively retransmit these packets to recover from loss locally.

The snoop protocol maintains the end-to-end semantics of the TCP connection and is effective in hiding random losses on wireless links from TCP congestion control. Unfortunately, snoop requires that the RTT at the wireless link should be small enough to allow multiple retransmissions on wireless links before the sender RTO triggers. Yet, snoop fails if IP payloads are encrypted.

End-to-End Approach

The end-to-end approach consists in distinguishing wireless random losses from congestion losses at end-systems, allowing congestion control to only respond to congestion losses. The end-to-end approach requires no explicit modifications and supports at intermediate routers, which are out of the control of the end users. Schemes in this approach can be deployed easily and gradually; hence, this approach has attracted extensive research attention.

ELN: In the explicit loss notification (ELN) scheme [25], the TCP sender is informed of the reason of packet losses by the receiver. If a packet loss is not due to congestion, future cumulative ACKs corresponding to this lost packet are marked in the ELN bit to identify that a wireless random loss has occurred. Upon receiving this ELN information, the sender performs retransmissions without invoking congestion control. One of the drawbacks of this scheme is the difficulty for the receiver transport layer to identify wireless losses.

WTCP: The wireless TCP (WTCP) [26] is designed to provide an efficient and reliable transport protocol for wireless wide area networks (WWANs) that exhibit low bandwidth, large RTT, asymmetric channel and non-congestion related packet loss. WTCP uses similar connection management and flow control mechanisms as the standard TCP, however, it adopts a different, rate-based congestion control mechanism. In WTCP, the receiver is responsible for computing the desired transmission rate based on the network condition on the forwarding path. WTCP uses inter-packet delay as the prime congestion signal for congestion control. Upon the arrival of each incoming packet, the receiver determines whether to ask the sender to increase, decrease or keep the sending rate to avoid congestions in the network.

WTCP identifies congestion losses from wireless random losses using a heuristic approach. When an out-of-order packet is received indicating there are several packets missing, the receiver computes the average inter-arrival delay for each pair of these missing packets, through dividing the packet absence period with the number of missing packets. If this value is close to the inter-arrival delay measured when there is no congestion, the receiver treats them as random losses and signals the sender to continue transmitting at the current rate. Otherwise, the receiver predicts that there was at least one congestion loss, and informs the sender to reduce the sending rate. WTCP adopts the selective acknowledgment (SACK) [27] to achieve reliable transmission while avoiding the usage of RTO timer. The authors argued that one of the main reasons for the sub-optimal performance of TCP in WWANs is the inaccurate RTO estimation. The receiver-generated cumulative ACKs also carry SACK information. WTCP also modifies the startup behavior to benefit short flows in large RTT WWAN's. WTCP attempts to use the "packet-pair" method [28] to compute the appropriate transmission rate for a connection immediately upon startup instead of going through the slow-start phase.

While WTCP follows the end-to-end design principle of TCP it requires dramatic changes to TCP at both end systems. It could however be used effectively only on the wireless part of the connection via a split-connection method as described in I-TCP [23]. Nevertheless, WTCP is more complicated than the conventional TCP and requires considerable additional workload at the receiver which may not be desirable in battery driven receivers such as cellular phone in cellular networks.

TCP Westwood family: TCP Westwood [29] is a sender-side modification of the TCP congestion control algorithm that takes advantage of an end-to-end bandwidth estimation to make TCP function better in lossy wireless networks. Namely, TCP Westwood is based not solely on packet loss, which itself is an ambiguous congestion indicator in presence of wireless links, but also on the bandwidth estimation at the time of loss. A TCP Westwood sender continuously estimates the source rate of the connection by averaging the rate of returning ACKs with a discrete low-pass filter. Upon detection of a packet loss, the $cwnd$ and $ssthresh$ are set to the estimated effective congestion window size, which relieves the congestion while keeping the link utilization high. The bandwidth estimator is the key component in the Westwood TCP design. It was reported that the original TCP Westwood bandwidth estimator (BE) [29] over-estimates the fair share of the bandwidth when competing with other flows that use TCP Reno. To solve this problem a rate estimator (RE) was introduced in [30]. RE tends to be closer to the achieved rate of a connection and works better when packet losses are mostly due to congestion, while BE provides better performance when packet losses are mostly due to link errors. A hybrid scheme combining the two estimators (called CRB) was also proposed in [31]. Yet another bandwidth estimator, called TIBET [32], was proposed to achieve accurate and fair bandwidth share estimation while being resilient to ACK compression and traffic clustering by monitoring the packet departures at the sender instead of the ACK arrivals.

Clearly the common goal with these estimators is to obtain the fair bandwidth share of a connection, which is a dynamic point and depends on the behavior of the TCP connection itself, thus is difficult to determine at the end system without a complete knowledge of the whole network. The options taken by these estimators are to estimate the achieved throughput, or the bottleneck link capacity, or even a value somewhere in between these two. None of these can be properly related to the fair bandwidth share, especially in presence of wireless random losses. Ironically the most aggressive estimator, BE, which overestimates the bandwidth share in normal cases, tends to have a better performance than other estimators in lossy wireless environments because of its aggressiveness. The performance of these congestion control protocols is still sub-optimal in combating wireless random losses.

NewReno-FF: NewReno-FF [33] was proposed based on the observation that if the connection is suffering congestion losses, the measured RTTs vary significantly; however, if there is only random losses, the RTTs vary little. In NewReno-FF, the flip flop (FF) filter is used to estimate the average and the variation of a connection's RTT. The flip flop filter uses two exponential weighted moving average (EWMA) filters on RTT samples: one is stable and the other is agile. The agile filter weights more on recently observed samples in contrast to the stable filter. The underlying idea is to employ an agile filter whenever possible but switch to the stable one when the RTT samples vary drastically. NewReno-FF estimates the ratio of RTT samples that deviate from the mean in a certain number of recent samples. If the ratio is larger than a predefined threshold, the packet loss during this period is treated as congestion loss and the normal NewReno congestion control operations are triggered; otherwise, the loss is treated as random loss, in which case no rate reduction is performed. It is argued that NewReno-FF can achieve 100% congestion loss prediction and high prediction accuracy of random losses. However, if link layer retransmission schemes or rate adaptation techniques are adopted by the wireless links large RTT variation would prevail without congestion.

TCP Veno: TCP Veno [34] is another sender side enhancement to TCP Reno for transmission over wireless networks. The idea is to use TCP Vegas congestion detection mechanism [35] to distinguish congestion loss from random loss. When a packet loss is detected, TCP Veno further checks the difference between the current flow rate and the expected flow rate to determine the type of loss.

To infer congestion build up, TCP Vegas [35] computes the difference between the *actual input rate* $(cwnd/\tau)$ and the *expected rate* $(cwnd/\tau_{min})$, where τ is the current estimated RTT and τ_{min} is the minimum RTT. When this difference is larger than a pre-specified threshold, TCP Vegas interprets that there is certain congestion along the path and reduces the sending rate. If the difference is less than another pre-specified threshold, it is inferred that little congestion exists along the connection path and TCP Vegas sender increases the sending rate. According to *Little's Law*, this difference essentially reflects the number of packets a connection has queued in the buffers along the path. TCP Vegas attempts to keep a small and constant flow queue size so as to achieve efficiency and avoid severe congestion at the intermediate routers. However, with the consideration of co-existence with conventional loss-based TCP, TCP Vegas also responds to packet losses the same as TCP Reno, thus equally suffer from wireless random losses.

TCP Veno follows the same mechanism as TCP Vegas to detect congestion level and pushes it further to infer loss type. When a loss is detected, if the estimated difference of the actual input rate and expected rate is beyond a certain threshold, it is probably a congestion loss, TCP Veno then halves the *cwnd* just as Reno does; otherwise, the loss is treated as a wireless random loss, thereafter, TCP Veno sets the *ssthresh* and *cwnd* to 4/5 of the current *cwnd*, and moves into the congestion avoidance phase. To deal with route changes, TCP Veno resets the minimum measured RTT whenever a packet loss is detected.

During the congestion avoidance phase, if the actual input rate is smaller than the expected rate by a predefined threshold, available bandwidth is not fully used; a Veno sender advances the *cwnd* by *1/cwnd* when each new ACK is received. Otherwise, the *cwnd* is advanced by *1/cwnd* every other newly received ACK.

It was shown that TCP Veno improves throughput when the random loss rate is moderate (in the order of 0.01) [34]. However, TCP Veno cannot perform well in high BER environments, as for each packet loss the congestion window is reduced somehow. When the loss is frequent, Veno cannot increase the congestion window at all. As a delay-based sender side modification of TCP, TCP Veno is also sensitive to the reverse traffic and delay variations contributed by factors other than queue dynamics, such as link layer retransmissions.

JTCP: Jitter-based TCP (JTCP) [36] is a jitter-based, end-to-end TCP modification to address random loss in wireless networks. JTCP makes use of the jitter ratio to determine the congestion level of the network thus to differentiate wireless random losses from congestion losses.

Based on information from the receiver, the JTCP sender estimates the inter-arrival jitter for a pair of packets as the difference between the inter-departure time of two packets and their inter-arrival time at the receiver. The average jitter ratio is calculated as the ratio between the sum of jitters for any two consecutive packets ACKed during a certain period and the duration of this period. Usually the computation time span is several RTTs. The average jitter ratio reflects the ratio of the queued packets in some sense. When the average jitter ratio is less than a pre-specified threshold, no congestion is inferred along the path. When a packet loss is detected, JTCP first checks whether the average jitter ratio is significant; if it is, the loss is considered to be due to congestion and the conventional TCP congestion control operations are performed. Otherwise, the loss is regarded as a random loss.

JTCP treats the inter-arrival jitter on the forwarding path as a congestion indication, hence is resilient to reverse traffic effects. However, this conjecture does not hold when the jitter is due to link layer retransmissions. This turns out to be a common drawback for delay-based loss discriminators.

ZBS: In NewReno-FF, TCP Veno and JTCP, the key idea is to rely on some loss discriminators to tell wireless random losses from congestion losses, and respond to them differently to improve efficiency. This technique can also be used in TCP-friendly rate control (TFRC) algorithms.

Cen et al. [37] extended TFRC with loss differentiation algorithms to improve the performance when there are wireless links in the path between the sender and the receiver. TFRC [38] is an equation-based rate control algorithm that ensures UDP traffic responsiveness to network congestions and only sends traffic at a TCP-friendly rate.

The hybrid loss discriminator, ZBS, proposed in [37] turned out to yield high wireless and congestion misclassifications [39]. Its accuracy also depends on the number of flows sharing the bottleneck.

Congestion predictors as loss discriminators: As pointed out in [37], many end-to-end schemes that detect congestion level in the network, can be used as loss discriminators to filter out wireless random losses from congestion losses. TCP Veno [34] uses the Vegas congestion predictor; while TFRC can rely on the ZBS congestion predictor [37]. In addition, other non-loss-based end-to-end congestion predictors can serve as loss discriminators, although they were not designed explicitly for improving congestion control in wireless environments [40–44]. These congestion predictors were designed based on the observation that as the sending rate increases, the throughput increases accordingly and the delay observed by the flow is low and somehow stable before reaching the knee point where the link is nearly saturated. Beyond the knee point, the throughput stabilizes and the delay increases sharply. CARD [41] monitors the normalized delay gradient of a flow at the sender for determining the knee point and controls the flow rate at this point. In Tri-s [42], the authors proposed to use the normalized throughput gradient instead of the normalized delay gradient to determine when the bottleneck link reaches saturation. In DUAL [43], the current RTT sample is compared with the average of the minimum and maximum RTT to determine congestion. In CIM [45], the authors compared the moving average of a small number of RTT samples with the moving average of a large number of RTT samples to determine congestion. TCP SantaCruz [40] is very similar to TCP Vegas [35]. TCP SantaCruz adopts a different technique to estimate the number of queued packets in the bottleneck buffer and attempts to maintain this operation point. The estimation is based on one-way trip time instead of the RTT; thus, it is more robust to reverse traffic congestion. Recently, Bhandarkar et al. [44] proposed to predict congestion based on an EWMA of the RTT. If the average RTT is larger than the sum of the minimum RTT and a predefined threshold, congestion is inferred along the path.

Evaluating the performance of these congestion predictors as loss discriminators, [46] reported that Vegas, CARD, and Tri-s perform poorly as wireless random loss predictors. We conjecture that all the congestion predictors cannot serve well as random loss predictors. The reason is simply that random losses occur regardless of network congestion states. If random losses occur when the congestion level is low, the congestion predictors can effectively detect them. However, if wireless random losses occur when the congestion is high, random losses are mixed with congestion losses, it is difficult for these congestion predictors to differentiate them.

Network Cooperation Approach

In the network cooperation approach, routers help in controlling the congestion passively by feeding back network condition information to the end-systems.

ECN: Packet loss is the only congestion indication in conventional TCP congestion control; hence, TCP performs poorly in wireless environments where non-congestion-related packet losses exist. Floyd et al. proposed Explicit Congestion Notification (ECN) [47] to decouple packet losses from congestion control. The idea is to "mark" a packet instead of dropping it when congestions is impending at the routers. Therefore, ECN marks serve as early congestion signals, and congestion packet losses may be avoided as well as the consequent retransmissions. Upon receiving an ACK with the ECN bit set, it is recommended by the ECN standard that the sender treats the marked packets just as lost and back off the congestion window [48]. As ECN requires modifications at routers and end-hosts, it is categorized as a network cooperation approach. It is worth noting that ECN has the potential to address the wireless random losses issue completely. If all the intermediate routers are ECN-enabled and the ECN marking schemes at the routers are robust enough to avoid any packet dropping due to buffer overflow, ECN marks can serve as the exclusive congestion signals replacing packet losses, thus wireless random losses do not impact TCP congestion control at all. However, neither of these two requirements is fulfilled in the current Internet.

TCP Jersey: TCP Jersey [49] combines the bandwidth estimator of TCP Westwood with congestion warnings (CW). The CW idea is similar to the ECN scheme. TCP Jersey estimates the available bandwidth at the sender by measuring the rate of the returning ACKs. To distinguish congestion losses from wireless random losses, TCP Jersey leverages on a non-standard-compliant usage of the ECN bit. When the average queue size is above a certain threshold at an intermediate router, this router marks all passing packets to warn the receiver of incipient congestion. By examining the marking history when a packet loss is detected, the ender infers whether it is a random loss or a congestion loss, as congestion losses usually occur preceded by massive consecutive marked packets. For random losses TCP Jersey retransmits the packet without invoking congestion control; otherwise, TCP Jersey acts the same as TCP Westwood. However, if a wireless random loss occurs when the congestion level is high, TCP Jersey won't be able to identify it from congestion losses. Furthermore, TCP Jersey does not follow the standard behaviors of ECN. Although TCP Jersey is originally proposed as an end-to-end scheme, we believe it is more appropriate as a network cooperation approach, as the deployment of TCP Jersey requires the support of intermediate routers.

TCP Casablanca: TCP Casablanca [39] proposes a simple biased queue management scheme that de-randomizes congestion losses and enables a TCP receiver to diagnose accurately the cause of a packet loss. Then the receiver informs the sender to react appropriately. The key idea of TCP Casablanca is, within the same TCP flow, to send packets with different drop preferences: a small proportion of selected packets are marked with lower priority to be dropped first by routers when congestion occurs. When congestion occurs, packets with the lower priority are dropped first.

When random losses occur, it is unlikely that only the lower priority packets are dropped. By examining the loss patten of the lower priority packets, the cause of the losses can be inferred with certain confidence. Note that TCP Casablanca requires changes at both the sender and the receiver, as well as an active queue management scheme at intermediate routers.

Summary

Table 12.1. Comparison of proxy-based, end-to-end, and network cooperation approaches.

| | proxy-based | | end-to-end | network cooperation |
	split-connection	snoop-like		
requires changes at	BS	BS	end hosts	routers, end hosts
requires per-flow state	at BS	at BS	no	no
deployment difficulties	medium	medium	low	high
maintains the end-to-end sematic of TCP	no	yes	yes	yes
handles encrypted IP payloads	decryption at BS	no	yes (except ELN)	yes
wireless scenarios targeted	last-hop wireless		wireless links along the path	

In this section we have surveyed a number of works intended to improve the performance of TCP in presence of wireless random losses. These works generally follow three approaches: proxy-based, end-to-end, and network cooperative approaches.

The proxy-based approach relies on a proxy agent, which is deployed on the edge between the wired and wireless networks, to make the wireless random losses transparent to TCP congestion control. Although the proxy-based approach can greatly improve TCP efficiency in wireless environment while leaving the current TCP/IP implementations in wired network untouched, the work load and complexity introduced to the proxy is heavy, for instance, per-flow sate information is required at the proxy, which limits the scalability of the proxy-based approach. Fortunately, in last-hop wireless scenarios the number of flows handled by each proxy is not large, thus the complexity on the proxy is still acceptable. In the proxy-based approach, the actions are to be taken by the network managers to improve services to their end-users.

In the end-to-end approach, the end systems differentiate wireless random packet losses from congestion losses purely based on the information available at the connection ends. According to the loss types, different congestion control actions are

Table 12.2. Summary of techniques and protocols for congestion control in presence of wireless random losses.

	Approach	Design scenario	Modification to TCP
ARQ	link layer retransmission (LLR)	link layer random loss	none
FEC	error correction coding scheme	high BER	none
ECN [47]	explicit congestion notification	general network	sender, receiver
ELN [25]	random loss notification	last-hop wireless	sender, receiver
I-TCP [23]	split connection	last-hop wireless, mobile	none
snoop [24]	TCP-aware LLR	last-hop wireless, mobile	none
W-TCP [26]	end-to-end	WWANs	sender, receiver
TCP Westwood [29–31]	end-to-end	wireless link along the path	sender
NewReno-FF [33]	end-to-end	wireless link along the path	sender
TCP-Veno [34]	end-to-end	wireless link along the path	sender
JTCP [36]	end-to-end	wireless link along the path	sender, receiver
ZBS [37]	end-to-end	wireless link along the path	n/a, TFRC receiver
congestion predictors: CARD [41]	end-to-end	early congestion prediction	sender
Tri-s [42]			sender
DUAL [43]			sender
TCP Vegas [35]			sender
CIM [45]			sender
TCP SantaCruz [40]			sender, receiver
end-AQM [44]			sender

taken to improve efficiency. Due to technical difficulties the accuracy and performance of these end-to-end solutions are not optimal. However, as only modifications to the end-systems are needed, the end-to-end approach has low deployment cost and gradual deployment potential, and is thus still promising and often favored in many cases.

The network cooperation approach refers to the scenario where the intermediate routers aid the end systems with extra information to improve congestion control efficiency in presence of wireless random losses. As the routers are involved, the scalability issue has to be taken seriously. None of the reviewed protocols in the network cooperation approach requires per-flow states on the routers. However, the deployment of network cooperation-based solutions involve the routers as well the end-systems.

A brief comparison among these approaches on their advantages and shortcomings is presented in Table 12.1, and a comparative study of the protocols reviewed in this section can be found in Table 12.2.

12.4.2 Mobility Support

In mobile wireless networks, MHs experience frequent temporary disconnections to the BS due to handoff procedures, as a MH moves from one BS to another, or simply due to radio signal obstruction. These temporary disconnections result in successive packet losses and can even trigger the RTO. In such scenarios TCP suffers severe performance degradation due to the reaction of TCP congestion control. The situation gets even worse when the MH is roaming among separate wireless systems where the MH is allocated with different IP address. With current TCP socket implementation, the change of the IP address of the MH will close the ongoing TCP connection and a new connection to the FH is required, which is certainly inefficient and results in a bad user experience.

Mobile IP: Mobile IP [19] was designed to avoid the need of the connection reestablishment. In Mobile IP design, the MH needs to register to a home network and get a permanent home IP address at the first place. Then after roaming to another network, the MH issues request to get another temporary IP address, called care-of address (CoA), from a foreign agent in the visited network. Before allocating this CoA to the MH, the foreign agent contacts the home agent located in the home network and informs the home agent about the CoA of the MH, to make sure the home agent knows the new location of the MH. When a node wants to communicate with this MH, the packets destined to the MH first reach the home network and are captured by the home agent. Then the home agent encapsulates these packets with the CoA and forwards them to the network the MH is visiting. This mechanism is usually referred to as "network tunnelling". At the foreign agent the encapsulated packets are decapsulated and relayed to the MH. The foreign agent continues serving the MH until the granted lifetime expires. If the MH wants to continue the service, it has to reissue the request.

Although the handoff-related connection re-establishment can be effectively avoided by techniques like Mobile IP, the temporary disconnections caused by handoffs can still hinder TCP congestion control severely. Snoop [24] aims to reduce handoff latency and packet losses to improve the performance of TCP by buffering packets destined to the MH at surrounding BSs, which forms a multicast group. Among these BSs, the MH selects the one with the best signal strength as the primary BS to relay the packets. There are a number of works in the literature on improving the performance of TCP in mobile wireless network with frequent handoffs.

M-TCP: M-TCP [50] was designed for the last-hop wireless scenario according to the split-connection approach described previously. The design goal in M-TCP is to handle disconnections and low-bandwidth. The basic idea is to allocate bandwidth to MHs at BS with a bandwidth management module. When MH disconnection occurs, BS stops the TCP sender at FH; when connection is resumed, BS asks the sender to send at its full rate.

M-TCP requires a three-level mobile network architecture. A MH connects to a BS, and several BSs are controlled by a supervisor host (SH), which functions like a gateway and maintains flow states. In this way, most of the time when a handoff occurs, there is no need to transfer the state information between BSs.

In contrast to I-TCP [23], M-TCP maintains the end-to-end semantic of TCP in a sense that the SH does not send back the corresponding acknowledgement to the FH until the data is successfully delivered to the MH. Another important design principle in M-TCP is that, after successfully delivering some data to the MH, the SH does not inform the FH about this directly; instead, it always keeps the last byte of the successful transmission unacknowledged to the FH sender. Assume that n bytes data have been successfully delivered to the MH. Then, the SH only sends back ACK to the FH to confirm that $n - 1$ bytes have been successfully received at the MH. The acknowledgement for the last byte is saved for later usage. A MH disconnection is detected if there are no packets received during a period; then, the SH sends back the ACK for the last byte with a zero advertised window in the TCP header to force the sender into *persist mode* to avoid unnecessary congestion window shrinks. When the wireless link is resumed, the MH notifies the SH by sending a greeting packet. The SH, in turn, informs the sender of this reconnection and allows the sender to resume its transmission from the frozen state. The adverse effects of disconnections on the performance of TCP are thus gracefully eliminated.

M-TCP performs well in mobile networks where there are frequent disconnections between the MH and the BS. However, M-TCP implicitly assumes that packets are reliably transferred over the wireless link from the BS to the MH even during the handoff period. This assumption may be difficult to achieve in practice. Furthermore, the complexity at the SH can be very high with M-TCP.

Freeze TCP: Freeze TCP [51] is a pure end-to-end solution to address the throughput degradation of TCP caused by frequent disconnections. M-TCP is quite effective to improve the performance of TCP in presence of frequent temporary disconnections in mobile networks; however, the split-connection approach cannot be applied if packets are encrypted or if the forward and reverse paths do not use the same gateway. In these scenarios, an end-to-end solution, like Freeze TCP, is preferred. In Freeze TCP, it is assumed that the MH has the ability to predict the impending disconnection by sensing the radio signal strength. In such events, the Freeze-

TCP receiver adopts a similar approach as M-TCP to force the sender into *persist mode* by sending back zero-window-ACKs (ZWA) about one RTT before handoff. In *persist mode*, the sender freezes all retransmission timers and sends zero-window-prob (ZWP) packets, whose inter-departure times are increased exponentially. On receiving a response with a positive advertised window, the sender exits from *persist mode* and resumes its transmission. However, due to the exponential back off of the ZWP interval, it is possible that after recovering from the disconnection the MH needs to wait for a long time for the ZWP from the sender. To tackle this problem, on detecting reconnection, the receiver can send three duplicate ACKs to the sender acknowledging the last received packet right before the disconnection, so that the transmission can be resumed quickly.

The main advantage of Freeze TCP is that it improves the performance of TCP without any modifications at intermediate nodes. However, the actual performance improvement depends largely on the prediction accuracy of impending disconnection at MH. Freeze TCP may also set the sending rate inappropriately after the restoration of the connection since the available bandwidth of the connection may have changed.

RCP: Reception Control Protocol (RCP) [52] is a receiver-centric transport protocol. RCP is a TCP clone in the general behaviors. The RCP receiver is responsible for congestion control, flow control, and reliability, and guides the sender accordingly. As a receiver-centric transport protocol, RCP has advantages over sender-centric ones in handling handoff disconnections. Since the MH knows when a handoff occurs and can accurately control which and how much data should be sent by the sender, it can stop the sender just before handoff and resume the sending right after handoff. When the MH is equipped with multiple wireless interfaces and has access to different wireless networks, a receiver-centric transport protocol can also easily select a proper interface without involving the sender.

Nevertheless, a receiver-centric transport protocol introduces significant work load at the MHs, which are usually limited in terms of computational power and energy.

pTCP: When the MH is equipped with multiple wireless interfaces that have access to different wireless networks, two type of handoffs may occur: horizontal handoff and vertical handoff. A horizontal handoff occurs between BSs in the same network; a vertical handoff happens between two separate access networks. To enable a seamless usage of heterogeneous wireless networks, parallel TCP (pTCP) [53], a multi-state transport layer solution was proposed. It is remarkable that pTCP does not need Mobile IP [19] support from the underlying network. For each active interface used in a pTCP connection, a TCP-virtual (TCP-v) pipe is created. A pTCP socket can be addressed through any of its component TCP-v pipe. pTCP dynamically adds or deletes virtual pipes in a connection depending on the connectivity. As a multi-state transport protocol, pTCP allows multiple pipes to co-exist in a connection, and hence soft handover of the transport layer states are possible. Moreover, congestion control operations are carried out by TCP-v, thus within one pTCP connection different congestion control protocols can be used to achieve good throughput in different wireless networks. When traversing from one access network to another, a pTCP connection selects the proper interface (TCP-v) for data transmission.

pTCP provides an effective way to switch to different access networks with multiple wireless interfaces. Although information can be shared among interfaces, determining the transmission rate after handoff is still a challenge, in particular for horizontal handoffs, as with vertical handoff the MH must reprobe the bandwidth in any case. Simply restoring the transmission state to the one prevailing before handoff may not be appropriate as the network condition may have changed.

Summary

In this section we have reviewed some congestion control protocols that were introduced to mitigate the negative effects of handoff in mobile wireless networks on the performance of TCP. A summary of these protocols is listed in Table 12.3.

Table 12.3. Summary of techniques and protocols to provide smooth handoff in mobile wireless networks.

	Approach	Design scenario	Handoff detection
Mobile IP [19]	agent-based tunnelling	IP alteration during roaming	n/a
M-TCP [50]	snoop-like	horizontal handoff	at BS
Freeze TCP [51]	handoff-aware TCP receiver	horizontal handoff	at MH
RCP [52]	receiver-centric congestion control	horizontal handoff	at MH
pTCP [53]	multi-state transport layer	multiple wireless access interface	at MH

12.4.3 High Bandwidth-Delay Product in Satellite Networks

Generally speaking, satellite networks exhibit high bandwidth-delay products. Due to the slackness of TCP, many high-speed versions of TCP were proposed, like HS-TCP [54], S-TCP [55], TCP FAST [56], and so on. Most of them were, however, designed for wired networks with high bandwidth-delay products and do not take into account the specific requirements and constraints of satellite networks. Satellite-specific congestion control enhancements are reviewed in this section.

TCP-Peach: TCP-Peach [22] was particularly designed for satellite communication. TCP-Peach keeps the congestion avoidance and the fast retransmit mechanisms of TCP Reno, but replaces *slow-start* and *fast recovery* with *sudden start* and *rapid recovery*. In *sudden start* and *rapid recovery*, the sender probes the available bandwidth with low-priority dummy packets. TCP-Peach assumes that routers along the path support priority queueing. A successfully delivered dummy packet indicates that there exist unused network resources and the transmission rate can be increased accordingly. In *sudden start*, the effective congestion window size is obtained in one RTT bandwidth probing at the beginning of a connection, instead of going through the round-by-round exponential-increase as in *slow-start*. After sending the first data packet, the sender sends out $(rwnd - 1)$ dummy packets in a RTT, where $rwnd$ is the maximum allowed congestion window size advertised by the receiver. For each ACK for the dummy packets, the sender increase $cwnd$ by one. In *rapid recovery*, unnecessary rate reduction is avoided for random losses: assuming when three duplicate ACKs are received the window size is W, the $cwnd$ is first set to $W/2$, then W dummy packets are also sent. For the first $W/2$ dummy packets, $cwnd$ does not increase, but for each ACK of the subsequent dummy packets, $cwnd$ is increased by one. If the three duplicate ACKs are due to a random loss, this *rapid recovery* operation can effectively capture back the available bandwidth. In TCP-Peach+ [57], dummy packets are replaced by the actual data packets with lower priority to improve the throughput.

TCP-Peach can effectively utilize the expensive satellite links with the help of the priority queuing on the satellite link. However, due to the requirement of this priority queuing TCP-Peach cannot be deployed in the Internet at large. Therefore, it is more appropriate to follow the split-connection approach to make TCP-Peach the congestion control protocol on the satellite links.

XCP: eXplicit Control Protocol (XCP) [58] is a router-centric congestion control protocol for high bandwidth-delay product networks. XCP takes an open loop control operating at the network level and generalizes the ECN scheme. Instead of one bit congestion indication used by ECN, XCP utilizes precise congestion signals, where routers explicitly tell the sender the state of congestion and how much to react to it. In XCP the efficiency and fairness control are separated from each other. Based on the deviation between the incoming rate and the link capacity, and the queue occupancy, an XCP router determines the aggregate feedback which is then fairly allocated among the flows through a per-packet based feedback scheme. XCP tends be able to effectively utilize bandwidth and fairly allocate the bandwidth among competing flows. One major advantage of XCP is that per-flow state is not required at the intermediate routers, so that the scalability is not an issue. Moreover, XCP can almost avoid all congestion losses, thus packet losses are not necessarily congestion signals any more. However, XCP requires a revolutionary change to the Internet architecture and implementations, thus its deployment in the Internet is still questionable.

In [59], the performance of XCP in satellite networks was studied. It is argued that XCP can be deployed in satellite networks in a split-connection approach, where the gateways between the satellite network and the wired network are equipped with performance enhancement proxies. It is shown that XCP can help to utilize the expensive satellite bandwidth effectively.

REFWA: The recursive, explicit, and fair window adjustment scheme (REFWA) [21] achieves the max-min fairness principle and improves the system efficiency in LEO networks. The fairness issue in LEO networks arises mainly due to the fact that flows may pass different numbers of satellite links, thus have significantly different RTTs, while the TCP throughput is inversely proportional to the RTT. REFWA promotes the max-min fairness principle in LEO networks and relies on intermediate satellite routers to compute and feedback a feasible window size to different flows by rewriting the advertised window field in the ACK packets. No modification is needed at end-systems. The intermediate satellite routers are capable of estimating the flow RTTs and compute the feedback. With proper feedback, RTT unfairness among flows can be mitigated; in the mean time, the convergence speed can be also improved when flows join and leave the network, while avoiding sudden changes of the queuing delay and potential excessive packet drops.

The RTT estimation scheme in REFWA uses a composite hop count from DATA and ACK packets that transit in the satellite. This gross estimation assumes that the TTL field is reset at the ingress of the satellite network and that its initial value is known and the same everywhere. This is a drawback when the TCP flow passes through not only the LEO satellite network but also through some wired links, unless performance enhancement proxies are deployed at the edge between the satellite network and wired links to split the connection or reset the TTL count. Moreover, the scalability of REFWA is another issue as it maintains per-flow state information.

Summary

In this section we have reviewed some congestion control enhancements for satellite networks. Table 12.4 summarizes these works.

Table 12.4. Summary of congestion control protocols for satellite network (BDP: bandwidth-delay product; $awnd$: advertised window).

	Approach	Design scenario	Support from satellite
TCP Peach [22]	router-assisted	satellite network	priority queuing
XCP [58] (in [59])	router-centric congestion control	high BDP network	congestion control module
REFWA [21]	router-assisted	LEO satellite network	$awnd$ rewriting

12.4.4 Wireless Links with Rate-Adaptation

In general, when the available bandwidth on a link changes dynamically, the performance of TCP is impaired. Consider a saturated bottleneck link: if there is a sudden reduction of the link nominal capacity, the original traffic volume will greatly overwhelm the link, leading to a sudden increase of the queue size and even bulk packet losses; while a sudden increase of the link nominal capacity can cause link underutilization. Furthermore, the sudden increase and decrease of the queue size caused by the link rate variation also leads to a large variation in RTT measurements; hence, the RTO cannot be estimated accurately. As a consequence, TCP flows may wait unnecessarily long for the timeout or trigger packet retransmission and slow-start due to false timeout. These phenomena impact the performance of TCP negatively.

Such varying link nominal capacity scenarios are becoming more and more customary in today's wireless networks, including 3G and especially WLAN. The so-called rate-adaptation (or link-adaptation) mechanisms are widely used in IEEE 802.11 to adjust the data transmission rate based on wireless media conditions to maximize the capacity given a reasonable SNR level. It is interesting and important to investigate how this emerging rate-adaptation feature impacts the performance of TCP.

RA-snoop: Rate Adaptive snoop (RA-snoop) [60] considers bandwidth variations as well as loss characteristics to improve the performance of TCP over rate-adaptive wireless links. RA-snoop is an extension of snoop. To effectively utilize the proxy buffer, RA-snoop caches packets selectively based on wireless channel conditions. Packets are only cached when the link condition is bad and the transmission is likely to be corrupted; RA-snoop retransmits the packets locally over the wireless link if these packets have not been successfully delivered, which is detected through duplicate ACKs. In addition, RA-snoop is designed to calculate the feedback window size based on the bandwidth-delay product estimation and the queue level, then convey this feedback back to the TCP source through the advertised window field in the ACK packets. As a result, TCP can adapt effectively to variable bandwidth. For implementation, RA-snoop needs to estimate each flow's RTT, the number of active flows, as well as an EWMA of the varying link capacity. The feedback window size of each flow is calculated as a fraction of the total available bandwidth estimated as a composite of the varying link capacity and the deviation of the current queue size from a target queue size.

RA-snoop helps TCP to effectively adapt to variable bandwidth. However, RA-snoop needs to maintain per-flow state information and also introduces considerable measurement workload on the proxy.

12.4.5 Congestion Control in Mobile Ad-hoc Networks

Mobile ad-hoc networks are easy to deploy in adverse environments where it is impossible or costly to deploy infrastructure-based networks. The topology of mobile ad-hoc networks is highly volatile, thus many new challenges come forth in the design of congestion control schemes for these dynamic networks. For example, an established link may be teared down due to high mobility; the capacity of a wireless link is shared with neighboring nodes while the nodes may join and leave the neighborhood dynamically. Due to such common challenges in wireless environments, and more importantly the intrinsic complexity of mobile ad-hoc networks, till now mobile ad-hoc networks are still at the research stage, where prototypes are usually constructed based on IEEE 802.11. In such dynamic networks, congestion control is one of the major problems that need to be addressed before real deployment. Many approaches have been proposed to enhance congestion control in mobile ad-hoc networks. Generally, they can be divided into two categories: improving the performance of TCP with sophisticated mechanisms to distinguish congestion packet losses from non-congestion packet losses and constructing new congestion control protocols by exploiting lower-layers information. We review here some of the proposed solutions and classify them according to the following issues: 1) whether they address disconnections due to frequent route changes or failures; 2) whether they handle persistent packet reordering due to route changes or multi-path routing; 3) whether they improve the fairness by a better sharing of the wireless medium; and finally, 4) whether they require network cooperation.

Frequent Route Changes and Failures

Route changes may occur when nodes move, join or leave a mobile ad-hoc network. In a route recovery or re-discovery procedure, packets are dropped due to link breaks. The performance of TCP may suffer significantly due to these non-congestion packet loss. To deal with this issue, TCP needs to distinguish packet loss due to route failures and network congestion, either by feedback mechanisms or inference.

TCP-Feedback: Chandran et al. proposed a router-assisted scheme, TCP-Feedback (TCP-F) [61], to address route failures in mobile ad-hoc networks to improve TCP throughput. Once a route disruption is detected, for example when the next-hop node is unreachable, a route failure notification (RFN) packet is sent back to the source along the downstream path. Upon receiving the RFN packet, each intermediate node disables the original upstream route and checks if an alternate route exists. If yes, the RFN packet is discarded and the traffic for the original route is redirected at this intermediate node; otherwise, the intermediate nodes relays the RFN packet back towards the source. When the source receives the RFN packet, it suspends the TCP connection activities and freezes all the timers and states. When an intermediate node between the break point and the source finds a new route to the destination, it sends a route re-establishment notification (RRN) packet to the source. The connection is re-activated upon either the snooze state timeout or the reception of such a RRN packet. Similar to Freeze-TCP, TCP-Feedback may set the sending rate inappropriately after the restoration of the connection since the available bandwidth of the connection may have changed. Moreover, TCP-Feedback does not handle the wireless random losses explicitly.

ATCP: Ad-hoc TCP (ATCP) [62] was designed for mobile ad-hoc networks to address random packet losses due to high BER, route changes, network partitions, and packet reorderings. ATCP is implemented as a thin layer between TCP and IP layer. ATCP has four states: *normal state, loss state, congested state*, and *disconnected state*. These states are defined to differentiate packet losses due to high BER/packet reordering, true network congestion, or route changes. It relies on ECN to distinguish congestion losses from random error losses. Only on congestion losses, congestion control operations are performed. Upon the reception of an ECN-set ACK, ATCP enters the *congested state* and passes this ACK to TCP sender to invoke necessary congestion control operations. ATCP leaves the *congested state* until a new packet is transmitted by TCP sender. If three duplicate ACKs are received but the ECN flag is not set, ATCP treats it as a random loss or a packet reordering – which may be due to route reamputation or even multi-path routing; ATCP enters the *loss state* and puts the TCP sender into *persist mode* and the "lost" packet is simply retransmitted at the ATCP layer. Until a new ACK is received ATCP leaves *loss state* and removes the TCP sender from the *persist mode*. When there is a route change or network partition, routers generate ICMP messages of "Destination Unreachable" in response to a packet transmission and send these ICMP messages back to the sender. Upon receiving such a message, the ATCP enters the "disconnected state" and puts the TCP sender into *persist mode* until a new route has been found. Implemented as an intermediate layer between TCP and IP layer, ATCP has the advantage that the original TCP implementation is left unchanged. To infer the cause of current packet loss, TCP relies on ACK, whose delivery is also vulnerable to route changes. This can impair the efficiency of ATCP.

Persistent Packet Reordering

In mobile ad-hoc networks, packets, which are originally sent in order, may follow different paths to the destination, when route changes occur or multi-path routing is deployed. As a result, these packets may arrive out-of-order at the destination. Note that TCP treats out-of-order packets as suspicious packet losses and may reduce the congestion window size by half at the sender although there is no congestion or packet loss in the network. Therefore, in mobile ad-hoc networks TCP needs to change its behavior when detecting out-of-order packets.

DSACK: The Dynamic SACK extension [63] to the TCP SACK option [27] was proposed to improve the robustness of TCP on packet reordering. The spurious retransmission inferred from DSACK is helpful in controlling sender behaviors correctly to improve the performance of TCP. The original DSACK proposal does not specify how the TCP sender should respond to DSACK notifications. In [64], a number of extensions to DSACK notifications were proposed. The simplest one restores the sender's congestion window to its value prior to the spurious retransmission detected through DSACK. Another extension is to adjust the DUPACK threshold ($dupthresh$). The proposed $dupthresh$ adjustment mechanisms include: 1) increment $dupthresh$ by a constant; 2) set the new value of $dupthresh$ to the EWMA of the current $dupthresh$ and the number of DUPACKs that caused the spurious retransmission; and 3) set $dupthresh$ to an EWMA of the number of DUPACKs received at the sender. There are also other schemes for improving TCP congestion control in presence of persistent packet reordering such as those proposed in [65, 66]. However, as they were not originally proposed explicitly for mobile ad-hoc networks, we skip them here.

TCP-DOOR: TCP-DOOR [67] attempts to improve the performance of TCP by detecting and responding to out-of-order packet deliveries in order to avoid unnecessary congestion control operations. TCP-DOOR assumes that packet recording is most likely due to route changes in mobile ad-hoc networks. When either the TCP sender or the receiver detects DATA/ACK packets reordering, route changes occur and the sender should respond with two possible operations: temporarily disabling congestion control and instantly recovering congestion states during congestion avoidance. In the first operation, the TCP sender maintains its state variables such as RTO and the congestion window size constant for a certain time period. In the second operation, before detecting out-of-order packet delivery, if the TCP sender has already backed-off its congestion window size and entered the congestion avoidance state, and the time is less than another pre-specified threshold, the sender should recover immediately to the state prior to the congestion avoidance phase. The main reason for the second operation is that the detection of out-of-order packet deliveries implies that a route change has just occurred. However, TCP-DOOR cannot address the out-of-order deliveries due to multi-path routing.

Fair Sharing Wireless Medium among Neighbors

Neighborhood RED: TCP exhibits a serious unfairness problem in ad-hoc networks due to the combination of MAC-inherent problems such as medium contention, the hidden terminal problem, and the exposed terminal problem. These problems are likely to occur in co-located nodes in a neighborhood. Xu et al. [68] proposed a scheme, neighborhood RED (NRED), to improve TCP fairness from the perspective of a neighborhood. By definition, a node's neighborhood consists of the node itself and the nodes which can interfere with this node. The key idea of NRED is that, since the wireless medium is shared by neighbor nodes, these neighbor nodes should share a "common" queue instead of a distributed queue on each node. In this "common" queue, an active queue management scheme, such as RED, can be applied to enforce the same packet loss probability for each flow. Nodes may exchange distributed queue sizes among themselves to detect an early congestion; however, this signaling overhead is very high. In NRED, the channel utilization is passively monitored instead at each node. It is believed that when the queues at its neighboring nodes are busy the channel utilization around a node is likely to increase. Thus, if the channel utilization, which is estimated at a node through monitoring radio states, exceeds a certain threshold, an early congestion is determined. Then this node calculates the packet dropping probability using the RED algorithm and sends this probability in an Neighborhood Congestion Notification (NCN) packet to its neighbors. Upon receiving such a NCN notification, the neighbors drop some packets in their queues when necessary. Simulation studies show that NRED improves TCP fairness to some extent in ad-hoc networks. However, the aggregate throughput in the network is actually reduced. Thus NRED achieves a tradeoff between efficiency and fairness in ad-hoc network.

Network-Cooperative Design

As wireless medium is shared by neighboring nodes, MAC layer contention is the direct cause of network congestion. In multi-hop wireless networks, congestion control can be enhanced by taking into account MAC layer information. This motivates a systematic design of congestion control protocols, combining transport layer and MAC layer, not just patching TCP.

RAIN: RAIN [69] is a new reliable wireless architecture for multi-hop wireless networks. RAIN pushes congestion control down to the link layer, where congestion control is combined with medium contention control. The transport layer only ensure reliability by detecting and recovering packet loss due to routing failures. RAIN also reorganizes the buffers in the protocol stack. RAIN coalesces the network layer transit traffic buffering into the link layer. Whenever a wireless router receives a frame, the contention and congestion controller (C3) at the link layer first checks the length of the transit traffic queue. If the queue length is above a small threshold (e.g., one packet), a contention and congestion control signal is sent (e.g., piggy-backed to the MAC ACK frame), to the upstream node before the frame is forwarded to the network layer for routing. The upstream router, which receives such a negative acknowledgement, freezes the transmission for certain time and then retransmits the frame. With this back-pressure mechanism, network congestion is successfully pushed back to the sender. With a safe link layer implemented in a RAIN network, packet losses due to contention or congestion are highly uncorrelated. Therefore, in RAIN, ReTCP is proposed as the transport protocol. ReTCP removes congestion control and flow control from TCP, and adds the delayed ACK adaptation feature.

ATP: Ad-hoc Transport Protocol [70] is a rate-based network-cooperative scheme, similar to XCP, for congestion control in mobile ad-hoc networks. In ATP, congestion control and reliability control are decoupled. Intermediate nodes measure the packet delivery delay and provide it as congestion signals for end hosts. The packet delivery delay is the sum of the queuing delay and the MAC transmission delay. This delay is piggy-backed to the data packets and updated by successive nodes if the piggy-backed delay is less than the delivery delay on the current node. Upon receiving a data packet, the receiver applies an EWMA low-pass filter to compute the average of the delivery delay and feed it back to the sender. With this feedback the sender computes the bottleneck capacity as the inverse of the delivery delay and adjusts its sending rate accordingly by linearly increasing it if the sending rate is 10% less than the measured bottleneck capacity, or by setting it to the measured bottleneck capacity if the current sending rate is larger than a pre-specified threshold; otherwise, the sending rate is sustained. ATP achieves reliability using the SACK option.

Summary

In this section, we have reviewed the congestion control problem in mobile ad-hoc networks. We have discussed several major reasons for the poor performance of TCP in these networks, e.g., route changes and failures and packet reordering. For each reason, we investigated some typical solutions. Due to the complexity of mobile ad-hoc networks, congestion control is difficult and still a great challenge to be addressed.

Conclusion

In this chapter, we have studied some major challenges to the conventional TCP congestion control in the emerging wired-cum-wireless Internet, namely, wireless random packet losses, mobility support, high bandwidth-delay product in some wireless networks, rate-adaptive wireless links, and efficiency and fairness issues in mobile ad-hoc networks. We have further reviewed the start-of-the-art progress on congestion control in the wired-cum-wireless Internet. Given the specific design scenarios, the proposed schemes achieve their design goals. However, none of them is suitable for addressing all these issues. Instead of identifying one solution which is better than other solutions, we intend to provide the readers with main ideas and possible tools to enhance congestion control in the wired-cum-wireless environment.

The conventional wisdom behind the TCP/IP architecture is that the lower layers only provide the best-effort packet delivery service while the transport layer addresses the efficiency and fairness issues. However, in wireless networks the link characteristics are very different from that in wired networks. Therefore, this layered design may not achieve a good performance. A revisit of the architecture design appears to be necessary. On the other hand, the TCP/IP architecture has been serving as the foundation of the Internet from the beginning. Turning over the TCP/IP architecture would almost lead to rebuild the Internet from scratch, which is unlikely to happen. To improve the performance of congestion control in wired-cum-wireless networks leaving the TCP/IP architecture untouched, there are two folds of works to be done: 1) enhance TCP properly to address the challenges in different applications; 2) improve the lower layer design to make the packet delivery interface to TCP in wireless networks similar to that in wired networks.

Most of the works reviewed in this chapter follow the first direction. For example, some differentiate random losses from congestion losses; some avoid unnecessary sending rate shrinking in presence of temporary disconnections due to handovers; some modify the congestion control algorithm to improve delivery efficiency in high bandwidth-delay product networks. Even though some solutions are not purely deployed on the transport layer, their implementations are still practical. However, as proposed by these works, congestion control protocols are customized to address particular challenges in different applications. With so many flavors of congestion control protocols coexisting in the Internet, the compatibility becomes crucial and should be examined carefully. Thus in designing congestion control enhancements for different applications, the performance gain over conventional TCP should not be the only consideration; instead, other issues, for example, deployment cost and TCP friendliness, are also important. In evaluating different congestion control protocols in wired Internet, research and standardization efforts have been going on in establishing a common benchmark suit, defining performance metrics, and identifying simulation and testbed scenarios [71, 72]. We envision that a similar TCP evaluation suite should also be established for the wired-cum-wireless Internet.

On the other hand, in viewing the limitation and the precondition of TCP congestion control, wireless technologies and lower network layers should take the performance of TCP into consideration. For example, the Neighborhood RED [68] emulates AQM behavior of a wired link among wireless neighboring nodes to achieve better MAC layer fairness. A counter example is the uplink and downlink unfairness problem in IEEE 802.11 WLAN caused by DCF. In DCF, an access point needs to compete with other mobile hosts to access the media. When there are n greedy mobile hosts transmitting data to the access point, the chance for this access point to access the channel is $1/(n+1)$; thus, the uplink and downlink bandwidth ratio in the WLAN is n. This unfairness problem is rooted at the MAC layer but is experienced at the transport layer. In wireless networks, where the communication condition is harsh, pushing congestion control exclusively to the transport layer does not seem to be wisdom any more. At least the lower layers should help to provide TCP with a comfortable operation environment like that in wired networks, instead of raising new challenges. In this way the TCP/IP architecture can still stay alive with only minor changes.

Another emerging issue that may impact congestion control is the bandwidth mismatch between the wired network and the wireless network. In the wired Internet, the link capacity and reliability improve significantly over the years. With the development of cheaper optical fiber and the switching capability due to more pervasive deployment of optical burst switching and 100 Gigabit Ethernet, the Internet core is much more powerful than ever before. On the other hand, despite the variety of wireless Internet access technologies, the last-mile link capacity at the edge of the Internet does not exhibit the same increasing rate. For example, the old wired standard ADSL can deliver 8 Mbps to the customer premises and the latest standard, ADSL2+, can deliver up to 24 Mbps, depending on the distance to the ISP access multiplexer. However, the commonly deployed IEEE 802.11b can only provide a maximum shared transmission rate of 11 Mbps and a maximum achievable throughput of 5Mbps due to the IEEE 802.11 MAC design. In LANs, Ethernet cards with 100 Mbps and even 1Gbps are quite common nowadays, while in WLAN the maximum shared bit rate is of 54 Mbps with IEEE 802.11g/a up to now. This bandwidth mismatch partially arises from the willingness that end-users make the trade-off between the capacity of broadband wired access and the convenience and mobility provided by wireless access networks. Nevertheless, it has been reported that in the Internet core, congestion seldom happen, the bottlenecks are usually at the Internet edge [73]. This mismatch in the capacity and link characteristics between the high-speed wired backbone and the interconnected wireless access edge could affect the user experience seriously and requires careful network planning and even new techniques to achieve good network efficiency. Techniques such as deploying content proxies at the edge of the network to improve the performance of the network, are becoming a common practice.

The Internet is evolving quickly. However, if the evolution is still going to occur in an ad-hoc fashion, unexpected challenges in the existing architecture will keep emerging. The Internet standardization entities should take responsibility to define a clear developing path for the Internet. New technologies should be designed with a joint consideration of the TCP/IP architecture.

References

1. J. Postel, "Transmission control protocol," RFC 793, September 1981.
2. K. C. Claffy, "Internet traffic characterization," Ph.D. dissertation, University of California, San Diego, 1994.
3. "Cooperative association for Internet data analysis," 2007, http://www.caida.org.
4. K. Pentikousis, "TCP in wired-cum-wireless environments," *IEEE Communications Surveys*, vol. 3, no. 4, pp. 2–14, Fourth Quarter 2000.
5. X. Chen, H. Zhai, J. Wang, and Y. Fang, "A survey on improving TCP performance over wireless networks," in *Resource Management in Wireless Networking, Network Theory and Applications*, M. Cardei, I. Cardei, and D.-Z. Du, Eds.   Springer US, 2005, vol. 16, pp. 657-695.
6. Y. Tian, K. Xu, and N. Ansari, "TCP in wireless environments: Problems and solutions," *IEEE Communications Magazine*, vol. 43, no. 3, pp. S27-S32, March 2005.
7. B. Sardar and D. Saha, "A survey of TCP enhancements for last-hop wireless networks," *IEEE Communications Surveys & Tutorials*, vol. 8, no. 3, pp. 20-34, Third Quarter 2006.
8. K.-C. Leung and V. Li, "Transmission control protocol (TCP) in wireless networks: issues, approaches, and challenges," *IEEE Communications Surveys & Tutorials*, vol. 8, no. 4, pp. 64-79, Fourth Quarter 2006.
9. Institute of Electrical and Electronics Engineers (IEEE), "Part 11: Wireless lan medium access control (MAC) and physical layer (PHY) specifications," IEEE Standard 802.11, 1999.
10. ——, "Part 16: Air interface for fixed broadband wireless access systems," IEEE Standard 802.16, 2004.
11. International Telecommunication Union (ITU), "International mobile telecommunications - 2000 (IMT-2000)," ITU standard IMT-2000. [Online]. Available: http://www.itu.int/home/imt.html
12. J. Bicket, D. Aguayo, S. Biswas, and R. Morris, "Architecture and evaluation of an unplanned 802.11b mesh network," in *MobiCom '05: Proceedings of the 11th annual international conference on Mobile computing and networking*.   New York, NY, USA: ACM, 2005, pp. 31-42.
13. Bluetooth Special Interest Group (SIG), "Bluetooth core specifications, core specification v2.0 + EDR," Bluetooth standard, 2004.
14. "Satellite Internet service providers." [Online]. Available: http://www.getisp.info/satellite-internet.html
15. V. Jacobson, "Modified TCP congestion avoidance algorithm," 1990. [Online]. Available: ftp://ftp.ee.lbl.gov/email/vanj.90apr30.txt
16. S. H. Low, "A duality model of TCP and queue management algorithms," *IEEE/ACM Trans. Netw.*, vol. 11, no. 4, pp. 525-536, 2003.
17. S. Floyd and T. Henderson, "The NewReno modification to TCP's fast recovery algorithm," RFC 2582, April 1999.
18. A. Medina, M. Allman, and S. Floyd, "Measuring the evolution of transport protocols in the Internet," *SIGCOMM Comput. Commun. Rev.*, vol. 35, no. 2, pp. 37-52, 2005.
19. C. E. Perkins, "Mobile IP," *IEEE Communications Magazine*, vol. 35, no. 5, pp. 84-99, 1997.
20. N. Ghani and S. Dixit, "TCP/IP enhancements for satellite networks," *IEEE Communications Magazine*, vol. 37, no. 7, pp. 64-72, 1999.

21. T. Taleb, N. Kato, and Y. Nemoto, "REFWA: an efficient and fair congestion control scheme for LEO satellite networks," *IEEE/ACM Transaction on Networking*, vol. 14, no. 5, pp. 1031-1044, 2006.
22. I. F. Akyildiz, G. Morabito, and S. Palazzo, "TCP-Peach: A new congestion control scheme for satellite IP networks," *IEEE/ACM Transactions on Networking*, vol. 9, no. 3, pp. 307-321, Jun 2001.
23. A. Bakre and B. R. Badrinath, "I-TCP: Indirect TCP for mobile hosts," in *Proc. IEEE ICDCS*, 1995I-TCP.
24. H. Balakrishnan, S. Seshan, E. Amir, and R. H. Katz, "Improving TCP/IP performance over wireless networks," in *Proc. ACM MobiCom*, 1995.
25. H. Balakrishnan, V. N. Padmanabhan, S. Seshan, and R. H. Katz, "A comparison of mechanisms for improving TCP performance over wireless links," *IEEE/ACM Transaction on Networking*, vol. 5, no. 6, pp. 756-769, 1997.
26. P. Sinha, T. Nandagopal, N. Venkitaraman, R. Sivakumar, and V. Bharghavan, "WTCP: a reliable transport protocol for wireless wide-area networks," *Wireless Network*, vol. 8, no. 2/3, pp. 301-316, 2002.
27. M. Mathis, J. Mahdavi, S. Floyd, and A. Romanow, "TCP selective acknowledgment options," RFC 2018, October 1996.
28. S. Keshav, "A control-theoretic approach to flow control," *ACM SIGCOMM Computer Communication Review*, vol. 21, no. 4, pp. 3-15, 1991.
29. C. Casetti, M. Gerla, S. Mascolo, M. Sansadidi, and R. Wang, "TCP Westwood: End-to-end congestion control for wired/wireless networks," *Wireless Networks*, vol. 8, pp. 467-479, 2002.
30. S. Mascolo, L. A. Grieco, R. Ferorelli, P. Camarda, and G. Piscitelli, "Performance evaluation of Westwood+ TCP congestion control," *Performance Evaluation*, vol. 55, no. 1-2, pp. 93-111, 2004.
31. M. Gerla, B. K. F. Ng, M. Y. Sanadidi, M. Valla, and R. Wang, "TCP Westwood with adaptive bandwidth estimation to improve efficiency/friendliness tradeoffs," *Computer Communications*, vol. 27, no. 1, pp. 41-58, 2004.
32. A. Capone, L. Fratta, and F. Martignon, "Bandwidth estimation schemes for TCP over wireless networks," *IEEE Transactions on Mobile Computing*, vol. 3, no. 2, pp. 129-143, 2004.
33. D. Barman and I. Matta, "Effectiveness of loss labeling in improving TCP performance in wired/wireless networks," in *Proc. IEEE ICNP*. Washington, DC, USA: IEEE Computer Society, 2002, pp. 2-11.
34. C. P. Fu and S. C. Liew, "TCP Veno: TCP enhancement for transmission over wireless access networks," *IEEE Journal on Selected Areas in Communications*, vol. 21, no. 2, pp. 216-228, Feb 2003.
35. L. Brakmo, S. O'Malley, and L. Peterson, "TCP Vegas: New techniques for congestion detection and avoidance," in *Proc. ACM SIGCOMM*, 1994.
36. E. H.-K. Wu and M.-Z. Chen, "JTCP: jitter-based TCP for heterogeneous wireless networks," *IEEE Journal on Selected Areas in Communications*, vol. 22, no. 4, pp. 757-766, 2004.
37. S. Cen, P. C. Cosman, and G. M. Voelker, "End-to-end differentiation of congestion and wireless losses," *IEEE/ACM Transaction on Networking*, vol. 11, no. 5, pp. 703-717, 2003.
38. S. Floyd, M. Handley, J. Padhye, and J. Widmer, "Equation-based congestion control for unicast applications," in *Proc. ACM SIGCOMM*. New York, NY, USA: ACM, 2000, pp. 43-56.

39. S. Biaz and N. H. Vaidya, ""De-randomizing" congestion losses to improve TCP performance over wired-wireless networks," *IEEE/ACM Transaction on Networking*, vol. 13, no. 3, pp. 596-608, 2005.

40. C. Parsa and J. Garcia-Luna-Aceves, "Improving TCP congestion control over internets with heterogeneous transmission media," in *Proc. IEEE ICNP*, 1999.

41. R. Jain, "A delay-based approach for congestion avoidance in interconnected heterogeneous computer networks," *ACM SIGCOMM Computer Communication Review*, vol. 19, no. 5, pp. 56-71, 1989.

42. Z. Wang and J. Crowcroft, "A new congestion control scheme: Slow start and search (Tri-S)," *ACM SIGCOMM Computer Communication Review*, vol. 21, no. 1, pp. 32-43, 1991.

43. ——, "Eliminating periodic packet losses in the 4.3-Tahoe BSD TCP congestion control algorithm," *ACM SIGCOMM Computer Communication Review*, vol. 22, no. 2, pp. 9-16, 1992.

44. S. Bhandarkar, A. L. N. Reddy, Y. Zhang, and D. Loguinov, "Emulating AQM from end hosts," in *Proc. ACM SIGCOMM.* New York, NY, USA: ACM, 2007, pp. 349-360.

45. J. Martin, A. Nilsson, and I. Rhee, "Delay-based congestion avoidance for TCP," *IEEE/ACM Transaction on Networking*, vol. 11, no. 3, pp. 356-369, 2003.

46. S. Biaz and N. H. Vaidya, "Distinguishing congestion losses from wireless transmission losses: A negative result," in *Proc. IC3N.* Washington, DC, USA: IEEE Computer Society, 1998, p. 722.

47. S. Floyd, "TCP and explicit congestion notification," *ACM SIGCOMM Computer Communication Review*, vol. 24, no. 5, pp. 8-23, 1994.

48. K. Ramakrishnan, S. Floyd, and D. Black, "The addition of explicit congestion notification (ECN) to IP," RFC 3168, September 2001.

49. K. Xu, Y. Tian, and N. Ansari, "TCP-Jersey for wireless IP communications," *IEEE Journal on Selected Areas in Communications*, vol. 22, no. 4, pp. 747-756, May 2004.

50. K. Brown and S. Singh, "M-TCP: TCP for mobile cellular networks," *ACM SIGCOMM Computer Communication Review*, vol. 27, no. 5, pp. 19-43, 1997.

51. T. Goff, J. Moronski, D. S. Phatak, and V. Gupta, "Freeze-TCP: A true end-to-end enhancement mechanism for mobile environments," in *Proc. IEEE INFOCOM*, 2000.

52. K.-H. Kim, Y. Zhu, R. Sivakumar, and H.-Y. Hsieh, "A receiver-centric transport protocol for mobile hosts with heterogeneous wireless interfaces," *Wireless Network*, vol. 11, no. 4, pp. 363-382, 2005.

53. H.-Y. Hsieh, K.-H. Kim, and R. Sivakumar, "An end-to-end approach for transparent mobility across heterogeneous wireless networks," *Mobile Networks and Applications*, vol. 9, no. 4, pp. 363-378, 2004.

54. S. Floyd, "HighSpeed TCP for large congestion windows," RFC 3649, Dec. 2003.

55. T. Kelly, "Scalable TCP: Improving performance in high-speed wide area networks," *ACM SIGCOMM Computer Communication Review*, vol. 33, pp. 83-91, Apr. 2003.

56. C. Jin, D. X. Wei, and S. H. Low, "FAST TCP: Motivation, architecture, algorithms, performance," in *Proc. IEEE INFOCOM*, Hong Kong, Mar. 2004.

57. I. Akyildiz, X. Zhang, and J. Fang, "TCP-Peach+: Enhancement of TCP-Peach for satellite IP networks," *IEEE Communications Letters*, vol. 6, no. 7, pp. 303-305, 2002.

58. D. Katabi, M. Handley, and C. Rohrs, "Congestion control for high bandwidth-delay product networks," in *Proc. ACM SIGCOMM.* New York, NY, USA: ACM, 2002, pp. 89-102.

59. A. Kapoor, A. Falk, T. Faber, and Y. Pryadkin, "Achieving faster access to satellite link bandwidth," in *Proc. IEEE INFOCOM*, 2005.

60. J.-C. Moon and B. G. Lee, "Rate-adaptive snoop: A TCP enhancement scheme over rate-controlled lossy links," *IEEE/ACM Transaction on Networking*, vol. 14, no. 3, pp. 603-615, 2006.

61. K. Chandran, S. Raghunathan, S. Venkatesan, and R. Prakash, "A feedback based scheme for improving TCP performance in ad-hoc wireless networks," in *Proc. IEEE ICDCS*. Washington, DC, USA: IEEE Computer Society, 1998, p. 472.

62. J. Liu and S. Singh, "ATCP: TCP for mobile ad hoc networks," *IEEE Journal on Selected Areas in Communications*, vol. 19, no. 7, pp. 1300-1315, July 2001.

63. S. Floyd, J. Mahdavi, M. Mathis, and M. Podolsky, "An extension to the selective acknowledgment (SACK) option for TCP," RFC 2883, July 2000.

64. E. Blanton and M. Allman, "On making TCP more robust to packet reordering," *ACM SIGCOMM Computer Communication Review*, vol. 32, no. 1, pp. 20-30, 2002.

65. M. Zhang, B. Karp, S. Floyd, and L. Peterson, "RR-TCP: A reordering-robust TCP with DSACK," in *Proc. IEEE ICNP*. Washington, DC, USA: IEEE Computer Society, 2003, p. 95.

66. S. Bohacek, J. P. Hespanha, J. Lee, C. Lim, and K. Obraczka, "A new TCP for persistent packet reordering," *IEEE/ACM Transaction on Networking*, vol. 14, no. 2, pp. 369-382, 2006.

67. F. Wang and Y. Zhang, "Improving TCP performance over mobile ad-hoc networks with out-of-order detection and response," in *Proc. ACM MobiHoc*. New York, NY, USA: ACM, 2002, pp. 217-225.

68. K. Xu, M. Gerla, L. Qi, and Y. Shu, "TCP unfairness in ad hoc wireless networks and a neighborhood RED solution," *Wireless Network*, vol. 11, no. 4, pp. 383-399, 2005.

69. C. Lim, H. Luo, and C.-H. Choi, "RAIN: A reliable wireless network architecture," in *Proc. IEEE ICNP*, 2006.

70. V. Anantharaman, K. Sundaresan, H.-Y. Hsieh, and R. Sivakumar, "ATP: A reliable transport protocol for ad hoc networks," *IEEE Transactions on Mobile Computing*, vol. 4, no. 6, pp. 588-603, 2005.

71. S. Floyd and E. Kohler, "Tools for the evaluation of simulation and testbed scenarios," July 2007, working in progress. [Online]. Available: http://tools.ietf.org/html/draft-irtf-tmrg-tools-04

72. S. Floyd, "Metrics for the evaluation of congestion control mechanisms," Oct. 2007, working in progress. [Online]. Available: http://tools.ietf.org/html/draft-irtf-tmrg-metrics-11

73. N. Hu, L. E. Li, Z. M. Mao, P. Steenkiste, and J. Wang, "Locating internet bottlenecks: algorithms, measurements, and implications," in *Proc. ACM SIGCOMM*, 2004, pp. 41-54.

13

Adaptive Solutions for Quality-Oriented Multimedia Streaming in Heterogeneous Network Environments

Gabriel-Miro Muntean

School of Electronic Engineering, Dublin City University, Ireland
munteang@eeng.dcu.ie

13.1 Introduction

As the broadband connectivity coverage reaches more than 95% of households in the developed world [1] providing support for bandwidth intensive applications, streamed high quality multimedia-related services (e.g. digital and interactive TV, Video on Demand, gaming, videoconferencing, etc.) are getting more popular among the users. At the same time, advantages such as flexibility of viewer location, mobility, cost and convenience of deployment are determining a shift of the regular homeowner or business manager interest from wired towards wireless solutions [2].

Moreover, based on the latest development of wireless technology, an increasing number of multimedia-based service users prefer to have access to these services anywhere, anytime and from heterogeneous devices such as desktops, laptops, tabletPCs, PDAs, wide screen TV monitors, etc [3]. However, many of these services are offered from a single point of service such as the broadband IP connection, TV antenna, DVD player or computer hard-drive and therefore local wireless connectivity is required for their distribution. Although the envisaged local distribution of multimedia-based services can be deployed in various scenarios, the term "in-home" is used for the local content deployment in this chapter.

A typical architecture for the emerging wired-cum-wireless multimedia-based service distribution is presented in Fig. 13.1. It involves a media server which delivers multimedia content over a hierarchical-designed broadband IP network infrastructure to local points of attachments. These end-connectivity points can be localized at the level of residential houses, residential block or business premises. Flexibility and convenience as well as the cost of deployment have determined a definite trend towards the last leg distribution of multimedia content using wireless solutions. In this context, Fig. 13.2 illustrates a wireless-based solution for in-home multimedia distribution. This distribution may not be restricted to a single house, as in the picture and it could involve a number of residential units.

High quality multimedia streaming has high bandwidth requirements and is highly sensitive to transmission delays and errors. Wireless networks allow for much lower delivery rates than wired networks (e.g. wireless IEEE 802.11b network - up to

E. Hossain (ed.), *Heterogeneous Wireless Access Networks*,
DOI: 10.1007/978-0-387-09777-0_13, © Springer Science+Business Media, LLC 2008

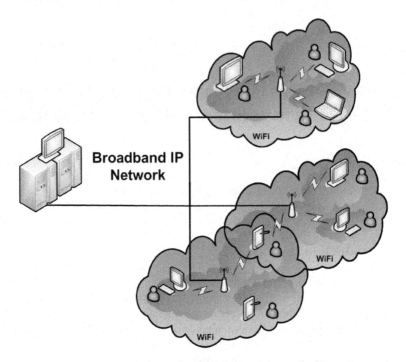

Fig. 13.1. Wired-cum-wireless architecture for distribution of multimedia content to heterogeneous devices.

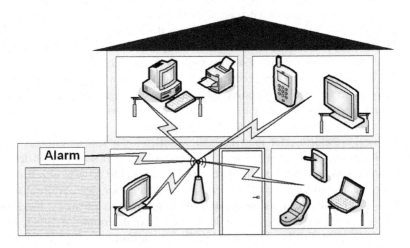

Fig. 13.2. In-home multimedia-based service distribution over wireless local area network.

11 Mbps; IEEE 802.11g - up to 54 Mbps in theory and only half that in practice), are error-prone and are affected by contention between stations for access to the shared medium, back-off mechanisms, collisions, signal attenuation with distance, signal interference, etc. Therefore wireless multimedia transmissions are often associated with lower user perceived quality. Significant efforts were put by the research community in proposing solutions to improve Quality of Service (QoS) when delivering multimedia over both wired and wireless networks [4–8]. Among these solutions most successful were the adaptive schemes [9], which adjust the content encoding rate and/or multimedia transmission rate to existing network conditions minimizing the delay and loss, which eventually determines increases in the user perceived quality.

The large majority of researchers who have studied and proposed techniques for adaptation of multimedia content to the network delivery conditions, have focused on network QoS-related aspects only. Unlike them, this book chapter presents a series of four solutions proposed to complement the network-based multimedia streaming adaptation in order to respond to other significant issues. These solutions offer support for user Quality of Experience (QoE)-oriented multimedia adaptation in loaded network delivery conditions, user-dependent region of interest multimedia content adaptation during streaming, distribution of differentiated multimedia quality streams when delivered over the same wireless network and device battery power level-based adaptation of multimedia content.

The book chapter is structured as follows. Next section briefly describes the proposed adaptive multimedia streaming solutions as well as the motivations of proposing the solutions. The related work section follows, summarizing the most significant research findings which have been presented in the literature in relation to adaptive multimedia streaming in general and quality-oriented, region of interest-based, differentiated quality and power-based adaptive streaming in particular. Sections 13.4, 13.5, 13.6 and 13.7 present the principles and major benefits of the four adaptive multimedia streaming solutions proposed by the author. The last section concludes the book chapter and presents possible future research avenues.

13.2 QoE-Oriented Adaptive Multimedia Streaming

13.2.1 Quality-Oriented Adaptation Scheme (QOAS)

Academic and industrial research aims to propose solutions for achieving and maintaining high user perceived quality while streaming multimedia over diverse network types and with various load levels. As user QoE is difficult to assess in-service during streaming, research has focused on easier-to-measure network performance-related QoS parameters. Most important of these parameters such as packet loss rate, delay, delay jitter for example can be monitored and based on their values and variation trends, adaptive decisions can be taken. Based on these decisions, the adaptive multimedia streaming schemes adjust the multimedia throughput, increasing or decreasing the transmission and/or multimedia content encoding rates. Consequently many of these solutions have shown improvements in the QoS levels and some in the resulting end-user QoE. However, the content adjustment policies of these QoS-based adaptive schemes are not directly related to the end-user perceived quality and there is only an indirect and often not causal relationship between the monitored QoS parameters, adaptation solutions and user QoE.

The author proposed the Quality Oriented Adaptation Scheme (QOAS) [10–12], which unlike other adaptive schemes, includes directly in the adaptive process an estimation of end-user perceived quality level. User QoE is monitored by the QOAS client and a feedback loop is employed to enable the server to take adjustment decisions based on most up to date information on the effect these decisions have on viewer perceived quality. As a direct consequence, QOAS's end-to-end content adaptation mechanism offers multimedia streaming with increased user QoE in loaded network delivery conditions.

13.2.2 Region Of Interest Adaptive Scheme (ROIAS)

As most adaptive multimedia solutions, QOAS affects equally the whole viewing area of video frames when adjusting the stream quality as part of the adaptive multimedia streaming process. However research has shown [13] that there are regions of the multimedia display image on which the viewers are more interested in than on others. It can be demonstrated that an area of maximum user interest (AMUI) exists and around it concentric multimedia display areas can be defined. These areas which are named regions of interest (ROI) attract lower user interest as their location is farther from the AMUI.

Based on these issues, the author proposed the Region Of Interest-based Adaptive Scheme (ROIAS) [14] for multimedia streaming that when performing adaptive streaming-related adjustments, selectively affects the quality of those regions of the image the viewers are the least interested in. As the quality of the regions the viewers are the most interested in will be affected by very little change (or will remain unchanged), ROIAS provides higher overall end-user perceived quality than any of the existing adaptive solutions that adjust equally the whole image area.

13.2.3 Priority-Based Differentiated Quality of Service Adaptive Wireless Multimedia Streaming Solution (pDiffQoS)

Available bandwidth limitation and variability is a well-known problem in wireless networks and equally affects the quality of all transmissions. However users have different requirements in relation to different services and they use many devices with various characteristics. For example, high definition television requires multimedia content to be streamed at a much higher bitrate than standard definition television to achieve the same level of expected user perceived quality. This is mainly due to the difference in screen size and resolution of the displaying devices which act as limiting factors.

However, existing wireless streaming solutions do not account for these differences, treating all the flows equally. Despite such an equalitarian solution being politically correct, the user perceived quality distribution is not fair among the competing sources, as their requirements differ. Consequently the viewers of multimedia content on high demanding devices will be more affected in their perceived QoE than those who view low complexity content on devices with basic characteristics.

Following this observation, the author has proposed the priority-based Differentiated Quality of Service adaptive wireless multimedia streaming solution (pDiffQoS) [15], an application-level adaptive scheme for in-home wireless delivery of multimedia-based services. pDiffQoS maintains high end-user perceived quality at heterogeneous client devices by adapting multimedia content bitrate to existing network conditions while taking into consideration their potentially different priorities. By using pDiffQoS, client device-priority will determine the end-user QoE level achieved during multimedia delivery in loaded network conditions.

13.2.4 Battery Power-Aware Adaptive Mechanism (BAM)

There are many technical difficulties when performing wireless multimedia streaming mainly due to the variability or high level of error rate, limitation and variability of available bandwidth, etc. Apart from these, streaming to mobile devices bring forward new challenges due to the limited processing power and battery power of the devices. Among all these issues, although battery power has significant importance as it enables the functionality of all mobile devices, it was the least researched and has experienced the least improvement in the past years.

As the latest complex applications and networking solutions drain more power, the batteries deplete faster. This depletion often occurs before an ongoing communication session has completed or before user interaction with an application has ended, triggering significant user dissatisfaction. The battery depletion process is accelerated when both a complex application and network communication are involved at the same time, as is the case when streaming multimedia. As most users prefer to have access to rich media services at lower quality levels than not to have access at all, multimedia adaptation based on current battery power level is envisaged.

In this context the author proposed the Battery Power-aware Adaptive Mechanism (BAM) for wireless multimedia streaming to mobile devices [16, 17]. This adaptive solution employs different individual power-saving schemes in the three major stages of the wireless multimedia streaming process to mobile devices: data transmission and reception, decoding and multimedia (dis)/playing. The mechanism introduces a step-wise solution to trade multimedia quality against battery power, eventually extending the life span of the current device battery cycle. As a direct consequence, BAM increases both user QoE and viewer satisfaction.

13.3 Related Work

13.3.1 Adaptive Multimedia Streaming Solutions

Important research effort has been put in order to propose more effective adaptive multimedia streaming solutions which adjust the streamed content to existing network load conditions. They aim at providing higher perceived quality for multimedia viewers by increasing QoS levels. Various adaptive schemes were designed and most were tested via either simulations or prototyping in wired best effort IP networks.

The Loss-Delay-based Adaptation Algorithm (LDA+) [18] is an AIMD network condition-based adaptive algorithm which uses RTP for data delivery and RTCP for feedback. It adapts very well in highly loaded network conditions, but it is often too aggressive with competing adaptive traffic. The equation-based TCP Friendly Rate Control (TFRC) [19] uses a TCP rate equation-based model to limit the aggressiveness of the adaptation to changing traffic conditions and prevent data starvation of competing adaptive traffic. However the competitiveness limitation is not best suited for multimedia streaming applications. TCP Friendly Rate Control with Compensation (TFRCC) [20] extends TFRC by providing better QoS support for multimedia streaming in short term while still providing good network fairness in long-term.

Other solutions include Rate Adaptation Protocol (RAP) [21] - a TCP-like acknowledgement-based AIMD scheme, Layered Quality Adaptation (LQA) [22] - algorithm based on layered encoding, Quality-Adaptive Media Streaming by Priority Drop [23] - adaptive approach based on voluntary priority drop of packets and Media and TCP-Friendly Rate-based Congestion Control (MTFRCC) [24] - a utility-based adaptive congestion control mechanism.

13.3.2 Wireless Multimedia Streaming Solutions

More recently the increase in the number of WLANs deployed has determined an increase in wireless multimedia streaming activities. Consequently, lately different adaptive solutions were proposed in order to respond to wireless-delivery specific challenges.

TFRC Wireless (TFRCW [25] enhances the TFRC adaptation mechanism by considering the particularity of wireless loss. TFRCW employs a Loss Discrimination Algorithm (LDA) which distinguishes between congestion-based and random wireless medium loss and improves the congestion control. Video Transport Protocol (VTP) [26] uses along LDA an estimation of the Achieved Rate instead of a multiplicatively decreased rate when loss occurs, achieving a higher consequent video quality. TCP Friendly Rate Control in Ad-hoc Wireless Networks (ADTFRC) [27] extends TFRC for ad-hoc wireless networks. ADTFRC involves multiple metrics which add complexity, but allow to estimate more accurately the network state and via adaptation to achieve higher throughput.

Other wireless adaptive multimedia delivery solutions include MULTFRC [28] - an adaptive solution which uses multiple parallel streams to increase the competitiveness of the current transmission, Optimum Adaptation Trajectory (OAT) [29] - a solution which assumes there is an optimum adaptation path in the resolution-frame rate bi-dimensional space to maximise user perceived quality and AMC-FGS [30] - an Adaptive Motion-Compensation Fine-Granular-Scalability solution for wireless video delivery.

13.3.3 Region of Interest Video Encoding

Region of interest-based adaptive streaming is a novel multimedia delivery approach proposed by the author and little existing research has been previously reported closely related to this. However, there is an important body of work in relation to ROI video encoding which refers to encoding different regions of the video clip frame with various qualities depending on user interest in the content presented in those regions of the display area. Authors of [31] provided a comprehensive introduction to the area of ROI video encoding and a novel Pixel Shader Algorithm which can be applied on both still images and video content in real-time. The implication is that this algorithm can be used for real-time video delivery and in particular for wireless multimedia streaming.

The authors of [32] considered the issue of perceptual disruptions in ROI video encoding. The results of their research indicate that by using ROI-based encoding, multimedia clips may not satisfy some viewers as this encoding process could introduce considerable perceptual distraction that can interrupt normal attentive processes. Moreover, [33] suggested that when adapting a high-resolution window at the point-of-gaze and degraded resolution in peripheral areas, participants took longer to identify a visual target, than when a low resolution was uniformly displayed across the whole display area. Therefore by using such a ROI-based encoding solution, a negative impact on some user tasks may be recorded.

A very important issue in ROI video encoding research is the determination of viewer ROIs. In general, the determination of ROIs is usually performed either by applying human vision models such as [34] and [35] in order to detect the most relevant areas of the video frame to the viewers or by empirically finding user ROIs via eye-tracking experiments [13]. Additionally, it is also important to find the best balance between quality and ROI location based on user interest in the video content. Although there is not any widely accepted procedure to determine the best balance between the quality and ROI location, most proposals put forward for multimedia clips vary either the stream's bitrate or the frame rate [13, 35–37].

13.3.4 Power Saving Solutions

There has been an intense research activity in relation to the proposal of different individual power-saving schemes in the three major stages of the wireless multimedia streaming process to mobile devices: data transmission and reception, decoding and multimedia (dis)/playing, respectively.

The data transmission and reception stage of the wireless multimedia streaming to mobile devices process is related to sending and receiving of data. In general, power saving in this stage involves more or less the wireless network interface card

(WNIC). Best known energy-aware solutions include an application-specific server side traffic shaping mechanism proposed by Chandra and Vahdat [38], a buffer-based energy efficient CPU scheduler for mobile devices described by Bae et al. [39], a scheme that uses traffic shaping with added benefit of a proxy scheduling algorithm introduced by Zhang and Chanson [40] and a client side energy prediction scheme is described by Wei et al [41]. Other power saving schemes for the WNIC have been designed at different levels of the OSI network model. For example, work proposed for the MAC layer includes [42–44] as well as the legacy PSM in the IEEE 802.11 standard [45]. Some solutions proposed for the network layer were described in [46, 47] whereas Stemm and Katz [48] proposed a transport layer solution. The work proposed in this book chapter is a solution that involves both transport and application layers.

The decoding stage is when the device receives the data and decodes it to a playable format. However, the encoding of multimedia data will impact the amount of battery power required for this stage. Power saving solutions which have been proposed in this stage generally focus on making the decoding of the data more efficient. Solutions include two schemes based on reducing the number of memory and bus accesses by high level language optimization presented in [49], a mechanism to reduce the multimedia decoding power by using feedback control proposed in [50], a scheme which can lower the supply voltage and reduce the power consumption detailed in [51] and the results of an investigation of the effect of dynamic voltage scaling on the trade off between energy consumption and high picture quality in multimedia decoding presented in [52]. The playing stage involves all activities related to displaying of the media to the user. Depending on the media type, this stage will involve the screen, the speakers or a combination of the two. The majority of power saving in this area has focused on the screen of the device, in particular, the backlight as in [53] and [54]. There is no known research on the effect of the speakers on the battery apart from that reported by the author in [16].

13.4 Quality-Oriented Adaptation Scheme (QOAS)

Quality Oriented Adaptation Scheme (QOAS) [10–12] is a feedback-based adaptive multimedia streaming solution which by using an innovative approach succeeded to achieve higher end-user perceived quality than other similar solutions.

As any other adaptive scheme for multimedia streaming, QOAS relies on the fact that random losses have a greater impact on the end-user perceived quality than a controlled reduction in quality [55]. However, the novel aspect that QOAS introduces is that it actively uses an estimation of viewer QoE in the feedback loop, helping in the adaptive optimisation process. QOAS uses an end-to-end sender-driven adaptation mechanism which controls the adjustment of both the quality of the streamed multimedia content and the transmission rate so that it maximises the end-user perceived quality in existing delivery conditions. This intra-stream QOAS adaptation mechanism controls the quality, and consequently, the quantity of streamed multimedia-related data and is based on information received from the client.

The QOAS-based system architecture includes a server and a client on which QOAS server and client adaptive applications are deployed. These applications unidirectionally transmit video data and bidirectionally exchange control packets through an IP-based delivery network. The QOAS client continuously monitors some transmission parameters and estimates the end-user perceived quality, and its Quality of Delivery Grading Scheme (QoDGS) regularly computes Quality of Delivery scores (QoDscores) that reflect the multimedia streaming quality in current delivery conditions. These grades are sent as feedback to the QOAS server, whose Server Arbitration Scheme (SAS) analyzes them and proposes adjustment decisions to be taken in order to increase the end-user perceived quality in existing conditions.

The QOAS adaptation principle is schematically described in Fig. 13.3. For each QOAS-based multimedia streaming process, a number of different quality states are defined at the server (e.g. the experimental tests have involved a five-state model). Each such state is then assigned to a different stream quality. The stream quality versions differ in terms of compression-related parameters (e.g. resolution, frame rate) and therefore have different bandwidth requirements. They also differ in expected end-user perceived quality. The difference between the average bitrates of these different quality streams is denoted as the "adaptation step". During data transmission, the client-located QoDGS computes QoDscores that are sent via feedback to the QOAS server. QOAS server-located SAS analyses these scores and makes suggestions regarding server quality state, based on which the QOAS server dynamically varies its quality state. When the delivery conditions cause excessive delays and/or loss, the client reports a decrease in end-user quality and the server switches to a lower quality state, reducing the bit-rate of the streamed multimedia. Consequently, this may reduce the delays and the loss, increasing the end-user perceived quality. If the QOAS client reports improved streaming conditions, the server increases the quality of the delivered stream. These switching to higher and lower quality states, respectively, are performed gradually with the granularity of the QOAS adaptation step. The smaller the adaptation step, the less noticeable to the viewer is the effect of the bitrate modification. However, the higher the adaptation step, the faster is the convergence of the algorithm to the bitrate best suited in existing network conditions. Experimental tests performed have suggested that a 0.5 Mbps adaptation step is not noticed by the viewers if the frequency of the quality state changes remains low.

QOAS client-located Quality of Delivery Grading Scheme (QoDGS), described in detail in [10], evaluates the effect of the delivery conditions on end-user perceived quality. It monitors both short-term and long-term variations of packet loss rate, delay and delay jitter, which have the most significant impact on the received quality [56, 57] and estimates the end-user perceived quality. The end-user quality is estimated using the no-reference moving pictures quality metric [58], which maps the joint impact of bitrate and data loss on video quality onto the ITU-T R P.910 five-point grading scale [59].

QOAS server-located Server Arbitration Scheme (SAS) assesses the values of a number of consecutive QoDscores received as feedback in order to reduce the effect of noise in the adaptive decision taking process. Based on these scores SAS suggests adjustment decisions. This process is asymmetric, requiring fewer QoDscores

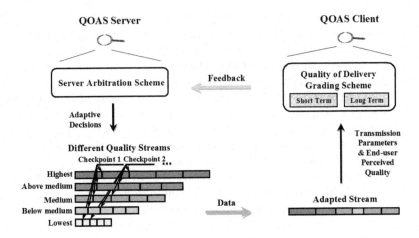

Fig. 13.3. Schematic illustration of QOAS's adaptation principle.

to trigger a decrease in the server's quality state than for an increase. This ensures a fast reaction during bad delivery conditions and helps to eliminate its cause. The increase is performed only when there is enough evidence that the network conditions have improved. This asymmetry helps also to maintain system stability, by reducing the frequency of quality variations. When QOAS is used to stream multimedia content to multiple viewers over a hierarchical network infrastructure like the one presented in Fig. 13.1, QOAS inter-stream adaptation mechanism complements the intra-stream adaptation and aims for a finer adjustment in the overall adaptation process. The inter-stream adaptation is responsible for preventing QOAS-based adaptive processes from reacting simultaneously to variations in the delivery network. It selectively allows some of the QOAS-based sources of multimedia data to react to the received feedback, in a step-by-step process, achieving near optimal link utilisation and long-term fairness between the clients.

QOAS was tested via both simulations and prototyping showing very significant positive results in terms of a very good balance between the viewers' need for high quality of multimedia service and the network operators' and service providers' goal of achieving high infrastructure utilisation and simultaneously serving a higher number of customers.

Extensive objective tests using simulation models have tested QOAS stand-alone and in comparison with other solutions such as TFRC [19], LDA+ [18], an ideal adaptive scheme and a non-adaptive mechanism in different delivery conditions [10, 11]. These objective tests have assessed the effect on QOAS performance of background traffic of various types and sizes and with different variation patterns commonly found in multi-service IP networks. QOAS showed very good performance in terms of end-user perceived quality, loss rate, throughput, and link utilization, and was very close to the performance of a hypothetical ideal adaptive scheme.

Subjective testing involving a prototype system, multimedia sequences representing different classes of multimedia clips in terms of motion content and type and a large number of participants have complemented the objective tests [12]. The subjective tests have assessed end-user perceived quality when using QOAS for multimedia streaming in high loaded delivery conditions and regardless of the motion content of the streamed clips, QOAS approach was highly graded by the test subjects which have appreciated it in comparison with a non-adaptive streaming approach in identical delivery conditions. These results highly recommend QOAS as a very efficient solution for delivering multimedia-based services to remote viewers at very good quality even in most variable and highly loaded network conditions.

13.5 Region Of Interest-Based Adaptive Scheme (ROIAS)

The Region of Interest-based Adaptive Scheme (ROIAS) [14] for multimedia streaming is an unicast rate-based adaptive solution for delivering high quality multimedia. ROIAS's goal is to increase the end-user perceived quality when viewing remotely streamed multimedia sequences in highly loaded delivery conditions by taking into consideration viewer's interest in certain regions of the multimedia frames and consequently differentiating their tratment during the adaptation process.

Fig. 13.4 presents schematically ROIAS architecture, which includes ROIAS client and server-located components. These components are involved in a bi-directional exchange of video data and control packets via the delivery network. In a similar fashion with QOAS, ROIAS client monitors some transmission-related parameters and regularly computes quality of delivery scores, which are sent as feedback to the server. The server analyses these scores and proposes content-related adjustment decisions in order to increase user QoE in existing delivery conditions. Consequently the transmitted quantity of multimedia data and eventually user perceived quality may vary during the streaming process.

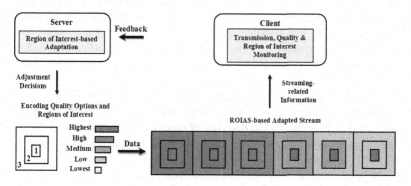

Fig. 13.4. Illustration of ROIAS's adaptation principle.

Existing adaptive multimedia streaming schemes involve content modifications that affect equally the whole viewing area of the multimedia frames being transmitted. However as eye-tracking research has shown [13], there are some regions within multimedia streams' frames the viewers are more interested in than in others. Consequently ROIAS enhances the classic network condition-based adaptive solution for streaming multimedia with a novel approach.

When required to reduce the quantity and consequently the quality of transmitted multimedia-related information in order to meet the available bandwidth constraints, ROIAS affects the streamed data in terms of some compression-related parameters such as resolution and frame-rate differently based on the ROIs and on the user interest level on them. As result those ROIs the user is highly interested in are transmitted at high quality, whereas those on which the user interest is lower are streamed at lower quality, saving bandwidth.

ROIAS's server side component maintains a viewer region of interest model that is updated regularly by the feedback. Based on the information from this model, ROIAS selectively adjusts the quality of ROIs the viewer is the least interested in when transmission-related quality adaptations are required to be performed. As the quality of the regions the viewers are the most interested in will le less affected by the quality adaptation, ROIAS will provide much higher overall end-user perceived quality than any of the existing adaptive solutions. Based on the multimedia stream resolution and on eye-tracking research results, the ROIAS server defines the overall multimedia viewing area and, within this area, a number of different regions of interests (ROIs). The placement of these ROIs is highly dependent on the multimedia sequence content and may also vary within the same multimedia stream from one scene to another. Following the observation that most professionally captured multimedia content includes the areas of highest user interest approximately in the middle of the viewing rectangle, in order for the proposed solution to be independent from the delivered content, ROIAS considers only concentric ROIs and associates the highest user interest to the ROI closest to the centre of the image. ROIAS orders these ROIs based on the decreasing user interest on them. For example Fig. 13.4 illustrates three such ROIs. The server also introduces a number of different potential multimedia quality levels which could be applied on the variously defined ROIs. Fig. 13.4 exemplifies five possible quality levels: highest, high, average, low and lowest. The different quality versions of the same content are to be obtained in real time by adjusting some compression-related parameters such as resolution and frame rate.

The ROIAS server has an associated finite state model. Each server state indicates what multimedia quality level is to be associated with each of the ROIs in terms of the pre-defined quality levels. The ROIAS server model gracefully degrades the quality of neighboring ROIs in order to offer a smooth transition from higher to lower quality regions and consequently to maintain high viewer QoE.

The ROIAS client monitors multimedia delivery in terms of loss, delay, jitter and estimated end-user perceived quality. It grades short-term and long-term Quality of Delivery grades (QoDscores) which are sent to the server via feedback. The same mechanism is employed by QOAS.

During transmission the server dynamically varies its state according to the client feedback. For example in highly loaded delivery conditions, when the client reports a decrease in end-user quality due to packet loss, the server switches to a lower quality state, which requires the reduction in the quantity of data sent. This reduction is achieved by employing the proposed ROIAS and affects the most the areas of least user interest. As a consequence of step-wise ROIAS-based adaptation, stream average rate reduction is achieved to a level that significantly lowers the loss rate. Consequently the stream's end-user perceived quality increases, in spite of transmitting less information. In improved conditions, the server gradually increases the quality of the delivered stream and if the loss rate is kept low, this determines an increase in the end-user perceived quality. As a direct consequence of the ROIAS-based multimedia stream adaptation the viewers will receive an "adapted" stream as illustrated in Fig. 13.4. This is performed in order to potentially reduce the loss rate and therefore increase the overall end-user perceived quality.

ROIAS was assessed via both objective and subjective testing [14]. Simulation results involving multiple clients streaming multimedia simultaneously using ROIAS showed how ROIAS performed much better in terms of average client throughput, loss and estimated viewer QoE than when a non-adaptive streaming scheme was used. Subjective testing involving clips with different levels of motion content showed how ROIAS-based adaptive multimedia streaming was very well received by the tests subjects. Other ROIAS tests are in progress.

13.6 Priority-Based Differentiated Quality of Service Adaptive Wireless Multimedia Streaming Solution (pDiffQoS)

The proposed Priority-based Differentiated Quality Adaptive Multimedia Streaming Scheme (pDiffQoS) [15] is an adaptive solution for multimedia streaming over in-home wireless local area networks which does not treat all traffic equally unlike most multimedia streaming solutions. Instead pDiffQoS enables the delivery of multimedia-based services of different quality to clients that have assigned different priorities by the viewer. This implies a decrease in transmitted quality and thus quantity of data transmitted for clients that have low priority, allowing the other streams to take advantage of the bandwidth resources available according to their associated priorities. This allows a fair distribution of end-user perceived quality even in loaded delivery network conditions as it is done according to the importance of the streams to their viewers.

As a direct consequence of pDiffQoS principle, in high loaded network delivery conditions lower priority clients have their transmission rate decreased more than the higher priority ones. When network resources become available again, pDiffQoS increases the quality of all multimedia streams. However this quality upgrade process is such performed that the higher priority clients will benefit more and faster than the ones that have lower priorities.

The high-level architecture of the system that enables in-home adaptive multimedia streaming with differentiated QoS includes an Intelligent in-Home Access

Point Server (IHAPS) that acts as a local proxy server and a number of Multimedia-enabled Clients (MC) to which multimedia-based services are delivered via the in-home WLAN. pDiffQoS has both client and server components deployed at the level of MC and IHAPS respectively. They exchange video data and control packets via the wireless network.

The pDiffQoS clients continuously monitor the quality of delivery in terms of throughput, loss rate and estimation of end-user perceived quality. Values of these metrics are regularly compared to session specific targets and determine the values of the QoDscores that reflect multimedia streaming quality in current delivery conditions. These scores along with the client priority are sent as feedback to the pDiffQoS server, which analyses them and takes content adjustment decisions in order to increase the viewer QoE according to the client associated priority.

The pDiffQoS server associates an internal state to each multimedia streaming process. This server state is related to the stream transmitted quality: the higher the quality server state, the higher the quality of the transmitted stream and therefore the larger the size of data to be streamed. Also the lower the server state, the lower the quality and consequently the smaller is the amount of information to transmit. During streaming, based on client feedback, the server changes its quality state upgrading or downgrading the quality of streamed multimedia. This results in adaptation that helps improving streaming quality. The quality of live-streamed multimedia can be varied dynamically based on client feedback by changing encoding parameters. The resulted stream can differ in resolution, frame rate, etc. For pre-recorded transmissions, different quality versions with the same content must be available for the server that dynamically switches between them based on feedback, during the streaming. Fig. 13.5 exemplifies a situation with five server quality states: high, above medium, medium, below medium and low. It also presents possible different quality streams associated to the same multimedia content.

Fig. 13.5 also presents an illustration of three resulted adapted streams as effect of possible pDiffQoS adaptations performed during streaming processes initiated by clients with different priorities on a three-point scale. It could be seen that the high priority client C1 receives a stream at a higher average quality than client C2 with normal priority and this streamed multimedia is at a higher quality than the one delivered to the viewer C3 with the lowest priority.

Client priorities are assigned statically by the users of the in-home systems based on their subjective assessment (e.g. living room home theatre has always the highest priority, while the game consoles may have the lowest). However pDiffQoS could also operate if these priorities are dynamically modified based on the efficient usage of resources for example.

The pDiffQoS bitrate adaptation is carried out using the Greediness Control Algorithm (GCA) [60, 61], a specially designed upper transport-level prioritization mechanism for multimedia delivery in wireless networks. It performs variable rate adaptation according to both receiver feedback and priority level based on user interest in the multimedia stream. GCA extends the TFRC solution stardardised also by the IETF [19] by introducing two parameters that allow the streaming applica-

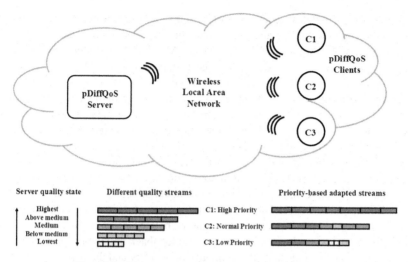

Fig. 13.5. Illustration of DiffQoS's adaptation principle.

tion to tune the aggressiveness of the rate adaptation and as a result, introduce true prioritization to the media streaming process.

Simulation results have shown that pDiffQoS combined with GCA enable fair prioritisation of the media flows based on the viewer assigned stream priorities. The results also showed that this form of prioritisation increases the overall user QoE achieved when streaming was performed to a number of heterogeneous devices operating within the same in-home wireless network.

13.7 Battery Power-based Adaptive Mechanism (BAM)

The Battery Power-based Adaptive Mechanism for wireless multimedia streaming to mobile devices (BAM) [16, 17] aims at maintaining a good balance between multimedia streaming quality and the duration of streaming as supported by the client device battery power level. BAM's goal is to enable an increase in the duration of the streaming session by reducing the quality of the multimedia stream, reducing therefore the quantity of data to be streamed, decoded and displayed by the client device. This will have a positive effect on the power drained from the device battery, extending its life and consequently increasing the duration it can support multimedia streaming.

The proposed BAM, which involves a client-server approach, has its architecture outlined at block level in Fig. 13.6. BAM's major modules at server side include a battery power model, an adaptation module and a streaming unit. The battery power model makes estimations regarding remote mobile device battery depletion time. This model considers information on current client applications usage and their complexity, wireless network interface usage as well as information on the usage of other battery power-draining components such as CPU, memory, display and speakers.

Regular feedback updates the model with most up-to-date information from the client side. The adaptation module analyses battery power consumption in all multimedia streaming phases: data transmission/reception, decoding and (dis)playing and based on the output of the power model, will suggest adaptive measures to be taken in order to slow down the battery depletion process. Among the adaptive measures to be taken are modifications in wireless data delivery manner, change in the way in which multimedia is (dis)played to the viewer and improvement of the data encoding-decoding process. These measures are to be taken in order to minimise their effect on the multimedia viewer perceived quality. The streaming unit is in charge with streaming the adapted content to the client device over the wireless network.

As Fig. 13.6 shows BAM's client major modules are the receive, decode and display module in charge with processing multimedia stream and making it available to the viewer, transmission information monitoring module in charge with streaming process monitoring and assessment of its quality, battery information module responsible with battery power level measurement and feedback unit in charge with informing regularly the server about both the quality of delivery and the battery power level at the client device.

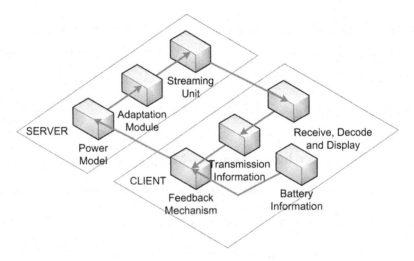

Fig. 13.6. BAM architecture.

BAM-based adaptive multimedia streaming process starts with a client requesting a multimedia stream from the BAM server. BAM client also provides information about its current battery state, such as current voltage and estimated remaining life in current device usage conditions. BAM server receives this request and before streaming the media content to the client, compares the media content playtime, at its current quality, with the received battery lifetime information. This comparison is carried out using the power model. If the media content playtime is higher than the battery lifetime remaining, then the adaptation module will suggest some power saving techniques to be applied in order to enable non-interrupted streaming of the whole clip. If the estimated client device battery life covers the clip duration, the adaptation module will not affect the multimedia content. The streaming unit delivers the adapted multimedia content to the client.

A number of power-save mechanisms can be applied incrementally by the BAM server-located adaptation module in order to carry out the power-based adaptation. If the estimation is that multimedia streaming will still not be completed in the new conditions, the next power save mechanism will be implemented, otherwise no further power save mechanisms will be deployed. The first power mechanisms employed can be those related to the wireless reception stage such as the AB-PSM [62], as this multimedia streaming stage drains the most power from battery and significant savings can be obtained with very little influence on end-user perceived quality. The next power saving mechanisms which can be applied are those from the decoding stage when content bitrate adjustment can be applied with different target bitrates. Last set of power saving mechanisms can be applied in the (dis)playing stage, affecting the screen luminosity and speakers' volume levels.

According to BAM, multimedia streaming server will have associated a state which indicates which power save mechanism is applied in each of the three streaming stages: reception, decoding and (dis)playing. Fig. 13.7 shows how different indexes are used for each stage to indicate the power save scheme employed as follows: the reception stage uses i, the decoding stage uses j and the playing stage uses k. Consequently the power save state of the BAM multimedia streaming system can be represented by the triplet $\{i, j, k\}$.

BAM client receives streamed encoded multimedia data, decodes it and (dis)plays the media content. At periodic intervals the client feedback mechanism sends updated battery information to the server. Once the server receives this feedback, it follows the same sequence as before, comparing the power readings and adapting the media content if necessary. The presence of the server-located battery power model provides a backup that enables the adaptive streaming process to continue to work as the model estimates the battery life span given certain device-processing load. If feedback is available, it will increase the accuracy of the power-based adaptation process.

As it is expected that the power-based adaptation carried out on the multimedia streaming process will influence the end-user quality, end-user perceived quality boundaries will be set such as overall streamed multimedia will always be at least at an acceptable quality level.

Emulation-based testing results showed how BAM-based multimedia streaming utilises more efficiently the available battery power, obtaining an important extension of mobile devices' battery life with little effect on user perceived quality.

Conclusion

The large majority of researchers who have studied and proposed techniques for adaptation of multimedia content to the network delivery conditions, have focused on

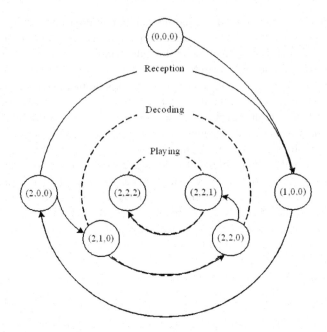

Fig. 13.7. BAM's adaptation state diagram.

network QoS-related aspects only. Unlike them, this book chapter presents a series of four solutions proposed to complement the network-based multimedia streaming adaptation in order to respond to other significant issues. These solutions offer support for user Quality of Experience (QoE)-oriented multimedia adaptation in loaded network delivery conditions, user-dependent region of interest multimedia content adaptation during streaming, distribution of differentiated multimedia quality streams when delivered over the same wireless network and device battery power level-based adaptation of multimedia content. Special sections were allocated to describe each of these schemes which have showed very good testing results: Quality Oriented Adaptive Scheme (QOAS), Region-Of-Interest Adaptive Scheme (ROIAS), Priority-based Differentiated Quality of Service (pDiffQoS) in wireless local area networks and Battery power Adaptive Mechanism for wireless multimedia streaming to mobile devices (BAM).

Acknowledgment

The author wishes to acknowledge the collaboration with Dr. Gheorghita Ghinea, Janet Adams, Edward Casey, Eric Vinouze and Timothy Noel Sheehan and to thank them for their contribution to the research reported in this book chapter.

References

1. IDC, "Top 10 Predictions for the Western European Broadband Market in 2008," *IDC*, December 2007, http://www.the-infoshop.com/study/id58981-european-broadband_toc.html

2. IDC, "Worldwide Telecommunications Infrastructure 2008 Top 10 Predictions," IDC, December 2007, http://www.the-infoshop.com/study/id58679-worldw-telecom_toc.html

3. C. Ngo, "A service-oriented wireless home network," *IEEE Consumer Communications and Networking Conference*, Jan. 2004, pp. 618-620.

4. B. Braden, D. Clark, and S. Shenker, "Integrated services in the Internet architecture: An overview," *IETF RFC 1633*, June 1994, http://www.ietf.org/rfc/rfc1633.txt

5. S. Blake et al., "An architecture for differentiated services," *IETF RFC 2475*, Dec. 1998, http://www.ietf.org/rfc/rfc2475.txt

6. E. C. Rosen, A. Viswanathan, and R. Callon, "MPLS architecture", *IETF RFC 3031*, Jan. 2001, http://www.ietf.org/rfc/rfc3031.txt

7. D. Hutchison, A. Mauthe, and N. Yeadon, "Quality-of-service architecture: Monitoring and control of multimedia communications," *IEEE Electronics and Comms. Eng. Journal*, vol. 9, no. 3, 1997, pp. 100-106.

8. D. Wu, Y. T. Hou, W. Zhu, Y.-Q. Zhang, and J. M. Peha, "Streaming video over the Internet: Approaches and directions," *IEEE Trans. on Circuits and Systems for Video Technology*, vol. 11, no. 3, 2001, pp. 282-300.

9. D. Wu, T. Hou, and Y.-Q. Zhang, "Scalable video coding and transport over broadband wireless networks," in *Proceedings of IEEE*, vol. 89, no. 1, Jan. 2001, pp. 6-20.

10. G.-M. Muntean, P. Perry, and L. Murphy, "A new adaptive multimedia streaming system for all-IP multi-service networks," *IEEE Transactions on Broadcasting*, vol. 50, no. 1, Mar. 2004, pp. 1-10.

11. G.-M. Muntean, "Efficient delivery of multimedia streams over broadband networks using QOAS," *IEEE Trans. on Broadcasting*, vol. 52, no. 2, June 2006, pp. 230-235.

12. G.-M. Muntean, P. Perry, and L. Murphy, "Subjective assessment of the quality-oriented adaptive scheme," *IEEE Trans. on Broadcasting*, vol. 51, no. 3, Sept. 2005, pp. 276-286.

13. S. R. Gulliver and G. Ghinea, "Stars in their eyes: What eye-tracking reveals about multimedia perceptual quality," *IEEE Transactions on Systems, Man and Cybernetics*, Part A, vol. 34, No. 4, July 2004, pp. 472-482.

14. G.-M. Muntean, G. Ghinea, and T. N. Sheehan, "Region of interest-based adaptive multimedia streaming scheme," *IEEE Trans. on Broadcasting*, vol. 54, no. 2, June 2008.

15. G.-M. Muntean, "Providing differentiated end-user quality for in-home wireless multimedia streaming," in *Proc. of IEEE Int. Conf. on Telecommunications*, Funchal, Portugal, May 2006.

16. J. Adams and G.-M. Muntean, "Power save adaptation algorithm for multimedia streaming to mobile devices," *IEEE International Conference on Portable Information Devices*, Orlando, Florida, USA, 2007.

17. J. Adams and G.-M. Muntean, "Adaptive-buffer power save mechanism for mobile multimedia streaming," in *Proc. of IEEE International Conference on Communications (ICC'07)*, Glasgow, Scotland, UK, 2007.

18. D. Sisalem and H. Schulzrinne, "The loss-delay based adjustment algorithm: A TCP-friendly adaptation scheme," in *Proc. of ACM NOSSDAV*, Cambridge, UK, July 1998, pp. 215-226.

19. M. Handley, S. Floyd, J. Padhye, and J. Widmer, "TCP friendly rate control (TFRC): Protocol specification," *IETF* RFC 3448, 2003, http://www.ietf.org/rfc/rfc3448.txt

20. P. Zhu, W. Zeng, and C. Li, "Joint design of source rate control and QoS-aware congestion control for video streaming over the Internet," *IEEE Trans. on Multimedia*, vol. 9, no. 2, Feb. 2007, pp. 366-376.

21. R. Rejaie, M. Handley, and D. Estrin, "RAP: An end-to-end rate-based congestion control mechanism for realtime streams in the Internet," in *Proc. of IEEE INFOCOM*, 1999, pp. 1337-1345.

22. R. Rejaie, M. Handley, and D. Estrin, "Layered quality adaptation for Internet video streaming," *IEEE JSAC, Special Issue on Internet QoS*, vol. 18, no. 12, Dec. 2000, pp. 2530-2543.

23. C. Krasic, J. Walpole, and W. Feng, "Quality-adaptive media streaming by priority drop," in *Proc. of ACM NOSSDAV*, Monterey, CA, USA, June 2003, pp. 112-121.

24. J. Yan, K. Katrinis, M. May, and B. Plattner, "Media- and TCP-friendly congestion control for scalable video streams," *IEEE Trans. on Multimedia*, vol. 8, no. 2, Apr. 2006, pp. 196-206.

25. S. Cen, P. C. Cosman, and G. M. Voelker, "End-to-end differentiation of congestion and wireless losses," *IEEE/ACM Transactions on Networking*, vol. 11, no. 5, 2003, pp. 703-717.

26. G. Yang, L.-J. Chen, T. Sun, M. Gerla, and M. Y. Sanadidi, "Smooth and efficient real-time video transport in presence of wireless errors," *ACM Transactions on Multimedia Computing, Communications and Applications (TOMCAPP)*, vol. 2, no. 2, May 2006, pp. 109-126.

27. Z. Fu, X. Meng, and S. Lu, "TCP friendly rate adaptation for multimedia streaming in mobile ad-hoc networks," *IEEE MMNS*, Belfast, Northern Ireland, Sept. 2003.

28. M. Chen and A. Zakhor, "Multiple TFRC connections-based rate control for wireless networks," *IEEE Trans. on Multimedia*, vol. 8, no. 10, Oct. 2006, pp. 1045-1061.

29. N. Cranley, L. Murphy, and P. Perry, "Optimum adaptation trajectories for streamed multimedia," *ACM Multimedia Systems Journal*, Springer-Verlag, vol. 10, no. 5, 2005, pp. 392-401.

30. M. van der Schaar and H. Radha, "Adaptive motion-compensation fine-granular-scalability (AMC-FGS) for wireless video," *IEEE Trans. Circuits Syst. Video Technol.*, vol. 12, no. 6, pp. 360-371, June 2002.

31. A. T. Duchowski and A. Coltekin, "Foveated gaze-contingent display for peripheral LOD management, 3D Visualisation, and stereo imaging," *ACM Transactions on Multimedia Computing, Communications and Applications*.

32. L. C. Loschky and G. S. Wolverton, "How late can you update gaze-contingent multi-resolutional displays without detection?," *ACM Transactions on Multimedia Computing, Communications and Applications*

33. E. M. Reingold and L. C. Loschky, "Reduced saliency of peripheral target in gaze-contingent multi-resolutional displays: Blended versus sharp boundary windows," in *Proc. of Eye Tracking Research and Applications Symposium*, New Orleans, Louisiana, USA, pp. 89-93, 2000.

34. W. Lai, X.-D. Gu, R.-H. Wang, W.-Y. Ma, and H.-J. Zhang, "A content-based bit allocation model for video streaming," in *Proc. of IEEE International Conference on Multimedia and Expo (ICME'04)*, Taipei, Taiwan, June 2004, pp. 1315-1318.

35. W. Osberger and A. J. Maeder, "Automatic identification of perceptually important regions in an image using a model of the human visual system," in Proc. of *14th Interna-*

tional Conference on Pattern Recognition, Brisbane, Australia, August 1998, pp. 701-704.

36. W. Lai, X-D. Gu, R-H. Wang, L-R. Dai, and H.-J. Zhang "A region based multiple frame-rate tradeoff of video streaming," in *Proc. of IEEE International Conference on Image Processing (ICIP'04)*, Singapore, October 2004, pp.2067–2070.

37. A. C.-W. Wong, and Y.-K. Kwok, "On a region-of-interest based approach to robust wireless video transmission," *IEEE International Symposium on Parallel Architectures, Algorithms and Networks (ISPAN'04)*, Hong Kong, China, May 2004, pp 385-390.

38. S. Chandra and A. Vahdat, "Application-specific network management for energy-aware streaming of popular multimedia formats," in *Proc. of General Track, USENIX Annual Technical Conference*, pp. 329-342, 2002.

39. G. Bae, J. Kim, D. Kim, and D. Park, "Low-power multimedia scheduling using output pre-buffering," in *Proc. of IEEE International Symposium on Modeling, Analysis, and Simulation of Computer and Telecommunication Systems*, pp. 389-396, 2005.

40. F. Zhang and S. Chanson, "Proxy-assisted scheduling for energy-efficient multimedia streaming over wireless LAN," in *Proc. of 4th International IFIP-TC6 Networking Conference, LNCS*, vol. 3462, pp. 980-991, 2005.

41. Y. Wei, S. M. Bhandarkar, and S. Chandra, "A client-side statistical prediction scheme for energy aware multimedia data streaming," *IEEE Trans. on Multimedia*, vol. 8, no. 4, pp. 866-874, 2006.

42. R. Krashinsky and H. Balakrishnan, "Minimizing energy for wireless web access with bounded slowdown," in *Proc. of Annual International Conference on Mobile Computing and Networking (MOBICOM)*, pp. 119-130, 2002.

43. B. Prabhakar, E. Biyikoglu, and A. El Gamal, "Energy-efficient transmission over a wireless link via lazy packet scheduling," in *Proc. of IEEE INFOCOM*, vol. 1, pp. 386-394, 2001.

44. L. Y. Zhang, Y. Ge, and J. Hou, "Energy-efficient real-time scheduling in IEEE 802.11 wireless LANs," in *Proc. of International Conference on Distributed Computing Systems*, pp. 658-667, 2003.

45. IEEE 802.11, (2000), "IEEE 802.11: Wireless LAN medium access control (MAC) and physical layer (PHY) specification," *IEEE*, 2000.

46. Q. Li, J. Aslam, and D. Rus, "Online power-aware routing in wireless ad-hoc networks," in *Proc. of Annual International Conference on Mobile Computing and Networking (MOBICOM)*, pp. 97-107, 2001.

47. Y. Xu, J. Heidemann, and D. Estrin, "Geography-informed energy conservation for ad hoc routing," in *Proc. of International Conference on Mobile Computing and Networking (MOBICOM)*, pp. 70-84, 2001.

48. M. Stemm and R. Katz, "Measuring and reducing energy consumption of network interfaces in hand-held devices," *IEICE Transactions on Communications*, vol. E80-B, pp. 1125-1131, August 1997.

49. P. Pakdeepaiboonpol and S. Kittitornkun, "Energy optimization for mobile MPEG-4 video decoder," *Int. Conference on Mobile Technology, Applications and Systems*, pp. 1-6, November 2005.

50. Z. Lu, J. Lach, M. Stan, and K. Skadron, "Reducing multimedia decode power using feedback control," in *Proc. of 21st International Conference on Computer Design*, pp. 489-496, 2003.

51. S. Lee, "Low-power video decoding on a variable voltage processor for mobile multimedia applications," *ETRI Journal*, vol. 27, no. 5, pp. 504-510, 2005.

52. M. Mesarina and Y. Turner, "Reduced energy decoding of MPEG streams," *ACM Multimedia Systems*, vol. 9, no. 2, pp. 202-213, 2003.

53. S. Pasricha, S. Mohapatra, M. Luthra, N. Dutt, and N. Venkatasubramanian, "Reducing backlight power consumption for streaming video applications on mobile handheld devices", in *Proc. of ACM/IEEE/IFIP Workshop on Embedded Systems for Real-Time Multimedia*, pp. 11-17, 2003.

54. H. Shim, N. Chang, and M. Pedram, "A backlight power management framework for battery-operated multimedia systems," *IEEE Design and Test of Computers*, vol. 21, no. 5, pp. 388-396, 2004.

55. G. Ghinea and J. P. Thomas, "QoS impact on user perception and understanding of multimedia video clips," in *Proc. of ACM Multimedia Conference*, Bristol, United Kingdom, 1998, pp. 49-54.

56. L. Zhang, L. Zheng, and K. S. Ngee, "Effect of delay and delay jitter on voice/video over IP," *ACM Computer Communications*, vol. 25, no. 9, June 2002, pp. 863-873.

57. J.-C. Bolot and T. Turletti, "Experience with control mechanisms for packet Video in the Internet," *ACM Computer Communication Review*, vol. 28, no. 1, January 1998, pp. 4-15.

58. O. Verscheure, P. Frossard, and M. Hamdi, "User-oriented QoS analysis in MPEG-2 video delivery," *Journal of Real-Time Imaging*, vol. 5, no. 5, October 1999, pp. 305-314.

59. ITU-T Recommendation P.910, "Subjective Video Quality Assessment Methods for Multimedia Applications," September 1999.

60. E. Casey and G.-M. Muntean, "A priority-based adaptive scheme for wireless multimedia delivery," in *Proc. of IEEE International Conference on Multimedia and Expo (ICME'06)*, Toronto, Canada, July 2006.

61. E. Casey and G.-M. Muntean, "TCP compatible greediness control for wireless multimedia streaming," in *Proc. of IEEE 65th Vehicular Technology Conference (VTC'07)*, Dublin, Ireland, April 2007.

62. J. Adams and G.-M. Muntean, "Adaptive-buffer power save mechanism for mobile multimedia streaming," in *Proc. of IEEE International Conference on Communications (ICC'07)*, Glasgow, Scotland, UK, June 2007.

14

Differentiated Pricing Policies in Heterogeneous Wireless Networks

Shamik Sengupta[1] and Mainak Chatterjee[2]

[1] Department of Electrical and Computer Engineering
Stevens Institute of Technology
Hoboken, NJ 07030
Shamik.Sengupta@stevens.edu

[2] School of Electrical Engineering and Computer Science
University of Central Florida
Orlando, FL 32816
mainak@eecs.ucf.edu

14.1 Introduction

With the technological advancements and economic changes, wireless service providers (WSPs) are finding it difficult to maintain a steady customer–base with just a single type of network. In the near future, it is anticipated that the wireless service providers (WSPs) will use a multitude of access technologies, operating on both licensed and unlicensed bands, to serve an increasing number of subscribers. It is likely that numerous types of access networks will prevail to support wireless services that have varied QoS requirements, e.g., real time video and telephony are more delay sensitive than non-real time services such as file downloads. Even for a particular kind of network i.e., cellular, wireless local area networks (WLAN) or wireless metropolitan area network (WMAN), it is not clear which technology will ultimately emerge as the global standard. For example, both CDMA (code division multiple access) or GPRS (GSM based general packet radio service) systems are widely deployed as far as cellular networks are concerned. To harness the wide variability of coverage, bandwidth, and reliability offered by different technologies operating at different spectrum bands, WSPs are planning to deploy heterogeneous access networks in an overlaid fashion. These heterogeneous networks would be capable of providing different sets of services governed by their corresponding quality–of–service (QoS) capabilities. The most common example seen today is the accessibility of Wi-Fi hotspots on top of third generation ($3G$) cellular services [1]. It is envisaged that multi-modal mobile terminals will have the capability to seamlessly switch between different access networks to support varying QoS and network connectivity constraints [2, 3]. A simple generalized illustration of such a heterogeneous wireless network can be seen in Fig. 14.1 where the mobile terminal can connect to the backhaul Internet through three different access networks.

E. Hossain (ed.), *Heterogeneous Wireless Access Networks*,
DOI: 10.1007/978-0-387-09777-0_14, © Springer Science+Business Media, LLC 2008

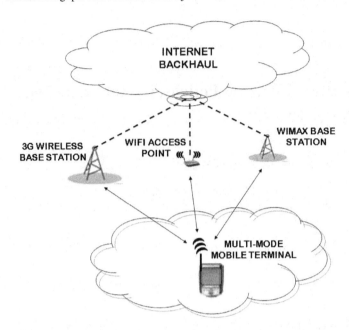

Fig. 14.1. A heterogeneous wireless networking scenario.

Even with the deployment of multiple access networks, the WSPs are finding it difficult to attract customers. This is primarily because of the presence of multiple WSPs in any geographic region which is forcing a competitive environment where each WSP is trying to maximize its profit. As far as the end users are concerned, there is still a strong association with a single WSP i.e., a user usually subscribes to one provider for a period of time (e.g., 1-2 years) and gets the services as per the contractual agreement. However, it is anticipated that in the near future, the concept of service brokers, technically known as Mobile Virtual Network Operators (MVNO) [4], will evolve that will act as an interface between the providers and the users [5]. These service brokers will allow end users more freedom to move from long-term service provider agreements to more opportunistic service models.

With such loose association between the users and the WSPs, the first question that arises is 'how or which wireless service provider should a user select for a particular service'? Second, 'what price must the WSPs charge such that they are able to attract the users and maximize their profit'? In light of these new developments, it is important to investigate the economic issues that has a profound impact on the service quality and the prices paid by the end users. By introducing the providers and users in a market like environment, it becomes convenient to leverage the concept of prices to regulate the demands of users who consume resources (bandwidth). This chapter discusses the popular approaches that have been adopted for pricing of wireless data services. In particular, the focus is on overlaid heterogeneous networks that are deployed by competing service providers who provide heterogeneous services.

14.2 Service Pricing

Similar to any market, the wireless services market will constitute services that are sold by the WSPs and bought by the end users. Thus, determination of the correct prices becomes a pivotal element for admission control and QoS provisioning. As far as the prices for the various services are concerned, the wireless industry simply followed the charging model for voice services, i.e., free or flat-rate. One example of flat-rate pricing could be paying a certain amount of money every month and having unlimited usage facility. (Some of the variations of flat-rate pricing can be found in [6, 7].) As a result, wireless users community have exhibited selfishness towards resource usage. However, the extensive increase in wireless service access and the corresponding quality of service degradation have demonstrated the need for a more dynamic pricing model. Also, the current practice of offering different contract plans (based on minute usage) for voice services will no longer be valid for data services. The same notion of resource sharing in voice networks can not be used because packet data systems are usually aimed at maximizing the throughput. The goal is to allocate each user the maximum data rate possible based on the application needs and wireless channel conditions while maintaining fairness among users. This new way of looking at the resource allocation brings forth the requirement for proper pricing schemes for wireless services.

Several pricing mechanisms so far have been proposed; three popular ones are listed below.

- **Paris Metro Pricing:** Paris metro pricing, more popularly known as PMP, was proposed by Odlyzko [8] to study differentiated services based on pricing only. However, like any other first generation proposal, PMP was the simplest differentiated services system and was intended for efficient congestion control. The main idea of PMP was to statically partition the main network into several logically separate channels that differ in price. Each of the channels would offer different expected QoS to the users. The key idea behind PMP is that the channel that is more highly priced would attract less traffic and therefore be able to deliver a higher level of QoS. PMP is mostly designed to be acceptable to users, who have a strong preference for flat-rate pricing.
- **Smart Market:** In the Smart market pricing, users no longer pay for their entire session but provide a bid price for each packet they wish to submit to the network [9]. The price to send a packet therefore dynamically varies with respect to the congestion in the network. However, the users do not actually pay their own bid submitted but the highest bid price of the user losing in the smart market network. An example on such bidding system an be found in [31]. However, the biggest disadvantage of smart market pricing mechanism is its high complexity even though QoS differentiation is not considered. Other risks with such market mechanism are the dangers associated with an incorrectly implemented auction-bidding model. For example, New Zealand followed smart market mechanism for allocating spectrum using second price auctions, i.e., the winner of the auction pays the second highest bid. In one case, a firm bid $NZ 100,000 but paid

the second highest bid of $NZ 6. A university student bid $NZ 1 for a T.V. license and won the auction but paid nothing because no one else bid for the license [10].

- **Threat Strategy Pricing:** Economics of network pricing with multiple competing Internet Service Providers (ISPs) are investigated in this approach [11]. In this model, Internet is separated into two tiers: *Local ISPs* which are co-located in a small geographical region and compete for the same customer base; *Transit ISPs* which carry traffic between local ISPs and customer base. The pricing mechanism is investigated from the standpoint of local and transit ISPs and customer pricing with differentiated QoS between the tiers. To devise the equilibrium pricing, threat strategy is used extensively, i.e., any selfish behavior from any of the tier would lead to the loss of revenue of other tiers and thus resulting in punishing strategy taken by the affected tiers to the selfish tier.

A major limitation with most of the pricing mechanisms is that they do not take into account the differentiated nature of quality of service and quality of experience (QoE) perceived by users for different applications. Users prefer static and low-fee pricing schemes, where they have limited control over the QoS they receive, to usage-based pricing schemes; while for a guaranteed QoS system, users do not mind approaching the usage-based pricing. Thus no service and QoS differentiation lead to the (in)famous tragedy of the commons [12]. Moreover, competition among multiple wireless service providers with heterogeneous networks and services poses a challenge in following the above-mentioned pricing schemes as these pricing mechanisms do not incorporate the interaction between multiple competitive service providers. As a result, it is important to deviate from the concept of flat pricing or variation of flat pricing towards a more dynamic competitive market based pricing where the service providers cater to end-users through overlaid heterogeneous networks.

The traditional concept of per–service static pricing is no longer considered from WSPs' perspective. In contrast, it is assumed that service providers will have more freedom in terms of choosing the price to be charged to their end–users. Thus in this new model, a WSP has freedom to change (increase or decrease) the price of a service continuously for a user depending on changing load, revenue etc. Note that with this freedom gained, a WSP can become malicious also to start with a low price to admit users and then increase the price in the midway. Thus it is necessary for end-users also to develop their strategies. End-users are anticipated to have more freedom in dynamically connecting to any service provider for any service through their multi-mode terminals [4].

14.3 Pricing Policies for Heterogeneous Networks

Several interesting investigations on multi-access networks have been conducted that deal with pricing issues. Convergence of cellular networks with broadcast networks was addressed in [13] with multi-radio functionalities. This architecture assumes the cooperation of existing radio networks to combine their spectrum-efficient capabilities leading to a fully coordinated system. In [14], Shabany *et al.* presented a resource allocation scheme for data traffic under heterogeneous cellular CDMA framework. Zemlianov *et al.* assumed the existence of orthogonal overlaid technologies in a wireless multi-provider setting for studying multi-access network architecture using cooperative decision making in [15, 16]. In particular, cellular and WLANs are considered where users are vertically transferred from one network to another based on the load of each network. Cooperation among radio nodes and/or networks were explicitly or implicitly assumed in all the above mentioned works and decisions were based on cooperative approaches.

In reality, entities who are involved in such pricing game approaches and decision making have always been greedy and selfish [17, 18] and they always act in a non-cooperative way. In contrast to the cooperative framework, multi-radio channel allocation problem using non-cooperative game theory was discussed in [30]. Non-cooperative, selfish behavior among radio nodes were considered in this work and pareto-optimal solutions were achieved. In [19], an integrated admission and rate control framework for CDMA based wireless data networks was proposed. The providers defined the admission criteria as the outcome of the game and the Nash equilibrium was reached using pure strategy. Users were categorized into multiple classes and were offered differentiated services based on the price they pay and the service degradation they can tolerate. However, dynamic pricing was not explored in most of these research. Shakkottai *et al.* presented a pricing policy for multiple competing ISPs using threat strategy in [11]. Local and transit ISPs play non-cooperative game with each other and common customer base to achieve Nash equilibrium in this game. In [20], a non–cooperative game for pricing Internet services were studied but concluded with an unfair Nash equilibrium where future upgradation of the networks were not discussed. Musacchio *et al.* studied the economic interests of a wireless access point owner and his paying client, and modeled their interaction as a dynamic game [21]. Economic equilibrium strategies were proposed in this work with differentiated QoS based services. Resource allocation and base-station assignment problems for the downlink in CDMA networks were studied based on dynamic pricing in [22]. Revenue maximization and pricing problems under heterogeneous service requirements were discussed in [12, 22, 23]. Mandjes proved the necessity and importance of differentiated pricing for heterogeneous services through the analysis of tragedy of the commons [12].

14.4 Internetworking System Architecture

To understand the complex nature of heterogeneous wireless networks a brief summary of the system architecture for heterogeneous inter-networking is presented in this section. Such architecture includes the general approach the system should adopt for enabling seamless roaming and functionalities for end–users across services, networks and providers in terms of pricing and differentiated QoS. The approach is assessed with regard to what requirements it places on the heterogeneous architecture, and what assumptions are made about underlying technology.

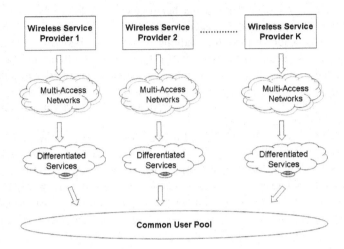

Fig. 14.2. Competition among multiple wireless service providers with heterogeneous network model.

A generalized heterogeneous network model is shown in Fig. 14.2. Also, a generic abstraction of "always greedy and profit seeking" providers and users are considered. "Always greedy and profit seeking" means that the providers and users are rational entities and thus would only follow the profit-earning strategies. Every provider is in a market-like scenario competing with other providers trying to maximize its own revenue by providing *heterogeneous services* through *heterogeneous networks* while competing for a common pool of users. Before proceeding any further, a formal discussion of *"heterogeneity"* of networks and services is necessary.

14.4.1 Heterogeneity of Network

To exploit the relative advantages of different kinds of technologies, different access networks are deployed. For example, cellular networks provide better coverage while Wi-Fi technology provides higher data rate. These technologies offer their advantages in terms of bit-rate and coverage. The frequency spectrum at which a technology operates also affects the QoS capabilities; for example, high frequencies impose restrictions on the coverage area but can support very high speed data transfer. Going down the hierarchy of the frequency spectrum, higher coverage with low speed data transfer is obtained. With respect to Fig. 14.3, the heterogeneous network technologies are shown as Access Network $1, 2, \cdots, X$. Each of these access networks is again capable of supporting a variety of services.

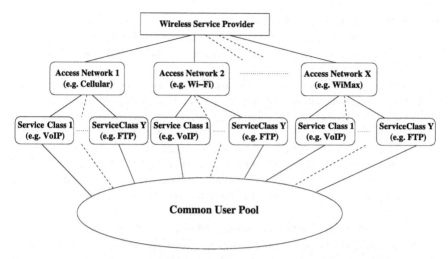

Fig. 14.3. Detailed heterogeneous access architecture under one wireless service provider.

14.4.2 Heterogeneity of Service

Heterogeneous services mean different QoS enabled wireless applications, such as telephony, video, file transfer etc. as shown in Fig. 14.3. Service classes are shown as service class 1, $\cdots$, service class Y for generalization under every access network. However, for simplicity, all services can be categorized into two broad classes each having different quality of service requirements [29].

- **Class A – Voice/Video service:** In this class, real-time services such as voice/video services are considered. These services are time-critical and therefore delayed packets are of no significance. Users pay for the duration they are connected to the network and do not pay if negotiated QoS is not met at any point during the session. This generic model can be extended to include any popular real-time services such as, voice, streaming video, VoIP, and teleconferences.
- **Class B – Data service:** In this class, the elastic services (popularly known as data services) that do not have strict delay requirement are considered. Packets are not discarded even if they are delayed. Services such as web downloads, FTP, emails are typical examples for this class.In contrast to the non-elastic model, users obtain full utility only when 100% of the data is downloaded successfully; else the utility becomes 0.

14.4.3 Internetworking Architecture

In order to provide a convenient and dynamic access of heterogeneous technologies, seamless inter-working and proper integration of the networks are very important. The key functions that must be supported by the providers to provide an efficient architecture are as follows.

- **Coupling among networks:** There are two generic coupling approaches toward the integration of different heterogeneous networks. They are *loose coupling* and *strong coupling* [32]. In the loose coupling architecture, the networks are deployed as a complement to each other and they are operated by the service provider individually and independently. The advantage of loose coupling is the flexibility of management and operation of the multiple networks. For example, in the inter-networking of $3G$ and IEEE 802.11, service provider operates both these networks separately; no $3G$ functionalities are involved during the IEEE 802.11 access and vice-versa. However the biggest disadvantage of loose coupling is the authentication procedure. End–users connected through IEEE 802.11 network need to go through authentication procedure all over again while attempting to connect to $3G$ even for the same provider thus increasing the access latency.

  In contrast to loose coupling, in strong coupling, service provider operates over the networks in a more coherent manner. Thus the need for separate access procedures are eliminated for authentication procedure. This gives the service providers and end-users the advantage of lower hand-over latency for switching from one network to another network. However, the flexibility to manage different networks independently and separately is lost and development of resource sharing mechanism across multiple networks becomes dynamic and more complex.

- **Network association:** Each service provider is required to provide information to end users about the types of networks available and the roaming agreement under the same service provider. This information advertisement is necessary as the user device (multi-mode terminal) needs to decide whether to connect to a particular network or not depending on the roaming agreement and hand-over latency.

- **Resource sharing across networks:** Each service provider required to provide information about current resource sharing across the networks. Depending on the coupling and/or existing load, the resource sharing mechanisms can be static or dynamic in nature. In the static resource sharing, resource is shared across networks in equal or fixed ratios. In dynamic resource sharing, providers follow users' distribution closely across networks and assign resources accordingly. Depending on the duration and periodicity of the assignment, the complexity of the dynamic resource sharing can range from very simple to very complex.

- **Pricing:** An important component for designing successful inter-networking architecture is pricing. The service provider must design proper pricing mechanisms to successfully compete with other service providers. The pricing can be different across networks and services and could depend on the existing load and QoS of the networks. Note that, the prices offered by the providers cannot be too high as that might repel the end-users. At the same time, it cannot be too low as that would attract too many users deteriorating QoS resulting in churning of users and subsequent loss for the provider.

14.4.4 Decision Problems

With regard to pricing, there are major decisions that need to be resolved both by the WSPs in the presence of other competitive WSPs and by the end users in deciding which WSP to pick for their service.

As a service provider, the first decision problem is to find a price for a service without knowing what prices are being simultaneously offered by its competitors. Note that providers have the knowledge of the distribution of advertised price but not the exact price at any instant. The optimization is to find different strategies for different services and find the price such that the provider is able to entice the end users and still sustain profit. With fixed resources and operational costs, prices exhibited cannot be too low, as that would attract too many users leading to degradation in performance because of resource sharing. At the same time, prices can not be too high because that will lead to user churn [19]. When a session is admitted, the WSP must then decide on the best access network to use to serve the session. Identification of the access network and allocation of resources for the services should be such that number of users being served can be increased without violating the QoS expected by the users.

As a user, the decision problem is to select the best service provider for the particular service request. All WSPs advertise their prices for the user request. The end-user will either reject all the prices and will not receive any service or the end-user might connect to the *best* service provider according to some criteria.

14.4.5 Problem Formulation Using Game Theory

The interaction (decision problems presented in the earlier section) of the service providers and the end users can be modeled formally using non-cooperative games. The game starts with a user making a request for an application service (voice/video or data). Each service requires a specific expected QoS in terms of bandwidth. Users in turn pay for the services. The consistent objective of any service provider in this game is to maximize profit for the service provided. Similarly, the users try to maximize their benefits from the service for the price they pay.

The duration of the game is the entire time the user is connected to a service provider and the game ends once the user or the provider disconnects. The exact duration of the game (how much time user would like to get the service) is not known to provider but only a distribution might be known.

With respect to end-users' and WSPs' best response, it is necessary to introduce the concept of *sub-games* in a game. Sub-games are defined as the discretized smaller unit time epochs of the whole game played between end-user and WSP. The assumption is that within a sub-game there is no change of pricing or strategy from any of the end–users or WSPs. But across the sub-games there might be changes as new users might be admitted or users may finish their session or may also churn out. Thus across sub-games there might be a change in the load of the WSPs as a result of which the WSPs might decide to reassign their prices of the services.

14.5 Game Analysis for Differentiated Services

Once the problems are identified and the game is formalized, the game needs to be solved for both service providers and end-users. Solving a game means predicting the strategy of each player (service providers and users) considering the information the game offers and assuming that the players are rational. One can see that if the strategies from the players are mutual best responses to each other, no player would have a reason to deviate from the given strategies and the game would reach a steady state. By definition this equilibrium is known as the Nash equilibrium [24]. Next, the utility functions for voice/video and data services are provided and the dominant strategies for both users and service providers are analyzed.

14.5.1 Utility Functions for Voice/Video Services

An utility function is a mathematical characterization that represents the benefits obtained and cost incurred by the players playing the game. Let there be $\mathcal{M}$ service providers that are trying to cater to a common pool of $\mathcal{N}$ users and one of the $\mathcal{N}$ users is requesting for a non–elastic voice/video service.

The methodology of mapping the user satisfaction derived from the bandwidth received is well established. A variant of the Sigmoid functions [25] is usually followed to model the non-elastic service user's satisfaction. In Fig. 14.4, an example of the user satisfaction is presented. The normalized user satisfaction is modeled as

$$US(b) = \frac{1}{1 + e^{(\frac{(b_{max}+b_{min})}{2} - b)}} \tag{14.1}$$

where $US(b)$ is the user satisfaction perceived for bandwidth b, b_{min} is the minimum bandwidth required to maintain the service, while b_{max} is the maximum bandwidth above which the user perceives no significant improvement in the QoS.

The utility function of the user i served by provider j can be expressed as

$$U_i(b_{ij}) = \frac{a_{ij}}{1 + e^{(\frac{(b_{max}+b_{min})}{2} - b_{ij})}} \tag{14.2}$$

where $U_i(b_{ij})$ is the utility perceived, and b_{ij} is the bandwidth received by ith user from jth provider. The coefficient a_{ij} is a positive parameter that indicates the relative importance of empirical benefit and acts as a weightage factor. It is a simple scaling parameter that maps user satisfaction to a dimension equitable to the price paid to the WSP as the cost of the service. Note that, a_{ij} could be even a function of bandwidth perceived; for simplicity, a_{ij} is a constant.

User satisfaction for non-elastic service is a non-decreasing function of the bandwidth received. However the satisfaction and thus the utility remains almost close to zero unless a minimum bandwidth (b_{min}) is received. This is because all real time services require a minimum amount of resource to sustain the application at the minimum QoS level. On the other hand, with a bandwidth more than maximum needed

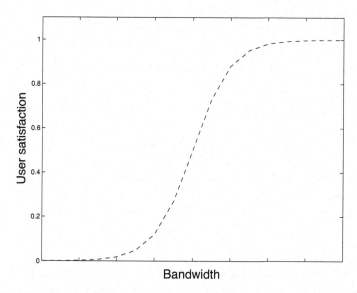

Fig. 14.4. User satisfaction as a function of bandwidth for non-elastic service.

for the service (b_{max}), the improvement in the service quality is almost not recognizable. Thus at very low and very high bandwidth, the marginal utility is almost 0. On the contrary, when the allocated bandwidth lies between the minimum and the maximum, the marginal utility changes significantly.

If the user satisfaction model is known, the price that must be advertised by the providers for every service request can be evaluated. As the duration of the voice/video service requested by the user is not known beforehand to the providers, it is not possible for the providers to set the price for the entire service duration. Thus it is necessary to divide the service pricing in smaller unit time periods, defined earlier as sub-games. The advertised price of the service from the providers can be then expressed as per unit of bandwidth, per unit time period.

If user i is connected to provider j for L time epochs, the cumulative utility obtained by user i can be given by

$$\sum_{l=1}^{L} \frac{a_{ij}}{1 + e^{\left(\frac{(b_{max}+b_{min})}{2} - b_{ij}(l)\right)}} \tag{14.3}$$

where $b_{ij}(l)$ is the bandwidth consumed in lth time period.

Net utility: There is a cost incurred by user i for this game. The cost is the price paid to the provider for obtaining the required amount of bandwidth. The price per unit of bandwidth in lth time epoch set by service provider j, $1 \leq j \leq \mathcal{M}$, is $p_j(l)$. The cumulative cost component for L time periods is

$$m_j = \sum_{l=1}^{L} p_j(l) b_{ij}(l). \tag{14.4}$$

Thus, the net utility from the user's perspective for L time epochs is given by

$$U_{ij} = \sum_{l=1}^{L} \frac{a_{ij}}{1 + e^{\left(\frac{(b_{max}+b_{min})}{2} - b_{ij}(l)\right)}} - \sum_{l=1}^{L} p_j(l) b_{ij}(l). \tag{14.5}$$

Recall, the dimension of the constant a_{ij} is adjusted so as to make both the right hand side terms of equation (14.5) dimensionally equatable.

The net utility for service provider j due to user i, if user i is connected for L time periods, is given by

$$V_j = \sum_{l=1}^{L} m_j(l) = \sum_{l=1}^{L} p_j(l) b_{ij}(l). \tag{14.6}$$

Strategies for Nash (Price) Equilibrium for Non-Elastic Service: To investigate the existence of Nash equilibrium, it is required to consider if the strategies taken by user and service provider are dominant best response to each other or not, and if they are, then whether deviating from those strategies unilaterally will have any impact or not. In this regard, it is also important to find why discretization of the game into sub-games is necessary for voice service.

Let the user i request for voice/video service. The 1st sub-game is considered, where providers advertise a price for the first time for this user i. Being rational, the aim of user i is to shortlist the service providers providing the user with positive net utility such that

$$\frac{a_{ij}}{1 + e^{\left(\frac{(b_{max}+b_{min})}{2} - b_{ij}(l=1)\right)}} \geq m_j(l=1) \quad \text{where } j \in \mathcal{M}. \tag{14.7}$$

This is the necessary condition for the user towards choosing the best response but not the sufficient. User i's aim would be to maximize the net utility. Thus the dominant best response from the user i's perspective is to choose the provider that not only provides positive net utility but also maximizes the net utility of all the providers. Let the provider j be the desired service provider in this case. For generalization purpose, it can be assumed that user i has been connected to the provider j for n^* consecutive sub-games, where, $n^* < n$ and n is the expected number of sub-games user intends to connect for voice/video service. Note that, n is not known to provider at the time of connection. Then net utility obtained by user i over these n^* consecutive sub-games is given by

$$\sum_{l=1}^{n^*} U_{ij}(l). \tag{14.8}$$

Suppose, at this point, provider j changes its price strategy (i.e., increase the price), such that, user i's net utility is not positive for $(n^* + 1)$th sub-game. If user i still decides to connect to provider j at the $(n^* + 1)$th sub-game, the net utility of user i will then be given by

$$\sum_{l=1}^{n^*} U_{ij}(l) + U_{ij}(l = n^* + 1). \tag{14.9}$$

Since $U_{ij}(l = n^* + 1)$ is a negative quantity, it is evident that user i's dominant best response would be to withdraw from the game with provider j at n^*th sub-game itself, i.e., to disconnect from the service and leave with net utility $\sum_{l=1}^{n^*} U_{ij}(l)$.

As far as the best response of provider j is concerned, the 1st sub-game of this session (game) is considered. Suppose, provider j advertises a price $\hat{p} = p_j(l = 1)$ per unit bandwidth for the 1st sub-game and user i connects to this provider. This implies that the price advertised by provider j maximizes the net utility for user i. Then, for the existence of Nash equilibrium, it is required to check if provider j or user i wants to change their strategies unilaterally. In this regard, without loss of generality, it can be assumed that all other service providers and users keep their strategies unchanged.

As user i has accepted the connection with provider j, the provider knows for sure that if price $\hat{p}$ is charged for subsequent sub-games, user i would continue to play the game (i.e., remain connected). Thus, at equilibrium, the lower bound on the pricing can be given by $\hat{p}$. As a greedy player, provider j's strategy would be to charge higher and thus maximize its expected net utility over the entire service. The expected net utility of the provider for the voice service can be then given by,

$$\sum_{l=1}^{n} m_j(l) P(U_{ij}(l) \geq 0) P(U_{ij}(l) > U_{ik}(l)), \quad \forall k \in \mathcal{M}, k \neq j \tag{14.10}$$

where $m_j(l)$ is the revenue generated from user i at the lth sub-game. $P(U_{ij}(l) \geq 0)$ denotes the probability of generating positive net utility for user i and $P(U_{ij}(l) > U_{ik}(l))$ denotes the probability of generating maximum net utility for user i with provider j than any other provider. Equation (14.10) being strictly concave and continuously differentiable upto 2nd order, it is clear that a maximum point exists.

As provider now knows the lower bound of the price charged at equilibrium, the natural inclination of the provider would be to charge the user with a non-decreasing pricing sequence over the consecutive sub-games. Let the non-decreasing pricing sequence at equilibrium be given by

$$p_j(l = 1) \leq p_j(l = 2) \leq \cdots \leq p_j(l = n) \tag{14.11}$$

which would maximize the expected net utility of the provider j as was given in equation (14.10). Provider's best response would be to choose the maximum price from this non-decreasing pricing sequence given in equation (14.11). Let the maximum price be denoted by $\hat{p}_{max}$. It is evident from equation (14.10) that $\hat{p}_{max}$ is upper bounded by the condition $P(U_{ij}(l) \geq 0)$ and $P(U_{ij}(l) > U_{ik}(l))$. Provider's best response would then be to charge $\hat{p}_{max}$ from the very 1st sub-game. But earlier, it has been seen that provider has charged $\hat{p}$ in the 1st sub-game, which is the lower bound of the prices charged. Thus it is clear that $\hat{p}_{max} = \hat{p}$, i.e., at equilibrium provider's best response would be to adhere to the price agreement made at the time of admission for any user.

Once, the best responses for both the user and the provider are found, it is rather easy to show that both players are affected by deviating from the equilibrium unilaterally. If user i deviates from its dominant strategy unilaterally, the net utility obtained by the user will not be maximized (as explained in equation (14.9)). Similarly, if provider j deviates from its best response by charging some other price than the non-decreasing pricing scheme, then it will not increase the expected net utility which proves the existence of Nash equilibrium.

14.5.2 Utility Functions for Data Services

In contrast to voice/video services, in data services, a user gets utility only when the whole file is downloaded completely (e.g., web download, FTP etc.); else the utility becomes 0. In such a scenario, the user satisfaction is modified and expressed as

$$US(b) = \begin{cases} k_s \, log(1 + b), & 0 < b < b_{max} \\ 1, & b \ge b_{max}. \end{cases} \qquad (14.12)$$

The parameter k_s is used for normalization and is defined such that

$$US(b_{max}) \cong 1. \qquad (14.13)$$

The normalized user satisfaction presented for elastic service can be modeled as in Fig. 14.5.

The utility function for user i from provider j can then be given by

$$\begin{cases} a_{ij}k_s log(1 + b_{ij}) & \text{if transaction is complete,} \\ 0 & \text{if transaction is incomplete} \end{cases} \qquad (14.14)$$

where a_{ij} is a simple scaling parameter that maps user satisfaction to a dimension equitable to the price paid to the WSP as the cost of the service.

Net utility: The net utility for user i for data service upon completion can be expressed as

$$U_{ij} = a_{ij}k_s log(1 + b_{ij}) - \hat{m}_j \qquad (14.15)$$

where $\hat{m}_j$ is the total price charged.

Strategies for Nash (Price) equilibrium for data service: To investigate the existence of Nash equilibrium for data service, the analysis can be done in similar manner as the voice/video service, i.e., the session (game) is discretized into sub-games and user i uses the same best response strategy proposed for the voice/video service, i.e., negotiate price at the beginning of each sub-game.

Suppose, user i requests for a data service at the beginning of the 1st sub-game and after QoS negotiation and price agreement, service provider j is selected as the desired provider. Let the expected number of game sessions to complete the entire

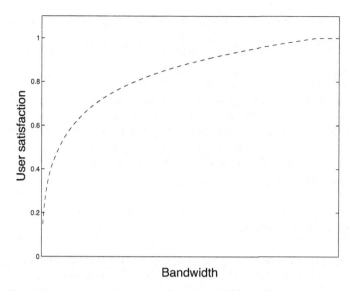

Fig. 14.5. User satisfaction as a function of bandwidth for elastic service.

data service be n $(n > 1)$. Then, at the 1st game session, user i would pay $m_j(l = 1)$ and will receive bandwidth $b_{ij}(l = 1)$. At the end of the 1st sub-game, the cost incurred by user i would be $m_j(l = 1)$, whereas, the utility obtained is still 0; thus conflicting the nature of a rational player. At this point, service provider j will become malicious and will charge a high price for bandwidth provided from 2nd sub-game onwards. If user i quits the game at this session and tries to connect to some other provider, the net utility perceived so far will be negative as the service has not been completed thus making the user a loser in the game. If user persists to be in the game with the losing strategy, even then the net utility over the n sub-games will be negative as provider continues to charge at a high rate. Thus it is clear that discretization of prices over sub–games will not be a choice for end–users in this case. Though discretization was necessary for voice services, for data services discretization in terms of pricing is not at all desired to prevent provider to be malicious. So user's best response would be to quit the data service game if it is discretized pricing. Thus from user i's perspective, the equilibrium strategy would be to negotiate price at the beginning for the entire data service if and only if net utility for the entire data service is positive and maximized. Following this strategy will enable the user to quit the game at any point of time if price advertised by the provider does not maximize his net utility. This prevents provider to be malicious as user quitting the game means provider would also loose the game obtaining zero revenue though bandwidth is wasted. Once the equilibrium strategy from user's point of view is established, it is easy to define the dominant strategy from service provider's point of view. As user would pay only once, service provider's aim is to maximize the expected net utility

$$\hat{m}_j P\big(a_{ij} \ k_s \ log(1 + b_{ij}) \geq \hat{m}_j\big) P\Big(\big(a_{ij}k_s log(1 + b_{ij}) - \hat{m}_j\big)$$

$$> \big(a_{ik}k_s log(1 + b_{ik}(l)) - \hat{m}_k\big)\Big) \text{s. t.} \forall k \in \mathcal{M}, k \neq j \quad (14.16)$$

where $\hat{m}_j$ is the price charged for the entire service (game). $P\big(\sum_{l=1}^{n} a_{ij}k_s log(1 + b_{ij}) \geq \hat{m}_j\big)$ denotes the probability that the service provider has to provide positive net utility to the user i with the one time charge $\hat{m}_j$ and the second probability in the above equation denotes the probability that the provider has to provide the maximized net utility for this service than any other provider. Equation (14.16) being concave and continuously differentiable upto 2nd order, the maximizing point and the Nash equilibrium exist.

14.6 Network Selection

Once a session is admitted, the service provider has to find the best network to serve the session. For generalization, each service provider is equipped with $\mathcal{X}$ heterogeneous networks, whereas each of these networks provide $\mathcal{Y}$ different services as was shown in Fig. 14.3. Then total bandwidth capacity of a service provider can be given by $\sum_{x=1}^{\mathcal{X}} \sum_{y=1}^{\mathcal{Y}} B_{xy}$, where B_{xy} is the bandwidth capacity of service y under network x.

Moreover, let the existing load (bandwidth already assigned to the existing users) under this service provider be $\sum_{x=1}^{\mathcal{X}} \sum_{y=1}^{\mathcal{Y}} B_{xy}^*$, where B_{xy}^* is the bandwidth allocated from service y under network x. Thus, a new parameter called QoS affect ratio is defined and denoted as

$$Q = \frac{\sum_{x=1}^{\mathcal{X}} \sum_{y=1}^{\mathcal{Y}} B_{xy}^*}{\sum_{x=1}^{\mathcal{X}} \sum_{y=1}^{\mathcal{Y}} B_{xy}} \quad (14.17)$$

Note that, $0 \leq Q \leq 1$ is directly related to the relative usage of the bandwidth. As is evident from equation (14.17), there is a chance that QoS might degrade as more and more users are allocated bandwidth by this service provider. On the other hand, QoS will improve if the bandwidth capacity of the service provider is also improved.

It is clear that provider's sole interest will be to increase the usage efficiency, as revenue earned by service provider is directly proportional to the existing usage efficiency, provided QoS of the existing load is maintained. In other words, the game from the service provider's point of view can be formalized as follows:

$$maximize \ R = maximize \ f\Big(\sum_{x=1}^{\mathcal{X}} \sum_{y=1}^{\mathcal{Y}} B_{xy}^*\Big) \quad (14.18)$$

where R is the revenue earned by the service provider and $f(\cdot)$ is an increasing function. Thus a service provider's best bet would be to admit the incoming service request to a particular chosen network which would introduce minimum relative increase in QoS affect ratio.

Best response strategy from service provider: When a new service request arrives, provider's aim is to serve the request to maximize the revenue, but at the same time maintain the QoS of the on–going services. The important question that arises is, which network (among the multiple overlaid networks) to select for servicing a particular service request such that the relative increase in QoS affect ratio is minimum. Note that, a service request can potentially be served by more than one network. For example, a user trying to make a voice call can use the Wi-Fi network or the cellular network.

Let the user i request for service y from provider j. The aim of the provider would be to admit this request so that expected revenue can be increased by

$$m_{ij}P(\text{user } i \text{ connects to provider } j) \qquad (14.19)$$

where m_{ij} is the price paid by the user i to provider j. As a result, provider's intention would be to follow a pricing mechanism such that probability of user i connecting to provider j is maximized. If the provider j admits this service request and allocates bandwidth b_{ij} then the relative increase in QoS affect ratio in service y under network x is given by

$$\frac{d}{db_{ij}}\left(Q_{xy}\right) = \frac{1}{B_{xy} - B_{xy}^*} \qquad (14.20)$$

where Q_{xy} denotes the QoS affect ratio of service y under network x. Similarly, relative increase in QoS affect ratio in network x considering all services, $y \in \mathcal{Y}$, is given by

$$\frac{d}{db_{ij}}\left(Q_x\right) = \frac{1}{\sum_{y=1}^{\mathcal{Y}} B_{xy} - \sum_{y=1}^{\mathcal{Y}} B_{xy}^*}. \qquad (14.21)$$

The ratio of relative increase in QoS affect ratio for a particular service y under network x, and all services, $y \in \mathcal{Y}$ under network x is given by

$$Q_x^r = \frac{\sum_{y=1}^{\mathcal{Y}} B_{xy} - \sum_{y=1}^{\mathcal{Y}} B_{xy}^*}{B_{xy} - B_{xy}^*}. \qquad (14.22)$$

A provider's aim would be to select the network x from $\mathcal{X}$, that would have the least value of Q_x^r so that users experience least QoS degradation with more users being served.

14.7 Proof of Concept

A system is considered where a wireless service provider has two access networks – 3G network based on CDMA/HDR [26, 27] and 802.11 based Wi-Fi network. The coverage area of the $3G$ network is large but the perceived QoS degrades as user goes far from the $3G$ base station. On the other hand, Wi-Fi network coverage is provided by Wi-Fi access point (AP) that provides uniform coverage for a small area. Users are randomly distributed over the service area and they are equipped with multi-mode terminals capable of connecting to any of the two networks provided it is inside the coverage area of that particular network. Note that $3G$ channels are more expensive because of the licensing fee for $3G$ spectrum as opposed to the unlicensed Wi-Fi band.

14.7.1 Network Layout

In the time-slotted CDMA/HDR system, the base station transmits to one user at full power using one of the 11 pre-defined modulation and coding schemes. Thus, the cell can be assumed as being divided into 11 concentric rings, each receiving a unique data rate [26].

It can be argued that proper utilization of the radio resources depends on the placement of the APs. In this regard, two different kinds of placement are considered: *ideal* placement and *random* placement. It is assumed that in the ideal placement, the distribution of users' locations are known and more number of APs are placed at the high density areas. Since the ideal placement is not possible in every premise due to incomplete information of user's profiles, random placement is the alternative solution where APs are placed uniform randomly within the $3G$ cell. For simulation purpose, the radius of the $3G$ cell and the Wi-Fi cell are considered as 2000 and 100 meters respectively. The data rates and corresponding distances from Table 14.1 as given in [28]. Number of users (N) is assumed to be 1000.

Table 14.1. Rates and ring radius.

Ring k	Data rate (Kbit/sec)	Radius (m.) ($\alpha = 4$)
11	38.4	2820
10	76.8	2370
9	102.6	2210
8	153.6	2000
7	204.8	1860
6	307.2	1680
5	614.4	1410
4	921.6	1280
3	1228.8	1190
2	1843.2	1070
1	2457.6	1000

14.7.2 Benefit Perceived by Users

For the users, the benefit is expressed in terms of blocking probability, expected delay, and perceived bandwidth. The blocking probability is defined as the probability with which a user's request is rejected i.e., the request is denied. Both cases are considered–without and with APs. The queue size is assumed to be 15.

In Figs. 14.6 and 14.7, the blocking probability and expected delay experienced per user are presented with and without APs. As evident from the figures, blocking probability and delay decrease with introduction of APs. Of course, the cost of deploying the APs is not considered here.

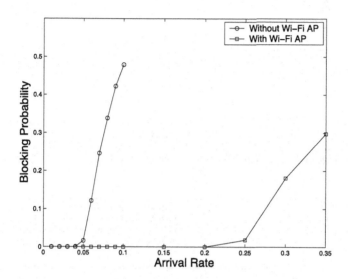

Fig. 14.6. Blocking Probability without and with inter-networking.

In Fig. 14.8, the perceived system bandwidths of the systems are compared for with and without inter-networking. The round-robin scheduling scheme is considered for fairness criteria in the inter-networked system and compared to both round robin and opportunistic scheduling in non-inter-networked system. (A simple opportunistic scheme is followed, where priorities are given to the users with better signal strength.) With increase of Wi-Fi APs, more and more number of users come under the direct influence of Wi-Fi coverage area and get the highest data rate. After a certain point – when all the users have the option of getting the highest rate for the service requested, the system performance reaches its maximum value.

14.7.3 Benefit Perceived by Providers

As discussed in equation (14.18), it is assumed that the revenue generated by each provider is directly proportional to the bandwidth usage efficiency. The capacity of the $3G$ network is assumed to be fixed and thus the spectrum maintenance cost due to the $3G$ network is assumed constant. E_{3G} is the cost for the $3G$ network. As far as the cost of the Wi-Fi APs are concerned, the cost is mainly due to deployment,

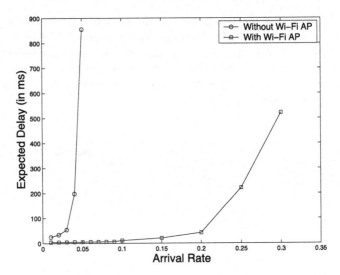

Fig. 14.7. Expected delay per user without and with inter-networking.

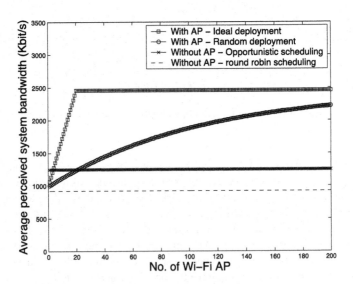

Fig. 14.8. Perceived system bandwidth: Simulation result.

infrastructure maintenance, and cost due to the backhaul link allocation. For simplicity, the cost incurred due to each of the APs is fixed and is given by e_{AP}. Then the provider's net utility without inter-networking is

$$V_j[3G] = f(B_{3G}^*) - E_{3G} \tag{14.23}$$

where B^*_{3G} is the allocated bandwidth from the service provider only from the $3G$ network. In contrast to equation (14.23), the net utility for the service provider with $3G$ and Wi-Fi inter-networked is

$$V_j[AP] = f(B^*_{AP}) - E_{3G} - Le_{AP} \qquad (14.24)$$

where B^*_{AP} is the allocated bandwidth from the service provider with $3G$ and Wi-Fi inter-networked and in the presence of L APs. The relative increase in net utility in the inter-networked system from the service provider's point of view is given by

$$\frac{(f(B^*_{AP}) - E_{3G} - Le_{AP}) - (f(B^*_{3G}) - E_{3G})}{f(B^*_{3G}) - E_{3G}}. \qquad (14.25)$$

Note that, relative increase in net utility depends on $f(B^*_{AP})$ and the ratio $\frac{E_{3G}}{e_{AP}}$. With increase in the allocated bandwidth in the inter-networked system and increase in the ratio, relative increase in net utility is also increased.

In Fig. 14.9, the net utility obtained by the service provider is presented with increase in number of Wi-Fi APs, keeping the capacity and thus the cost of the $3G$ network fixed. In addition to the above parameters, the cost due to the $3G$ network is assumed as $E_{3G} = 1000$ unit. The cost incurred due to each AP is $e_{AP} = 20$ unit.

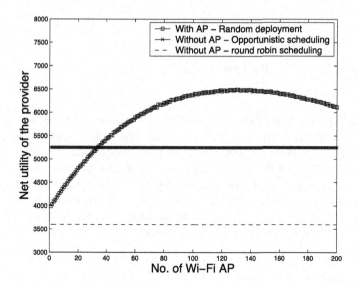

Fig. 14.9. Net utility of the provider.

14.7.4 Per-User Utility

Per-user perceived utility is studied for both with and without inter-networked system. The same parameters are used as mentioned earlier but the number of APs used are 100 and 200. The result obtained is shown in Fig. 14.10. As expected, with the increase in number of active users, the per-user perceived utility decreases because of resource sharing, but the rate of decrease is lesser for the case with inter-networked system. With more number of APs, this decreasing rate can be further reduced.

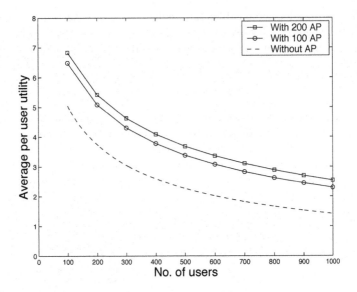

Fig. 14.10. Per-user perceived utility.

So far, the radii of the cell and the APs were considered constant. In Fig. 14.11, the radius of the Wi-Fi coverage area is kept constant (100 m) and the radius of the cell is varied from 1000 m to 3000 m so as to get a varying radius ratio. It is observed that when the radius of the $3G$ cell is small, throughput with and without APs does not vary much because users are situated very near to the base station and get very high rate. But when the $3G$ cell radius increases, more and more number of users get lesser signal strength and hence lesser data rate, resulting in decreased system performance for system without inter-networking.

Conclusion

The continuous growth of QoS enabled wireless services along with the deployment of overlaid access networks is creating a market for short-term wireless services. The accessibility to different wireless networks serviced by different service providers and competitive service offerings are making dynamic trading of wireless services at a finer granularity. A framework based on dynamic games is presented and the interaction between providers and users in a market-based environment is formulated. Different pricing approaches have been investigated and a new dynamic

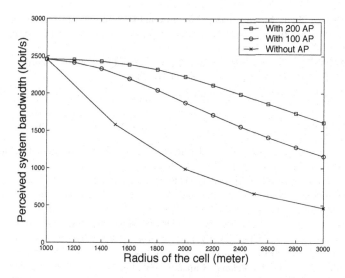

Fig. 14.11. Perceived system bandwidth due to effect of radius ratio.

pricing mechanism has been presented for both voice and data services that have varying QoS requirements. It is shown that if users and providers comply with the proposed pricing strategies, Nash (price) equilibrium can be achieved maximizing both users' and providers' net utilities. A case study is also presented where a wireless service provider provides service through two overlapped networks. Simulation experiments demonstrate that dynamic pricing mechanism achieves better net utility for both providers and users.

References

1. CDMA Development Group, "CDMA2000 and WiFi: Making a business case for interoperability," Sep. 2003, http://www.cdg.org.
2. J. Mitola and G.Q. Maguire Jr., "Cognitive radio: Making software radios more personal," *IEEE Personal Communications*, vol. 6, no. 4, pp. 13-18, 1999.
3. [Online] GTRAN, "GTRAN dual-mode 802.11/CDMA wireless modem," http://www.gtranwireless.com.
4. [Online] http://en.wikipedia.org/wiki/MVNO.
5. P. J. Seok and K.S. Rye, "Developing MVNO market scenarios and strategies through a scenario planning approach," in *Proc. of 7th International Conference on Advanced Communication Technology (ICACT'05)*, vol. 1, pp. 137-142.
6. R. Cocchi, S. Shenker, D. Estrin, and L. Zhang, "Pricing in computer networks: Motivation, formulation, and example," *IEEE/ACM Transactions on Networking*, vol. 1, no. 6, Dec. 1993.
7. C. Parris, S. Keshav, and D. Ferrari, "A framework for the study of pricing in integrated networks," Institute of Computer Science, Berkeley, California, TR-92-016, Mar. 1992.

8. A. Odlyzko, "Paris Metro pricing: The minimalist differentiated services solution," in *Proc. of Seventh International Workshop on Quality of Service (IWQoS)*, pp. 159-161, 1999.

9. J. K. Mackie-Mason and H.R. Varian, "Economic FAQs about the Internet," *Journal of Economic Perspectives*, vol. 8, no. 3, pp.24-36, 1994.

10. J. McMillan, "Selling spectrum rights," *Journal of Economic Perspectives*, vol. 8, no. 3, pp. 145-162, 1994.

11. S. Shakkottai and R. Srikant,"Economics of network pricing with multiple ISPs," in *Proc. of IEEE INFOCOM*, vol. 1, pp. 184-194, 2005.

12. M. Mandjes, "Pricing strategies under heterogeneous service requirements," in *Proc. of IEEE INFOCOM'03*, vol. 2, pp. 1210-1220, 2003.

13. R. Keller, T. Lohmar, R. Tonjes, and J. Thielecke, "Convergence of cellular and broadcast networks from a multi-radio perspective," *IEEE Personal Communications*, vol. 8, no. 2, pp. 51-56, 2001.

14. M. Shabany, K. Navaie, and E.S. Sousa, "Downlink resource allocation for data traffic in heterogeneous cellular CDMA networks," in *Proc. of Ninth International Symposium on Computers and Communications*, vol. 1, pp. 436-441, 2004.

15. A. Zemlianov and G. de Veciana, "Capacity of ad hoc wireless networks with infrastructure support," *IEEE Journal on Selected Areas in Communications*, vol. 23, no. 3, pp. 657-667, March 2005.

16. A. Zemlianov and G. de Veciana, "Cooperation and decision-making in a wireless multi-provider setting," in *Proc. of IEEE INFOCOM*, vol. 1, pp. 386-397, 2005.

17. C. K. Chau and K. M. Sim, "Analyzing the impact of selfish behaviors of Internet users and operators," *IEEE Communications Letters*, vol. 7, no. 9, pp. 463-465, Sept. 2003.

18. W. Wang and B. Li, "Market-driven bandwidth allocation in selfish overlay networks," in *Proc. of IEEE INFOCOM*, vol. 4, pp. 2578-2589, 2005.

19. H. Lin, M. Chatterjee, S. K. Das, and K. Basu, "ARC: An integrated admission and rate control framework for CDMA data networks based on non-cooperative games," in *Proc. of 9th Annual International Conference on Mobile Computing and Networking (MobiCom 2003)*, San Diego, pp. 326-338, 2003.

20. L. He and J. Walrand, "Pricing Internet services with multiple providers," in *Proc. of Allerton Conference 2003*.

21. J. Musacchio and J. Walrand, "WiFi access point pricing as a dynamic game," *IEEE/ACM Transactions on Networking*, vol. 14, no. 2, pp. 289-301, April 2006.

22. S. C. M. Lee, J. W. J. Jiang, J. C. S. Lui, and C. D.-Ming, "Interplay of ISPs: Distributed resource allocation and revenue maximization," in *Proc. of 26th IEEE International Conference on Distributed Computing Systems*, 2006.

23. X. Wang and H. Schulzrinne, "Pricing network resources for adaptive applications," *IEEE/ACM Transactions on Networking*, vol. 14, no. 3, pp. 506-519, 2006.

24. J. F. Nash, "Equilibrium points in N-person games," Proc. of the National Academy of Sciences, vol. 36, pp. 48-49, 1950.

25. [Online] http://en.wikipedia.org/wiki/Sigmoid_function

26. P. Bender, P. Black, M. Grob, R. Padovani, N. Sindhushyana, and S. Viterbi, "CDMA/HDR: A bandwidth efficient high speed wireless data service for nomadic users," *IEEE Communications Magazine*, vol. 38, issue 7, pp. 70-77, 2000.

27. 3GPP TR25.848 V0.5.0, "Physical Layer Aspects of UTRA High Speed Downlink Packet Access(HSDPA)," ANNEX B, TSGR1#18(01)186, Jan. 2001.

28. T. Bonald and A. Proutiere, "Wireless downlink data channels: User performance and cell dimensioning," in *Proc. of ACM MobiCom*, pp. 339-352, 2003.

29. S. Faizullah and I. Marsic, "Charging for QoS in internetworks," in *Proc. of IEEE Global Telecommunications Conference*, vol. 3, pp. 1957-1962, 2001.
30. M. Felegyhazi, M. Cagalj, and J.-P. Hubaux, "Multi-radio channel allocation in competitive wireless networks," in *Proc. of 26th IEEE International Conference on Distributed Computing Systems*, July 2006.
31. H. Kikuchi, "(M+1)st-price auction protocol," in *Proc. of the 5th International Conference on Financial Cryptography*, pp. 351-363, 2002.
32. ETSI, "Requirements and architectures for intenvorking between HIPEKLANI3 and 3rd generation cellular systems," Tech. rep. ETSI TR 191 957, Aug.2001.

15

Content Discovery in Heterogeneous Mobile Networks

Diego Borsetti[1], Claudio Casetti[1], and Carla-Fabiana Chiasserini[1], and Luigi Liquori[2]

[1] Dipartimento di Elettronica, Politecnico di Torino, Italy
{borsetti, casetti, chiasserini}@tlc.polito.it
[2] INRIA Sophia Antipolis Méditerranée, France
Luigi.Liquori@inria.fr

15.1 Introduction

Today, it is not uncommon for the inhabitants of major cities to have a large selection of network technologies, both wired and wireless, at their fingertips. Hence, to have an unprecedented number of services readily accessible.

Whereas the availability of wired technologies is limited to one at a time (e.g., an office LAN, a residential xDSL), as far as wireless technologies are concerned, people are usually exposed to several of them simultaneously (e.g., GSM/GPRS and UMTS cellular networks, WiFi, Bluetooth). Although users commonly rely on one technology at a time for a specific service, considerable research effort has been devoted to inter-operability among technologies, with the aim of granting more bandwidth to users and, potentially, lower fees.

Among recently-developed mobile applications (e.g., the Google Mobile suite, the Traffic Message Channel or the Intermodal Journey Planner), a widespread interest is surrounding those that enable the collection and sharing of various information contents for users on the move. The growth of such services will foster the creation of "virtual communities" of users sharing similar interests and goals: as way of example, students on a campus who share class material, tourists browsing through local attractions listings, car drivers seeking and exchanging traffic information, or pedestrians accessing their domestic or working environment via hand-held devices.

Taking into account the previous remarks, it is clear that a fundamental problem in wireless networks is the discovery and sharing of information among users, which can rely on more than one communication technology. While in the context of wired networks, many algorithms and protocols have been presented for content discovery, much fewer proposals target heterogeneous wireless environments.

In this chapter, we first review some of the most relevant solutions presented for content discovery, in both wired and wireless contexts (Sections 15.2 and 15.3). We focus on the well-known *publish/subscribe* asynchronous messaging paradigm and discuss the benefits of designing an overlay network relying on an underlay wireless network. We then introduce a possible network architecture, composed on the one

E. Hossain (ed.), *Heterogeneous Wireless Access Networks*,
DOI: 10.1007/978-0-387-09777-0_15, © Springer Science+Business Media, LLC 2008

hand of infrastructured nodes, such as WLAN access points and cellular system base stations, and on the other hand of mobile nodes that are equipped with multi-interface wireless terminals, namely, pedestrian users as well as vehicles (Section 15.4). We devise how an overlay network can be created and efficiently implemented on such a communication network using the publish/subscribe paradigm and its content-based routing algorithms. We describe the semantics and the interaction among the logical network entities as well as the possible interface selection strategies (Section 15.5). Finally, we take a heterogeneous wireless network, including the IEEE 802.11 technology and the UMTS cellular network as a case study, and we show some performance results of the publish/subscribe approach in such a scenario (Sections 15.6 and 15.7).

15.2 Content Discovery in Dynamic Networks

The explosive growth of the information technology fosters the design of communication networks that let users discover and share information, anywhere and anytime. In a wireless, urban environment, information sought by users can be acquired either from server nodes belonging to the backbone infrastructure or from other users, by exploiting cooperative mechanisms [1, 2]. Such a communication system, however, poses several technical challenges, which have been recently addressed both within research projects and in the scientific literature.

Examples of projects focusing on the support of innovative services within an urban environment are IntelCities [3] and WikiCity [4]. IntelCities is a European project that develops efficient and innovative e-government services for citizens. WikiCity, instead, is aimed at distributing and processing real-time data coming from electronic devices scattered throughout the city and connected to the Internet infrastructure.

In the literature, solutions proposed for wired peer-to-peer (P2P) networks, such as Gnutella, Freenet, Fastrack, or eDonkey, do not apply to mobile scenarios because of the dynamic nature of the network where users continuously move and contents may appear and disappear. A promising approach is instead the publish/subscribe messaging paradigm [5], an asynchronous, many-to-many communication model designed to distribute information to a large number of users adopting either a content-based or a topic-based routing. According to the publish-subscribe messaging paradigm, users can be both information providers (publishers) and information consumers (subscribers). Consumers subscribe to the publish/subscribe system and specify the type of information that they are interested in, while producers publish data to the system. The system disseminates the information to all (if possible) the consumers that are interested in receiving it, according to the interests they declared. Having decoupled users in publishers and subscribers allows for a great flexibility and scalability in presence of networks with a dynamic network topology (i.e., with high node churning). The selection of messages for reception and processing is called filtering. There are two kinds of filtering, namely topic-based or content-based, each associated to a specific routing. In a nutshell, topic-based routing broadcasts all published messages (on some topic) to all users subscribed to that topic; content-based routing delivers to subscribers only messages that exactly match subscriber-defined attributes or contents. Many publish/subscribe systems have recently been developed for wired networks (e.g., Xnet [6, 7], Siena [8] or Gryphon [9]), as well as for wireless networks [10–18]. In particular, the work in [11] specifies the way consumers and producers are matched together, i.e., by applying a cross-layer approach that leverages some routing-specific metrics, such as hop count or node traffic load. In [12], the cooperative downloading strategy, named SPAWN, addresses peer discovery and content selection. Peer discovery uses a centralized approach, as in Bit-Torrent [19], and Slurpie [20], as well as a distributed one that leverages the broadcast nature of the wireless medium and allows nodes to overhear information about the content availability at neighbors. Content selection is determined by a proximity-driven piece selection strategy, where proximity estimation is based on hop count. The work in [14] combines the publish/subscribe approach with a content-based routing scheme. In [16], instead, the authors present a dynamic publish/subscribe system for mobile peer-to-peer environments, which integrates an extended on-demand multicast routing protocol and content-based messaging.

A valuable solution to effectively implement the publish/subscribe approach is to define an overlay network, which is built on top of the physical one, such that two neighbor nodes in the overlay may be many links apart in the physical network. In general, overlay networks are capable of providing a rich spectrum of services through the use of aggregated computational power, storage and contents [21–23]. The main idea is therefore to achieve information discovery and retrieval via a seamless, geographically distributed, open-ended network of bounded services owned by the overlay components.

In the field of wireless networks, the publish/subscribe paradigm and overlay networks were exploited together in [15, 24, 25]. In [24, 25], service discovery protocols are based on the deployment of a virtual backbone of directories within an infrastructure-less network. Each node composing the backbone performs service discovery in its proximity, while global service discovery is provided by the cooperative action of the directories. In [15], an overlay network is conceived to operate in a mobile ad hoc network: the overlay network that routes events from publishers to subscribers dynamically adapts itself to the changing topology by means of cross-layer interactions.

In the context of wired computer networks, an interesting solution, named Arigatoni [26], was designed to provide a large variety of services through a multi-layer overlay network. In a nutshell, in a multi-layer overlay network, the responsibility assigned to network nodes differs. Super-peers, called Brokers, act as servers for a subset of peers (named colony). Ordinary peers, called Agents, submit queries to their Broker and receive results from it. Brokers are also connected to each other; they route messages across the overlay network, submit, delegate, and answer queries on behalf of the Agents in their colony. This structure is replicated *recursively*, creating an *n-layer topology*. Arigatoni provides service discovery with variable guarantees in a virtual organization, where peers can dynamically appear, disappear, and self-organize. Furthermore, it is a fully–programmable overlay network: it dictates how and where services are declared, discovered and orchestrated (via a simple business language) in the overlay, allowing peers to ask, provide and use global services and resources.

Thanks to the above features, Arigatoni appears to be a suitable choice for information delivery and sharing in a mobile environment. We therefore define a urban mobile network scenario, whose nodes are equipped with multiple wireless interfaces and, inspired by Arigatoni. We devise an overlay network for information discovery and retrieval in such an environment. For the sake of completeness, some further details on the Arigatoni solution are provided below.

15.3 Arigatoni in a Nutshell

What follows is a short overview of the activity of the main entities and of the protocols involved in Arigatoni (the interested readers can refer to [26–28]).

15.3.1 Functional Units

Two main logical entities (the Agent and the Broker) and two basic protocols (a registration and a service discovery protocol) are the core of the Arigatoni overlay network.

The Agent is a computing device with wired/wireless connectivity capabilities. It does not necessarily need to be a high-end device, such as a supercomputer; on the contrary, it may have limited storage and computation capabilities, and few basic installed applications (a simple editor, one or two compilers, an email client, a mini browser). Agents are organized in Colonies, led by a Broker. Unlike the Agent, though, the Broker is required to be a mid- to high-end device, equipped with a high–speed wired/wireless connection and a service table, crucial to perform the publish/subscribe content-based routing. Given the hierarchical overlay, colonies may recursively be embedded into super-colonies, each led by a super-Broker.

The Agent

It should be able to work in *local mode* for all the tasks that it can manage locally or in *colony mode*, by first registering itself to one or many colonies of the overlay, and then by asking and serving colony-originated requests via the Brokers. The tasks of an Agent can thus be summarized as follows:

- discovering the address of one or more Brokers, acting as colony leaders, upon its arrival in a "connected area";
- registering to one or many Brokers, thus entering the virtual organization;
- requesting and offering services to other Agents, through its own Broker;
- connecting directly to other Agents in a peer-to-peer fashion, and exchanging services between each others. Note that an Agent can also be a service provider. This symmetry is one of the key features of Arigatoni.

The Broker

It requires higher capabilities than an Agent to store and manage the content-based routing table of the colony it leads. Such table is essential to route queries, and the Broker must also efficiently match and filter the routing table against a received query. The tasks of a Broker are

- discovering the address of another Broker, and possibly embedding its colony into the other Broker's;
- registering/unregistering Agents in its colony and updating the internal content-based routing table accordingly (who offers what within the colony, its address, and other geographical information);
- receiving Agents service requests, discovering the services that satisfy an Agent request in its local colony, according to its content-based routing table, or delegating the request to its direct super-Broker;
- in case the Agent request can be satisfied, forwarding, in a service response, all the information necessary to allow the requesting agent to communicate directly with the agent offering the service;
- in case the agent request cannot be satisfied, notifying the requesting agent, after a fixed timeout period, that its service request could not be served.

There are mostly two mechanisms of service discovery, namely:

- the process of a Broker finding and negotiating services to serve an Agent request in its own colony;
- the process of an Agent discovering a Broker, upon physical/logical insertion in a colony.

The first discovery is processed by Arigatoni's service discovery protocol, while the second is processed out of the Arigatoni overlay, using well-known network protocols like DHCP, SLP in Bluetooth or Active/Passive Scanning in WiFi.

The Service Discovery Protocol (RDP)

It is used by a Broker to find and negotiate services to serve agent requests in its own colony. RDP allows the request for multiple services and service conjunctions (i.e., each agent may offer several services at the same time). The RDP protocol allows Agents to

- ask to a Broker a request for a service set S;
- reply to a Broker the availability to offer a service set S'.

The colony's Broker handles the service request received through RDP and looks up the service set in its routing table, filtering S against the offered set S'. If a match is found, the Broker returns to the requesting agent the address of the agent matching, partly or fully, the request. From then on, the two agents interact in a peer-to-peer fashion, without further intervention of the Broker using a simple coordination language (e.g., the BPEL language [29]).

Each Broker maintains a content-based *routing table* locating the *services* that are registered in its colony. The table carries one entry for each member matching the ID of the Agent with the set of services it can offer.

The Virtual Intermittent Protocol (VIP)

It manages peers' participation in Arigatoni's colonies. The protocol deals with the *dynamic topology* of the overlay, by allowing individuals to login/logout to/from a colony. Registration is the act through which an agent becomes member of a colony. An Agent is allowed to unregister when it has no pending service requests or offers. Agents that abruptly unregister or behave as "free riders" (using other Agents' services without offering or giving theirs in return) are tagged as unfair and may be forcefully unregistered from a colony by its Broker.

An Agent registers to a colony with a list of services. If a Broker accepts an individual in its colony, then it sends a service update to its direct super-Broker in order to communicate the availability of the new services in its colony. This message is then propagated from Broker to Broker until the root (if any) of the multi-layer overlay is reached. This means a high degree of node churning forces routing tables to be *faulty* until all service updates are properly propagated. As such, service registration in an overlay network computer is an activity that must be taken seriously into account.

15.4 A Mobile Heterogeneous Network Scenario

Let us consider an urban area in which a mobile network is deployed by using an ad hoc communication technology (e.g., WiFi). A further coverage of the network area is obtained through a cellular network, such as GSM/GPRS or UMTS. Such mobile network is populated by both mobile users, e.g., pedestrians with hand-held devices, cars equipped with browsing/computational capabilities, public-transportation vehicles and roadside infrastructures such as bus stops. All devices have multiple wireless interfaces. Depending on their mobility, they may also be equipped with a wired interface. Such is the case of wireless Points of Access (PoA), which are installed at bus stops, in order to provide connectivity either to users waiting for a bus or to the bus itself (hence to its passengers). PoAs are thus equipped with a wired and one or more wireless interfaces. In such setting, devices carried by cars and pedestrians play the role of Mobile Agents; PoAs, such as roadside infrastructures and public transportation vehicles (buses, trams, cabs etc.), act as Brokers. In the urban mobile and dynamic scenario described above, it is possible to define an overlay network based on the publish/subscribe messaging paradigm and content-based routing. We describe this overlay network, named Arimove, in the next section.

15.5 Content Discovery in Heterogeneous Wireless Network: The Arimove system

Arimove, firstly introduced in [30], provides an efficient mapping between physical devices in the wireless underlay network and virtual entities in the overlay network. Below, we present an overview of the overlay network followed by a description of the main entities and their interactions.

15.5.1 Overlay Overview

A Broker is implemented at a fixed infrastructure in the urban environment. A suitable choice could be a bus stop, since metropolitan transportation companies are likely to bundle bus stops with electric power and, in some cases, even with fixed network connectivity. A Broker colony is composed of Mobile Agents that have registered to it when they were within radio range of the PoA installed at the bus stop. Therefore, a colony is a logical entity, whose members may be physically located anywhere within the area where Arimove is deployed.

However, to take into account the high mobility of the scenario and enhance its performance in terms of load balancing and service response time, we introduce an additional, Arimove-specific entity, the *Mobile Broker* (mB). This unit may be a public transport vehicle equipped with a scaled-down Broker-like wireless device. Every Mobile Broker is *associated* to (i.e., it has the same identity of) a single Broker. Clearly, at the underlay level, connectivity between the Mobile Broker and the associated Broker may at times be severed.

The main aim of the Mobile Broker is to introduce the novel concept of *colony–room*: a small subset of Mobile Agents with a wireless connection to the Mobile Broker (e.g., pedestrian/vehicles around a bus/cab or traveling along the same direction of the bus/cab during a traffic jam). In addition, thanks to its mobility, the Mobile Broker can collect registrations from Mobile Agents that were too far from the PoA of the associated Broker, and, therefore, might had never had the chance to register to it.

The Mobile Broker collects (un)registrations, service requests and service offers from the Agents within the colony–room. When a wireless connection has been established between the Mobile Broker and a roadside PoA (not necessarily corresponding to the associated Broker), the data path to the associated Broker is again available and an information exchange takes place resulting in the updating of each other's data. Specifically, the following actions occur. Firstly, the associated Broker merges the Mobile Broker's routing table with the one it currently carries. Then, the associated Broker handles the registration/discovery information and generates the appropriate responses. Finally, depending on the response time, the responses are returned to the Mobile Broker before it leaves the wireless PoA coverage, or the next time it connects to a PoA.

Fig. 15.1 illustrates the relationships among overlay and underlay entities. A central coordination entity is located at a headquarter (HQ), in our case corresponding to the local transportation authority building. The coordination entity plays the role of a super-Broker and it is provided with a wired connection to each of the 4 road-side PoA at bus stops (B_1 to B_4). Mobile Brokers (mB_1 and mB_3) shuttle between bus stops, each carrying a different Broker association (to B_1 and B_3), while Mobile Agents (portable devices in the figure) are either connected to Brokers or Mobile Brokers, depending on their mobility. A base station (BS), belonging to a cellular network system, provides a wireless coverage on the whole area.

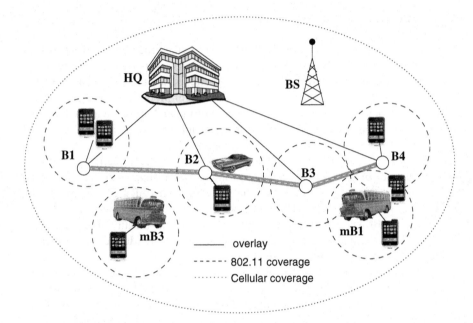

Fig. 15.1. An example of heterogeneous networking scenario.

15.5.2 Overlay Entities

We can now summarize the different activities of the three main entities in Arimove, i.e., Mobile Agents, Brokers and Mobile Brokers.

Mobile Agents

Their activity is carried out through the following main operations.

- *Broker Discovery*: if the underlay is a broadcast network, such as a 802.11 WLAN, HELLO messages are issued by one or more Brokers (and/or Mobile Brokers)[3]; their reception allows an Agent to perform the choice of the Broker to

---

[3]Namely, HELLO messages from a broker may be sent on the broadcast channel of a cellular network, but it entails the direct cooperation of the cellular carrier/operator.

which to register using information stored in the HELLO message itself. Broker discovery in a non-broadcast network may be performed through the use of a directory service offered by a well-known host readily available on the network.

- *(Un)Registration*: the Agent sends a service registration (SREG) message of the VIP protocol trying to register to a new Broker or to unregister from the current Broker.
- *Service Request/Response*: these tasks require that the Agent is already registered to a Broker and that it is part of a colony. It can send a service request (SREQ) of the RDP protocol to its Broker or a service response (SRESP) offering some services. The service will be then exchanged in peer-to-peer fashion.

Brokers

Their activity is carried out through the following main operations.

- *Colony Health*: in a broadcast underlay, periodically broadcasting of HELLO messages to let Mobile Agents discover them;
- *Colony (Un)Registration*: to and from a higher-level Broker;
- *Colony Management*: the Broker interacts with colony members, (un)registering and handling service requests through the VIP and RDP protocols.

Mobile Brokers

The activity of a Mobile Broker is carried out through the following three main operations.

- *Broker Association*: the process of associating to a specific Broker; the Broker for which the Mobile Broker acts as colony-room is chosen according to some policy (see below) and the association is held throughout the Mobile Broker's route.
- *Colony-Room Advertising*: periodical broadcasting of HELLO messages along its whole route; HELLO messages forward information about the associated Broker for which they are acting as colony-room.
- *Relaying*: relaying VIP and RDP messages from (to) Agents inside the colony-room to (from) the associated Broker. Relaying may occur at once if a wireless connection to a PoA exists, or it may be deferred until the wireless connection is re-established.

15.5.3 Membership Policies

As a byproduct of the Arimove overlay, members of a colony will be geographically distributed, although this distribution should be carefully planned by a Broker (accepting or refusing registration requests) for load balancing purposes.

VIP registration policies are usually not specified in the protocol itself; thus every Broker is free to choose its acceptance policy. Different self-organization policies may be used to address issues such as load-balancing and colony specialization.

Possible policies therefore are: (i) *mono-thematic*: a Broker accepts an Agent in its colony only if it offers services that the colony already owns in large quantities, so as to increase its specialization; (ii) *balanced*: a Broker accepts an Agent in its colony only if it offers services that the colony lacks, with the aim of evening out its service offer; (iii) *unbalanced*: a Broker unconditionally accepts all Agent registrations. Membership to a colony is also affected by the Mobile Broker association. The choice of which Broker is associated to a Mobile Broker can be performed by a specific load balancing algorithm that is run periodically, i.e., when the public transportation vehicle leaves the deposit and sets off on its route. One possible aim of the load balancing algorithm is to let the Mobile Broker collect Agents for a Broker with a scarcely populated colony at the time of the Mobile Broker departure; other aims, such as specializing the colony population, can be also envisaged.

15.5.4 Overlay Service Discovery

Service discovery over a MANET, hence in Arimove, plays a crucial role in the successful retrieval of information. There are two main concerns regarding the issue of service discovery. The first one is to *expedite* the selection of the Mobile Agent providing the service, given that Agents are nominally free to roam in and out of their own Broker's underlay reach. Therefore, if a service is advertised by more than one Agent, and a request for that service is pending, a Broker should be given the opportunity to hand the service request over to the Agent that is more likely to be within its reach. The second concern is the *suitability* of the match that the Broker is about to create. Indeed, finding a "good" Agent carrying the requested service must also account for its ability to establish an effective communication channel with the Agent requesting the service. The mobility of both Agents (the requesting one and the potential provider) must be accounted for, e.g., by selecting Agents that are either within radio range of each other, or that are likely to remain within some PoA coverage for enough time. In publish/subscribe jargon, the "filtering" matching function between potential subscriber and publishers must take also into account the "reachability" at the underlay network level.

From a practical standpoint, service discovery at an Arigatoni Broker is carried out through a table that mainly records colony member IDs and their service lists. In Arimove, however, the table information for each Agent is integrated by a *liveliness* field, indicating the time elapsed since the last contact from that Agent, and by a *mobility* field, that can be used to pinpoint the position of the Agent and to infer the direction of its movement (the latter information is provided by the Agent in its last message sent to its Broker).

In order for these additional table parameters to be maintained up-to-date, Agents are expected to interact with their own Broker on a regular basis. Such interaction, in the form of a refresh SREG, should not be limited to the period when the Agent is within its own Broker radio range. Rather, it should also be promoted when the Agent is within *any* Broker range (whence it will be relayed to the Agent's Broker). The refresh SREG will therefore be issued by the Agent upon hearing a HELLO message

Table 15.1. Comparison of the three underlay networks under study in terms of various performance metrics.

Underlay	Overlay connection	Capacity	Energy	QoS	Pricing
IP	AP to HQ	high	low	high	low
UMTS	Mobile Agent to AP	med	med	med	high
WiFi	Mobile Agent to mB and AP	low	high	low	low

from a Broker[4]. The rationale is to let the Agent's Broker know that the Agent is within coverage of whatever wireless technology is used by any Brokers, hence it is readily accessible if a service is requested. It should be noted that an Agent may choose to refrain from sending refresh SREG if the underlay is charging a per-access fee. For each Broker, the content-based routing table has the following form:

Agent ID	Services	Liveliness	Mobility
$user$	$\{S_i\}^{i=1\ldots n}$	t	(x, y, θ, v)
...	...	...	...

where $\{S_i\}^{i=1\ldots n}$ is the set of services it can offer, t is the time passed since the Agent has sent a message, and (x, y, θ, v) is a quadruple denoting physical position, direction and speed (all those information easily retrievable by a GPS module, or computed through it). The table is updated according to the dynamic registration and unregistration of Agents in the overlay. When an Agent asks for a service, then the query is *filtered* against the routing tables of its own Broker; in case of a *filter-failure*, the Broker forwards the query to its direct super-Broker.

15.6 A Case Study: IEEE 802.11 and UMTS Coexisting Technologies

The underlay network we envision hinges on different PoAs. As already mentioned, a bus stop is a suitable place for a WiFi hot spot, but it clearly is not a viable choice to encase a UMTS base station (commonly termed a *Node B*). Both types of PoAs are however expected to route incoming traffic to an IP-based fixed network that guarantees connectivity among the different elements of the architecture. Also, the two wireless technologies have different features in terms of QoS, energy consumption and, mostly, pricing. Table 15.1 compares the three underlay networks in terms of logical connectivity (from-to), bandwidth capacity, energy consumption, quality of service, and pricing.

Therefore, a strategy should be devised in order to select the "appropriate" wireless underlay network. In many cases, one may need to switch from WiFi to UMTS

---

[4]Broadcast storms are prevented by forcing a latency period between consecutive SREG from the same Agent.

and vice-versa because of signal loss, radio hardware problems, pricing issues, or required quality of service (QoS). Moving to one underlay network wireless to another is usually transparent to the Arimove overlay network. Indeed, in either case, RDP and VIP traffic is routed toward the overlay peers. Switching may be automatic when both wireless networks are available, or may be suggested or imposed by the Broker or by the Mobile Agent. Moreover, in the registration phase, the Mobile Agent and the Broker may agree on which underlay to use, whether it is WiFi, UMTS, or a smooth combination of both.

At the time of the first VIP registration, the Mobile Agent communicates its physical position and speed (x, y, θ, v) to the Broker. The Broker and the user negotiate the choice of the underlay (e.g., WiFi connection in a downtown zone until the available bandwidth is not dangerously low). The rationale of the strategy is simple and obeys to two (natural) rules:

1. When a Mobile Agent moves to one coverage area to another, then either it continues to stay connected with the Broker (in the abandoned area) using UMTS, or it is invited to change colony, i.e., the Broker sends to it a SREG logout unless it has some pending requests to serve or receive.
2. When a Mobile Agent moves from one WiFi zone to another *in the same area*, then it essentially has two ways to stay connected to the Arimove virtual organization: either using ad hoc MANET protocols at the cost of poorer performance, or switching to the UMTS network until another WiFi zone is detected. The choice of the underlay protocol is in principle left to the mobile user unless the Broker forces the choice of one network instead of another.

This simple strategy induces the following considerations: moving from one area to another "forces" the Mobile Agent to change broker, while moving in the same area "advises" the Mobile Agent to stay in the colony it is connected to. The overlay therefore tries to limit as much as possible high node churning that, as is well-known, contributes to slow down overlay performance.

15.7 Performance Evaluation

Below, we show the performance of the Arimove architecture obtained under the case study network introduced in Section 15.6. We first describe the simulation scenario and then present the results.

15.7.1 Simulation Scenario

We implemented Arimove in the Omnet++ [31] simulator, coding the overlay part and exploiting the existing wireless underlay network modules. In the underlay, we used IEEE 802.11 at the MAC layer and the DYMO routing protocol (an AODV-like reactive routing protocol). The UMTS RACH (Random Access Channel) and dedicated channel were also simulated.

We tested the performance of Arimove in a mobile environment. We considered 3 Mobile Brokers and 120 Mobile Agents. We assumed that the nodes are randomly scattered in a 1-km-wide square city section and move around according to the random walk mobility model. Every node is assigned a speed of v m/s, randomly selected in a specified interval (see below).

Upon entering the topology, a node acting as Mobile Agent owns a set of S unitary services (e.g., files, traffic information, point of interests) randomly chosen from a set of C services. A service is included in a node subset with probability p, equal for all services and all nodes, so that each node owns an average of S $= p \cdot C$ services. In our simulations, we choose $C = 20$ while p is a varying parameter.

A Mobile Agent issues an SREQmessage for a service it is missing. The inter-request time of each missing service is supposed to be exponentially distributed with parameter λ. It entails that the larger the number of services a node is lacking, the more frequently it issues an SREQ message. To further clarify, using the above notation, p is inversely proportional to the average SREQ rate.

If an SREQ is successful (i.e., the Broker returns to the requesting Agent the address of another Agent who is willing to provide the requested service), the two peers establish a connection and a download is started. We will assume that a peer engaged in a download (either as sender or receiver) rejects any SREQ or any further connection request *coming from the same type of underlay* until the download is complete. In our simulations, the size of a downloaded file is set to 500 Kbytes.

The simulated city topology features six 802.11 PoAs, each corresponding to a Broker. A single UMTS cell covers the entire simulated area. The wireless channel is considered as always reliable, and no obstacles (such as high-rise buildings) are assumed to be present. Brokers apply the unbalanced acceptance policy and filter the routing table against a received SREQby using the liveliness information only.

15.7.2 Simulation Results

Although the number of parameters that can potentially be analyzed is quite high, we focus on few selected sets of results that highlight the performance of the system in presence of either 802.11 coverage or UMTS coverage, as well as in a scenario featuring both technologies in the same area.

We initially address the success probability of a service request in an 802.11-only coverage scenario, i.e., the probability that a service request is positively answered by a Mobile Agent either in the requester's colony or in another colony (after delegation to the super-Broker). Fig. 15.2 carries a comparison of success probabilities for an Arimove system as a function of p. The top plot shows results for two different inter-request times (the node speed is set between 0 and 1 m/s), while the bottom plot allows us to examine different node speeds (with a 600 s inter-request time). As expected, higher content availability entails more successful retrievals, regardless of request rate or node speed. As expected, a longer inter-request time, hence less congested underlay network, results in a higher retrieval success probability. Also, it is quite meaningful to underscore that higher speeds negatively affect the retrieval due to nodes leaving the 802.11 coverage more frequently and thus being unable to complete a request/response cycle that they may have initiated.

We point out that some underlay-related events are responsible for missing responses: for example, a miscue from the routing protocol may lead an Agent to wrongly believing that it is within radio range of a Broker (or that a Broker may be reachable in multi-hop fashion through other nodes); an SREQ will therefore be issued but never delivered to the Broker, negatively affecting the success probability.

Another metric of interest is the average time after which a response to an SREQ is returned to the requesting Agent (Fig. 15.3). Again, we first look at an 802.11 scenario. As can be expected, some of the trends already observed for the success probability are found in the response times as well. However, all curves peak around $p = 0.4$. As can be recalled, p and SREQ rates are inversely proportional, so values lower than $p = 0.4$ counterbalance the scarcity of resources with a high SREQ message rate that "keeps alive" many routing paths; this avoids the high overhead of the path setup process typical of reactive routing protocols (such as DYMO). In addition to it, a smaller success probability results in fewer peer-to-peer downloads overloading the network. On the contrary, values higher than $p = 0.4$ exhibit a lower discovery time thanks to the wealth of services available throughout the network.

When the attention shifts to a scenario featuring UMTS-only coverage, the underlay becomes less of an issue. The probability of successful retrieval is very close to 1, the only failures being due to connection requests towards Agents that have already become engaged in a download with another Agent. Similarly, the discovery time is very low, as shown in Fig. 15.4, since the only significant latency comes from RACH access (and no multihop connection, along with its routing overhead, is involved).

Finally, we look at a scenario that combines a spotty 802.11 coverage and an "umbrella" UMTS coverage. Clearly, if the UMTS operator charges a per-access fee, while the 802.11 access is free or available at reasonably hourly fees, Agents will strive to connect to the latter. Therefore, in areas where both coverages are available, Agents will try and connect through the 802.11 underlay. If an Agent cannot connect to an 802.11 PoA (either directly or through a multihop path), it will use UMTS to access the Arimove system. In our simulations, we considered one of the setups used in previous tests, (i.e., Agent speed between 1 and 2 m/s, $p = 0.6$) and analyzed three cases, each with a different 802.11 coverage of the overall area (either 25% or 75%).

Interestingly, the probability of resorting to the UMTS network is directly mapped onto the coverage percentage for higher loads (i.e., inter-request times equal to 100 ms), as can be seen in Fig. 15.5. Therefore, if, on average, 75% of the Agents are outside the 802.11 PoA coverage (802.11 coverage = 25% in the figure), the probability of switching onto UMTS matches such figure. However, as the load decreases, the number of Agents who successfully complete an SREQ cycle on 802.11 becomes larger, since multihop connectivity even *outside* the radio range of a PoA is no longer hindered by high traffic on the underlay network.

Fig. 15.6 presents the probability of successful retrieval observed for SREQ cycles that are performed on 802.11 only (lower curves, with 25% and 75% cover-

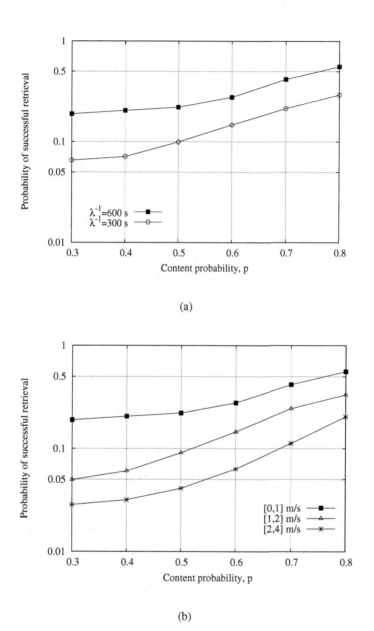

(a)

(b)

Fig. 15.2. Success probability of a service request as a function of the initial content availability at each user. Results are derived for different request rates (top) and different user speeds (bottom) for 802.11 scenario.

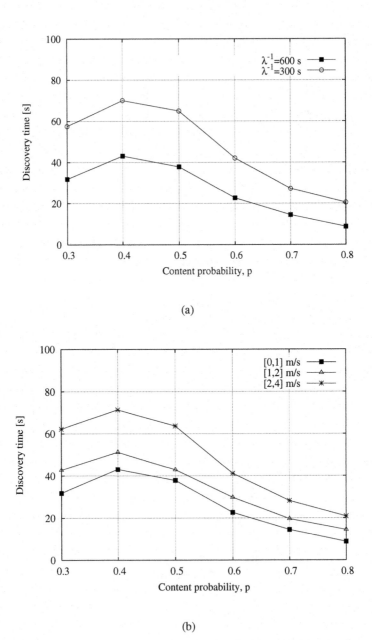

(a)

(b)

Fig. 15.3. Average time to satisfy a service request as a function of the initial content availability at each user. Results are derived for different request rates (top) and different user speeds (bottom) for 802.11 scenario.

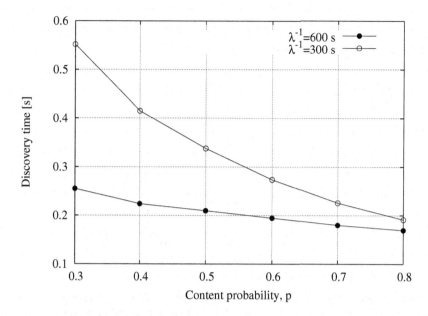

Fig. 15.4. Average time to satisfy a service request as a function of the initial content availability at each user. Results are derived for different request rates for UMTS scenario.

age), and the one observed for all SREQ cycles, i.e., including those that are routed over the UMTS network (higher curves). In the 802.11-only message exchanges, beside the usual increase in success probability as the inter-request time increases, we also observe that smaller coverages yield a better success rate than larger coverages. This is due to fewer downloads on the underlay network (most occur over UMTS, therefore they do not hinder SREQs on 802.11). Clearly, the very high completion probability under UMTS positively affects the overall performance.

The last set of results, shown in Fig. 15.7, attains to the average time to satisfy a service request, computed over all SREQ cycles. Depending on the coverage, the use of UMTS rather than 802.11 boosts the performance, lowering the discovery time for 802.11 coverage at 25% (when the majority of Agents is forced to use UMTS).

Conclusion

This chapter addressed the problem of content discovery in heterogeneous mobile networks. It focused on a network architecture, composed on the one hand of infrastructured nodes, such as WLAN access points and cellular system base stations, and on the other hand of mobile nodes, that are equipped with multi-interface wireless terminals. The study adopts the well-known *publish/subscribe* paradigm and designs an overlay network, relying on the underlay wireless network, to implement such a

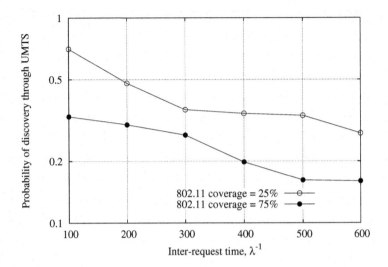

Fig. 15.5. Probability that a SREQ has to be routed over UMTS due to unavailability of an 802.11 route to the Broker, as a function of inter-request time, for 802.11/UMTS mixed scenario and different 802.11 coverages.

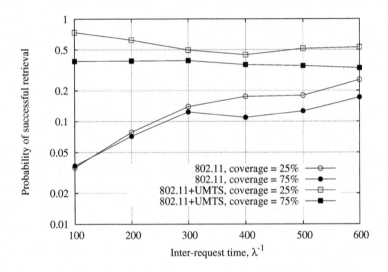

Fig. 15.6. Probability that a SREQ issued is successful, as a function of inter-request time in 802.11/UMTS mixed scenario and with different 802.11 coverages.

messaging paradigm. The chapter described the semantics and the interaction among the logical network entities, as well as the possible interface selection strategies. Fi-

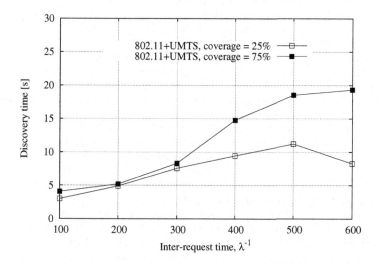

Fig. 15.7. Average time to satisfy a service request as a function of inter-request time, for 802.11/UMTS mixed scenario and different 802.11 coverages.

nally, the benefits of using multiple wireless technologies for content retrieval in a mobile environment were highlighted through some performance results. The results were obtained for a case study including WiFi hot spots and UMTS base stations as network infrastructure entities.

Acknowledgment

This work is supported by AEOLUS FP6-IST-FET Proactive, and by the *Regione Piemonte* through the project VICSUM.

References

1. F. H. P. Fitzek and M. Katz, *Cooperation in wireless networks: Principles and applications – Real egoistic behavior is to cooperate!*, Springer, 2006.
2. M. Dohler, D.-E. Meddour, S.-M. Senouci, and A. Saadani, Cooperation in 4g: hype or ripe. *IEEE Technology and Society Mag.* in press.
3. The InterCity project. http://intelcities.iti.gr/intelcities.
4. The WikiCity project. http://senseable.mit.edu/wikicity.
5. P. Th. Eugster, P. Felber, R. Guerraoui, and A. M. Kermarrec, The Many Faces of Publish/Subscribe. *Computing Survey*, 35(2):114–131, 2003.
6. R. Chand and P. Felber, "A scalable protocol for content-based routing in overlay networks," in *Proc. of NCA*, 2003.

7. R. Chand and P. Felber, "XNet: A reliable content-based publish/subscribe system," in *Proc. of SRDS: Symposium on Reliable Distributed Systems*, 2004.
8. A. Carzaniga, D. S. Rosenblum, and A. L. Wolf, "Design and evaluation of a wide-area event notification service," in *ACM TOCS*, 19(3), 2001.
9. G. Banavar, T. Chandra, B. Mukherjee, J. Nagarajarao, R. E. Strom, and D. C. Sturman, "An efficient multicast protocol for content-based publish-subscribe systems," in *Proc. of ICDCS*, 1999.
10. L. Iftode, C. Borcea, N. Ravi, and T. Nadeem, "Exploring the design and implementation of vehicular networked systems," Technical Report DCS-TR-585, Rutgers.
11. A. Varshavsky, B. Reid, and E. de Lara, "A cross-layer approach to service discovery and selection in MANETs," in *Proc. of IEEE MASS*, Washington, DC, 2005.
12. A. Nandan and et al., "Co-operative downloading in vehicular ad-hoc wireless networks," in *Proc. of WONS '05*, pp. 32-41, St. Moritz, Switzerland, January 2005.
13. M. Fiore, C. Casetti, and C.-F. Chiasserini, "Efficient retrieval of user contents in MANETs," in *Proc. of IEEE Infocom '07*, May 2007.
14. D. Lundquist and A. Ouksel, "An efficient demand-driven and density-controlled publish/subscribe protocol for mobile environments," in *Proc. of ACM International Conference on Distributed Event-based Systems*, pages 26-37, Toronto, Canada, 2007.
15. M. Avvenuti, A. Vecchio, and G. Turi, "A cross-layer approach for publish/subscribe in mobile ad hoc networks," in *Proc. of MATA*, pp. 203-214, 2005.
16. E. Yoneki and J. Bacon, "Dynamic publish/subscribe in mobile peer-to-peer systems," in *OTM Workshops*, pp. 1-2, 2004.
17. R. Meier and V. Cahill, "Steam: Event-based middleware for wireless ad hoc network," in *ICDCS Workshops*, pp. 639-644, 2002.
18. R. Meier, V. Cahill, A. Nedos, and S. Clarke, "Proximity-based service discovery in mobile ad hoc networks," in *DAIS*, pp. 115-129, 2005.
19. BitTorrent, Inc. http://www.bittorrent.com/.
20. R. Sherwood, R. Braud, and B. Bhattacharjee, "Slurpie: A cooperative bulk data transfer protocol," in *Proc. of IEEE Infocom'04*, Hong Kong, Mar. 2004.
21. Globus Alliance. http://www.globus.org/.
22. JXTA Community. http://www.jxta.org/.
23. OSGi Alliance. Open Services Gateway Initiative. http://www.osgi.org/.
24. U. C. Kozat and L. Tassiulas, "Network layer support for service discovery in mobile ad hoc networks," in *Proc. of IEEE Infocom'03*, pp. 1965-1975, San Francisco, CA, April 2003.
25. F. Sailhan and V. Issarny, "Scalable service discovery for MANET," in *Proc. of IEEE PerComm*, Kauai Island, Hawaii, Mar. 2005.
26. L. Liquori and M. Cosnard, "Logical networks: Towards foundations for programmable overlay networks and overlay computing systems," in *TGC, Trustworthy Global Computing*, Lecture Notes in Computer Science. Springer, 2008.
27. D. Benza, M. Cosnard, L. Liquori, and M. Vesin, "Arigatoni: Overlaying Internet via low level network protocols," in *JVA, John Vincent Atanasoff International Symposium on Modern Computing*, pp. 82-91, IEEE, 2006.
28. R. Chand, M. Cosnard, and L. Liquori, "Powerful resource discovery for Arigatoni overlay network," *Future Generation Computer Systems*, 24(1):31-38, 2008.
29. D. Jordan and J. Evdemon, *Web Services Business Process Execution Language Version 2.0*, OASIS Web Services Business Process Execution Language (WSBPEL) TCO.
30. L. Liquori, D. Borsetti, C. Casetti, and C. Chiasserini, "Overlay networks for vehicular networks," Research report, INRIA Sophia-Antipolis Méditerranée, 2008.
31. OMNET++ Discrete Events Simulator, http://www.omnetpp.org.

16

Wireless Networks Test-beds: When Heterogeneity Plays with Us

Alessio Botta[1], Antonio Pescapé[1], and Roger Karrer[2]

[1] Universitá di Napoli Federico II, Napoli, Italy
{a.botta,pescape}@unina.it
[2] Deutsche Telekom Laboratories, Berlin, Germany
roger.karrer@telekom.de

16.1 Introduction

The Internet has been designed for heterogeneity. In particular, Clark formulates in the design goals of the DARPA Internet that the network must support (i) multiple types of services and (ii) accommodate a variety of physical networks. These design goals are the most important goals besides the interconnection of existing networks and survivability [1]. Moreover, they have led to two design principles: the end-to-end argument and layering. These principles have coined the Internet architecture and were among the key enablers of the stunning success of the Internet. In particular, they have shaped the architecture of the Internet into the well-known hourglass (see Fig. 16.1).

However, is the heterogeneity envisioned four decades ago still the same heterogeneity we experience today? We argue that the notion and the challenges of heterogeneity have significantly changed over time. In particular, the heterogeneity targeted in the early days focused on *co-existence*, i.e. the ability to seamlessly connect different network technologies and shield the upper layer protocols and end systems from the details of the underlying technologies and protocols. In contrast, today, we are challenged to make the heterogeneous technology *concurrently collaborate*. In particular, in the wake of the fixed-mobile convergence, networks suddenly face the challenge to either dynamically choose one of the available technologies or even to concurrently use multiple technologies. For example, modern cities typically provide multiple wireless access technologies, such as GSM, 3G, WLAN oder even WiMAX. All these technologies are concurrently available and modern devices are even equipped with multiple radios to take advantage of the concurrent availability of the heterogeneous technology.

The concurrent availability of heterogeneous resources puts forward a set of unprecedented challenges. A first challenge is to decide who controls the resources. Given a modern device with multiple radios that can be used in parallel, some instance has to decide which and how many resources should be used. Should the end system control the resources? Technically, an end-system approach is able to take

E. Hossain (ed.), *Heterogeneous Wireless Access Networks*,
DOI: 10.1007/978-0-387-09777-0_16, © Springer Science+Business Media, LLC 2008

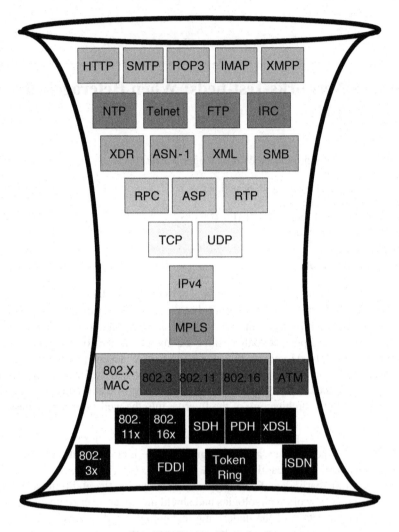

Fig. 16.1. Internet "hourglass".

the entire end-to-end path into account, including the non-wireless access network as well as potentially multiple providers, whereas a network provider only has information about the technology he deployed. Moreover, an end system may also take end system resources into account, such as battery life. Economically, the end user ultimately needs to be informed about the costs of using multiple technologies. On the other hand, to make such decisions, an end system needs information about the network, including the availability of the different technologies as well as the actual resource usage.

The second challenge is to maximize and manage the usage of the different resources. The availability of multiple technologies in the new heterogeneity allows to exploit the features of the different technologies and consider and arrange them according to some specific metrics: short-delay paths may be used for delay-sensitive applications such as VoIP (Voice over IP), high bandwidth paths for bulk traffic. Again, the question of who controls the resources influences the ultimate outcome.

Finally, the third challenge is to implement the cooperation. In particular, the end-to-end argument, the layering principle and the economical separation between ISPs and end users prevent an easy information exchange among the different layers that hide the technological diversity as well as among networks and end systems. Thus, from a design perspective, it is far from obvious how such an implementation could be done with the Internet protocol stack. Should layering, cross-layer implementation and intermediate layers be considered harmful, or should they be considered as necessary steps towards efficiency for a future development of the Internet?

To sum up, we notice that the challenges raised here are fundamental problems that touch to the very core principles of the Internet design, such as the end-to-end argument [1]. Solutions should therefore be considered only in this entire context. Moreover, it is likely that the road towards solutions will have to consider the tussles raised by the competing and conflicting demands, preferences and needs of the different stakeholders [2].

Our work focuses on shedding light on these questions by planning, deploying and experimentally evaluating test-beds. We argue that many of these questions will ultimately be decided by convincing arguments that are supported by hard facts from real data. Measurements provide insight into the real benefits an operator or an end system may gain. The deployment yields detailed numbers on the deployment costs and ultimately on the incentives for an operator to invest into enhancing collaboration. Ultimately, test-beds contribute to the debate on the fundamental principles of systems and networks design.

This chapter first gives an overview of the heterogeneity in the Internet today. Section 16.2 thereby emphasizes the tremendous *heterogeneization* of the Internet along various dimensions. The discussion highlights the challenges and the need for a clear structure in the control plane to monitor and manage the heterogeneous devices. Then, Section 16.3 discusses the impact of the heterogeneity on the Internet architecture and the protocols. The section digs into the fundamentals of the Internet architecture and shows that addressing the heterogeneity requires a fundamental consideration, potentially even a re-thinking, of the design principles of the current Internet. We emphasize the need for test-beds to address these challenges, in particular to verify that novel approaches are feasible and comply with the requirements, such as scalability. Section 16.4 then discusses two heterogeneous test-beds at work: the Magnets test-bed in Berlin and the small scale test-bed at the University of Napoli. We describe such test-beds and show how they allow us to tackle the above challenges. Finally, Section 16.4.2 concludes this article.

16.2 Dimensions of Heterogeneity

One of the greatest achievements of the Internet has been to expand and evolve in spite of the increasing heterogeneity. Heterogeneity has steadily increased over the past decades for a number of reasons and in different dimensions.

A key technical driver for heterogeneity has been the miniaturization of the integrated circuits. In the first years of the Internet, only a few devices were built to send and receive packets over the Internet. With the increasing miniaturization, entire operating systems with complete TCP/IP stacks fit on pocket devices and even tiny sensor nodes. Therefore, the Internet today consists of a plethora of devices with a wide variety of physical capabilities, ranging from low-speed battery-conserving sensor nodes to Gigabit routers.

The technical advances have been joined by a rapid decrease in costs. The Commodore 64, the first computer sold for the mass market, contained only 64 kB of memory - a size that seems ridiculously small when comparing it to today's portable devices. Similarly, CPU speed, disk space and network interfaces have rapidly increased their capabilities while the prices constantly dropped. In 2000, Vint Cerf wrote "By 2020, so many appliances, vehicles, and buildings will be online that it is likely there will be more Internet devices than people online at any given moment" [3]. If any part of this statement is wrong, it will be the time frame by when this vision is achieved. This means that now, and ever more in the future, a plethora of devices will exist that are able to communicate.

Similar to the devices, but at a much lower speed, did heterogeneity increase at the physical and data-link layer. Quite interestingly, though, heterogeneity is limited in the wired Internet where Ethernet has largely triumphed over competing technologies such as ATM. In the wireless world, in contrast, we are still in the infancy of the technological deployment. Significant improvements at both devices and antenna technology will therefore continue to change. Today, WLAN has established itself as the dominant technology for wireless communication, but it is unclear how WLAN will compete with WiMAX. Finally, WLAN is currently also challenged by the advances and the integration of cellular technologies, such as UTMS and HSDPA.

Third, the heterogeneity of the end-host devices has an immediate impact also on the operating systems. Different appliances have often different operating systems. The OS are typically tailored to the capabilities of the end systems and therefore differ for PCs and handheld devices. It is this optimization that causes the heterogeneity: a wirleess device will e.g. also send data via TCP-IP, yet the sent traffic pattern may differ from the traffic sent from a PC because it optimizes its resources differently. For example, the traffic pattern of a battery-powered device may significantly differ from a wired-powered PC.

Fourth, heterogeneity exists due to the wide variety of applications. While the Internet has originally been dominated by file transfer data, applications and protocols such as the World Wide Web [4] and email have tremendously increased the popularity of the Internet. Today, virtually any imaginable application has been ported to the packet-based Internet, even phone calls are increasingly replaced by VoIP, e.g. via Skype. Finally, two of the greatest surprises are the network games and the file sharing applications. In the first Internet years, network games were predicted to have a very small share of users and traffic [4]. On the contrary, the last games released for network-equipped console games have caused an observable increment in overall Internet traffic [5]. As for the file sharing applications, it is almost known that nowadays they generate a large percentage of all the Internet traffic [6]. Similar evolutions have been noticeable with respect to content: while the Internet was originally text-based, we see an increasing *mediazation* of the content, from text to pictures to multimedia content. The latest push has come from the Web 2.0 that simplifies the exchange of content.

Finally, heterogeneity comes from the different protocols that implement the various functionality. We hereby distinguish three types of heterogeneity. First, different protocols have been specified to implement different functionality. The prominent example here are TCP and UDP, which present a stream and a packet-based transport-layer interface to the application. Moreover, the transport-layer interface shows an interesting evolution. On the one hand, TCP is becoming the dominant transport-layer protocol, and it is even used for e.g. real-time streaming of multimedia content, even though the retransmissions and the strong reaction to congestion make it difficult for TCP to maintain the required streaming rate. UDP, in contrast, is increasingly blocked by firewalls to prevent security exploitations such as DDoS attacks. On the other hand, we see a push towards diversification of TCP, away from the point-to-point protocol towards multipoint-to-point communication. This diversification is motivated by the increasing server-side replication of data (e.g. in Content Distribution Networks and even peer-to-peer networks) as well as client-side multi-homing. Protocols such as Stream Transmission Control Protocol (SCTP [8]) or Structured Streams [9] emphasize the need for enhanced communication support at the transport layer. Thus, in the future we will see an increasing heterogeneity of TCP-friendly protocols that open multiple streams in parallel.

The second level of heterogeneity comes from the different flavors of a protocol, e.g. TCP. Over the past years, many TCP variants have evolved, such as TCP Reno, Tahoe, NewReno, WestWood, FAST, BIG, etc. These variants were the response to the increasing heterogeneity of the lower layers: some TCP variants target high-speed wired networks, others target wireless networks. While all flavors have their pros and cons, we largely ignore today how the different variants inter-operate, e.g. in the case that they are concurrently deployed in wired-cum-wireless networks. Finally, we notice an increasing "heterogeneity" in protocols, especially at the application layer, due to security constraints. In particular, today's firewalls increasingly block potentially suspicious ports. As a result, applications "hijack" ports and protocols to tunnel content through the firewalls. HTTP is one of the most (ab)used protocols for this purpose today because port 80 is most frequently open. Similarly, Skype is known to actively search for holes in the firewall. In the future, as long as the binding between port and service identification prevails, we expect that the raising security concerns lead to more heterogeneity in (ab)used application-layer traffic.

16.3 Impact on Architecture and Protocols

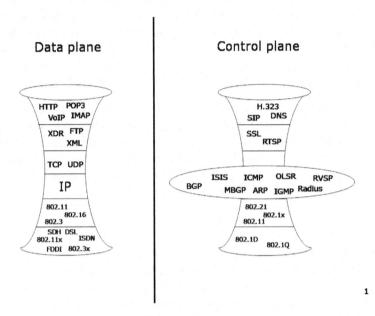

Fig. 16.2. IP waist.

The heterogeneity at both the application layer and the lower layers raises questions how the Internet architecture should evolve in the future. At the data plane, the hourglass model has shown to have many advantages. However, at the control plane, the picture looks quite different, as Fig. 16.2 shows. The constant demand for new services has led to an almost inverted hourglass shape, with the thick waist around the network layer. Moreover, cross- and inter-layer protocols have started to blur the layer boundaries. While some of the problems can only be solved with a clean slate approach, e.g. splitting locators and ids as well as services and ports, other issues concern the layering structure of the control plane. We identify two challenges.

The first challenge is to address the question whether layers are necessary, and how strict they need to be. Or, in other words, is the layering principles a necessary precondition to ensure the future evolution of the Internet? Or are they even preventing evolution today where the Internet challenges are no longer just technical, but also economical and social? Just consider security problems: how many problems arise because the information exchange among the layers does not exist? Similarly, to which degree will it be possible to organize and optimize multiple heterogeneous (wireless) access technologies without an integration at the control plane?

To give a brief example, consider routing in wireless mesh networks. Routing is typically addressed at the network layer with IP. However, in wireless mesh networks, the discussion is ongoing if routing in the mesh should be performed at layer

3 or, as the 802.11s standard prescribes, at layer 2? The advantages of a layer 2 approach are that the entire mesh cloud is visible as a single "node". It integrates well with the broadcast properties of the mesh and promises high performance. In contrast, a layer 3 approach reveals the mesh topology, allows a re-usage of IP-layer mechanisms such as IP multicast and mobile IP. Finally, layer 2.5 approaches promise a combination of the advantages, but also require reprogramming. Thus, this example shows that the traditional distribution of functionality onto the different layers may be subject to change in the future. Similar considerations are possible for network coding approaches [10], multipath [11] and security.

The second challenge is the distribution of control. As outlined above, it is unclear how the control of network resources should be divided in the future: end systems or the network (i.e. ISPs). Consider the case of multi-homing. Should multi-homing be implemented in the end device or inside the network? Both possibilities have pros and cons. Or is there even a third, compromising option, e.g. in the form of middleboxes that provide a limited support of customization for end systems but are still under the control of ISPs?

All these questions and the subsequent decisions must be supported by test-bed implementations and evaluations. Only test-beds provide the necessary power to assess the performance gain and the implementation overhead. By deploying and experimenting with these parameters, vital insight can also be gained on the incentives for either solution. For example, by deploying a wireless test-bed in a city, both ISPs and users can be integrated into the test-bed and their needs and their willingness to cooperate can be investigated under real conditions.

16.4 Heterogeneous Wireless Test-beds at Work

This section describes two test-beds and some experiences we had with them. We aim at providing some information useful to set up heterogeneous test-beds in reality, to perform measurements on them, and to interpret the outcomes of the experiments in order to shed light on the potential and limitations of current heterogeneous networks.

The first test-bed we describe is located in Berlin, Germany. This project is called Magnets and has a lot of interesting features which span from being a joint research-operational network to mixing different access network technologies, from being located in the center of a very big city to having a multi-hop Wireless Wide Area Network (WWAN) as a backbone, from having such a wireless backbone made with off-the-shelf components to being able to reach more than 60 Mbps of throughput. We describe the complete plan and then provide some details regarding the different components: the wireless backbone, the wireless mesh networks that will be interconnected, and the points of integration with other technologies (i.e. GPRS, UMTS, and WiMAX). Also, we present the experimentations we have performed providing some interesting results.

The second test-bed has a smaller scale with respect to the first. However, it comprises a large mix of different devices, operating systems, and access networks. The smaller scale improves the ability to control the environment, the behavior, the measurements and thus, in turn, the predictability of the experiments and the interpretation of the results. Therefore, while Magnets allows to assess what a real user would experiment, the small scale test-bed allows to go in deep into the root causes of the observed behaviors. We present the architecture of this network, our measurement methodology, and some obtained results.

16.4.1 Large Scale Test-bed: Magnets, a Next Generation Access Network

Magnets is designed as a next generation wireless infrastructure. It consists of two main parts: a wireless mesh with 100 nodes and a high-speed wireless backbone with a raw end-to-end throughput of 108 Mbps. Besides the size of the network in terms of nodes and link speed, a key distinguishing characteristic of Magnets is its heterogeneity along several dimensions: it features multiple wireless interfaces with diverse link characteristics, nodes with varying degrees of processing and storage capabilities, and interconnection of multiple mesh networks with disparate routing protocols.

To further exploit this uniqueness, Magnets is designed with a three-fold goal. First, Magnets is designed as a semi-productive network. That is, the network is used as a test-bed, e.g. to experimentally evaluate protocols, but at the same time the network is integrated into the productive campus network of the TU Berlin. Therefore, Magnets will extend the Internet coverage of the students. This combination eventually allows us to also perform measurements of real user traffic and to evaluate protocols under realistic conditions. Second, with Magnets we will systematically assess ways to build wireless mesh networks. For example, the WiFi backbone is designed as a fully planned network, optimized for throughput. In the mesh, we encounter several constraining factors, such as the buildings, the density of already deployed access points. We will compare the capabilities of this mesh with other mesh networks in Berlin that are driven by communities and their structure is therefore unplanned.

The Magnets architecture consists of three parts: a high-speed wireless 802.11 backbone, an 802.11-based wireless mesh network and integration points to alternative technologies (GPRS, UMTS, and WiMAX). Next, we describe the three parts in more detail.

Backbone

The Magnets backbone is designed to interconnect 2 facilities in Berlin with a high-speed connection that is purely wireless. After a careful planning that involved network-specific parameters, such as finding buildings to provide line-of-sight, but also economical parameters such as the deployment costs or rent for space, we decided on a layout that consists of 5 nodes, as depicted in Fig. 16.3. The total distance between the two end points at the T-Labs and T-Systems is 2.3 Km. All nodes reside on top of high-rise buildings and have unobstructed line of sight. All transmissions are in the unlicensed spectrum (2.4 GHz and 5 GHz) range.

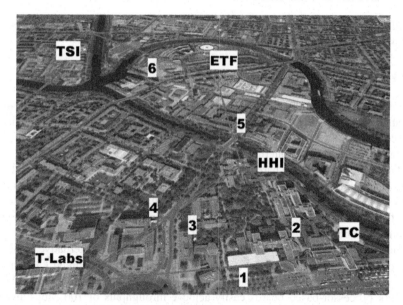

Fig. 16.3. Magnets WiFi backbone in the heart of Berlin.

The nodes are designed to ensure an efficient multi-hop communication. In particular, to avoid well-known performance and unfairness problems in multi-hop communication [16], we decided that each wireless link should be operated by an individual access point. Thus, a total of 12 WiFi access points (APs), suitable for outdoor usage, are mounted along the antennas to shorten the cable length between the antenna and the AP. While the APs all support 802.11a/g modes at 54 Mbps, with the option to improve to 108 Mbps via Super-A/G, the use of the frequency band is defined by the antennas. We decided to operate 8 APs at in the 2.4 GHz band and the rest in the 5 GHz to also have heterogeneity in the transmission frequencies, e.g. to observe the impact of interference in a dense urban area.

Since most nodes on the buildings consist of multiple access points, we decided to inter-connect them via a workstation. In addition to the pure connecting of the APs, these workstations can additionally be used to inject traffic and to monitor the forwarded traffic. To ensure that the workstation is not becoming the bottleneck, we equipped them with a fast 3 GHz processor and 1 GB of RAM.

This setup allows us to perform a range of measurements. First, we are able to observe the per-link characteristics. Since every node is physically located at a different environment, we are able to monitor the node and link performance as a function of the link distance, the capacity and the interference at the receiver. In terms of distance, the links vary from 330 m up to 920 m. We are able to assess short-term statistics, e.g. to assess the frequency and the impact of link-layer retransmissions, as well as long-term statistics, e.g. the evolution of the link speed over several days or even months. Moreover, we are able to monitor low-level information, such as link-layer retransmissions as well as end-to-end throughput, e.g. the performance of different TCP versions over multiple wireless hops and even over wired-cum-wireless connections.

WiFi Mesh

The WiFi mesh with 100 mesh nodes is deployed on the campus of the TU Berlin in collaboration with the IT department. The mesh will consist of 100 nodes deployed as a combination of in- and outdoor nodes. The mesh shall cover the entire campus area and thereby provide Internet access to the students.

For the selection of the hardware for the mesh nodes, we opted for two hardware platform: routerboards and Avila Gateworks. The routerboards provide maximum extensibility. The RB500, e.g., has the ability to attach external storage via a compact flash card, which is important to add management and measurement tools and eventually to collect traces. With the help of a daughterboard, up to 6 MiniPCI slots provide ample opportunities to attach WiFi cards or to insert alternative technologies on each node. The drawback of the Routerboards is the limited CPU speed. Therefore, the largest portion of the mesh will be built with Avila Gateworks. These network processor-based boards easily achieve throughputs of 100 Mbps and are therefore well suited for high capacity.

To perform experiments, we set up the nodes with 2 particularities. First, all nodes run openWRT, a Linux-based operating system that provides the flexibility to access kernel information. Moreover, we use Atheros cards with MadWiFi to get access to the MAC and PHY statistics. This software provides ample opportunities and flexibility to deploy and evaluate protocols at any layer. It allows experimental evaluation of benefits and drawbacks of cross-layer optimizations that have been proposed in the research literature [14]. Our main objective here is to shed practical, experimental light on the ongoing discussion. Second, we equip most nodes with 2 boards: a main board for data transmissions (Avila or Routerboard) and a secondary board for monitoring (mostly a cheaper ASUS board). The reason for the monitoring boards is that CPU, memory and network speed may cause limitations to concurrently transfer data and perform monitoring on the nodes. For example, running tcpdump on multiple interfaces may severely slow down the performance of the nodes.

In terms of deployment, we decided to build three different "mesh networks": a "smoke" test-bed, consisting of pairs of boards only, an indoor test-bed of 20 nodes and finally the outdoor test-bed of 100 nodes. The smoke test-bed is used for node configuration and testing. That is, before any software is deployed on a mesh, it must be tested and shown to be runnable at least on two nodes. While the smoke test does not guarantee that software does not crash deployed nodes, it reduces the risk that somebody has to climb and unscrew the mounted indoor or outdoor boxes.

Fig. 16.4 shows the deployment of the indoor test-bed. The figure shows that 5 mesh nodes are deployed on opposite sides of the T-Labs building and one node at the center. Generally, we made sure that two neighboring nodes are within range of each other, whereas two-hop neighbors have a limited or no connectivity due to glass and other obstructing material. The same applies to the connectivity among floors. Here, two nodes at the same physical location but on different floors are within range,

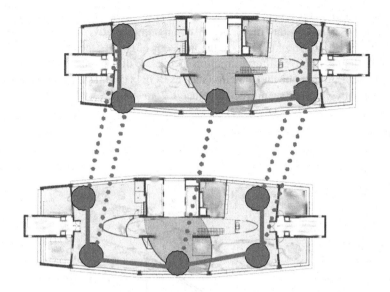

Fig. 16.4. Deployment of the indoor test-bed.

but nodes that are two floors apart or nodes at different floor locations have limited connectivity.

Heterogeneous Nodes

Besides the heterogeneity given by environmental factors (interference) at the different stations, the frequency range, the physical distance of the links and the node density, another degree of heterogeneity can be added to the network by augmenting with alternative wireless technologies. Taking advantage of the up to 6 Mini PCI slots of the nodes, alternative technologies can be added to the network, such as GPRS, UMTS, and WiMAX. By superimposing multiple technologies within the same area, we are able to address questions on how to operate, manage and optimize future 4G networks. Again, important here is the ability to gain first-hand experience in a semi-productive testbed with real user traffic. Issues such as TCP performance during horizontal and vertical handovers between multiple access technologies can be experimentally assessed and evaluated.

Constellation of Mesh Networks

Berlin is the center of the Freifunk community. Driven by the need to provide Internet connectivity in an area where DSL is not available because (ironically) fiber but not copper is available, the community stepped up to build the Freifunk network [3].

---

[3]http://www.olsrexperiment.de

Interesting research questions arise when we think about the options to inter-connect wireless mesh networks that are under different administrative authorities. For example, it will be intersting to observe the behavior of two meshes that run different routing protocols or have different routing metrics. Will it be necessary to develop novel protocols to separate the domains (such as BGP in the Internet), or is it possible to weave the two meshes seamlessly together, e.g. to simplify mobility?

Results

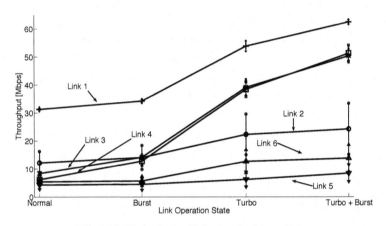

Fig. 16.5. Throughput of Magnets backbone links.

To provide initial insight into the heterogeneity we are able to observe with Magnets, we consider the per-link throughput of the backbone links. For this purpose, we measure each link individually. We generate UDP traffic using iperf and vary the mode of each AP between 802.11a/g and the options provided by Super-A/G (Turbo- and Burst mode). For 600 seconds, we generate traffic at 70 Mpbs, which lies above the saturation rate of the link. At the receiving workstation, we monitor the incoming packets with tcpdump. Then, to calculate the bandwidth, we sample the traces at 50ms intervals.

Fig. 16.5 shows the resulting throughput. The x-axis denotes the mode, the lines show the average throughput of the different links. Finally, the whiskers show the standard deviation. In normal 802.11 modes, the figure shows that link 1 outperforms the others with an average throughput of 31.3 Mbps. Moreover, the low standard deviation of 0.9 Mbps indicates that the link is very stable. Next, links 2-4 have an average throughput between 6.2 and 12.2 Mbps. These links operate in the 2.4 GHz range and the throughput degradation is attributed to interference. Finally, links 5 and 6 are the weakest links, with an average bandwidth of 4.3 and 5.4 Mbps, respectively. Link 5 has strong interference because the ETF building is lower than the others, and link 6 spans as much larger distance with 930m. Thus, we conclude that the link characteristics vary significantly even though they have been measured in the same test-bed.

Then, we assess the impact of *Turbo* and *Burst Mode* on the link performance. Even though the reference manual indicates a doubling of the throughput via *Turbo Mode* and an increase of 10 Mbps with *Burst Mode*, it is not obvious how these modes impact the link characteristic of *MagNets*. As we can see in Fig. 16.5, the *Turbo* and *Burst Mode* increase the throughput on link 1 significantly. Compared to the basic mode (31.3 Mbps), the throughput increases with *Burst Mode* to 34.2 Mbps. *Turbo Mode* boosts the throughput to an average of 53.8 Mbps. Finally, with both modes enabled, the average throughput reaches 62.4 Mbps! Thus, we conclude that link 1 matches the original specifications and expectations of *Turbo* and *Burst Mode*. Link 3 also shows throughput gains with *Turbo* and *Burst Mode*. The corresponding rates are 8.4, 14.2, 39.1, and 50.3 Mbps. Note here that the improvement with *Turbo Mode* is more than twice the base rate. All other links obtain results comparable to link 3. With the exception of links 5 and 6 that suffer from the above mentioned problems, we can state that the performance was significantly improved with *Turbo* and *Burst Mode* enabled. Therefore, we argue that the MagNets backbone is able to support a substantial amount of traffic.

16.4.2 Small Scale Test-bed: A Heterogeneous Network at the University of Napoli

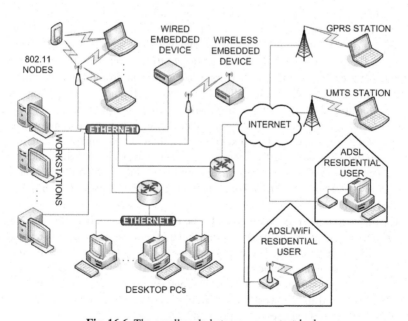

Fig. 16.6. The small scale heterogeneous test-bed.

In this section we present another example of a real test-bed useful to perform experiments aimed to uncover the potential and the problems of heterogeneous networks. Such test-bed is sketched in Fig. 16.6. As shown, it is composed of a number of heterogeneous wireless/wired networks. Over such test-bed a number of different configurations have been produced: we have varied several configuration parameters such as the operating system, end user device, access network, transport protocol, and traffic condition.

All these components constitute our definition of end-to-end path. In details, we define an end-to-end path (e2eP) as

$$e2eP = (S_{UD}, R_{UD}, S_{OS}, R_{OS}, S_{AN}, R_{AN}, Protocol, Bitrate) \qquad (16.1)$$

where UD stands for the User Devices (S_{UD} at sender side and R_{UD} at receiver side) (e.g. Laptop, Palmtop, Workstation, etc.); OS identifies the Operating Systems of each of the two users (e.g. Windows, Linux, Linux Familiar[4], etc.), S_{OS} at sender side and R_{OS} at receiver side; AN is the Access Networks (LAN, 802.11, ADSL, GPRS, etc.), S_{AN} at sender side and R_{AN} at receiver side; $Protocol$ identifies the protocol the users are communicating through (e.g. TCP, UDP, SCTP, etc.); and, finally, $Bitrate$ is that imposed by the application. By combining all these variables, our test-bed allows to set up about 350 different end-to-end paths.

Measurement Methodology

For the measurements we used an active approach and our tool called Distributed Internet Traffic Generator (D-ITG) [17]. D-ITG is able to generate a multitude of traffic patterns by combining pairs of PS (Packet Size) and IDT (Inter Departure Time). In this way it is possible to generate controlled yet realistic traffic. In this chapter, we present UDP Constant Bitrate (CBR) traffic profile obtained with constant PS and constant IDT. This allows to draw a reference curve for successive analysis and to reduce the number of variables. Thanks to its features, D-ITG can be used as an active measurement tool. It can measure and analyze one-way-delay (OWD), round-trip-time (RTT), packet loss rate, jitter, and throughput, using the various components of such platform: (i) sender, (ii) receiver, (iii) decoder, and (iv) log server. The experiments have been carried out by producing three traffic conditions named *Low*, *Medium*, and *High Traffic* [18]. The characteristics of such traffic are reported in Table 16.1.

The measurement stage has been performed between December 2003 and November 2004, in the day hours between 9:00 am and 6:00 pm. Such stage allowd to collect over 34 GB of traffic traces. The traces have been carefully inspected and sanitized detecting and removing samples affected by errors. At [7] we made freely available several archives containing the outcomes of measurements over real networks (not only those we used in this work). Each archive contains files with samples of QoS parameters measured over several end-to-end paths.

---

[4]An open source porting of Linux for Palmtop devices

Table 16.1. Characteristics of measurement traffic.

Traffic Condition	IDT [s]	PS [Bytes]	Bit Rate [Kbps]
Low	$1/100$ s	$\in \{32, 64, ..., 1024, 1500\}$	$\in \{26.1, ..., 819.2, 1200\}$
Medium	$1/1000$ s	$\in \{64, ..., 512\}$	$\in \{512, ..., 4096\}$
High	$1/10000$ s	$\in \{64, 128\}$	$\in \{5120, 10240\}$

Results

For the purpose of this chapter, it is interesting to report here a comparison of what we achieved with different network configurations. In Fig. 16.7 we sketch a three-dimensional plot showing the average throughput, jitter, and round trip time we obtained in the *low* traffic condition with a PS equal to 256 bytes (i.e. with a generated bit-rate equal to 204.8 Kbps). For the sake of clarity, we have selected 6 network path that present different characteristics only in terms of AN. This allows us to exclude the other variables (OS, EuD, ...) from the possible explanation of the results. This is a very important point when performing measurement on a such heterogeneous test-beds: it is necessary to vary only one variable at a time.

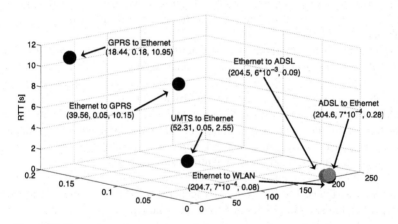

Fig. 16.7. Throughput, jitter, and round trip time over the small scale test-bed.

Fig. 16.7 shows a clear separation between slow (i.e. GPRS and UMTS) and fast (i.e. ADSL and WLAN) Access Networks. The very different results obtained by the slow AN have the effect of making the fast AN appear as a single point in the graph. However, differences are noted are between them. This figure allows to separate the ANs not only looking the the obtained throughput, which was expected, but also looking the the other statistics. This is useful because, in some cases, it is simple to obtain an estimate of the RTT and then of the jitter (e.g thanks to the TCP acknowledgments), while it can be difficult to estimate the throughput. Having such statistics and using these results, several applications can be devised. For example, in [12] and in [13] automatic identification of network characteristics was performed.

Conclusion

The Internet has evolved from a simple and purpose-specific network to the common infrastructure for global user communication. Its current shape was impossible to imagine for its designers. As a consequence, it has now become something very different from the initial plan. However, it still preserves some of the original protocols which causes different problems to both the users and the network administrators with respect to new services and applications.

To understand the benefit and the limitations of the current Internet, in this chapter we have analyzed the main causes of its heterogeneity. We have seen that such causes can be partitioned along different independent dimensions, and we have explored these dimensions. Moreover, we have identified the main challenges that this infrastructure poses with specific regard to the protocols.

Thanks to the use of two real life examples we have then observed how the heterogeneity can be studied. We have described a large and a small scale test-bed, both characterized by an high degree of heterogeneity. We presented some results obtained on the test-bed and showed how the obtained results, can be exploited for addressing the aforementioned issues.

Heterogeneity - in particular heterogeneity where multiple technologies can dynamically be chosen in parallel - raises fundamental questions: how to optimize a network, who decides, who pays. For this reason test-beds are needed to evaluate principles, to show tradeoffs and to create firm arguments.

Moreover, we believe that the heterogeneity may represent a problem; but, at the same time, it provides a great opportunity that must be exploited for network and service convergence: convergence of fixed, wireless and cellular technology increases the demand to support heterogeneity within a single Internet architecture.

Concluding, heterogeneity is a challenge, far from easy to solve at the Internet scale in a distributed fashion. Yet, heterogeneity needs to be addressed to simplify the use of different technologies and to provide unified services to users.

References

1. J. Saltzer, D. Reed, and D. Clark, "End-to-end arguments in system design," *ACM Transactions on Computer Systems*, 2(4):277-288, Nov 1984
2. D. Clark, J. Wroclawski, K. Sollins, and R. Braden, "Tussle in cyberspace: Defining tomorrow's Internet," *ACM Sigcomm*, Aug. 2002, Pittsburgh, PA.
3. Vinton Cerf (June 19, 2000), "Visions of the 21st Century: What Will Replace the Internet". TIME.com.
4. Kevin Kelly (August 2005), "We Are the Web," Wired.com.
5. Graeme Wearden (December 2004), "Does the 'Halo 2' effect threaten broadband?", CNET News.com.

6. Sean McCreary, kc claffy (2000), "Trends in wide area IP traffic patterns - A view from Ames Internet Exchange," ITC Specialist Seminar.

7. http://www.grid.unina.it/Traffic [Online]

8. R. Stewart, Q. Xie, K. Morneault, C. Sharp, H. Schwarzbauer, T. Taylor, I. Rytina, M. Kalla, L. Zhang, and V. Paxson, RFC 2960, "Stream Control Transmission Protocol," 2000.

9. B. Ford, "Structured Streams: a new transport abstraction," in *Proc. of ACM Sigcomm*, Aug. 2007, Kyoto, Japan.

10. S. Katti, S. Gollakota, and D. Katabi, "Embracing wireless interference: Analog network coding," in *Proc. of ACM Sigcomm*, Aug. 2007, Kyoto, Japan.

11. P. Key, L. Massoulie, and D. Towsley, "Combining multipath routing and congestion control for robustness," in *Proc. of CISS*, Mar. 2006.

12. A. Botta, A. Pescapé, and G. Ventre, "Identification of network bricks in heterogeneous scenarios," in *Proc. of First IEEE LCN Workshop on Network Measurements*, Nov. 2006, Tampa, Florida.

13. W. Wei, B. Wang, C. Zhang, J. Kurose, and D. Towsley. "Classification of access network types: Ethernet, wireless LAN, ADSL, cable modem or dialup?," in *Proc. of IEEE Infocom'05*, Miami, Mar. 13-17, pp. 1060-1071.

14. V. Srivastava and M. Motani, "Cross-layer design: A survey and the road ahead," *IEEE Communications Magazine*, vol. 43, no. 12, pp. 112-119, Dec. 2005.

15. National Science Foundation (January 2006), "GENI: Global Environment for Network Innovations - Conceptual Design and Project Execution Plan".

16. Z. Fu, P. Zerfos, H. Luo, S. Lu, L. Zhang, and M. Gerla, "The impact of multihop wireless channel on TCP throughput and loss," in *Proc. of IEEE Infocom'03*.

17. http://www.grid.unina.it/software/ITG [Online]

18. G. Iannello, A. Pescapé, G. Ventre, and L. Vollero, "Measuring quality of service parameters over heterogeneous IP networks," in *Proc. of International Conference on Networking (ICN'05)*, LNCS 3421, pp. 718-727, Apr. 2005, Reunion Island.

Index